NORTH-HOLLAND PERSONAL LIBRARY

STATISTICAL MECHANICS

STATISTICAL MECHANICS

An Advanced Course with Problems and Solutions

RYOGO KUBO

University of Tokyo

in cooperation with

HIROSHI ICHIMURA
Tokyo Institute of Technology

TSUNEMARU USUI
Kyoto University

NATSUKI HASHITSUME
Ochanomizu University

ELSEVIER

AMSTERDAM • BOSTON • HEIDELBERG • LONDON • NEW YORK
OXFORD • PARIS • SAN DIEGO • SAN FRANCISCO • SINGAPORE • SYDNEY • TOKYO

ELSEVIER B.V.
Sara Burgerhartstraat 25
P.O. Box 211, 1000 AE
Amsterdam, The Netherlands

ELSEVIER Inc.
525 B Street
Suite 1900, San Diego
CA 92101-4495, USA

ELSEVIER Ltd.
The Boulevard
Langford Lane, Kidlington,
Oxford OX5 1GB, UK

ELSEVIER Ltd.
84 Theobalds Road
London WC1X 8RR
UK

First edition (hardbound) 1965
Second impression (hardbound) 1967
Third impression (hardbound) 1971
Fourth impression (hardbound) 1974
Fifth impression (hardbound) 1978
Sixth impression (hardbound) 1981
Second edition (paperback) 1988
Third impression (paperback) 1993
Fourth impression (paperback) 1999
Fifth impression (paperback) 2004

Library of Congress Card Number 74-76060

ISBN: 978 0 44487 103 9

∞ The paper used in this publication meets the requirements of ANSI/NISO Z39.48-1992 (Permanence of Paper).

Transferred tp digital printing 2005

PREFACE TO ENGLISH EDITION

The original text of this volume is part of the book "Problems and Solutions in Thermodynamics and Statistical Mechanics", itself one of the "University Series" published by the Shokabo Publishing Company. At the request of the present publisher, the English edition is being published in two volumes, one on thermodynamics and the other on statistical mechanics. Considering the more urgent interest of university students in statistical mechanics, this volume has been translated and published first. The volume on thermodynamics is expected to be published within a year.

The translation was made from the Japanese text by the original authors, together with a few collaborators. As the editor of the original Japanese edition and of the English edition, I wish to express my deep appreciation to Drs. Masaji Kubo, Toshihiko Tsuneto and Satoru Miyake who did the translation work with the authors, and particularly to Professor Donald C. Worth of International Christian University, Tokyo, who kindly took the trouble of helping us with linguistic difficulties. The authors are also indebted to Miss N. Tokuda for the preparation of the manuscript.

1964 RYOGO KUBO

PREFACE TO JAPANESE EDITION

Thermodynamics and statistical mechanics are indispensable tools in studying the physics of the properties of matter. Statistical mechanics, together with quantum mechanics, provides a foundation for modern physics which aims at the thorough understanding of physical phenomena from the microscopic viewpoint of atomic physics. Fundamental knowledge and training in statistical mechanics are therefore of vital importance not only for students studying the physical properties of matter but also for those who study nuclear physics or even astrophysics. Outside the realm of physics, its importance is rapidly penetrating into chemistry, biology and into those vast areas of technology which owe their growth to the advances in modern physics.

Thermodynamics belongs completely to classical physics and is sometimes regarded as unimportant by students of physics who are over-occupied in learning modern physics. Even for students in chemistry, the present is different from the time some decades ago when physical chemistry was almost nothing but chemical thermodynamics. However, it must be stressed here that the usefulness and unique significance of thermodynamics as a fundamental science remain as basic today as they were in the latter half of the last century. Thermodynamics teaches us the value of a phenomenological approach. It avoids explicit use of physical images or models such as atoms and molecules. Instead it deals with relations between somewhat abstract quantities such as energy, entropy, free energy and so forth. Admittedly it does not give intuitive pictures as atomic theories do, which is one of the reasons why students find it difficult to gain sufficient understanding and familiarity to use thermodynamics in real problems. But the simplicity of the logic of thermodynamics sometimes makes us see more clearly into the nature of the basic physics of a given problem from very general principles. This is the great advantage of a phenomenological approach.

Obviously, however, it is impossible to explore more deeply the underlying atomic processes in a given physical phenomenon if we confine our attention to thermodynamics. Such progress is made possible only by quantum mechanics and statistical mechanics. Statistical mechanics provides us with a means to link the physical laws of the microscopic world to those of the macroscopic world. Without close cooperation with statistical mechanics, quantum mechanics itself would not be able to represent the physics of the real world. In this sense, statistical mechanics is indispensable as one of the keystones of modern physics.

Like any other science, statistical mechanics cannot be mastered easily just by learning its principles once. One has to think by oneself a great deal before one grasps the way in which to use the statistical approach in one's thinking, and to apply statistical mechanics to real physical problems. In statistical mechanics and in thermodynamics, there are certain aspects which are quite different from other fields of physics. We often meet students who find difficulty in mastering thermodynamics or statistical mechanics, lacking confidence in applying it to real problems, although they know the principles. Such difficulties are due to insufficient and inadequate training.

The purpose of the present book is to provide a guide for students studying and acquiring facility in thermodynamics and statistical mechanics. Thus it contains fundamental topics, examples and a fairly large number of problems with complete solutions. The fundamental topics are rather condensed, but still they cover all of the points which are basic. This book is meant to be readable without reference to other textbooks. By reading through these topics only, one would be able to obtain fundamental knowledge of thermodynamics and statistical mechanics. The examples are partly to supplement the fundamental topics, but they are primarily meant to show the reader how the principles are applied to physical problems.

The problems are classified into three groups, A, B and C, in order of increasing difficulty. If a reader has enough time he may go through all problems in each chapter. But, if not, it is recommended that he studies first the problems in group A throughout the whole book and then later comes back to try B and C. By just finishing group A problems, he will find himself to have obtained a much better understanding of physics. The number of group A problems is fairly large, so that he may even select about half of these and come back later to the other half. The subjects in the fundamental topics and examples which are marked by $^+$ are not needed in solving problems in group A.

In this book†, thermodynamics and statistical mechanical problems are mostly limited to those of equilibrium states. It might be desirable to include kinetic methods and extensions of thermodynamics and statistical mechanics

† The reader is reminded that this text is a translation of the Preface to the original Japanese edition, in which thermodynamics and statistical mechanics are contained in one volume.

which apply to non-equilibrium problems. We had, however, to content ourselves in treating such topics in a limited way only in the last chapter (Chapter 6 of the present English edition). This is because the whole volume had become much larger than the original plan and also because such non-equilibrium problems are certainly somewhat advanced.

As mentioned previously in this preface, quantum mechanics is the fundamental dynamics of the microscopic world. In this sense, statistical mechanics ought to be essentially quantum-statistics. However, since the present book devotes itself to clear understanding of the nature of statistical considerations, only an elementary knowledge of quantum mechanics is required in studying problems in groups A and B. Therefore, even those students who are not specializing in physics but have only an elementary background in quantum mechanics will not find any serious difficulty in starting to study this book.

What is most important in studying a physical problem is to grasp it as a problem in *physics.* Mathematical manipulations may sometimes be tedious and sometimes may require specialized techniques. Training in mathematical methods should not be ignored, but it would be a serious mistake if one was to be dazzled by the mathematics and to forget the physics. Teachers often meet students' papers in which the student seems to be in no doubt about the numerical answers although they are in error by two or three orders of magnitudes or are dimensionally incorrect. Professor H. Nagaoka (a pioneer physicist in Japan) was carrying out calculations on a blackboard in his class. He changed the sign of his answer saying "It is plus rather than minus. Isn't it?" Mathematical calculations may very often be in error. A physical mind is very important, for this can give you the right sign even when your calculation betrays you. An answer obtained by calculation is in many cases easily understood, at least qualitatively. It may not be guessed before making calculations, but one should not forget to think it over again in order to see if one can see some physical meaning contained in it. Such remarks are not given in each solution of the problems, so that we should like to emphasize here the importance of such reasoning.

Here and there between the pages some comments † are inserted under the title "Divertissement". While giving seminars to students we sometimes take a rest to drink a cup of tea and chat. We hope that the reader will spare a few minutes at these spots to listen to a chat from the authors, drinking tea or coffee or just smoking.

† These are revised in this English edition.

The fundamental topics were mostly written by R. K. Examples and problems were selected after repeated discussion by all the authors. The final check of the solutions was made by R. K. and the whole design of the book was made by N. H. The authors would appreciate it if readers would kindly point out any mistakes which may have escaped our notice.

Five years have passed since this book was originally planned, and two years since we started actually to write it. The undertaking proved to be much more difficult than we anticipated. The authors are particularly grateful to Mr. K. Endo, editor of Shokabo Publishing Company, for his continual encouragement and help.

January, 1961 RYOGO KUBO

CONTENTS

CHAPTER 1. PRINCIPLES OF STATISTICAL MECHANICS

CHAPTER 2. APPLICATIONS OF THE CANONICAL DISTRIBUTION

CHAPTER 6. FLUCTUATIONS AND KINETIC THEORIES

CHAPTER 1

PRINCIPLES OF STATISTICAL MECHANICS

Thermodynamics is a phenomenological theory based upon a few fundamental laws derived from empirical facts. In contrast to this, statistical mechanics aims to provide a deductive method which leads us from the microscopic physical world to the macroscopic world starting from the atomic or molecular structure of matter and the fundamental dynamical principles of the atomic world and combining with these the logic of probability theory. It answers the questions what are the physical laws of the microscopic world behind the thermodynamic laws, how the thermodynamics can be "explained" from such laws and why a specific physical system exhibits such thermodynamic characteristics. The fundamental principles of statistical mechanics involve, in fact, very profound and difficult questions if one meditates upon them, but it would not be very wise for the beginners to be too much concerned with such questions. The most important thing is to learn how one thinks in statistical mechanics and how one applies statistical considerations to physical problems.

Fundamental Topics

§ 1.1. MICROSCOPIC STATES

Microscopic and macroscopic states: A physical system which one observes usually consists of a great number of atoms or molecules and so has an enormously large number of degrees of dynamical freedom. But in the usual case, only a few physical quantities, say the temperature, the pressure and the density, are measured, by means of which the "state" of the system is specified. A state defined in this crude manner is called a *macroscopic state* (example: a thermodynamic state). On the other hand, from a dynamical point of view, each state of a system can be defined, at least in principle, as precisely as possible by specifying all of the dynamical variables of the system. Such a state is called a *microscopic state.*

Classical statistical mechanics and quantum statistical mechanics: The statistical mechanics based on classical mechanics is called *classical statistical mechanics* and that based on quantum mechanics is called *quantum statistical mechanics.* Since rigorous mechanics in the atomic world is quantum mechanics, rigorous statistical mechanics must be quantum statistical mechanics and so classical statistical mechanics may be said to be useful only as a certain approximation to quantum statistical mechanics. But the classical theory has even today a great value from theoretical and educational points of view because it makes us understand more clearly the basic ways of thinking in statistical mechanics.

Classical phase space: Let $(q_1, q_2, \dots q_f)$ be the generalized coordinates of a system with f degrees of freedom and $(p_1, p_2, \dots p_f)$ their conjugate momenta. A microscopic state of the system is defined by specifying the values of $(q_1, q_2, \dots q_f, p_1, p_2, \dots p_f)$. The $2f$-dimensional space constructed from these $2f$ variables as the coordinates is the *phase space* of the system. Each point in the phase space (phase point) corresponds to a microscopic state. Therefore the microscopic states in classical statistical mechanics make a continuous set of points in phase space.

If the Hamiltonian of the system is denoted by $\mathscr{H}(q, p)$, the motion of the system is determined by the canonical equation of motion

$$\dot{p}_j = -\frac{\partial \mathscr{H}}{\partial q_j}, \qquad \dot{q}_j = \frac{\partial \mathscr{H}}{\partial p_j}, \qquad (j = 1, 2 \dots, f). \tag{1.1}$$

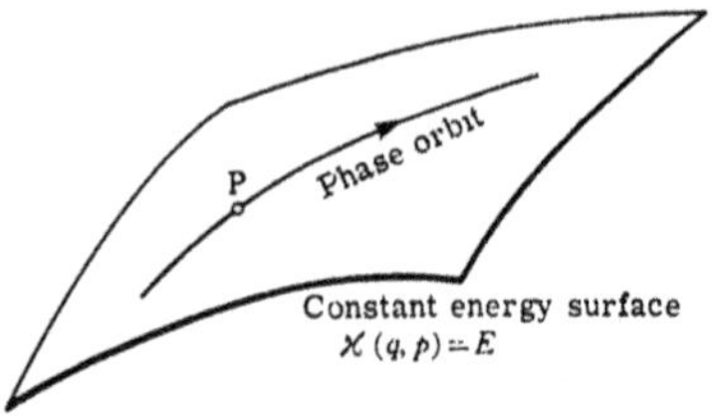

Fig. 1.1.

This determines the motion of the phase point, P_t, defining the state of the system at time t. This motion of P_t will be called the *natural motion* in the phase space. The trajectory of the phase point occurring during natural motion is called a *phase orbit*. For a conservative system, the energy is constant, i.e.

$$\mathscr{H}(q, p) = E. \tag{1.2}$$

Therefore the phase orbit must lie on a surface of constant energy (ergodic surface).

Quantum states: According to quantum mechanics, p and q cannot be specified simultaneously (the uncertainty principle of Heisenberg), so that classical phase space loses its rigorous meaning. In quantum statistical mechanics, a microscopic state is a state defined in a quantum mechanical sense. In particular, a stationary dynamical state of a system must be one of the quantum states determined by the equation

$$\mathscr{H}\varphi_l = E_l\varphi_l \qquad (l = 1, 2, \ldots). \tag{1.3}$$

Here $\mathscr{H}$ is the Hamiltonian of the system, E_l the energy of the quantum state l and φ_l is the wave function representing the quantum state l.

The set of microscopic states in quantum statistical mechanics is thus a discrete denumerable set of quantum states denoted by the quantum number l. (In statistical mechanics, one usually considers a system confined in a limited space, so that the quantum number l is usually discrete. A system with infinite extension is considered as the limit of one of finite extension.)

§ 1.2. STATISTICAL TREATMENT

Whenever a system is kept in equilibrium and remains constant according to macroscopic observations, it never stays constant from the microscopic point of view, and so one can never say precisely in which microscopic state the system is found. One can only define the probability for the set of all possible microscopic states of the system.

Fundamental assumption for observed values of physical quantities: Suppose a physical quantity A is observed for the system under consideration. A is a dynamical quantity from the microscopic point of view and is a function of microscopic states. The microscopic value of A is represented by $A(q, p) = A(\mathrm{P})$ in classical mechanics (P is a phase point) and by the expectation value†

$$A_l = \int \varphi_l^* A \varphi_l \,\mathrm{d}\tau \equiv \langle l | A | l \rangle \tag{1.4}$$

in the quantum state l in quantum mechanics. The observed value A_{obs} in the macroscopic sense must be a certain average of microscopic A: i.e.

$$A_{\mathrm{obs}} = \bar{A}. \tag{1.5}$$

Realization probability of a microscopic state: Let $\mathfrak{M}$ be the set of all possible microscopic states which can be realized by the system under a certain macroscopic condition. $\mathfrak{M}$ is classically a certain subspace of the phase space and quantum-mechanically it is a set of quantum states of the system. The probability that these microscopic states are realized is defined as the

† The integration in the following expression is carried out over the variables which are used to represent the wave function, say $q_1, q_2, \ldots q_f$. Here $\mathrm{d}\tau$ is a volume element of the space of these variables. Note that a quantum state corresponds to a phase orbit in classical mechanics and so A_l corresponds to the average taken over such an orbit.

probability that one of the microscopic states in the volume element $\Delta\Gamma$ of phase space is realized:

$$Pr(\Delta\Gamma) = \int_{\Delta\Gamma} f(\mathrm{P})\,\mathrm{d}\Gamma, \qquad (\Delta\Gamma \in \mathfrak{M}) \tag{1.6a}$$ †

or the probability that the quantum state l is realized:

$$Pr(l) = f(l), \qquad (l \in \mathfrak{M}) \tag{1.6b}$$

that is, by giving the probability density $f(\mathrm{P}) = f(q, p)$ or the probability $f(l)$. $f(\mathrm{P})$ and $f(l)$ are sometimes called simply the distribution functions.†† When the distribution functions are given, the average value (1.5) is explicitly written as

$$A_{\mathrm{obs}} = \bar{A} = \int_{\mathfrak{M}} A(\mathrm{P}) f(\mathrm{P})\,\mathrm{d}\Gamma, \tag{1.7a}$$

$$\bar{A} = \sum_{\mathfrak{M}} A_l f(l)\,. \tag{1.7b}$$

Statistical ensembles: In order to make the probabilistic idea as clear as possible, let us consider an hypothetical ensemble consisting of a great number of systems each of which has the same structure as the system under observation, and assume that the probability that a system arbitrarily chosen from this ensemble is found to be in a particular microscopic state is given by (1.6a) or (1.6b). For this hypothetical ensemble, (1.5) may be written as

$$A_{\mathrm{obs}} = \text{ensemble average of } A \equiv \bar{A}\,. \tag{1.8}$$

A statistical ensemble is defined by the distribution function which characterizes it. The most fundamental ensemble is the micro-canonical ensemble to be discussed later, but many other ensembles can be considered corresponding to various physical conditions (see (1.12) and (1.13)).

Ideal gas – Γ space and μ-space: So far the whole system in question is considered as the object of statistical treatment. This is the general standpoint of statistical mechanics established, in particular, by Gibbs. If the system under consideration is an ideal gas or a nearly ideal gas, it is possible to take each molecule as a statistical unit and regard the gas as a real ensemble

† A volume element of phase space is denoted by $\mathrm{d}\Gamma$: $\mathrm{d}\Gamma = \mathrm{d}q_1 \mathrm{d}q_2 \ldots \mathrm{d}q_f \mathrm{d}p_1 \mathrm{d}p_2 \ldots \mathrm{d}p_f$.

†† In mathematical probability theory, a distribution function is usually defined by

$$\int_{-\infty}^{x} f(x)\,\mathrm{d}x = F(x)$$

in the one-dimensional case, for example. The term "distribution function" in statistical mechanics is usually used in a loose way.

consisting of such units. This point of view was taken in the kinetic theory of gases which became the prototype of statistical mechanics. From this standpoint, the important thing is, in classical statistical mechanics, the distribution function of the position $\boldsymbol{x}$ and the momentum $\boldsymbol{p}$ of a molecule, i.e., the probability that a molecule chosen from the ensemble of gas molecules is found to have the coordinate and momentum values between $\boldsymbol{x}$ and $\boldsymbol{x} + d\boldsymbol{x}$, $\boldsymbol{p}$ and $\boldsymbol{p} + d\boldsymbol{p}$ is equal to

$$f(\boldsymbol{x}, \boldsymbol{p})\,d\boldsymbol{x}\,d\boldsymbol{p}\,. \tag{1.9}$$

Most of the properties of dilute gases can be derived from a knowledge of this distribution function. This is a distribution in a six-dimensional space, which is often called the *μ-space*. The phase space of the N molecules of the gas is called the *Γ space*.

Maxwell distribution: In a thermal equilibrium state at high temperatures, the distribution function f for a dilute gas is given by

$$f(\boldsymbol{p}) = \frac{1}{(2\pi mkT)^{\frac{3}{2}}} \exp\left\{ -\frac{1}{2mkT}(p_x^2 + p_y^2 + p_z^2) \right\} \tag{1.10}$$

where T is the absolute temperature, m the mass of a molecule, and k the Boltzmann constant. This Maxwell distribution can be derived by various methods. The most general derivation will be described later § 1.15, eq. (1.100).

§ 1.3. THE PRINCIPLE OF EQUAL WEIGHT AND THE MICROCANONICAL ENSEMBLE

When a system consisting of a great number of particles (more generally a system having a great number of degrees of freedom) is isolated for a long time from its environment, it will finally reach a thermal equilibrium state. In this case, the energy of the system is constant, so that it is presumed to be fixed at the value E with a certain allowance δE. This is the prescribed macroscopic condition. The set $\mathfrak{M}(E, \delta E)$ of the microscopic states to be considered under such conditions † is

classically: the shell-like subspace of the phase space between the two constant-energy surfaces for $\mathscr{H} = E$ and $\mathscr{H} = E + \delta E$;

and quantum-mechanically: the set of quantum states having the energy eigenvalues in the interval $E < E_l < E + \delta E$.

The principle of equal weight: In a thermal equilibrium state of an isolated

† Under certain circumstances, other constants of motion such as the total linear momentum or the total angular momentum may be prescribed. In such a case, $\mathfrak{M}$ is further restricted.

system, each of the microscopic states belonging to the set $\mathfrak{M}(E, \delta E)$ is realized with equal probability, namely:

(classically)

$$f(\mathrm{P}) = \text{constant} = \left[\int_{E<\mathscr{H}<E+\delta E} \mathrm{d}\Gamma\right]^{-1}, \qquad \mathrm{P}\in\mathfrak{M}(E, \delta E) \tag{1.11a}$$

(quantum-mechanically)

$$f(l) = \text{constant} = \left[\sum_{E<E_l<E+\delta E} 1\right]^{-1}, \qquad l\in\mathfrak{M}(E, \delta E). \tag{1.11b}$$

The microcanonical ensemble: A statistical ensemble defined by the principle of equal weight, or more precisely by the probability distribution given by (1.11a) or (1.11b), is called a microcanonical ensemble, and the distribution is called a *microcanonical distribution.* A microcanonical ensemble thus represents an isolated system which has reached thermal equilibrium.

Classical limit ($\delta E \to 0$): Using classical considerations, one may go to the limit $\delta E \to 0$ and take the set $\sigma(E)$ on the surface of constant energy E instead of $\mathfrak{M}(E, \delta E)$. Then one has, instead of (1.6a) and (1.11a)

$$Pr(\Delta\sigma) = \int_{\Delta\sigma} f(\mathrm{P})\,\mathrm{d}\sigma, \qquad \mathrm{P}\in\sigma(E) \tag{1.12a}$$

or,

$$f(P)\,\mathrm{d}\sigma = \frac{\mathrm{d}\sigma}{|\operatorname{grad}\mathscr{H}|} \bigg/ \int_{\mathscr{H}=E} \frac{\mathrm{d}\sigma}{|\operatorname{grad}\mathscr{H}|}, \tag{1.12b}$$

where $\mathrm{d}\sigma$ is a surface element on the constant energy surface, and

$$|\operatorname{grad}\mathscr{H}| = \left[\sum_j \left\{\left(\frac{\partial\mathscr{H}}{\partial p_j}\right)^2 + \left(\frac{\partial\mathscr{H}}{\partial q_j}\right)^2\right\}\right]^{\frac{1}{2}}.$$

Equation (1.7a) becomes

$$\bar{A} = \int_{\mathscr{H}=E} \frac{A(q,p)\,\mathrm{d}\sigma}{|\operatorname{grad}\mathscr{H}|} \bigg/ \int_{\mathscr{H}=E} \frac{\mathrm{d}\sigma}{|\operatorname{grad}\mathscr{H}|}. \tag{1.13}$$

Ergodic theorem: In classical mechanics, the dynamical states of an isolated system are represented by the motion of a phase point in phase space and so a dynamical quantity A is represented by a time-dependent quantity $A_t = A(\mathrm{P}_t)$ which changes in time according to the motion of the phase point. An observed value A_{obs} of A is therefore to be considered as a time average of A_t. Since A_{obs} remains constant for the thermal equilibrium state of the system, it may be an average over a sufficiently long period of time. In this way, it appears possible to justify the principle of equal weight as

$$\begin{aligned} A_{\text{obs}} &= \text{a long time average of } A_t \\ &= \text{the phase average (1.13) of } A(\text{P}). \end{aligned} \tag{1.14}$$

The second equality of the above equation,

$$A_{\text{time average}} = A_{\text{phase average}} \tag{1.15}$$

is called the ergodic theorem. This theorem has been studied as a mathematical problem. But there exist various arguments about its physical significance as to whether it really provides convincing grounds for the principle of equal weight.

The finite allowance of the energy δE: In quantum mechanics, the energy of a system has an uncertainty

$$(\delta E)_{\text{qu}} \sim h/t \tag{1.16}$$

determined by the length of the time of observation, t. Here h is the Planck constant. Therefore one has to choose $\delta E > (\delta E)_{\text{qu}}$.

δE may be small, but in this range there exist a great number of quantum states if the system has a macroscopic size so that statistical considerations become possible. (If the system stays certainly in one quantum state, there would be no need of a statistical treatment.)

NOTE: The existence of the allowance δE is necessary. But its ambiguity may cause concern for some readers. It can be seen, however, that the magnitude of δE does not affect thermodynamic properties in macroscopic systems (see § 1.6, (1.27)).

§ 1.4. THE THERMODYNAMIC WEIGHT OF A MACROSCOPIC STATE AND ENTROPY

Variables defining a macroscopic state: As the variables defining a macroscopic state of a system under consideration, one may choose the energy E (with an allowance δE), the numbers N_A, N_B ... of particles of various kinds existing in the system, the volume V of the box which contains the system, and other parameters x, ... to specify external forces, such as the electric field strength acting on the system. The Hamiltonian $\mathscr{H}$ of the system involves the variables N_A, N_B, ... V, x, ...

Thermodynamic weight: In quantum statistical mechanics, the total number $W(E, \delta E, N, V, x)$ of the possible quantum states for the set of prescribed values of the variables $E(\delta E)$, N_A N_B, ... V, x, ... is called the thermodynamic weight of that particular macroscopic state of the system. Namely (quantum-mechanically):

$$W(E, \delta E, N, V, x) = \sum_{E < E_l(N, V, x) < E + \delta E} 1 . \tag{1.17a}$$

The thermodynamical weight in classical statistical mechanics should be defined as the limit of quantum statistical mechanics. It is given by (classically):

$$W(E, \delta E, N, V, x) = \int\limits_{E < \mathscr{H}(N, V, x) < E + \delta E} \frac{\mathrm{d}\Gamma}{h^{3(N_A + N_B + \cdots)}\, N_A! N_B! \ldots} . \tag{1.17b}$$

The denominator dividing the volume element $\mathrm{d}\Gamma$ of phase space is derived from the correspondence of the classical phase space and quantum states (see § 1.5).

Statistical definition of entropy: The entropy defined by the Boltzmann relation

$$S(E, N_A, N_B, \ldots, V, x) = k \log W(E, \delta E, N, V, x) \tag{1.18}$$

is called the statistical entropy. Here k is the Boltzmann constant. One has to show the well-known thermodynamic equation

$$\mathrm{d}S(E, N_A, N_B, \ldots, V, x) = \frac{1}{T}(\mathrm{d}E + p\,\mathrm{d}V - \sum X\,\mathrm{d}x - \sum \mu\,\mathrm{d}N_i) \tag{1.19}$$

in order to convince oneself that the statistical entropy is identical with the thermodynamic entropy. This will be seen later.

NOTE: The allowance δE of the energy does not affect the numerical values of the entropy S (§ 1.6).

§ 1.5. NUMBER OF STATES AND THE DENSITY OF STATES

Number of states: The zero of the energy E is now so chosen that the energy levels of the system under consideration are non-negative:

$$0 \leqq E_1 \leqq E_2 \leqq E_3 \ldots .$$

The number of quantum states with energy values between 0 and E is denoted by $\Omega_0(E, N, V, x)$ and is called the number of states of the system:

$$\Omega_0(E, N, V, x) = \sum_{0 \leqq E_l \leqq E} 1 . \tag{1.20a}$$

Classically this is redefined by

$$\Omega_0(E, N, V, x) = \frac{1}{h^{3(N_A + N_B + \cdots)} \cdot N_A! N_B! \ldots} \int\limits_{\mathscr{H} \leqq E} \mathrm{d}\Gamma . \tag{1.20b}$$

State density:

$$\Omega(E,N,V,x) = \frac{\mathrm{d}}{\mathrm{d}E}\,\Omega_0(E,N,V,x) \tag{1.21}$$

is called the state density of the system. If δE is small enough, one may write

$$\Omega(E,N,V,x)\,\delta E = W(E,\delta E,N,V,x). \tag{1.22}$$

Classical number of states (Phase integral) (Correspondence of classical and quantum mechanics): Quantum mechanics reduces to classical mechanics in the limit of $h \to 0$, so that (1.20b) must be derived from (1.20a) by this limiting process. A general proof of this is somewhat too advanced to be given here (see problem 33, Chapter 2). Therefore only some explanation of (1.20b) will be given in the following. There are two elements in the correspondence of classical and quantum mechanics, which give rise to the two factors appearing in (1.20b).

(1) *Quantization of phase space:* For a system with f degrees of freedom, the set of microscopic states held in a volume element $\Delta\Gamma$ corresponds, in the limit $h \to 0$ (not considering another factor to be stated in (2)), to a set of

$$\Delta\Gamma/h^f \tag{1.23}$$

quantum states (as long as h remains finite, this correspondence is an approximation). This can be understood by means of the uncertainty principle

$$\Delta p \Delta q \sim h,$$

since the classical states in a cell of h per degree of freedom, or h^f per f degrees of freedom, merge into a single quantum state which cannot be further distinguished (see problems 7 and 8).

(2) *Indistinguishability of identical particles:* In quantum mechanics, *identical particles are indistinguishable* in principle. For instance, the state in which two identical particles have coordinates (p', x') and (p'', x'') respectively is distinguished classically from another state in which the second particle has coordinates (p', x') and the first particle has coordinates (p'', x''). However, these are simply the same single state from a quantum-mechanical point of view. This indistinguishability leads to the concept of Fermi and Bose statistics (§ 1.15), but in the limit $h \to 0$ it gives rise to the factor $1/N_A!\,N_B!\ldots$ in equation (1.20b). For example, the expression

$$\frac{1}{N!}\int\cdots\int_{\mathscr{H}\leq E} \frac{\mathrm{d}p_1\,\mathrm{d}x_1\ldots\mathrm{d}p_N\,\mathrm{d}x_N}{h^{3N}}$$

gives an approximation to the number of states below E for a system consisting of N identical particles. The indistinguishability of identical particles introduces the denominator $N!$ in the above expression because the $N!$ classical states† arising from a given phase point $\boldsymbol{p}_1, \boldsymbol{x}_1, \ldots \boldsymbol{p}_N, \boldsymbol{x}_N$ must be identified with each other by this principle (see the Note to Chapter 2, problem 33 for a more rigorous discussion).

NOTE: The denominator $N!$ was very difficult to understand before the principle of the indistinguishability of identical particles was introduced into quantum mechanics. In spite of this, the necessity for this denominator term had long been recognized in order to make the entropy defined by (1.18) an extensive quantity as it should be.

§ 1.6. NORMAL SYSTEMS IN STATISTICAL THERMODYNAMICS

Asymptotic forms of the number of states and state density of a macroscopic system: A system consisting of a great number of particles, or of a system with an indefinite number of particles but with a volume of macroscopic extension usually has a number of states $\Omega_0(E)$ which shows the following properties (in which case the system will be called *normal in the statistical-thermodynamic sense*):

(1) When the number N of particles (or the volume V) is large, the number of states $\Omega_0(E)$ approaches asymptotically to

$$\Omega_0 \sim \exp\left\{N\phi\left(\frac{E}{N}\right)\right\} \quad \text{or} \quad \exp\left\{V\psi\left(\frac{E}{V}\right)\right\}, \tag{1.24a}$$

$$\Omega_0 \sim \exp\left\{N\phi\left(\frac{E}{N},\frac{V}{N}\right)\right\} \quad \text{or} \quad \exp\left\{V\psi\left(\frac{E}{V},\frac{N}{V}\right)\right\}. \tag{1.24b}$$

If E/N (or E/V) is looked upon as a quantity of the order of $\mathrm{O}(1)$††, ϕ is also $\mathrm{O}(1)$ (the same holds for ψ), and

$$\phi > 0, \qquad \phi' > 0, \qquad \phi'' < 0. \tag{1.25}$$

(2) Therefore

$$\Omega = \mathrm{d}\Omega_0/\mathrm{d}E = \phi' \exp(N\phi) > 0,$$

$$\frac{\mathrm{d}\Omega}{\mathrm{d}E} = \left(\phi'^2 + \frac{\phi''}{N}\right)\mathrm{e}^{N\phi} \sim \phi'^2 \mathrm{e}^{N\phi} > 0. \tag{1.26}$$

---

† When some of $(\boldsymbol{p}_1, \boldsymbol{x}_1), (\boldsymbol{p}_2, \boldsymbol{x}_2) \ldots (\boldsymbol{p}_N, \boldsymbol{x}_N)$ coincide with each other, the number of classical states produced by the permutation of particle states is less than $N!$. But the chance for such coincidence is negligible in the limit of $h \to 0$.

†† One writes $y = \mathrm{O}(x)$ and $z = \mathrm{o}(x)$ if $\lim_{x\to\infty} y/x = \text{finite} \neq 0$ and $\lim_{x\to\infty} z/x = 0$.

When N (or V) is large, Ω_0 or Ω increases *very rapidly* with energy E. No general proof of these properties will be attempted here. If a system existed which did not have these properties, it would show a rather strange macroscopic behavior, very different from ordinary thermodynamic systems (see example 4, Chapter 1).

Entropy of a normal system: For the statistical entropy defined by (1.18), one finds the following from (1.24)–(1.26):

(1)
$$S = k \log\{\Omega(E)\delta E\} \simeq k \log \Omega_0(E) = kN\phi . \tag{1.27}$$

The error involved here is $o(N)$ (or $o(V)$), and so is negligible for a macroscopic system (for which N, V, or E is very large).

(2) The statistical temperature $T(E)$ is introduced by means of the definition,

$$\frac{\partial S}{\partial E} = \frac{1}{T} \tag{1.28}$$

$$T(E) = \frac{1}{k\phi'} > 0 . \tag{1.29}$$

By (1.24) and (1.25) it will be shown later that this temperature in fact agrees with the thermodynamic temperature (see § 1.9).

The allowance of the energy and the definition of entropy: By (1.24)–(1.26), the function $\Omega_0(E)$ is positive and increases monotonically with E. Therefore one has

$$\Omega(E)\delta E < \Omega_0(E) < \Omega(E)E ,$$

thus
$$S = k \log \Omega(E)\delta E < k \log \Omega_0(E) < k \log \Omega(E)E .$$

Also by (1.24) and (1.25) and using the fact that $E = O(N)$, one finds:

$$k\{\log \Omega(E)E - \log \Omega_0(E)\} = k \log E \cdot \phi' = O(\log N) = o(N) \quad (\text{or } o(V))$$

and

$$k\{\log \Omega(E)E - \log \Omega(E)\delta E\} = k \log E/\delta E = o(N)\dagger \quad (\text{or } o(V)).$$

Therefore (1.27) is seen to be valid.

§ 1.7. CONTACT BETWEEN TWO SYSTEMS

There can be various kinds of interactions between two systems in contact.

† If one supposes that $\log E/\delta E = O(N) = \alpha N$, then $\delta E = E \exp(-\alpha N)$. According to the uncertainty principle (1.16) the time of the observation t is then $t \sim h/\delta E = (h/E) \exp \alpha N$. If $\alpha = O(1)$, this t is astronomically long for a macroscopic system. Therefore, for a t of ordinary length, δE cannot be so small and thus one must have $\log E/\delta E = o(N)$ (namely $\alpha = o(1)$).

In phenomenological thermodynamics, these interactions are idealized as thermodynamic contacts, i.e., mechanical, thermal or material-transferring contacts. In statistical mechanics, correspondingly one considers the following types of contact:

(1) *Mechanical contact with a work source:* If the outside world (environment) is conceived simply as a source which exerts a force on the system under consideration, this mechanical (or electromagnetic) effect can be represented by the Hamiltonian $\mathscr{H}(q, p, x)$, in which x is the coordinate (for instance, the position of a piston in a box containing a gas), to describe the interaction between the system and the environment and is regarded as a variable of the system. Then

$$X = \frac{\partial \mathscr{H}(q, p, x)}{\partial x} \tag{1.30}$$

represents the force which the system exerts on the outside world.

(2) *Thermal contact between two systems:* When two systems with Hamiltonians $\mathscr{H}_{\mathrm{I}}$ and $\mathscr{H}_{\mathrm{II}}$ are in contact and are interacting with the interaction Hamiltonian $\mathscr{H}'$, then the total Hamiltonian for the composite system I + II is written as

$$\mathscr{H}_{\mathrm{I+II}} = \mathscr{H}_{\mathrm{I}} + \mathscr{H}_{\mathrm{II}} + \mathscr{H}' . \tag{1.31}$$

The two systems are said to be in thermal contact if the interaction $\mathscr{H}'$ satisfies the following two conditions:

a) $\mathscr{H}'$ is so small (the interaction is so weak) that each of the microscopic states of the composite system I + II, say l, is specified by giving the microscopic states l' and l'' of the subsystems I and II, and the energy E_l is to a good approximation the sum of the energies of the subsystems, namely

$$l = (l', l''), \qquad E_l = E^{\mathrm{I}}_{l'} + E^{\mathrm{II}}_{l''} . \tag{1.32}$$

b) On the other hand, the existence of $\mathscr{H}'$ allows a sufficiently frequent exchange of energy between the systems I and II. As a result, after waiting long enough one can certainly expect that the composite system I + II reaches a final state, whatever the initial state was. In this final state every microscopic state (l', l'') of the composite system is realized with equal probability in accordance with the principle of equal weight. This final state is called the statistical equilibrium of the two systems and corresponds to thermal equilibrium in thermodynamics. The two systems here may be locally separated, or they can be different sets of dynamical degrees of freedom.

(3) *Material-transferring contact:* In similar fashion to (2), if the interaction $\mathscr{H}$ between two systems allows an exchange of material particles and

still the microscopic states of the composite system can be represented by

$$(N, l) = (N'l', N''l''), \qquad E_l(N) = E^{\mathrm{I}}_{l'}(N') + E^{\mathrm{II}}_{l''}(N'') \qquad (1.33)\dagger$$

to a good approximation (namely the interaction is weak enough), then the interaction is called a material-transferring contact.

DIVERTISSEMENT 1

Maxwell's demon. Entropy never decreases spontaneously in an isolated system. Who in the world has ever seen water in a kettle boiling by itself taking heat from a block of ice on which the kettle has been placed? Who has ever seen two boxes filled with gas at the same temperature and pressure spontaneously producing an unbalance of temperature, one becoming automatically heated up and the other cooled down, when they are connected by an open window on the wall between the boxes? Of course, nobody. Wouldn't it be possible, however, to find an extremely clever creature who stands by the window watching gas molecules coming by and who opens the window only when hot molecules, namely those molecules having kinetic energy greater than the average, approach the window from one side or when cold molecules come from the opposite side? Maxwell imagined such a clever demon.

The second law of thermodynamics denies the existence of Maxwell's demon. You may be able to find a demon who starts to do an extremely fine job of selecting molecules passing through the window. But he will never be able to continue the work indefinitely. Soon he will become dazzled and get sick and lose his control. Then the whole system, the gas molecules and the demon himself, will again approach a final equilibrium, where the temperature difference the demon once succeeded in building up disappears and the demon will run a fever at a temperature equal to that of the gas. A living organism may look like a Maxwell's demon, but it is not. A living organism is an open system, through which material, energy, and entropy are flowing. But life itself cannot violate thermodynamic laws.

(4) *Pressure-transmitting contact:* When two systems are separated by a movable wall, then an exchange of volume is possible between the two systems. If the wall allows only volume exchange but not energy or particle exchange, this is an example of a purely mechanical contact, but it may also be regarded as an interaction of two subsystems for which the approximation

$$(V, l) = (V'l', V''l''), \qquad E_l(V) = E^{\mathrm{I}}_{l'}(V') + E^{\mathrm{II}}_{l''}(V'') \qquad (1.34)$$

is possible.

† (N', N'') is the partition of N particles to the subsystems I and II. $E_{l'}{}^{\mathrm{I}}(N')$ and $E_{l''}{}^{\mathrm{II}}(N'')$ denote the energies of quantum states of the subsystems containing N' and N'' particles respectively.

§ 1.8. QUASI-STATIC ADIABATIC PROCESS

Quasi-static adiabatic process in statistical mechanics†: This is a process involving a very slow change of a parameter x which determines a purely mechanical interaction of a system with an external work source, namely:

$$\mathscr{H}(q, p, x) \to \mathscr{H}(q, p, x + \Delta x), \qquad \mathrm{d}x/\mathrm{d}t \to 0. \tag{1.35}$$

Adiabatic theorem in dynamics: Dynamical quantities which are kept invariant in a (quasi-static) adiabatic process are called *adiabatic invariants*. It can be shown, in particular, that the number of states $\Omega_0(E)$ is an adiabatic invariant. Namely,

$$\Omega_0(E, x) = \Omega_0(E + \mathrm{d}E, x + \mathrm{d}x), \qquad (\mathrm{d}x/\mathrm{d}t \to 0), \tag{1.36}$$

where $\mathrm{d}E$ is the increment of energy of the system in the process involving an adiabatic increase $\mathrm{d}x$ of x. For this, it holds that

$$\mathrm{d}E = \langle X \rangle_{\text{time average}}\, \mathrm{d}x = \bar{X}\, \mathrm{d}x\,; \tag{1.37}$$

$\bar{X}$ here means the phase average of X, or the average of X over a microcanonical ensemble.

Adiabatic theorem in statistical mechanics: It is a rather debatable question whether the adiabatic theorem in pure dynamics retains its original significance for systems with enormously large number of degrees of freedom such as treated in statistical mechanics, but (1.36) and (1.37) will be assumed valid here without examining any detailed arguments about this question (see problem 34). Therefore one has, from (1.36) and (1.37),

$$\mathrm{d}S/\mathrm{d}x = 0, \qquad (\mathrm{d}x/\mathrm{d}t \to 0) \tag{1.38}$$

for a quasi-static adiabatic process. In other words,

$$\mathrm{d}S = \frac{\partial S}{\partial E}\mathrm{d}E + \frac{\partial S}{\partial x}\mathrm{d}x = 0, \qquad \mathrm{d}E = \bar{X}\mathrm{d}x \tag{1.39}$$

for an adiabatic reversible process. This equation gives the formulae

$$\bar{X} = \left(\frac{\partial E}{\partial x}\right)_S, \tag{1.40}$$

$$\bar{X} = -\left(\frac{\partial S}{\partial x}\right)_E \Big/ \left(\frac{\partial S}{\partial E}\right)_x = -T\left(\frac{\partial S}{\partial x}\right)_E \tag{1.41}$$

† In mechanics, this is often called simply an adiabatic process.

for the statistical mechanical force or the average of X. Here T is the statistical temperature (1.28). Also one has

$$dS = \frac{1}{T}(dE - \bar{X}dx). \tag{1.42}$$

§ 1.9. EQUILIBRIUM BETWEEN TWO SYSTEMS IN CONTACT

Partition of energy between two systems in thermal contact: The principle of equal weight (see § 1.3) can be applied to the composite system I + II represented by (1.32), whereby one obtains the probability that the subsystems I and II have the energies E_{I} and E_{II} ($E_{\text{I}} + E_{\text{II}} = E$) respectively. Let the state densities of I and II be denoted by Ω_{I} and Ω_{II} respectively and that of I + II by Ω, then one has

$$\begin{aligned}\Omega(E)\delta E &= \iint\limits_{E<E_{\text{I}}+E_{\text{II}}<E+\delta E} \Omega_{\text{I}}(E_{\text{I}})\Omega_{\text{II}}(E_{\text{II}})\, dE_{\text{I}}\, dE_{\text{II}} \\ &= \delta E \int \Omega_{\text{I}}(E_{\text{I}})\Omega_{\text{II}}(E - E_{\text{I}})\, dE_{\text{I}}.\end{aligned} \tag{1.43}$$

Therefore the probability that the system I has energy in the range between E_{I} and $E_{\text{I}} + dE_{\text{I}}$ is given by

$$f(E_{\text{I}})\, dE_{\text{I}} = \frac{\Omega_{\text{I}}(E_{\text{I}})\Omega_{\text{II}}(E - E_{\text{I}})\, dE_{\text{I}}\delta E}{\Omega(E)\delta E} \qquad \left(\int f(E_{\text{I}})\, dE_{\text{I}} = 1\right). \tag{1.44}$$

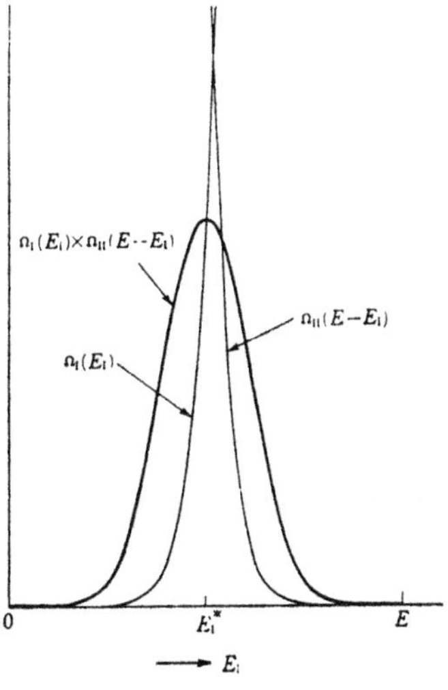

Fig. 1.2.

The most probable partition of energy: (1.44) gives the probability of energy partition when I + II are in statistical equilibrium (thermal equilibrium).

The function $\Omega_{\text{I}}(E_{\text{I}})$ increases very rapidly with E_{I} (see (1.26)) and $\Omega_{\text{II}}(E-E_{\text{I}})$ decreases very rapidly with E_{I}, so that this probability has a *very steep maximum* at a certain partition (E_{I}^*, $E_{\text{II}}^* = E - E_{\text{I}}^*$) (see Fig. 1.2). This is the partition to be expected almost certainly in the equilibrium state of I + II. This is determined by the condition

$$\Omega_{\text{I}}(E_{\text{I}})\Omega_{\text{II}}(E - E_{\text{I}})\,\mathrm{d}E_{\text{I}}\delta E = \text{max.} \tag{1.45}$$

which is, by (1.27), equivalent to

$$S_{\text{I}}(E_{\text{I}}) + S_{\text{II}}(E - E_{\text{I}}) = \text{max.} \tag{1.46a}$$

or to

$$\frac{\partial S_{\text{I}}}{\partial E_{\text{I}}} = \frac{\partial S_{\text{II}}}{\partial E_{\text{II}}}, \qquad (E_{\text{I}}^* + E_{\text{II}}^* = E)\,. \tag{1.46b}$$

This may be also expressed as

$$T_{\text{I}}(E_{\text{II}}^*) = T_{\text{II}}(E_{\text{II}}^*)\,, \qquad E_{\text{I}}^* + E_{\text{II}}^* = E \tag{1.47}$$

(see (1.28)).

Equilibrium of two systems with a material-transferring contact: When particle exchange is made possible between two systems I and II through a material-transferring contact, the probability of partition (N_{I}, N_{II}) of a certain kind of particles (with the total number N) is derived in the same way as (1.44) and is given by

$$f(E_{\text{I}}, N_{\text{I}})\,\mathrm{d}E_{\text{I}} = \frac{\Omega_{\text{I}}(E_{\text{I}}, N_{\text{I}})\Omega_{\text{II}}(E - E_{\text{I}}, N - N_{\text{I}})\,\mathrm{d}E_{\text{I}}\delta E}{\Omega(E, N)\delta E} \tag{1.48}$$

which represents the probability that the subsystem I has an energy E_{I} and N_{I} particles. In this expression

$$\Omega(E, N)\delta E = \delta E \sum_{N_{\text{I}}=0}^{N} \int \Omega_{\text{I}}(E_{\text{I}}, N_{\text{I}})\Omega_{\text{II}}(E - E_{\text{I}}, N - N_{\text{I}})\,\mathrm{d}E_{\text{I}} \tag{1.49}$$

is the thermodynamic weight for the composite system I + II. The quantities $\Omega_{\text{I}}(E_{\text{I}}, N_{\text{I}})\,\mathrm{d}E_{\text{I}}$ and $\Omega_{\text{II}}(E_{\text{II}}, N_{\text{II}})\,\mathrm{d}E_{\text{II}}$ refer, respectively, to system I having E_{I} and N_{I} and to system II having E_{II} and N_{II}, the energy allowance being $\mathrm{d}E_{\text{I}}$ and $\mathrm{d}E_{\text{II}}$ respectively. The most probable partition is obtained in the same way as (1.45), (1.46) and (1.47) from the conditions

$$\begin{gathered} S_{\text{I}}(E_{\text{I}}, N_{\text{I}}) + S_{\text{II}}(E_{\text{II}}, N_{\text{II}}) = \text{max.}, \\ E_{\text{I}} + E_{\text{II}} = E\,, \qquad N_{\text{I}} + N_{\text{II}} = N \end{gathered} \tag{1.50}$$

or by the conditions that the temperatures of the subsystems must be equal, (1.47), and that

$$\frac{\mu_{\text{I}}}{T_{\text{I}}}(E_{\text{I}}^*, N_{\text{I}}^*) = \frac{\mu_{\text{II}}}{T_{\text{II}}}(E_{\text{II}}^*, N_{\text{II}}^*), \qquad N_{\text{I}}^* + N_{\text{II}}^* = N\,. \tag{1.51}$$

Here the chemical potential μ is introduced by the definition

$$\frac{\partial}{\partial N} S(E, N) = -\frac{\mu}{T}. \tag{1.52}$$

Equilibrium of two systems exerting pressure upon each other: When two systems are in contact through a movable wall, so that an exchange of volume is made possible (§ 1.7,(4)), the condition to determine the most probable partition of volumes is given by

$$\frac{p_{\text{I}}}{T_{\text{I}}}(E_{\text{I}}^*, N_{\text{I}}^*, V_{\text{I}}^*) = \frac{p_{\text{II}}}{T_{\text{II}}}(E_{\text{II}}^*, N_{\text{II}}^*, V_{\text{II}}^*), \qquad V_{\text{I}}^* + V_{\text{II}}^* = V \tag{1.53}$$

where p is defined by

$$\frac{\partial}{\partial V} S(E, N, V) = \frac{p}{T}. \tag{1.54}$$

This is derived in the same way as before. (If heat exchange is possible, then one has also the condition (1.47).)

Additivity of entropy for systems in equilibrium: If the subsystems I and II are normal in a statistical-thermodynamic sense, the additivity of entropy holds for the composite system I + II, namely,

$$S_{\text{I}+\text{II}}(E, N, V) = S_{\text{I}}(E_{\text{I}}^*, N_{\text{I}}^*, V_{\text{I}}^*) + S_{\text{II}}(E_{\text{II}}^*, N_{\text{II}}^*, V_{\text{II}}^*) \tag{1.55}$$

where E_{I}^*, etc., represent the most probable partitions (all quantities which are not allowed to be exchanged by virtue of the nature of the contact are fixed). Equation (1.55) means that the entropy of the composite system is the sum of the entropies of the subsystems (example 5).

Increase of entropy by establishment of new equilibrium: Two separate systems I and II are given with the partition (E_{I}^0, N_{I}^0, V_{I}^0; E_{II}^0, N_{II}^0, V_{II}^0) which is different from (E_{I}^*, N_{I}^*, V_{I}^*; E_{II}^*, N_{II}^*, V_{II}^*). Then the two are brought into contact and are made to reach a new equilibrium. In this new equilibrium, the entropy of the total system becomes

$$\begin{aligned} &S_{\text{I}+\text{II}}(E, N, V) \\ &= S_{\text{I}}(E_{\text{I}}^*, N_{\text{I}}^*, V_{\text{I}}^*) + S_{\text{II}}(E_{\text{II}}^*, N_{\text{II}}^*, V_{\text{II}}^*) \\ &\geqq S_{\text{I}}(E_{\text{I}}^0, N_{\text{I}}^0, V_{\text{I}}^0) + S_{\text{II}}(E_{\text{II}}^0, N_{\text{II}}^0, V_{\text{II}}^0) = S_{\text{initial}}. \end{aligned} \tag{1.56}$$

If the systems I and II are separated again, they will almost certainly keep the partition $(E_{\mathrm{I}}^{*}, N_{\mathrm{I}}^{*}, V_{\mathrm{I}}^{*}; E_{\mathrm{II}}^{*}, N_{\mathrm{II}}^{*}, V_{\mathrm{II}}^{*})$. Therefore one has

$$S_{\mathrm{final}} = S_{\mathrm{I}}(E_{\mathrm{I}}^{*}, N_{\mathrm{I}}^{*}, V_{\mathrm{I}}^{*}) + S_{\mathrm{II}}(E_{\mathrm{II}}^{*}, N_{\mathrm{II}}^{*}, V_{\mathrm{II}}^{*}) \geqq S_{\mathrm{initial}}. \tag{1.56'}$$

In other words, the entropy has been increased as a result of bringing I and II into contact, and remains unchanged only when I and II were originally in equilibrium with each other (namely the initial partition is equal to the most probable partition after contact). When the two systems are separated after contact, the entropy of the two systems is almost certainly found to be greater than before contact, and they are almost certainly in equilibrium.

§ 1.10. FUNDAMENTAL LAWS OF THERMODYNAMICS

The first law of thermodynamics is, so to say, a self-evident consequence of the dynamical laws which satisfy the conservation of energy. The internal energy U is nothing but the energy E of the system, if the mechanical energy is not included, and is therefore a function of the state.

Entropy: By the definitions (1.27), (1.28), (1.52) and (1.54), the entropy S satisfies the relation

$$\mathrm{d}S = \frac{1}{T}(\mathrm{d}E + p\,\mathrm{d}V - \mathfrak{X}\,\mathrm{d}x - \mu\,\mathrm{d}N) \tag{1.57}$$

which gives the difference of entropy in the two neighboring equilibrium states (E, V, x, N) and $(E + \mathrm{d}E, V + \mathrm{d}V, x + \mathrm{d}x, N + \mathrm{d}N)$. When the system suffers a change between these two states in a quasi-static process, (1.57) can be written as

$$\mathrm{d}E = -p\,\mathrm{d}V + \mathfrak{X}\,\mathrm{d}x + \mu\,\mathrm{d}N + \mathrm{d}'Q, \tag{1.58}$$

with

$$\mathrm{d}'Q = T\,\mathrm{d}S \tag{1.59}$$

which represents the heat introduced into the system from the outside in the course of the process (because the first two terms represent the mechanical work given to the system from the outside, the third term is the energy increase brought about by material transfer, and so the last term must be the energy given from the outside in other forms, namely as heat flow).

Temperature: The temperature $T(E)$ defined by (1.28) coincides with the thermodynamic absolute temperature and accordingly statistical entropy coincides with the thermodynamic entropy (note that, however, statistical entropy has been defined without an arbitrary additive constant), because (i) $T(E)$ gives the condition for thermal equilibrium as shown by (1.47) and (ii) it becomes the integrating denominator for the differential form $\mathrm{d}'Q$. The first is a characteristic general property for the temperature and the second agrees with the definition of absolute temperature as introduced by the second law of thermodynamics.

The second law of thermodynamics: This law is interpreted by a probabilistic concept in statistical mechanics. When a certain inhibition (which, for instance, prohibits mutual contact of two systems) is removed and two systems are brought into contact, the entropy of the whole system will almost certainly increase (see (1.56′)).

Therefore a spontaneous decrease of entropy against the second law of thermodynamics is extremely improbable.

For any process in which the system under consideration comes into contact with an external environment, it very certainly holds that

$$\mathrm{d}S + \mathrm{d}S_{\mathrm{e}} \geqq 0 \,. \tag{1.60}$$

If the external environment is so large that its own change can be considered as quasi-static and its temperature is equal to the temperature of the system, then one may put $\mathrm{d}S_{\mathrm{e}} = -\,\mathrm{d}'Q/T$. Therefore the inequality

$$T\,\mathrm{d}S \geqq \mathrm{d}'Q \tag{1.61}$$

will be satisfied almost certainly for the entropy increase $\mathrm{d}S$ of the system and the heat $\mathrm{d}'Q$ given to the system in the process.

The third law of thermodynamics: Entropy is defined by (1.18) in an absolute manner. It is always non-negative by definition ($W \geqq 1$). For a real quantummechanical system the existence of the lowest ground state may be usually-assumed. If the density of the system remains finite, $\log \Omega(E)$ will approach, as $E \to 0$ (the lowest energy value), a value which is independent of the size, N or V, of the system. This means that the statement $S \to 0$ (as $E \to 0$) is a consequence of the quantum-mechanical definition (1.18) for real physical systems. It must be noted, however, that this does not necessarily assure that an actual experimental measurement will give $S \to 0$ as $T \to 0$. It may happen that the lowest state of the system is not attained in the course of the experiment because the motion of particles is extremely slowed down as one goes to $T \to 0$. When such *freezing* occurs, the observed entropy will approach a value not equal to zero (example: glass).

§ 1.11. THE MOST PROBABLE STATE AND FLUCTUATIONS

Probability and entropy: Let $\alpha(\alpha_1, \alpha_2, \ldots)$ be the parameters besides E, N, and V required to specify the macroscopic state of a system, and let $W(E, N, V, \alpha)$ be the thermodynamic weight of the state (E, N, V, α). Then the probability for the realization of the state (E, N, V, α) is, by the principle of equal weight,

$$P(\alpha) = \frac{W(E,N,V,\alpha)}{\sum_{\alpha} W(E,N,V,\alpha)} = Ce^{S(E,N,V,\alpha)/k}, \tag{1.62}$$

where

$$S(E,N,V,\alpha) = k \log W(E,N,V,\alpha) \tag{1.63}$$

is the entropy of the state (E, N, V, α).

Usually the most probable value α^* and the average value $\bar{\alpha}$ of α coincide with each other, because the maximum of the probability $P(\alpha)$ at $\alpha = \alpha^*$ is extremely steep when the system is large. The most probable value α^* is determined by the condition that $P(\alpha)$ is maximum or

$$S(E,N,V,\alpha) = \text{max.}, \qquad \alpha = \alpha^* \tag{1.64}$$

or

$$\frac{\partial S(E,N,V,\alpha_1^*,\alpha_2^*,\ldots)}{\partial \alpha_j^*} = 0, \qquad j = 1,2,\ldots,m. \tag{1.65}$$

Determination of most probable values under constraints: When certain constraints,

$$\Phi_n(\alpha_1,\alpha_2,\ldots) = 0, \qquad n = 1,2,\ldots,r \tag{1.66}$$

are imposed on the variables $\alpha_1, \alpha_2, \ldots \alpha_m$, the most probable values of α's are determined by the condition (1.66) and

$$\frac{\partial S}{\partial \alpha_j} + \sum_n \lambda_n \frac{\partial \Phi_n}{\partial \alpha_j} = 0, \qquad j = 1,2,\ldots \tag{1.67}$$

where $\lambda_1, \lambda_2, \ldots \lambda_r$ are Lagrange's undetermined multipliers.†

Fluctuations: By (1.62)

$$P(\alpha') = C' \exp\left[\frac{1}{k}\{S(E,N,V,\alpha^* + \alpha') - S(E,N,V,\alpha^*)\}\right] \tag{1.68}$$

gives the probability for a realization of deviations of α's from their most

† For the maximum probability one must have

$$\delta S = \sum_{j=1}^{m} \frac{\partial S}{\partial \alpha_j} \delta\alpha_j = 0 \tag{1}$$

subject to the conditions

$$\delta \Phi_k = \sum_{j=1}^{m} \frac{\partial \Phi_k}{\partial \alpha_j} \delta\alpha_j = 0 \qquad k = 1, \ldots, r. \tag{2}$$

With r initially undetermined multipliers $\lambda_1, \lambda_2, \ldots, \lambda_r$, the following equations must probable values. If the α's are continuous, (1.68) may be written as

$$P(\alpha_1', \alpha_2', \ldots)\,\mathrm{d}\alpha_1'\,\mathrm{d}\alpha_2' \ldots = C'\exp\left\{\frac{1}{2k}\sum\frac{\partial^2 S}{\partial\alpha_i^*\partial\alpha_j^*}\alpha_i\alpha_j\right\}\mathrm{d}\alpha_1'\,\mathrm{d}\alpha_2' \ldots \quad (1.69)$$

which can be used for deviations which are not too large †.

§ 1.12. CANONICAL DISTRIBUTIONS

Canonical distribution: When a system with volume V and $N_A, N_B \ldots$ particles is in equilibrium with a heat bath at temperature T, the probability of its microscopic states is given by

$$\text{(classically)} \qquad Pr(\mathrm{d}\Gamma) = \frac{1}{\prod N_A! h^{3N}} \frac{\mathrm{e}^{-\beta\mathscr{H}_N}\mathrm{d}\Gamma}{Z_N}, \qquad (1.70\mathrm{a})$$

$$\text{(quantum-mechanically)} \quad Pr(l) \equiv f(l) = \frac{\mathrm{e}^{-\beta E_{N,l}}}{Z_N}, \qquad (1.70\mathrm{b})$$

where

$$\beta = 1/kT, \qquad (k = \text{Boltzmann constant})$$

and $\mathscr{H}_N$ is the Hamiltonian of the system, $E_{N,l}$ being the energy of the l-th quantum state. This distribution is called a *canonical distribution*, and an ensemble

---

obviously hold

$$\delta S + \sum_{k=1}^{r} \lambda_k \delta\Phi_k = \sum_{j=1}^{m}\left(\frac{\partial S}{\partial\alpha_j} + \sum_{k=1}^{r}\lambda_k\frac{\partial\Phi_k}{\partial\alpha_j}\right)\delta\alpha_j = 0. \qquad (3)$$

Now $\lambda_1, \ldots \lambda_r$ are supposed to be chosen by the conditions

$$\frac{\partial S}{\partial\alpha_j} + \sum_{k=1}^{r}\lambda_k\frac{\partial\Phi_k}{\partial\alpha_j} = 0 \qquad j = 1, \ldots, r. \qquad (4)$$

Then equation (3) is reduced to

$$\sum_{j=r+1}^{m}\left(\frac{\partial S}{\partial\alpha_j} + \sum_{k=1}^{r}\lambda_k\frac{\partial\Phi_k}{\partial\alpha_j}\right)\delta\alpha_j = 0.$$

Since $m - r$ variables out of $\alpha_1, \ldots \alpha_m$ are independent by the constraints (1.66), $\delta\alpha_{r+1}, \ldots, \delta\alpha_m$ can be regarded as independent variations, so that the conditions

$$\frac{\partial S}{\partial\alpha_j} + \sum_{k=1}^{r}\lambda_k\frac{\partial\Phi_k}{\partial\alpha_j} = 0 \qquad j = r+1, \ldots, m \qquad (5)$$

must also hold. Equations (4) and (5) together yield equation (1.67). The α's and λ's are determined from $r + m$ equations (1.66) and (1.67).

† Generally speaking there exists a weight function $\omega(\alpha_1', \alpha_2', \ldots)$. The choice of ω depends on the nature of the variables α. If they are originally expressed in terms of dynamical variables (q, p), the explicit form of ω can be derived from the principle of equal weight.

defined by this distribution is called a *canonical ensemble*. The system considered here may be a small system with a rather few degrees of freedom, or it may well be a large macroscopic system. Z_N in (1.70a, b) is defined by

$$\text{(classically)} \qquad Z_N = \frac{1}{\prod N_A! h^{3N_A}} \int e^{-\beta \mathscr{H}_N} \, d\Gamma, \tag{1.71a}$$

$$\text{(quantum-mechanically)} \qquad Z_N = \sum e^{-\beta E_{N,l}}, \tag{1.71b}$$

or more generally by

$$Z_N = \int_0^\infty e^{-\beta E} \Omega(E) \, dE \tag{1.71c}$$

and is called the *partition function*.

NOTE: The factor $1/\prod N_A! h^{3N_A}$ is introduced in (1.71a), as was explained in § 1.5, in order to secure the correct correspondence between classical and quantum mechanics.

Derivation of canonical distribution: Let $\Omega(E)$ be the state density of the heat bath, E_t the total energy of the composite system (the system under consideration plus the heat bath), and E_l the energy of l-th quantum state of the system ($E_t = E + E_l$). According to the principle of equal weight, the probability that the quantum state l is realized is proportional to the number of microscopic states allowed, which is equal to $\Omega(E_t - E_l)\, \delta E$. Therefore

$$f(l) \propto \Omega(E_t - E_l)\delta E \propto \frac{\Omega(E_t - E_l)\delta E}{\Omega(E_t)\delta E}$$

$$= \exp\left\{\frac{S(E_t - E_l) - S(E_t)}{k}\right\}. \tag{1.72}$$

Since the heat bath is very large compared to the system considered it may be assumed that $E_t \gg E_l$, and so the exponent may be expanded as

$$S(E_t - E_l) - S(E_t) = -E_l \frac{\partial S}{\partial E} + \frac{1}{2} E_l^2 \frac{\partial^2 S}{\partial E^2} + \dots \Bigg|_{E = E_t}$$

$$= -\frac{E_l}{T}\left\{1 + \frac{E_l}{2CT} + \dots\right\}\Bigg|_{E=E_t}$$

where

$$T = (\partial S/\partial E)^{-1}$$

is the temperature of the heat bath and $C = \partial E/\partial T$ is its heat capacity. Because of the assumed large size of the heat bath,

$$E_l \ll CT$$

and the second term in the last brackets is so small that it may be ignored. Then (1.72) reduces to

$$f(l) \propto e^{-E_l/kT}.$$

§ 1.13. GENERALIZED CANONICAL DISTRIBUTIONS

T-μ distribution (Grand canonical distribution): When a system enclosed in a volume V is in contact with a heat bath at temperature T and with a particle source characterized by the chemical potentials μ_A, μ_B, ... for particles A, B, ..., then the number of particles it contains is also indeterminate. The probability that the system has the microscopic state which contains N_A, N_B, ... particles is given by

(classical) $$Pr(N, d\Gamma) = \frac{1}{\prod N_A! h^{3N_A}} \frac{e^{-\beta(\mathscr{H}_N - \Sigma N_A \mu_A)} d\Gamma}{\Xi} \tag{1.73a}$$

(quantum mechanical) $$Pr(N, l) = \frac{e^{-\beta(E_{N,l} - \Sigma N_A \mu_A)}}{\Xi} \tag{1.73b}$$

where Ξ is called the grand canonical partition function (or may be called the T-μ partition function) defined by

(classical)

$$\Xi = \sum_{N_A=0}^{\infty} \sum_{N_B=0}^{\infty} \cdots \frac{1}{\prod N_A! h^{3N_A}} \int e^{-\beta(\mathscr{H}_N - \Sigma N_A \mu_A)} d\Gamma, \tag{1.74a}$$

(quantum-mechanical)

$$\Xi = \sum_{N_A=0}^{\infty} \sum_{N_B=0}^{\infty} \cdots \sum e^{-\beta(E_{N,l} - \Sigma N_A \mu_A)}, \tag{1.74b}$$

or more generally by

$$\Xi = \left(\sum_{N_A=0}^{\infty} \sum_{N_B=0}^{\infty} \cdots \right) e^{\beta \Sigma N_A \mu_A} \cdot e^{\beta \Sigma N_B \mu_B} \cdots\cdots Z_N$$

$$= \left(\sum_{N_A=0}^{\infty} \sum_{N_B=0}^{\infty} \cdots \right) \lambda_A^{N_A} \cdot \lambda_B^{N_B} \cdots\cdots Z_N. \tag{1.74c}$$

Instead of the chemical potentials, the *absolute activities*

$$\lambda_A = e^{\beta \mu_A}, \ldots$$

are sometimes used.

If one wishes only the probability distribution of particle numbers, it is given by

$$Pr(N_A, N_B, \ldots) = \frac{e^{\beta(N_A \mu_A + N_B \mu_B + \ldots)} Z_N}{\Xi}$$

$$= \frac{\lambda_A^{N_A} \lambda_B^{N_B} \cdots\cdots Z_N}{\Xi}. \tag{1.75}$$

Derivation of T-μ distribution: Let N_t be the total number of particles of a given kind in the system considered and in the particle source, then the probability that the system has a partition of particles and energy (N, E) becomes

$$Pr(N, E) \propto \Omega(N_t - N, E_t - E)$$
$$\propto \exp\frac{1}{k}\{S(N_t - N, E_t - E) - S(N_t, E_t)\}.$$

The exponent may be expanded in the same way as for (1.72) since we may assume that $N_t \gg N$ and $E_t \gg E$. The higher order terms in the expansion can safely be omitted. Then this reduces to (1.75) with the chemical potentials defined by (1.52) (see example 6).

T-p distribution: When a system with given numbers of particles N_A, N_B, ... is in contact with a bath at the temperature T and pressure p through a movable wall, then its volume V is also indeterminate. The probability that the system is found to be in a microscopic state with a volume V is given by (classically)

$$Pr(\mathrm{d}V, \mathrm{d}\Gamma) = \frac{1}{\prod N_A! h^{3N_A}} \mathrm{e}^{-\beta(\mathscr{H}(V) + pV)} \frac{\mathrm{d}V\, \mathrm{d}\Gamma}{Y}, \tag{1.76a}$$

(quantum-mechanically)

$$Pr(\mathrm{d}V, l) = \frac{\mathrm{e}^{-\beta(E_l(V) + pV)}\mathrm{d}V}{Y}, \tag{1.76b}$$

where $\mathscr{H}(V)$ is the Hamiltonian of the system with the volume V specified, $E_l(V)$ is its energy quantum energy and Y is the *T-p partition function:*

(classically)

$$Y = \frac{1}{\prod N_A! h^{3N_A}} \int_0^\infty \mathrm{d}V \int \mathrm{d}\Gamma \mathrm{e}^{-\beta(\mathscr{H}(V) + pV)}, \tag{1.77a}$$

(quantum-mechanically)

$$Y = \int_0^\infty \mathrm{d}V \sum \mathrm{e}^{-\beta(E_l(V) + pV)}. \tag{1.77b}$$

These may be written as

$$Y = \int_0^\infty Z_N(V) \mathrm{e}^{-\beta pV} \mathrm{d}V. \tag{1.77c}$$

The distribution in volume only is given by

$$Pr(\mathrm{d}V) = Z_N(V) \frac{\mathrm{e}^{-\beta pV}\mathrm{d}V}{Y}. \tag{1.78}$$

This distribution is derived in the same way as the canonical or grand canonical distributions.

NOTE: The T-μ distribution and the T-p distribution, like the canonical distribution, can be applied irrespective of the size of the system under consideration.

§ 1.14. PARTITION FUNCTIONS AND THERMODYNAMIC FUNCTIONS

The microcanonical, canonical, T-μ (grand canonical) and T-p distributions are the distributions for given energy (E = constant), given temperature (T = constant), given temperature and chemical potential (T = constant, μ = constant) and given temperature and pressure (T = constant, p = constant), respectively. *If the system is macroscopic,* then the thermodynamic function (potential) for each of these prescribed conditions is derived from each partition function. This is summarized in the following table:

Distribution	Partition function	Thermodynamic function
Microcanonical	$\Omega(E, V, N)\delta E$ or $\Omega_0(E, V, N)$	$S(E, V, N) = k \log \Omega(E, V, N)\delta E$ or $= k \log \Omega_0(E, V, N)$
Canonical	$Z(T, V, N) = \sum_l e^{-E_l(V,N)/kT} = \int_0^\infty e^{-E/kT}\Omega(E, V, N)\,dE$	$F(T, V, N) = -kT \log Z(T, V, N)$ $\psi(T, V, N) = k \log Z(T, V, N)$
Grand canonical (T-μ distribution)	$\Xi(T, V, \mu) = \sum_{N=0}^{\infty} e^{N\mu/kT} Z(T, V, N) = \sum_{N=0}^{\infty} \lambda^N Z(T, V, N)$	$J(T, V, \mu) \equiv -pV = F - G = -kT \log \Xi(T, V, \mu)$ $q(T, V, \mu) = k \log \Xi(T, V, \mu)$
T-p distribution	$Y(T, p, N) = \int_0^\infty e^{-pV/kT} Z(T, V, N)\,dV$	$G(T, p, N) = -kT \log Y(T, p, N)$ $\Phi(T, p, N) = k \log Y(T, p, N)$

In statistical mechanics, the thermodynamic relations between thermodynamic functions are derived as the relations between certain average values obtained from probability laws suitable for the description of the given conditions. Well known transformations (Legendre transformations) for the thermodynamic functions are derived by approximating the partition function Z, Ξ, or Y by taking the maximum term in the sum or the integral.

This will be shown by the following argument for the canonical distribution (see also § 2.3). Consider the quantum-mechanical case: the generalized force X conjugate to a generalized coordinate x, involved in the Hamiltonian $\mathscr{H}(p, q, x)$ is defined by

$$X = \frac{\partial \mathscr{H}(q, p, x)}{\partial x}.$$

The quantum-mechanical expectation value of this force in the quantum state l is given by

$$X_l = \frac{\partial E_l(x)}{\partial x} \tag{1.79}$$

where the energy E_l is regarded as a function of the parameter x.† The averages of energy and force in the canonical distribution are given by

$$\bar{E} = \sum_l E_l \mathrm{e}^{-\beta E_l} \Big/ \sum_l \mathrm{e}^{-\beta E_l} = -\frac{\partial}{\partial \beta} \log Z(\beta, x), \tag{1.80a}$$

$$\bar{X} = \sum_l X_l \mathrm{e}^{-\beta E_l} \Big/ Z = \sum_l \frac{\partial E_l}{\partial x} \mathrm{e}^{-\beta E_l} \Big/ \sum_l \mathrm{e}^{-\beta E_l} = -\frac{\partial}{\partial x} \frac{1}{\beta} \log Z(\beta, x). \tag{1.80b}$$

† The energy $E_l(x)$ is determined by the eigenvalue equation,

$$\mathscr{H}(p, q, x)\varphi_l = E_l \varphi_l$$

where the eigenfunction φ_l depends on the parameter x. When differentiated by x, this becomes

$$\frac{\partial \mathscr{H}}{\partial x} \varphi_l + \mathscr{H} \frac{\partial \varphi_l}{\partial x} = \frac{\partial E_l}{\partial x} \varphi_l + E_l \frac{\partial \varphi_l}{\partial x}.$$

This is multiplied by φ_l^* from the left and is integrated over the variables (τ) of the wave function,

$$\int \varphi_l^* \frac{\partial \mathscr{H}}{\partial x} \varphi_l \, \mathrm{d}\tau + \int \varphi_l^* \mathscr{H} \frac{\partial \varphi_l}{\partial x} \mathrm{d}\tau = \frac{\partial E_l}{\partial x} \int \varphi_l^* \varphi_l \, \mathrm{d}\tau + E_l \int \varphi_l^* \frac{\partial \varphi_l}{\partial x} \mathrm{d}\tau$$

or

$$X_l + E_l \int \varphi_l^* \frac{\partial \varphi_l}{\partial x} \mathrm{d}\tau = \frac{\partial E_l}{\partial x} + E_l \int \varphi_l^* \frac{\partial \varphi_l}{\partial x} \mathrm{d}\tau.$$

Therefore,

$$X_l = \frac{\partial E_l}{\partial x},$$

where the relations

$$X_l = \int \varphi_l^* \frac{\partial \mathscr{H}}{\partial x} \varphi_l \, \mathrm{d}\tau \qquad \text{(definition of } X_l\text{)}$$

$$\int \varphi_l^* \varphi_l \, \mathrm{d}\tau = 1 \qquad \text{(normalization)}$$

$$\int \varphi_l^* \mathscr{H} \psi \, \mathrm{d}\tau = E_l \int \varphi_l^* \psi \, \mathrm{d}\tau \qquad \text{(Hermitian property)}$$

are used.

In particular, the average pressure is given by

$$p = \frac{\partial}{\partial V}\frac{1}{\beta}\log Z(\beta, V). \tag{1.80c}$$

If one now writes

$$-kT\log Z = F(T, V, x, \ldots), \tag{1.81}$$

equations (1.80a–c) can be written as

$$\frac{\partial}{\partial(1/T)}\frac{F}{T} = E, \qquad \frac{\partial F}{\partial x} = \bar{X}, \qquad \frac{\partial F}{\partial V} = -p. \tag{1.82}$$

This shows that F is nothing but the free energy of the system if $\bar{E}$, $\bar{X}$, and p are identified with the thermodynamic energy, force and pressure.

The above can be shown in another way. In the integral

$$\begin{aligned} Z(T, V, N) &= \int e^{-E/kT}\Omega(E, V, N)\,dE \\ &= \int \exp\left[-\frac{1}{k}\left\{\frac{E}{T} - S(E, V, N)\right\}\right]dE/\delta E, \end{aligned} \tag{1.83}$$

the exponential function varies with E extremely rapidly if E and S $(= k\log\Omega\cdot\delta E)$ are of the order N in magnitude (N is the total number of particles in the system, which is very large), so that an overwhelming contribution to the integral comes from the neighborhood of E^* which is the most probable value of E in the canonical distribution. E^* is determined by

$$E/T - S = \min, \quad \text{thus} \quad 1/T = \partial S/\partial E \qquad (E = E^*).$$

The exponent in the integral is expanded in this neighborhood as

$$\exp\left[-\frac{1}{k}\{E^*/T - S(E^*, V, N)\} - \frac{1}{2kT^2C}(E - E^*)^2 + \ldots\right], \tag{1.84}$$

where the relations

$$\partial^2 S/\partial E^2 = \partial T^{-1}/\partial E = -1/T^2C \qquad (C = \partial E/\partial T, \qquad E = E^*)$$

are used (E in these expressions should be put equal to E^*). To the extent that the difference $E - E^*$ is small, or more exactly, of the order $O(N^{\frac{1}{2}})$ in magnitude, the higher order terms in the expansion are of the order of $(E - E^*)^m \cdot O(N^{-m+1}) = (N^{-\frac{1}{2}m+1})$ and can be ignored as $N \to \infty$. Hence (1.83) is approximated by

$$Z \sim (2\pi kT^2C)^{\frac{1}{2}}\delta E^{-1}\exp[-k^{-1}\{E^*/T - S(E^*, V, N)\}],$$

the logarithm of which gives

$$F(T, V, N) = E - TS(E, V, N)$$

if the free energy F is introduced by (1.81). The above may be written as

$$-\frac{1}{T}F(T, V, N) = S(E, V, N) - \frac{E}{T}, \tag{1.85a}$$

where E^* is written simply as E. E on the right hand side of this equation is regarded as a function of V, T, and N by the relation

$$\frac{\partial S(E, V, N)}{\partial E} = \frac{1}{T}. \tag{1.85b}$$

Equations (1.85a) and (1.85b) show that $-F/T$ is the Legendre transform of S by which the independent variables are changed from E to $1/T$. Therefore one finds

$$-\frac{\partial}{\partial(1/T)}\left(\frac{F}{T}\right) = \left(\frac{\partial S}{\partial E} - \frac{1}{T}\right)\frac{\partial E}{\partial(1/T)} - E = -E.$$

The second and third equations of (1.82) are easily obtained from (1.85), (1.52) and (1.54).

§ 1.15. FERMI-, BOSE-, AND BOLTZMANN-STATISTICS

One-particle states and the states of a system of particles: Let us consider a system consisting of N particles of a certain kind. *If the interaction of the particles is weak enough,* each particle has its own motion which is independent of all others.† The quantum states allowed for this individual motion – *one-particle states* – are determined by the Schrödinger equation

$$\mathscr{H}_1(\boldsymbol{p}, \boldsymbol{x})\varphi_\tau(\boldsymbol{x}) = \varepsilon_\tau \varphi_\tau(\boldsymbol{x}) \tag{1.86}$$

and are represented by the wave functions φ_τ, the ε_τ's being the energies of these quantum states.

Since identical particles are indistinguishable in quantum-mechanics, each quantum state of the particle system is completely specified when the number

† Even for a system of particles with rather strong interaction, it often happens that the one-particle approximation works surprisingly well when it is properly modified. We shall not discuss such complicated cases here, but it should be recognized that the present approach is also basic for many advanced problems. (1.88) will, however, then cease to be valid.

occupying one-particle states is given precisely. That is, the set of *occupation numbers*

$$l \equiv \{n_\tau\} = (n_1, n_2, \ldots, n_\tau, \ldots) \tag{1.87}$$

gives the quantum numbers of the whole system. The energy of the system is then given by

$$E_l \equiv E_{(n)} = \sum_\tau \varepsilon_\tau n_\tau . \tag{1.88}$$

Fermi-statistics (F.D.) and Bose-statistics (B.E.) †: The occupation numbers, or the number of particles in each one-particle state, are strongly restricted by a general principle of quantum mechanics.†† There can be only the following two cases:

$$\begin{aligned} &\text{for F. D. (Fermi-statistics)} \quad n = 0 \text{ or } 1 \\ &\text{for B. E. (Bose-statistics)} \quad n = 0, 1, 2, \ldots . \end{aligned} \tag{1.89}$$

The difference between the two cases is determined by the nature of the particle. Particles which follow Fermi-statistics are called *Fermi-particles (Fermions)* and those which follow Bose-statistics are called *Bose-particles (Bosons)*.

Electrons(e), positrons(e^+), protons(P), and neutrons(N) are Fermi-particles, whereas photons are Bose-particles.

Generally, a particle consisting of an odd number of Fermi-particles (example: D = P + N + e) is a Fermi-particle, and a particle consisting of an even number of Fermi-particles (example: H = P + e) is a Bose-particle (a particle consisting of Bose-particles only is a Bose-particle). A Fermi-particle has half-integral spin and a Bose-particle has integral spin.

Fermi distribution and Bose distribution: When the particle system is in equilibrium (the total number of particles N is assumed very large, of course), the average occupation number for each one-particle state is shown to be (example 12, problem 3)

$$\bar{n}_\tau = \frac{1}{e^{(\varepsilon_\tau - \mu)/kT} + 1}, \qquad \text{(F.D.)} \tag{1.90}$$

$$\bar{n}_\tau = \frac{1}{e^{(\varepsilon_\tau - \mu)/kT} - 1}, \qquad \text{(B.E.)} \tag{1.91}$$

† These are often called Fermi-Dirac statistics and Bose-Einstein statistics.

†† The wave function of a system of identical particles must be either symmetrical (Bose) or antisymmetrical (Fermi) in permutation of the particle coordinates (including spin) (see, for example, Schiff: *Quantum Mechanics*).

where T is the temperature of the system and μ the chemical potential of the particle. The energy of the total system is given by

$$E = \sum_{\tau} \varepsilon_{\tau} \bar{n}_{\tau}, \tag{1.92}$$

while the total number of particles is

$$N = \sum \bar{n}_{\tau}. \tag{1.93}$$

Meaning of μ and T: μ and T appearing in equation (1.90) and (1.91) may be interpreted in different ways:

(1) If the system is isolated (microcanonical ensemble), equation (1.92) and (1.93) determine T and μ for prescribed E and N.

(2) When the system is in contact with a heat bath at temperature T (canonical ensemble), equation (1.92) gives the average energy and (1.93) determines μ for prescribed T and N.

(3) When the system is in contact with a bath at temperature T and with a particle source characterized by the chemical potential μ (grand canonical), (1.92) and (1.93) give the average E and N.

Thermodynamic functions: The upper sign is for F. D. and the lower sign is for B. E. (see example 12):

$$S = k \sum_{\tau} [-\bar{n}_{\tau} \log \bar{n}_{\tau} \mp (1 \mp \bar{n}_{\tau}) \log (1 \mp \bar{n}_{\tau})], \tag{1.94a}$$

$$F = E - TS, \qquad G = N\mu, \tag{1.94b}$$

$$\begin{aligned} J \equiv -pV = F - N\mu &= \pm kT \sum_{\tau} \log (1 \mp \bar{n}_{\tau}) \\ &= \pm kT \sum_{\tau} \log (1 \pm e^{(\mu - \varepsilon_{\tau})/kT}). \end{aligned} \tag{1.94c}$$

The classical limit (Boltzmann-statistics): When the particle density is so low and the temperature so high that the condition

$$\frac{N}{V} \ll \left(\frac{2\pi mkT}{h^2} \right)^{\frac{3}{2}} \tag{1.95}$$

is satisfied, we have

$$\bar{n}_{\tau} \ll 1, \qquad \varepsilon_{\tau} - \mu \gg kT. \tag{1.96}$$

Both F. D. and B. E. statistics are reduced to the classical limit, which corresponds to the classical approximation mentioned in § 1.5. This limiting law is called Boltzmann-statistics, and its distribution (the Boltzmann-distribution, sometimes called the Boltzmann-Maxwell distribution) is given by

$$\bar{n}_\tau = e^{(\mu - \varepsilon_\tau)/kT}. \tag{1.97}$$

For Boltzmann statistics, the *one-particle partition function* can be defined by

$$f = \sum_\tau e^{-\varepsilon_\tau/kT} \equiv \sum e^{-\beta\varepsilon_\tau}. \tag{1.98}$$

With this definition the following formulae are obtained (see problem 19):

$$\left.\begin{aligned} &N = \lambda f, &&\lambda = e^{\mu/kT}, \\ &E = -N\frac{\partial \log f}{\partial \beta}, && \\ &Z_N = \frac{f^N}{N!}, &&F = -NkT - NkT\log\frac{f}{N}, \\ &\Xi = e^{\lambda f}, &&pV = NkT. \end{aligned}\right\} \tag{1.99}$$

The Maxwell velocity distribution: The probability that a molecule (mass m) in an ideal gas at temperature T will be found to have a velocity in a range between (v_x, v_y, v_z) and $(v_x + dv_x, v_y + dv_y, v_z + dv_z)$ is given by

$$f(v_x, v_y, v_z)\,dv_x\,dv_y\,dv_z = \left(\frac{m}{2\pi kT}\right)^{\frac{3}{2}} e^{-\frac{m}{2kT}(v_x^2 + v_y^2 + v_z^2)}\,dv_x\,dv_y\,dv_z. \tag{1.100}$$

This is called the Maxwell (Boltzmann) velocity distribution law.

§ 1.16. GENERALIZED ENTROPY

It is possible to define the entropy for a statistical ensemble by

$$S = -k\sum_l f(l)\log f(l) \equiv -k\overline{\log f}, \tag{1.101}$$

where $f(l)$ is the probability of the realization of the quantum state l. It is recommended that the readers prove for themselves that this definition yields the correct expressions for statistical entropies for microcanonical, canonical, grand canonical and other ensembles. Equation (1.101) is related to general *H-theorems* which include time explicitly.

Divertissement 2

Statistical method. "And here I wish to point out that, in adopting this statistical method of considering the average number of groups of molecules selected according to their velocities, we have abandoned the strict kinetic method of tracing the exact circumstances of each individual molecule in all its encounters. It is therefore possible that we may arrive at results which, though they fairly represent the facts as long as we are supposed to deal with a gas in mass, would cease to be applicable if our faculties and instruments were so sharpened that we could detect and lay hold of each molecule and trace it through all its course."

"For the same reason, a theory of the effects of education deduced from a study of the returns of registrars, in which no names of individuals are given, might be found not to be applicable to the experience of a schoolmaster who is able to trace the progress of each individual pupil."

"The distribution of the molecules according to their velocities is found to be of exactly the same mathematical form as the distribution of observations according to the magnitude of their errors, as described in the theory of errors of observation. The distribution of bullet-holes in a target according to their distances from the point aimed at is found to be of the same form, provided a great many shots are fired by persons by the same degree of skill."

– J. Clerk Maxwell, "Theory of Heat," 1897 –

Examples

1. The pressure which a gas exerts on the walls of a vessel can be regarded as the time average of the impulses which the gas molecules impart on the wall when colliding with and recoiling from it. From this point of view, calculate the pressure p and show that

$$p = \tfrac{2}{3} n\bar{\varepsilon} \qquad \text{(Bernoulli's formula)},$$

where n is the average number of molecules per unit volume and $\bar{\varepsilon}$ the mean kinetic evergy per molecule.

SOLUTION

Let the vessel be a cube with edges of length l. Suppose that the wall is perfectly smooth so that a molecule colliding with it is reflected in a completely elastic way. Let us take the axis of an orthogonal coordinate system x, y, z parallel to the edges. We shall neglect collisions of molecules with each other. The momentum components of a given molecule do not change their magnitude as a result of collisions with the walls. Therefore, this molecule collides $|p_x|/2ml$ times per unit time with one of the walls prependicular to the x-axis, where m is the mass of the molecule. As a result of each collision the wall receives a momentum $2|p_x|$ directed along the outward normal of the wall (perfect reflection). Hence the time average of the momentum, in other words, the sum of the momentum imparted to the wall per unit time, is equal to $2\,|p_x|\cdot|p_x|/2ml = p_x^2/ml$. Since the sum of these momenta yields a force, the contribution of this molecule to the pressure is given by this sum divided by the area of the wall l^2 ($V = l^3$ is the volume of the vessel). Therefore, adding together the contributions from all the molecules, we obtain the pressure exerted by all the molecules:

$$p = \frac{1}{3V}\sum_{i=1}^{N}\frac{p_{ix}^2 + p_{iy}^2 + p_{iz}^2}{m} = \frac{2}{3V}\sum_{i=1}^{N}\frac{p_i^2}{2m} = \frac{2}{3}\frac{N}{V}\bar{\varepsilon} = \frac{2}{3}n\bar{\varepsilon}. \tag{1}$$

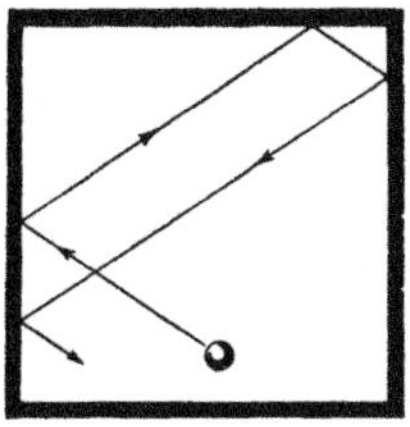

Fig. 1.3.

Alternative solution

(Method using the virial theorem). Denoting the position vector of the i-th molecule by $\boldsymbol{x}_i$, one has

$$\frac{\mathrm{d}}{\mathrm{d}t}\sum_{i=1}^{N}\boldsymbol{p}_i\cdot\boldsymbol{x}_i = \sum_{i=1}^{N}\dot{\boldsymbol{p}}_i\cdot\boldsymbol{x}_i + \sum_{i=1}^{N}\boldsymbol{p}_i\cdot\dot{\boldsymbol{x}}_i = \sum_{i=1}^{N}\boldsymbol{F}_i\cdot\boldsymbol{x}_i + 2\sum_{i=1}^{N}\varepsilon_i \tag{2}$$

where $\boldsymbol{F}_i$ is the force acting on the i-th molecule and ε_i its kinetic energy. When the time average is taken of both sides, the left hand side vanishes, since

$$\overline{\frac{\mathrm{d}}{\mathrm{d}t}A(t)} = \lim_{t\to\infty}\frac{1}{t}\int_{t_0}^{t_0+t}\frac{\mathrm{d}}{\mathrm{d}t}A(t)\,\mathrm{d}t = \lim_{t\to\infty}\frac{1}{t}\{A(t_0+t) - A(t_0)\} \tag{3}$$

is clearly equal to 0 for any time-dependent quantity $A(t)$ which remains finite and bounded at any instant. In the present case the quantity varies with the motion of the gas molecules but always stays finite (since the momenta as well as the coordinates remain finite). Therefore, one gets from (2)

$$-2\sum_{i=1}^{N}\varepsilon_i = \sum_{i=1}^{N}\overline{\boldsymbol{F}_i\cdot\boldsymbol{x}_i}. \tag{4}$$

The right hand side is called the *virial*. If the intermolecular force is negligible,

F_i is simply the force exerted on the molecule by the walls. Its time average per unit area being the pressure p, the virial due to this force is given by

$$\sum_{i=1}^{N} \overline{F_i \cdot x_i} = - \int_{\text{wall}} p\mathbf{v} \cdot \mathbf{x} \, \mathrm{d}S \tag{5}$$

where $\mathbf{v}$ is the outward normal to the walls and the integral is evaluated over the walls. By means of Gauss' theorem

$$\int \mathbf{a} \cdot \mathbf{v} \, \mathrm{d}S = \int \operatorname{div} \mathbf{a} \, \mathrm{d}V$$

one has

$$\int \mathbf{v} \cdot \mathbf{x} \, \mathrm{d}S = \int \operatorname{div} \mathbf{x} \, \mathrm{d}V = 3 \int \mathrm{d}V = 3V. \tag{6}$$

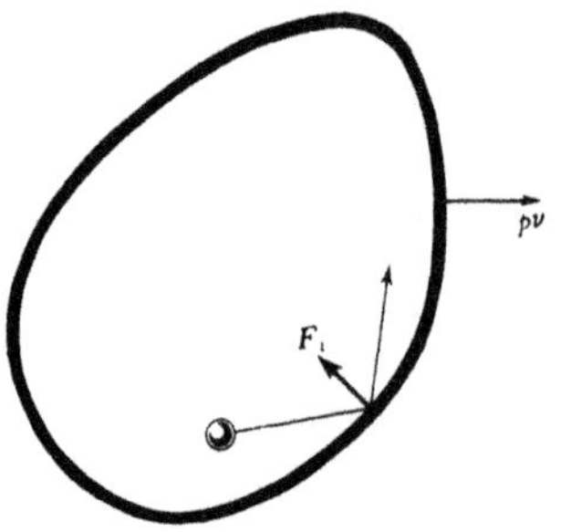

Fig. 1.4. Calculation of the virial due to the force exerted by the wall.

From (4), (5) and (6) one gets

$$2N\bar{\varepsilon} = 3pV, \qquad \text{or} \quad p = \tfrac{2}{3} n \bar{\varepsilon}.$$

NOTE: The virial theorem (4) is valid also in the case where one must take into account the intermolecular force explicitly. If F_{ij} is the force the i-th molecule exerts on the j-th, one has $F_{ij} = F_{ji}$ because of the law of action and reaction. The virial due to the intermolecular force is

$$\Phi = \sum_i \sum_j \overline{x_i \cdot F_{ij}} = \tfrac{1}{2} \sum_{(i \neq j)} \sum \overline{(x_i - x_j) \cdot F_{ij}}. \tag{7}$$

From (4) one gets

$$p = \tfrac{2}{3} n \bar{\varepsilon} + \frac{\Phi}{3V}. \tag{8}$$

When the range of the intermolecular force is much smaller than the mean intermolecular distance, Φ is almost zero and Bernoulli's formula holds. In the general case one has to use (8) (see problem 6 in Chapter 2).

2. An ideal gas consisting of N mass points is contained in a box of volume V. Find the number of states (phase integral) $\Omega_0(E)$ (1.20b) classically and, using it, derive the equation of state. (*Hint:* The volume of a unit sphere in a n-dimensional space, C_n, is equal to $\pi^{\frac{1}{2}N}/\Gamma(\frac{1}{2}n + 1)$.)

SOLUTION

The energy of this system is

$$\mathscr{H} = \sum_{i=1}^{3N} p_i^2/2m \tag{1}$$

where we have labelled the momenta conjugate to the Cartesian coordinates of each mass point with running index number: $p_1, \ldots, p_{3N}$. The number of states is given by (1.20b) where the integral over the coordinates gives V^N,

$$\Omega_0(E, N, V) = \frac{V^N}{h^{3N}N!} \int \cdots \int_{\sum_1^{3N} p_i^2 \leq 2mE} \mathrm{d}p_1 \ldots \mathrm{d}p_{3N}. \tag{2}$$

Since this integral is equal to the volume of a $3N$-dimensional sphere of radius $(2mE)^{\frac{1}{2}}$ it is proportional to $(2mE)^{3N/2}$, the proportionality coefficient being the number C_{3N} given in the hint. Hence,

$$\Omega_0(E, N, V) = \frac{V^N}{h^{3N}N!} \frac{(2\pi mE)^{3N/2}}{\Gamma(\frac{3}{2}N + 1)}. \tag{3}$$

By (1.27) the entropy becomes

$$S(E, N, V) = k \log \Omega_0 = Nk\left\{\log\frac{V}{N} + \tfrac{3}{2}\log\frac{2E}{3N} + \log\frac{(2\pi m)^{\frac{3}{2}}\,\mathrm{e}^{\frac{5}{2}}}{h^3}\right\}, \tag{4}$$

where use has been made of Stirling's formula

$$\log N! \simeq N \log(N/\mathrm{e}) = N \log N - N. \tag{5}$$

From (1.28) and (1.54) we get

$$\frac{1}{T} = \left(\frac{\partial S}{\partial E}\right)_{V,N} = Nk\frac{3}{2E}, \quad \text{or} \quad kT = \frac{2E}{3N} = \tfrac{2}{3}\bar{\varepsilon}, \tag{6}$$

$$\frac{p}{T} = \left(\frac{\partial S}{\partial V}\right)_{E,N} = Nk/V, \quad \text{or} \quad pV = NkT. \tag{7}$$

NOTE 1: There are a number of ways of calculating the volume of a n-dimensional sphere. We shall give only one of them. If the volume of a n-dimensional sphere of radius r is $V_n(r) = C_n r^n$, the surface area of the sphere is equal to $nC_n r^{N-1}$. Using this relation we evaluate

$$I_n = \int_{-\infty}^{\infty} \dots \int_{-\infty}^{\infty} \exp\{-a(x_1^2 + \dots + x_n^2)\} \, dx_1 \dots dx_n$$

by two different methods. First, because of the well-known formula,

$$\int_{-\infty}^{\infty} \exp(-ax^2) \, dx = (\pi/a)^{\frac{1}{2}},$$

we get $I_n = (\pi/a)^{\frac{1}{2}n}$.

Secondly, by dividing the $x_1, \dots x_n$ space into spherical shells we can perform the integration:

$$I_n = \int_0^{\infty} e^{-ar^2} nC_n r^{n-1} \, dr$$

$$= \tfrac{1}{2}nC_n \int_0^{\infty} e^{-ay} y^{\frac{1}{2}n-1} \, dy = \tfrac{1}{2}nC_n \Gamma(\tfrac{1}{2}n) \, a^{-\frac{1}{2}n} = C_n \Gamma(\tfrac{1}{2}n + 1) \, a^{-\frac{1}{2}n}$$

$$\left(\int_0^{\infty} e^{-x} x^{m-1} dx = \Gamma(m)\right).$$

Equating the two expressions we have

$$C_n = \pi^{\frac{1}{2}n} / \Gamma(\tfrac{1}{2}n + 1).$$

NOTE 2: *Stirling's formula*

$$\Gamma(x + 1) = x! \sim x^{x+\frac{1}{2}} e^{-x} \sqrt{2\pi}, \tag{8a}$$

or

$$\log x! \sim x \log x - x + \dots \tag{8b}$$

is an asymptotic formula valid for $x \gg 1$, which is frequently used in statistical mechanics.

3. The energy level of an oscillator with frequency ν is given by

$$\varepsilon = \tfrac{1}{2}h\nu, \tfrac{3}{2}h\nu, \dots, (n + \tfrac{1}{2}) h\nu, \dots .$$

When a system consisting of N almost independent oscillators has the total energy

$$E = \tfrac{1}{2}Nh\nu + Mh\nu \qquad (M \text{ is an integer})$$

(i) find the thermodynamic weight W_M, and (ii) determine the relation between the temperature of this system and E.

SOLUTION

(i) If the quantum number of the i-th oscillator is denoted by n_i, the statement that the total energy of the system is equal to $\frac{1}{2}Nh\nu + Mh\nu$ implies that

$$n_1 + n_2 + \ldots + n_N = M .$$

Therefore, the thermodynamic weight W_M of a macroscopic state with the total energy E is equal to the number of ways of distributing M white balls among N labelled boxes. A box may be empty since $n_i = 0$ is possible. As is evident from Fig. 1.5, one can get this number by finding the number of permutations of placing in a row all the white balls together with $(N - 1)$ red balls that designate the dividing walls. If one labels all the balls with the running numbers, 1, 2, ..., $M + N - 1$, the number of permutations is $(M + N - 1)!$. When one erases these numbers there appear indistinguishable distributions, the number of which is equal to the number of permutations among the (numbered) balls with the same color, $M!(N - 1)!$. Therefore,

$$W_M = \frac{(M + N - 1)!}{M!(N - 1)!} . \tag{1}$$

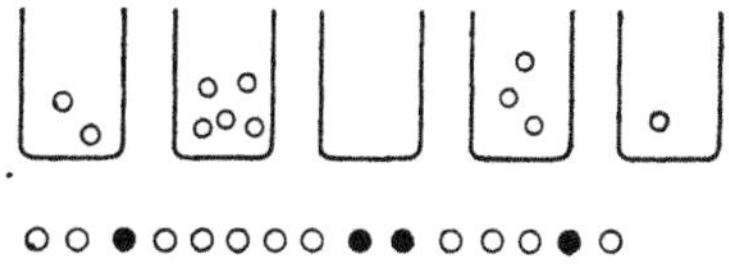

Fig. 1.5.

(ii) From (1.18) the entropy is

$$S = k \log W_M .$$

Substituting (1) and using Stirling's formula under the assumption, $N \gg 1$, $M \gg 1$, one has

$$S = k\{(M + N)\log(M + N) - M\log M - N\log N\} .$$

From (1.28)

$$\begin{aligned} \frac{1}{T} &= \frac{\partial S}{\partial E} = \frac{\partial S}{\partial M}\frac{\partial M}{\partial E} \\ &= k\log\left(\frac{M + N}{M}\right)\cdot\frac{\partial M}{\partial E} = \frac{k}{h\nu}\log\frac{M + \frac{1}{2}N + \frac{1}{2}N}{M + \frac{1}{2}N - \frac{1}{2}N} \\ &= \frac{k}{h\nu}\log\left(\frac{E/N + \frac{1}{2}h\nu}{E/N - \frac{1}{2}h\nu}\right). \end{aligned} \tag{2}$$

Or, inversely,

$$\frac{E + \frac{1}{2}Nh\nu}{E - \frac{1}{2}Nh\nu} = e^{h\nu/kT}.$$

Solving this for E, one gets

$$E = N\left\{\tfrac{1}{2}h\nu + \frac{h\nu}{e^{h\nu/kT} - 1}\right\}. \tag{3}$$

In Fig. 1.6, $E/Nh\nu$ is plotted against $kT/h\nu$.

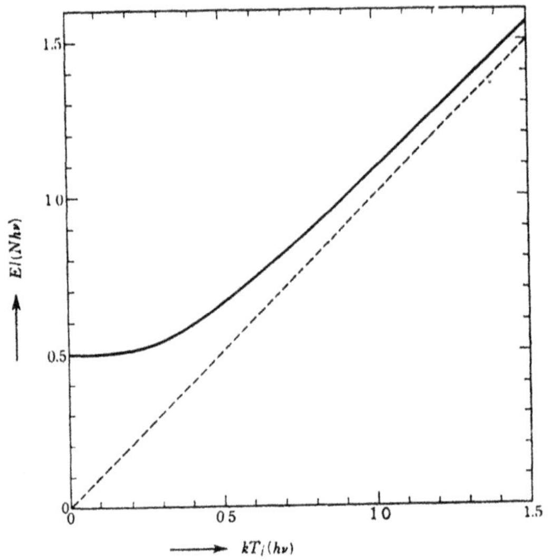

Fig. 1.6.

4. There is a system consisting of N independent particles. Each particle can have only one of the two energy levels, $-\varepsilon_0$, ε_0. Find the thermodynamic weight W_M of a state with the total energy $E = M\varepsilon_0 (M = -N, \ldots, N)$, and discuss the statistical-thermodynamic properties of the system for the range $E < 0$, deriving especially the relation between temperature and the energy as well as the specific heat.

SOLUTION

If N_- particles are in the state with energy $-\varepsilon_0$ and N_+ particles in the state with energy ε_0, the energy of the total system is

$$E = M\varepsilon_0 = -N_-\varepsilon_0 + N_+\varepsilon_0, \qquad M = N_+ - N_-. \tag{1}$$

Since $N = N_+ + N_-$, one has

$$N_- = \tfrac{1}{2}(N - M), \qquad N_+ = \tfrac{1}{2}(N + M). \tag{2}$$

Now, there are $N!/(N_-!N_+!)$ possible ways of choosing N_- particles out of N to occupy the state with $-\varepsilon_0$, each of which gives a different microscopic state with the energy E. Hence the thermodynamic weight is

$$W_M = \frac{N!}{[\tfrac{1}{2}(N - M)]!\,[\tfrac{1}{2}(N + M)]!}. \tag{3}$$

According to (1.18) the entropy of the system is

$$\begin{aligned} S(E) &= k \log W_M \\ &\simeq k\{N \log N - \tfrac{1}{2}(N - M)\log\tfrac{1}{2}(N - M) - \tfrac{1}{2}(N + M)\log\tfrac{1}{2}(N + M)\} \\ &= -\,k\{N_- \log(N_-/N) + N_+ \log(N_+/N)\} \end{aligned} \tag{4}$$

where Stirling's formula has been used. Defining the temperature by (1.28), one has

$$\frac{1}{T} = \frac{1}{\varepsilon_0}\frac{\partial S}{\partial M} = \tfrac{1}{2}\,\frac{k}{\varepsilon_0}\log\frac{N - M}{N + M}. \tag{5}$$

As one can see from this equation or from the relation between S and E (Fig. 1.7), $T < 0$ for $M > 0$ ($E > 0$), so that this system is *not normal* in the sense of statistical mechanics. Since, however, it is normal in the range $M < 0$ ($E < 0$), one can discuss the property of this system within this range.

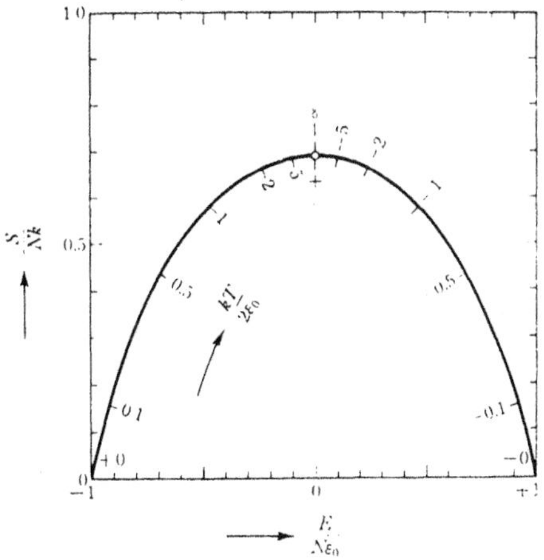

Fig. 1.7.

From (5) one gets

$$\frac{N_-}{N_+} = \frac{N-M}{N+M} = e^{2\varepsilon_0/kT},$$

or

$$\left.\begin{aligned} \frac{N_-}{N} &= \frac{e^{\varepsilon_0/kT}}{e^{\varepsilon_0/kT}+e^{-\varepsilon_0/kT}}, \\ \frac{N_+}{N} &= \frac{e^{-\varepsilon_0/kT}}{e^{\varepsilon_0/kT}+e^{-\varepsilon_0/kT}}. \end{aligned}\right\} \tag{6}$$

Therefore,

$$E = -(N_- - N_+)\varepsilon_0 = -N\varepsilon_0 \tanh(\varepsilon_0/kT). \tag{7}$$

Equations (6) give the probabilities of finding any one particle in the states $-\varepsilon_0$ and $+\varepsilon_0$, respectively, and have the form of a canonical distribution (1.70). From (7) the specific heat becomes

$$C = \mathrm{d}E/\mathrm{d}T = Nk\left(\frac{\varepsilon_0}{kT}\right)^2 \Big/ \cosh^2 \frac{\varepsilon_0}{kT}. \tag{8}$$

Figs. 1.8 and 1.9 show the graphs of E versus T and C versus T, respectively. NOTE: Putting $2\varepsilon_0 = \Delta E$, one rewrites (8) as

$$C = Nk\left(\frac{\Delta E}{kT}\right)^2 e^{\Delta E/kT}/(1+e^{\Delta E/kT})^2. \tag{9}$$

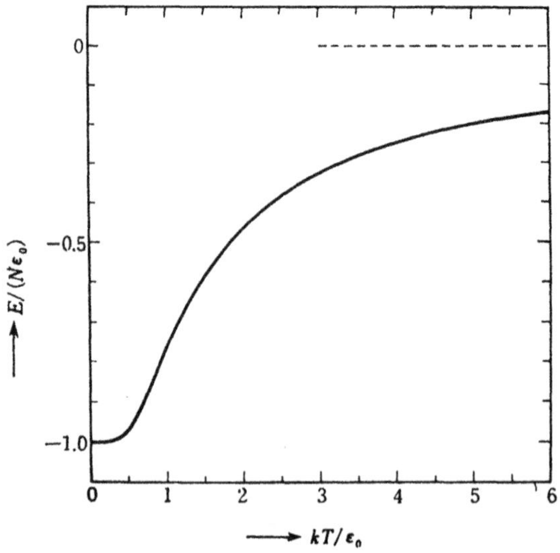

Fig. 1.8.

The specific heat of this form is called the *Schottky specific heat*. When a body contains a substance having the excitation energy ΔE, the specific heat anomaly with a peak as shown in Fig. 1.9 is actually observed.

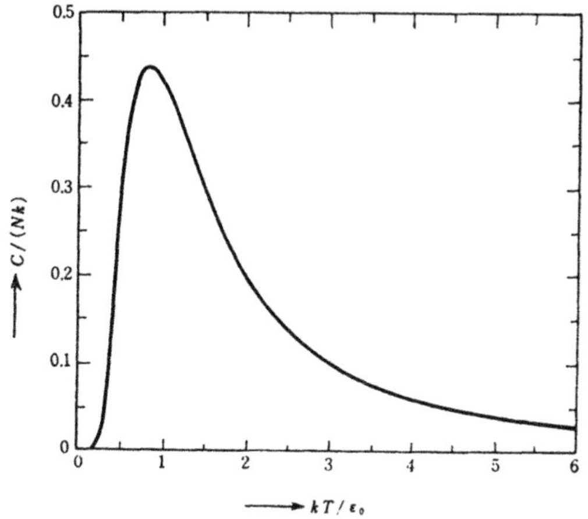

Fig. 1.9.

5. Suppose that two systems, which are normal in the sense of statistical thermodynamics, have the number of states of the form (1.24a). Prove the following statements concerning the thermal contact of the two system. (i) If their initial temperatures are T_1 and T_2 ($T_1 > T_2$), heat will flow from 1 to 2 upon contact and the entropy will increase by the amount $d'Q(T_2^{-1} - T_1^{-1})$. (ii) When both systems are in thermal equilibrium with each other, the entropy of the composite system S_{1+2} is equal to the sum of the entropies of the subsystems S_1 and S_2. (iii) The fluctuations in the energy of subsystem 1 or subsystem 2 when in a state of thermal equilibrium, $\overline{(E_1 - E_1^*)^2}$ or $\overline{(E_2 - E_2^*)^2}$, are given by $kT^2/(C_1^{-1} + C_2^{-1})$, where E_1^*, E_2^* are the mean values (the most probable values) of the energies of 1 and 2, and C_1 and C_2 are the respective specific heats.

SOLUTION

We suppose that the numbers of states $\Omega_1^0(E_1)$, $\Omega_2^0(E_2)$ are respectively of the form,

$$\Omega_1^0(E_1) \sim \exp\{N_1\phi_1(E_1/N_1)\}, \qquad \Omega_2^0(E_2) \sim \exp\{N_2\phi_2(E_2/N_2)\} \quad (1)$$

where N_1, N_2 are the number of particles in each system, both of them being very large. The probability of a state in which 1 and 2 have energies approximately equal to E_1 and E_2 respectively is given by

$$W(E_1, E_2) = \Omega_1(E_1)\Omega_2(E_2)\delta E_1 \delta E_2$$
$$\sim \exp\{N_1\phi_1(E_1/N_1) + N_2\phi_2(E_2/N_2)\}\cdot\phi_1'(E_1/N_1)\phi_2'(E_2/N_2)\delta E_1\delta E_2, \quad (2)$$

where the widths of the energy uncertainty in each system are denoted by δE_1 and δE_2, and $\mathrm{d}\Omega_1^0(E)/\mathrm{d}E = \Omega_1(E)$. Since N_1 and N_2 are large, the exponential function dominates in (2), the rest of the factors ϕ_1', ϕ_2' being of negligible influence (their influence is of the order of $1/N$).

(i) It is probable that upon thermal contact between 1 and 2 there occurs an energy flow between the systems and, as a result, they will go into a state with larger $W(E_1, E_2)$. In other words, $W(E_1, E_2)$ almost certainly increases as a result of thermal contact. Hence

$$k \log W(E_1, E_2) = S(E_1, E_2) = kN_1\phi_1(E_1/N_1) + kN_2\phi_2(E_2/N_2) \quad (3)$$

also increases. The amount of its increase due to the change $\mathrm{d}E_1 = -\mathrm{d}E_2 = -\mathrm{d}'Q$ is, according to (1.27) and (1.29),

$$\mathrm{d}S(E_1, E_2) = \{k\phi_1'(E_1/N_1) - k\phi_2'(E_2/N_2)\}\,\mathrm{d}E_1 = \mathrm{d}'Q\,\{T_2^{-1} - T_1^{-1}\}.$$

If $T_1 > T_2$, we must have $\mathrm{d}'Q > 0$ in order that $\mathrm{d}S(E_1, E_2) > 0$. That is to say, it is almost certain that heat flows from 1 to 2.

(ii) When the systems 1 and 2 are in thermal equilibrium, the total number of microscopic states (thermodynamic weight) of the compound system 1 + 2 having a total energy of approximately E can be obtained by integrating (2):

$$W(E) = \delta E \int \Omega_1(E_1)\Omega_2(E - E_1)\,\mathrm{d}E_1$$
$$= \int \exp\left\{N_1\phi_1\left(\frac{E_1}{N_1}\right) + N_2\phi_2\left(\frac{E_2}{N_2}\right)\right\}\phi_1'\left(\frac{E_1}{N_1}\right)\phi_2'\left(\frac{E_2}{N_2}\right)\mathrm{d}E_1\delta E \quad (4)$$
$$(E_2 = E - E_1).$$

Since under the condition $E_1 + E_2 = E =$ constant, the exponential function in the integrand has its maximum value at E_1^*, E_2^* determined by

$$N_1\phi(E_1/N_1) + N_2\phi_2(E_2/N_2) = \text{max.}, \qquad \text{or} \quad \phi_1'(E_1^*/N_1) = \phi_2'(E_2^*/N_2). \quad (5)$$

Corresponding to (1.46) and (1.47), we can expand this function around the maximum in the following way:

$$\exp\left\{N_1\phi_1\left(\frac{E_1}{N_1}\right)+N_2\phi_2\left(\frac{E_2}{N_2}\right)\right\}=\exp\left[N_1\phi_1\left(\frac{E_1^*}{N_1}\right)+N_2\phi_2\left(\frac{E_2^*}{N_2}\right)+\right.$$

$$\left.+\frac{1}{2}\left\{\frac{1}{N_1}\phi_1''\left(\frac{E_1^*}{N_1}\right)+\frac{1}{N_2}\phi_2''\left(\frac{E_2^*}{N_2}\right)\right\}(E_1-E_1^*)^2+\ldots\right]$$

$$=\exp\left[N_1\phi_1\left(\frac{E_1^*}{N_1}\right)+N_2\phi_2\left(\frac{E_2^*}{N_2}\right)-\frac{(E_1-E_1^*)^2}{2\Delta^2}+\ldots\right], \tag{6}$$

where

$$\Delta^2=-\left[\frac{1}{N_1}\phi_1''\left(\frac{E_1^*}{N_1}\right)+\frac{1}{N_2}\phi_2''\left(\frac{E_2^*}{N_2}\right)\right]^{-1}>0 \tag{7}$$

(using the condition (1.25)).
If $O(N_1)=O(N_2)=O(N)$, $\Delta^2=O(N)$. Since the higher order terms in [ ] are proportional to $N^{-(m-1)}(E_1-E_1^*)^m$ $(m>3)$, they are of the order $O(N^{-\frac{1}{2}m+1})\cdot\varepsilon^m$ if we put $E_1-E_1^*=N^{\frac{1}{2}}\varepsilon$. Hence we may neglect them (except the term $m=2$) in the limit $N\to\infty$. Because

$$\frac{E_1}{N_1}=\frac{E_1^*}{N_1}+\frac{E_1-E_1^*}{N_1}=\frac{E_1^*}{N_1}+\frac{\varepsilon}{N_1^{\frac{1}{2}}}\to\frac{E_1^*}{N_1}$$

the change in the values of the factors ϕ_1', ϕ_2' in (4) can also be neglected. Therefore, in the limit $N\to\infty$, (4) becomes asymptotically equal to

$$W(E)\simeq\delta E\exp\left\{N_1\phi_1\left(\frac{E_1^*}{N_1}\right)+N_2\phi_2\left(\frac{E_2^*}{N_2}\right)\right\}\cdot\int\exp\left\{-\frac{1}{2\Delta^2}(E_1-E_1^*)^2\right\}\mathrm{d}E_1$$

$$=\delta E\sqrt{2\pi}\,\Delta\exp\left\{N_1\phi_1\left(\frac{E_1^*}{N_1}\right)+N_2\phi_2\left(\frac{E_2^*}{N_2}\right)\right\}. \tag{8}$$

Taking the logarithm of this expression, we get

$$S_{1+2}(E)=k\log W\sim k\left\{N_1\phi_1\left(\frac{E_1^*}{N_1}\right)+N_2\phi_2\left(\frac{E_2^*}{N_2}\right)\right\} \tag{9}$$

$$=S_1(E_1^*)+S_2(E_2^*)$$

since $\log\Delta=O(\log N)$ may be neglected. The right-hand side is the sum of the entropies when 1 and 2 have the energies E_1^* and E_2^* respectively.

(iii) The integrand of (4) is the probability of 1 and 2 having energies E_1 and E_2, respectively, in a state of thermal equilibrium. The probability of 1 having energy in the range E_1 and $E_1 + \mathrm{d}E_1$ is equal to

$$P(E_1)\,\mathrm{d}E_1 = C\exp\left\{N_1\phi_1\left(\frac{E_1}{N_1}\right) + N_2\phi_2\left(\frac{E-E_1}{N_2}\right)\right\}\phi_1'\left(\frac{E_1}{N_1}\right)\phi_2'\left(\frac{E-E_1}{N_2}\right)\mathrm{d}E_1, \tag{10}$$

where C is the normalization constant. According to what we have said earlier, when $\mathrm{O}(N_1) = \mathrm{O}(N_2) = \mathrm{O}(N)$, we may put

$$P(E_1)\,\mathrm{d}E_1 = C\exp\left[-\frac{1}{2\Delta^2}(E_1 - E_1^*)^2\right] \tag{11}$$

asymptotically in the limit $N \to \infty$. Here Δ is given by (7). Since one gets

$$\phi''\left(\frac{E_1}{N_1}\right) = \frac{-N_1}{kT^2}\bigg/\frac{\partial E_1}{\partial T} = -\frac{N_1}{kT^2 C_1} \quad \left(\begin{matrix}C_1 = \text{the specific heat}\\ \text{of the system 1}\end{matrix}\right), \tag{12}$$

it can be rewritten as

$$\Delta^2 = kT^2\left(\frac{1}{C_1} + \frac{1}{C_2}\right)^{-1} \tag{13}$$

Equation (11) holds when $E_1 - E_1^*$ is of the order $\mathrm{O}(N^{\frac{1}{2}})$ and shows that $\overline{(E_1 - E_1^*)^2} = \Delta^2$. (When $E_1 - E_1^* = \mathrm{O}(N_1)$, it will no longer be of the Gaussian type. The probability that such a fluctuation will occur is, however, extremely small and need not be considered.)

6. Show that a system in contact with a heat and particle source has particle number N and energy E with the probability given by (1.73a, b).

SOLUTION

Let $\Omega(N_s, E_s)$ be the density of states of the heat-particle source. We shall denote by N_t and E_t the total number of particles and the total energy of the compound system composed by the system under consideration and this heat-particle source, respectively. The probability $Pr\,(N,E)$ that the system under consideration is in a microscopic state with N and E is proportional to the thermodynamic weight of a state of the heat-particle source with $N_t - N$ and $E_t - E$:

$$\begin{aligned} Pr(N,E) &\propto \Omega(N_t - N, E_t - E)\,\delta E_t \\ &\propto \exp\frac{1}{k}\{S(N_t - N, E_t - E) - S(N_t, E_t)\}\,. \end{aligned} \tag{1}$$

Since the heat-particle source is a very large system, we may put $N_t \gg N$ and $E_t \gg E$. Hence

$$S(N_t - N, E_t - E) - S(N_t, E_t) =$$

$$= -N\left(\frac{\partial S}{\partial N_t}\right)_{E_t} - E\left(\frac{\partial S}{\partial E_t}\right)_{N_t} + \frac{1}{2}\left\{N^2 \frac{\partial^2 S}{\partial N_t^2} + 2NE\frac{\partial^2 S}{\partial N_t \partial E_t} + E^2 \frac{\partial^2 S}{\partial E_t^2}\right\} + \dots$$

$$= \frac{\mu}{T}N - \frac{E}{T} + \frac{1}{2}\left\{N\Delta\left(-\frac{\mu}{T}\right) + E\Delta\left(\frac{1}{T}\right)\right\} + \dots . \tag{2}$$

Here μ and T are the chemical potential and the temperature of the heat-particle source with N_t and E_t, and Δ indicates the variations due to the deviation of N and E from N_t and E_t. Since these changes are $O(N/N_t)$ and $O(E/E_t)$, we can neglect them as long as $N \ll N_t$, $E \ll E_t$. Since the probability of having $N \sim N_t$, $E \sim E_t$ is, at any rate, extremely small, we can use this approximation. Therefore, from (1) we get

$$Pr(N, E) \propto \exp\left(\frac{\mu N - E}{kT}\right).$$

Normalizing this, we obtain (1.73a). (The same argument applies to the case where many kinds of particles are present.)

7. There are more than two systems A, B, C, ... , which are almost independent of each other. Suppose that they interact with each other weakly, so that they can be regarded as a compound system $A + B + C \dots$ Show that the partition function $Z_{A+B+\dots}$ and the free energy $F_{A+B+\dots}$ are given by

$$Z_{A+B+\dots} = Z_A \cdot Z_B \dots, \qquad F_{A+B+\dots} = F_A + F_B + \dots$$

respectively, where Z_A, Z_B, ... are the partition functions of the individual systems.

SOLUTION

If one denotes the energy of the quantum state l of the system A by $E_{A,l}$ ($l = 1, 2, \dots$), one has, according to (1.71b),

$$Z_A = \sum_l e^{-\beta E_{A,l}}. \tag{1}$$

Similarly, if one writes the energy of the quantum state m of the system B as $E_{B,m}$ ($m = 1, 2, \dots$), one has

$$Z_B = \sum_m e^{-\beta E_{B,m}}. \tag{2}$$

We have assumed that these systems are in contact with a heat reservoir at the same temperature.

Since the subsystems are almost independent, the quantum state of the compound system $A + B + \dots$ is determined if one specifies the quantum states of the individual subsystems: it is specified by a set of the quantum numbers $(l, m, \dots)$, where $l, m, \dots$ take the values $1, 2, \dots$ independently, and its energy is approximated by $E_{A,l} + E_{B,m} + \dots$. Therefore, the partition function defined by (1.71b) can be transformed as follows:

$$Z_{A+B} = \sum_l \sum_m \dots e^{-\beta(E_{A,l}+E_{B,m}+\dots)} = \sum_l \sum_m \dots e^{-\beta E_{A,l}} e^{-\beta E_{B,m}} \dots$$
$$= \sum_l e^{-\beta E_{A,l}} \sum_m e^{-\beta E_{B,m}} \dots .$$

Hence one has

$$Z_{A+B\dots} = Z_A Z_B \dots . \tag{3}$$

According to the table in § 1.14 the free energy is given by $F = -kT \log Z$. Substituting Z_A, Z_B, $\dots$, $Z_{A+B+\dots}$, one gets from (3)

$$F_{A+B+\dots} = -kT \log Z_{A+B+\dots} = -kT \log Z_A Z_B \dots$$
$$= -kT \log Z_A - kT \log Z_B \dots = F_A + F_B + \dots .$$

8. Apply the canonical and the T-p distribution in classical statistical mechanics to an ideal gas consisting of N monatomic molecules and derive the respective thermodynamic functions.

SOLUTION

Let us calculate them according to § 1.14. The Hamiltonian of this system is

$$\mathscr{H} = \sum_{i=1}^{3N} \frac{p_i^2}{2m},$$

where we have adopted a Cartesian coordinate system and used the same notation as in example 2. Let us first apply the canonical distribution. According to (1.71a) the partition function is

$$Z = \frac{1}{h^{3N} N!} \int e^{-\beta \mathscr{H}} d\Gamma, \qquad d\Gamma \equiv \prod_{i=1}^{3N} dq_i dp_i .$$

Since the Hamiltonian is a function only of the momenta, the integration over the coordinates can readily be performed, yielding V^N (V is the volume of the vessel). The integral over the momenta is equal to the product of the integrals over each momentum coordinate since the Hamiltonian is the sum of the terms for individual degrees of freedom. All the integrals being identical, it is equal to the $3N$-th power of the integral with respect to one degree of freedom, which is

$$\int_{-\infty}^{+\infty} \mathrm{e}^{-(\beta/2m)\,p^2}\,\mathrm{d}p = \sqrt{\frac{2m}{\beta}}\int_{-\infty}^{+\infty} \mathrm{e}^{-x^2}\,\mathrm{d}x = \sqrt{\frac{2\pi m}{\beta}}\,,$$

thus

$$Z = \frac{1}{h^{3N}N!}\,V^N(2\pi mkT)^{\frac{3}{2}N}\,. \tag{1}$$

The thermodynamic function corresponding to this is the Helmholtz free energy and is given by

$$F = -\,kT\log Z = -\,NkT\left\{\tfrac{3}{2}\log T + \log\frac{V}{N} + \log\frac{(2\pi mk)^{\frac{3}{2}}\,\mathrm{e}}{h^3}\right\} \tag{2}$$

where we have used Stirling's formula, $\log N! \approx N\log(N/\mathrm{e})$.

Next, the partition function for the T-p distribution is

$$Y(T,p,N) = \int_0^\infty \mathrm{e}^{-pV/kT}Z(T,V,N)\,\mathrm{d}V.$$

Substituting (1), one gets

$$Y = \frac{1}{h^{3N}N!}(2\pi mkT)^{3N/2}\int_0^\infty \mathrm{e}^{-pV/kT}V^N\,\mathrm{d}V.$$

Writing $p/kT = \alpha$, one can evaluate this integral as follows:

$$\int_0^\infty \mathrm{e}^{-\alpha V}V^N\,\mathrm{d}V = \left(-\frac{\partial}{\partial\alpha}\right)^N\int_0^\infty \mathrm{e}^{-\alpha V}\,\mathrm{d}V = \left(-\frac{\partial}{\partial\alpha}\right)^N\alpha^{-1} = N!/\alpha^{N+1}\,,$$

thus

$$Y(T,p,N) = \frac{1}{h^{3N}N!}(2\pi mkT)^{\frac{3}{2}N}N!\left(\frac{kT}{p}\right)^{N+1} \approx$$

$$\approx \frac{1}{h^{3N}}(2\pi mkT)^{\frac{3}{2}N}\left(\frac{kT}{p}\right)^N. \tag{3}$$

Here the exponent is approximated by $N+1 \to N$ because of $N \gg 1$. The corresponding thermodynamic function is the Gibbs free energy,

$$G(T,p,N) = -\,kT\log Y = -\,NkT\left\{\tfrac{5}{2}\log T - \log p + \log\frac{(2\pi m)^{\frac{3}{2}}k^{\frac{5}{2}}}{h^3}\right\}. \tag{4}$$

NOTE: From $p = -\,(\partial F/\partial V)_{T,N}$ or $V = (\partial G/\partial p)_{T,N}$ one can readily obtain $pV = NkT$. One can also confirm that the relation $G = F + pV$ holds for (2) and (4). The entropy can be obtained either from $S = -(\partial F/\partial T)_{V,N}$ or from $S = -(\partial G/\partial T)_{p,N}$, the results being identical with each other because of $pV = NkT$ (see example 2):

$$S = Nk\left\{\tfrac{3}{2}\log T + \log\frac{V}{N} + \log\left[(2\pi mk)^{\frac{3}{2}}\,e^{\frac{5}{2}}/h^3\right]\right\}$$

$$= Nk\{\tfrac{5}{2}\log T - \log p + \log[(2\pi mk)^{\frac{3}{2}}k\,e^{\frac{5}{2}}/h^3]\}. \tag{5}$$

According to $\mu = (\partial F/\partial N)_{T,V}$ or $G = N\mu$ the chemical potential is equal to

$$\mu = kT\log\frac{N}{V}\left(\frac{h^2}{2\pi mkT}\right)^{\frac{3}{2}} = kT\log\frac{p}{kT}\left(\frac{h^2}{2\pi mkT}\right)^{\frac{3}{2}}. \tag{6}$$

The internal energy is $U = -T^2\partial[F/T]/\partial T = \frac{3}{2}NkT$, and $C_V = \frac{3}{2}Nk$, $C_p = \frac{5}{2}Nk$.

9. Suppose there are two systems I and II in thermal contact with a heat bath at temperature T and there exists some mechanism enabling the two systems to exchange particles. Obtain the expression for the probability of having a distribution $(N_{\mathrm{I}}, N_{\mathrm{II}})$ of particles into I and II in terms of the partition functions $Z_{\mathrm{I}}(N_{\mathrm{I}}, T)$ and $Z_{\mathrm{II}}(N_{\mathrm{II}}, T)$, and derive the condition for finding the most probable distribution.

SOLUTION

Since the composite system I + II of the systems I, II is in contact with a heat bath, its microscopic states are realized according to the canonical distribution. The probability that the composite system I + II is in a state in which the system I has an energy in the range between E_{I} and $E_{\mathrm{I}} + dE_{\mathrm{I}}$ and a particle number equal to N_{I}, while the system II has an energy in the range $(E_{\mathrm{II}}, E_{\mathrm{II}} + dE_{\mathrm{II}})$ and a particle number N_{II}, is given by the sum of the elementary probability (1.70) over all the possible microscopic states subject to the given condition:

$$f(E_{\mathrm{I}}, N_{\mathrm{I}}, E_{\mathrm{II}}, N_{\mathrm{II}}, T)\,dE_{\mathrm{I}}\,dE_{\mathrm{II}} = \frac{1}{Z_{\mathrm{I+II}}}\,e^{-\beta(E_{\mathrm{I}}+E_{\mathrm{II}})}\,\Omega_{\mathrm{I}}(E_{\mathrm{I}}, N_{\mathrm{I}})\,dE_{\mathrm{I}}\,\Omega_{\mathrm{II}}(E_{\mathrm{II}}, N_{\mathrm{II}})\,dE_{\mathrm{II}} \tag{1}$$

($\beta = 1/kT$). $Z_{\mathrm{I+II}}$ is the partition function of the composite system I + II and, as one can see from (1.49), is equal to

$$Z_{\mathrm{I+II}}(N, T) = \sum_{N_{\mathrm{I}}=0}^{N} Z_{\mathrm{I}}(N_{\mathrm{I}}, T)Z_{\mathrm{II}}(N - N_{\mathrm{I}}, T) \tag{2}$$

where N is the particle number. Since we are concerned only with the probability of a particle distribution $(N_\text{I}, N_\text{II})$, we integrate (1) over the energies and obtain

$$f(N_\text{I}, N_\text{II}) = \frac{Z_\text{I}(N_\text{I}, T)Z_\text{II}(N - N_\text{I}, T)}{Z_{\text{I}+\text{II}}(N, T)}, \qquad (N_\text{II} = N - N_\text{I}). \tag{3}$$

The most probable distribution $(N_\text{I}^*, N_\text{II}^*)$ is given by the values of N_I and N_II which make this maximum,

$$Z_\text{I}(N_\text{I}, T)Z_\text{II}(N - N_\text{I}, T) = \text{max.} \tag{4}$$

Introducing the Helmholtz free energy for the respective systems by $F = -kT \log Z$, we have

$$F_\text{I}(N_\text{I}, T) + F_\text{II}(N - N_\text{I}, T) = \text{min.}, \tag{5}$$

or

$$\frac{\partial}{\partial N_\text{I}} F_\text{I}(N_\text{I}, T) = \frac{\partial}{\partial N_\text{II}} F_\text{II}(N_\text{II}, T), \qquad (N_\text{I} + N_\text{II} = N)$$

that is, in terms of the chemical potential $\partial F/\partial N = \mu$,

$$\mu_\text{I}(N_\text{I}, T) = \mu_\text{II}(N_\text{II}, T). \tag{6}$$

NOTE: One can get the Helmholtz free energy of the composite system I + II from the partition function (2). However, when the systems I, II are both sufficiently large (which condition is necessary for the most probable distribution $(N_\text{I}^*, N_\text{II}^*)$ to correspond to the values in the thermal equilibrium state), one may approximate it by the left hand side of (5) with $N_\text{I} = N_\text{I}^*$. This is what is meant by the additivity of the free energy. One can prove it in the same way as (ii) of example 5.

10. When a particle with spin $\frac{1}{2}$ is placed in a magnetic field H, its energy level is split into $-\mu H$ and $+\mu H$ and it has a magnetic moment μ or $-\mu$ along the direction of the magnetic field, respectively. Suppose a system consisting of N such particles is in a magnetic field H and is kept at temperature T. Find the

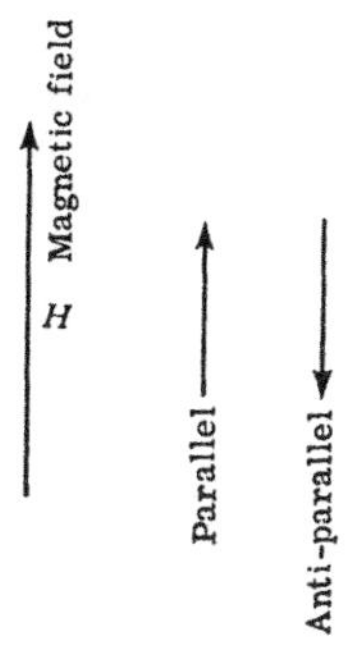

Fig. 1.10.

internal energy, the entropy, the specific heat and the total magnetic moment M of this system with the help of the canonical distribution.

Solution

Since the spins are independent of each other, the partition function of the total system Z_N is equal to the Nth power of the partition function for the

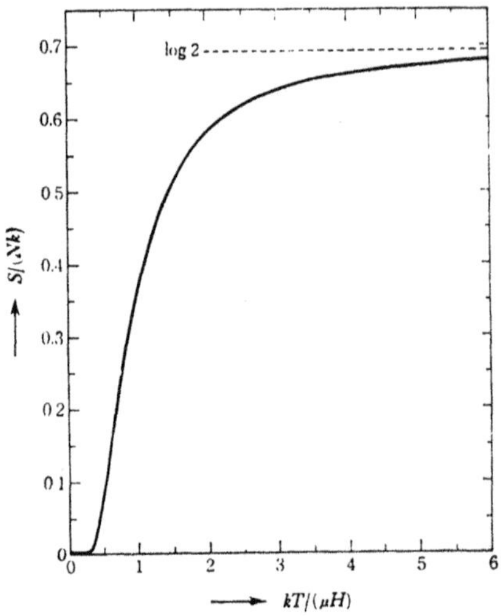

Fig. 1.11.

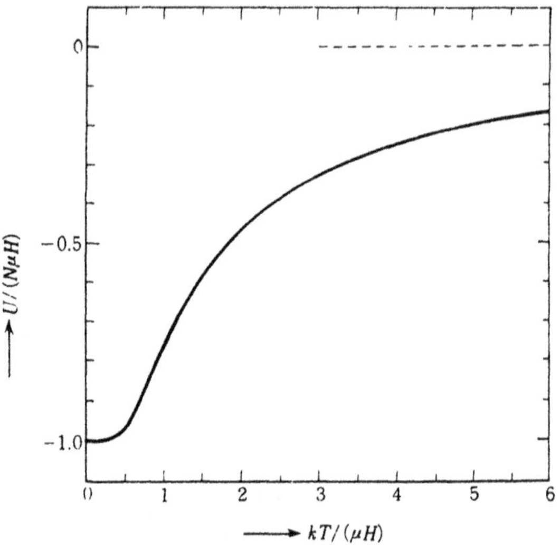

Fig. 1.12.

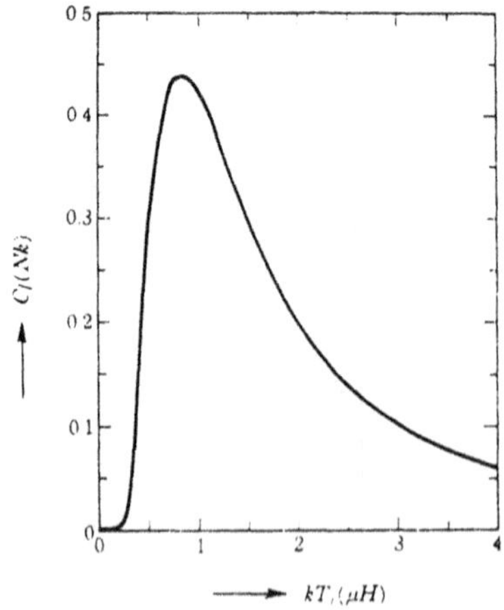

Fig. 1.13.

individual spin $$Z_1 = e^{\beta\mu H} + e^{-\beta\mu H} = 2\cosh(\mu H/kT)$$

(see example 7). Hence

$$Z_N = Z_1^N = [2\cosh(\mu H/kT)]^N . \tag{1}$$

The Helmholtz free energy is

$$F = -NkT\log\{2\cosh(\mu H/kT)\} \tag{2}$$

from which one gets

$$S = -\frac{\partial F}{\partial T} = Nk\left[\log\left\{2\cosh\left(\frac{\mu H}{kT}\right)\right\} - \frac{\mu H}{kT}\tanh\left(\frac{\mu H}{kT}\right)\right], \tag{3}$$

$$U = F + TS = -N\mu H\tanh\left(\frac{\mu H}{kT}\right), \tag{4}$$

$$M = -\frac{\partial F}{\partial H}\left(= N\bar{\mu} = \frac{N(\mu e^{\beta\mu H} - \mu e^{-\beta\mu H})}{(e^{\beta\mu H} + e^{-\beta\mu H})}\right) = N\mu\tanh\left(\frac{\mu H}{kT}\right), \tag{5}$$

$$C = \left(\frac{\partial U}{\partial T}\right)_H = Nk\left(\frac{\mu H}{kT}\right)^2 \Big/ \cosh^2\left(\frac{\mu H}{kT}\right). \tag{6}$$

(It is easier to calculate U by $-T^2\partial[F/T]/\partial T$. Also $U = -MH$ as it should be.)

These quantities are plotted in Figs. 1.11, 1.12 and 1.13.

DIVERTISSEMENT 3

Entropy and probability. In the central cemetery of the beautiful city of Vienna, travellers will see the grave of Ludwig Boltzmann (1844–1906) on which there remains forever his most precious bequest to mankind, namely the formula

$$S = k\log W .$$

When Boltzmann started his study, the kinetic theory of gases had been developed by such pioneers as Clausius and Maxwell to the stage that it could be regarded as so far very successful. Diffusion, heat conduction, viscosity and so forth had been treated with great success. For a believer in the kinetic theory of heat, the first law of thermodynamics was a simple consequence of the "Prinzip der lebendigen Kräfte" in mechanics, as Helmholtz pointed out, if the elementary processes were all purely mechanical. Boltzmann asked himself then the question, "Sollte also nicht auch der zweite Hauptsatz im Grunde ein rein mechanisches Prinzip sein?"

Through a series of elaborate studies of the kinetic theory of gases, in 1872 he reached the famous H-theorem, which states that the function

$$H = \iiint f(u, v, w)\log f(u, v, w)\,du \cdot dv \cdot dw$$

never increases by molecular collision, so that the equilibrium of a gas can be attained only when this function reaches its minimum. Later in a paper † published in 1877, he called the function $-H$ by the name "Permutabilitätsmass" and identified it, except for a constant factor, with the entropy. "Denken wir uns ein beliebiges System von Körpern gegeben, dasselbe mache eine beliebige Zustandsveränderung durch, ohne dass notwendig der Anfangs- und Endzustands Zustande des Gleichgewichtes zu sein brauchen; dann wird immer das Permutabilitätsmass aller Körper im Verlaufe der Zustandsveränderungen fortwährend wachsen und kann hochstens konstant bleiben solange sich sämtliche Körper während der Zustandsveränderung mit unendlicher Annährung im Wärmegleichgewichte befinden (umkehrbare Zustandsveranderungen)."

By this statement Boltzmann opened the way for statistical mechanics to get free of ties to the traditional method of kinetic theory and to construct itself on a more general and firm basis of probability theory combined with mechanics.

It appears that Boltzmann himself never explicitly wrote the formula (1.18) itself. Planck gave this formula in his famous lecture †† on "Wärmestrahlung", emphasizing then that this formula defines entropy absolutely leaving no ambiguity of an additive constant. This was possible for Planck who first introduced the quantum hypothesis in 1900.

11. Prove the following relation for the grand partition function $\Xi(T, \mu, V)$:

$$pV = kT \log \Xi .$$

SOLUTION

For simplicity, suppose there is only one kind of particles. From the definition of Ξ it follows that

$$kT \frac{\partial \log \Xi}{\partial V} = \sum_{N=0}^{\infty} \sum_{l} \left(- \frac{\partial E_{N,l}}{\partial V} \right) \frac{\exp[-\beta(E_{N,l} - N\mu)]}{\Xi} = \sum_{N=0}^{\infty} Pr(N) p_N \rightleftharpoons p \quad (1)$$

where p_N is the pressure in the state (N, T, V) and p is what one gets by taking the mean of p_N with the probability distribution $Pr(N)$. The latter is the pressure observed in the system in which the number of particles is not fixed. When the volume V is large, $\log \Xi$ is proportional to V, that is,

$$\partial \log \Xi / \partial V = \log \Xi / V . \quad (2)$$

If one grants (2), equation (1) gives the relation to be proved. Equation (2)

† L. Boltzmann, Über die Beziehungen zwischen dem II Hauptsatz der mechanischen Wärmetheorie und der Wahrscheinlichkeitsrechnung. Wein. Ber. 77 (1877) 373. See also: L. Boltzmann, Vorlesungen über Gastheorie, Bd. 1. (Barth, Leipzig, 1923) § 6.

†† Max Planck, Vorlesungen über die Theorie der Wärmestrahlung (Barth, Leipzig, 1923) pp. 119–123.

can be proved in the following way. Let us divide the volume V into n parts of equal volume V/n by inserting walls which can transmit both heat and particles. The total system is in contact with a heat-particle source with fixed μ and T. The macroscopic properties are not affected by the insertion of the walls. In statistical mechanics one can regard each part as independent, assuming the interaction between the system and the walls, as well as the interactions between the parts, to be weak. Therefore, when $N_1, N_2, \dots$ particles are distributed in the respective parts, the partition function of the whole system $Z(N_1, N_2, \dots, T, V)$ is equal to the product of the partition functions of each part (example 7):

$$Z(N_1, N_2, \dots, T, V) = Z(N_1, T, V/n)Z(N_2, T, V/n)\dots .$$

Substituting this into the definition (1.74c) we have

$$\begin{aligned}\Xi(\mu, T, V) &= \sum_{N=0}^{\infty} e^{\beta N\mu} \sum_{N_1+N_2+\dots=N} Z(N_1, N_2, \dots, T, V)\\ &= \sum_{N=0}^{\infty} \sum_{N_1+N_2+\dots=N} e^{\beta(N_1+N_2+\dots)\mu} Z(N_1, T, V/n)Z(N_2, T, V/n)\dots .\end{aligned}$$

Since the summation over the total number N is equivalent to summing over $N_1, N_2 \dots$ independently, we obtain

$$\Xi(\mu, T, V) = [\Xi(\mu, T, V/n)]^n,$$

or

$$\log \Xi(\mu, T, V) = n \log \Xi(\mu, T, V/n).$$

This is just the relation (2) (if we put $1/n = \alpha$, $\log \Xi(\mu, T, \alpha V) = \alpha \log \Xi(\mu, T, V)$).

12. Using the method of grand canonical ensemble, prove that the distribution function of an ideal quantum gas is given by

$$f(\varepsilon) = \frac{1}{e^{\beta(\varepsilon-\mu)} \mp 1}, \quad \begin{array}{l} - \text{ for Bose-Einstein statistics,} \\ + \text{ for Fermi-Dirac statistics} \end{array}$$

and derive the expression for the entropy (1.94a) as well as the formula giving the Helmholtz free energy.

SOLUTION

Let the energy of a quantum state τ of a particle be ε_τ ($\tau = 1, 2, \dots$) and denote the number of particles in this state by n_τ. A quantum state of the total system is specified by $\{n_\tau\} = (n_1, n_2, \dots)$. The energy of the total system is

$$E_n = \sum_\tau \varepsilon_\tau n_\tau . \qquad (1)$$

Since the total number of particles is N, one has

$$N = \sum_{\tau} n_{\tau}. \tag{2}$$

Hence, the partition function Z_N is

$$Z_N = \sum_{\{N_\tau\}} e^{-\beta \Sigma_\tau \varepsilon_\tau n_\tau}, \tag{3}$$

where the summation is to be taken under condition (2).

From (1.74c) one has the grand partition function

$$\Xi = \sum_{N=0}^{\infty} e^{\beta N \mu} Z_N = \sum_{N=0}^{\infty} e^{\beta N \mu} \sum_{\{n_\tau\}} e^{-\beta \Sigma_\tau \varepsilon_\tau n_\tau} = \sum_{N=0}^{\infty} \sum_{\{n_\tau\}} e^{\beta \Sigma_\tau (\mu - \varepsilon_\tau) n_\tau}.$$

In the last form, the summation over N allows us to get rid of the condition (2), and we can sum over n_τ independently:

$$\Xi = \sum_{n_1} \sum_{n_2} \ldots e^{\beta \Sigma_\tau (\mu - \varepsilon_\tau) n_\tau} = \prod_{\tau} \sum_{n_\tau} e^{\beta(\mu - \varepsilon_\tau) n_\tau}. \tag{4}$$

In carrying out the summation, one has to take into account the statistics.

In Bose-Einstein statistics, n_τ can take on the value of all positive integers as well as zero ($n = 0, 1, 2, \ldots$) so that

$$\sum_{n=0}^{\infty} e^{\beta(\mu - \varepsilon_\tau) n} = (1 - e^{\beta(\mu - \varepsilon_\tau)})^{-1},$$

where it is necessary that $\mu < \varepsilon_\tau$ (for all τ).

In Fermi-Dirac statistics, one has only the values $n_\tau = 0$ or 1, so that

$$\sum_{n} e^{\beta(\mu - \varepsilon_\tau) n} = 1 + e^{\beta(\mu - \varepsilon_\tau)}.$$

Assigning the $-$ and $+$ signs to the B. E. statistics and the F. D. statistics, respectively, we can write the above results in the form

$$\sum_{n} e^{\beta(\mu - \varepsilon_\tau) n} = \{1 \mp e^{\beta(\mu - \varepsilon_\tau)}\}^{\mp 1}, \tag{5}$$

thus

$$\Xi = \prod_{\tau} \{1 \mp e^{\beta(\mu - \varepsilon_\tau)}\}^{\mp 1}. \tag{6}$$

The probability that n_τ particles occupy a single particle quantum state τ is now obtained by adding (1.73b) for all the quantum states $\{n_\sigma\}$ with the same n_τ

$$Pr(n_\tau) = \frac{e^{\beta(\mu - \varepsilon_\tau) n_\tau} \prod_{\sigma}' \sum_{n_\sigma} e^{\beta(\mu - \varepsilon_\sigma) n_\sigma}}{\Xi},$$

where $\prod'$ in the numerator means the product over all σ excluding τ. On substitution of (4) this simplifies to

$$P_r(n_\tau) = e^{\beta(\mu-\varepsilon_\tau)n_\tau} \Big/ \sum e^{\beta(\mu-\varepsilon_\tau)n_\tau}. \tag{7}$$

Hence the mean number $\bar{n}_\tau$ of particles in the state τ is

$$\bar{n}_\tau = \sum_{n_\tau=0}^{\infty} n_\tau Pr(n_\tau) = \frac{\sum n_\tau e^{\beta(\mu-\varepsilon_\tau)n_\tau}}{\sum e^{\beta(\mu-\varepsilon_\tau)n_\tau}} = \frac{1}{\beta}\frac{\partial}{\partial\mu}\log \sum e^{\beta(\mu-\varepsilon_\tau)n_\tau}. \tag{8}$$

Using result (5) for the sum we get

$$\bar{n}_\tau = \mp \frac{1}{\beta}\frac{\partial}{\partial\mu}\log\{1 \mp e^{\beta(\mu-\varepsilon_\tau)}\} = \frac{e^{\beta(\mu-\varepsilon_\tau)}}{1 \mp e^{\beta(\mu-\varepsilon_\tau)}} = \frac{1}{e^{\beta(\varepsilon_\tau-\mu)} \mp 1}. \tag{9}$$

To obtain the thermodynamic function, it is convenient to use $J = -kT\log \Xi$ in the table of 1.14 (see example 11). From (6) one finds

$$J = -pV = -kT\log\Xi = \pm kT\sum \log\{1 \mp e^{\beta(\mu-\varepsilon_\tau)}\} \tag{10a}$$

$$= \mp kT \sum \log(1 \pm \bar{n}_\tau), \tag{10b}$$

where use has been made of

$$1 \pm \bar{n}_\tau = (1 \mp e^{\beta(\mu-\varepsilon_\tau)})^{-1}. \tag{11}$$

From the thermodynamic relation $dJ = -SdT - pdV - Nd\mu$ it follows that

$$S = -\partial J/\partial T = \mp k\sum\log\{1 \mp e^{\beta(\mu-\varepsilon_\tau)}\} - \frac{1}{T}\sum \frac{\mu-\varepsilon_\tau}{e^{\beta(\varepsilon_\tau-\mu)} \mp 1} \tag{12a}$$

$$= \frac{-J - N\mu + E}{T}. \tag{12b}$$

In (12b) we have used

$$N = \sum \frac{1}{e^{\beta(\varepsilon_\tau-\mu)} \mp 1}, \qquad E = \sum \frac{\varepsilon_\tau}{e^{\beta(\varepsilon_\tau-\mu)} \mp 1}. \tag{13}$$

The relation (12b) is obviously what one expects to get ($TS = pV - G + E$). On the other hand, from (11) and (9) we have

$$\bar{n}_\tau/(1 \pm \bar{n}_\tau) = e^{\beta(\mu-\varepsilon_\tau)}, \qquad \text{or} \quad \log\{\bar{n}_\tau/(1 \pm \bar{n}_\tau)\} = \beta(\mu - \varepsilon_\tau). \tag{14}$$

By means of (11) and (14) one can rewrite (12a) in the form

$$\begin{aligned} S &= \pm k\sum \log(1 \pm \bar{n}) - k\sum \bar{n}_\tau \log\{\bar{n}_\tau/(1 \pm \bar{n}_\tau)\} \\ &= k\sum[-\bar{n}_\tau \log \bar{n}_\tau \pm (1 \pm \bar{n}_\tau)\log(1 \pm \bar{n}_\tau)]. \end{aligned} \tag{15}$$

Since $J = -pV = F - G = F - N\mu$, the Helmholtz free energy is given as a function N, V, T by

$$\left.\begin{aligned} F = J + N\mu &= N\mu \pm kT \sum \log(1 \mp e^{\beta(\mu - \varepsilon_\tau)}), \\ N &= \sum \frac{1}{e^{\beta(\varepsilon_\tau - \mu)} \mp 1}. \end{aligned}\right\} \tag{16}$$

Problems

[A]

1. Explain Dalton's law for an ideal gas mixture on the basis of elementary kinetic theory of gas molecules.

2. Imagine a fictitious surface element in an ideal gas. Supposing that momentum transfer takes place through the surface element due to penetration by gas molecules, find the formula for calculating the pressure which both sides of the surface element exert upon each other (Lorentz's method). Assume that the gas molecules obey a Maxwellian velocity distribution.

3. A rarefied gas is contained in a vessel of volume V at pressure p. Supposing that the velocity distribution of molecules of the gas is Maxwellian, calculate the rate at which the gas flows out of the vessel into a vacuum through a small hole (of area A).

   Taking the wall with the hole as the y-z plane, find the velocity distribution in the x-direction of the gas molecules moving out of the hole.

4. There is a furnace containing a gas at high temperature. Through a small window of the furnace one observes, using a spectrometer, a spectral line of the gas molecules. The width of the observed spectral line is broadened (this is called Doppler broadening). Show that the relation between spectral line intensity I and wavelength λ is given by the following formula:

$$I(\lambda) \propto \exp\left\{-\frac{mc^2(\lambda - \lambda_0)^2}{2\lambda_0^2 kT}\right\}.$$

   Here T is the temperature of the furnace, c the velocity of light, m the mass of a molecule, and λ_0 the wavelength of the spectral line when the molecule is at rest.

5. A mass point with mass m moves within the range $0 \leqq x \leqq l$ and is reflected

by walls at $x = 0$ and l. (i) Illustrate the trajectory of this mass point in the phase space (x, p), (ii) find the volume of the phase space $\Gamma_0(E)$ with energy smaller than E and (iii) show that $\Gamma_0(E)$ is kept constant when the wall at $x = l$ is moved slowly (adiabatic invariance). (iv) Going over to quantum mechanics, find the number $\Omega_0(E)$ of quantum states with energy below E and compare it with $\Gamma_0(E)$.

6. Find the number of quantum states for a particle contained in a cubic box with edge length l, and compare it with the volume in classical phase space. Obtain also the density of states.

7. What does a surface of constant energy look like in the phase space of an oscillator of frequency ν? Find the volume $\Gamma_0(E)$ in the phase space with energy below E. Then find the number of quantum states $\Omega_0(E)$ with energy below E for this oscillator, and show that when E is large we have

$$\Gamma_0(E)/h \sim \Omega_0(E).$$

8. (i) When a system of N oscillators ($N \gg 1$) with total energy E is in thermal equilibrium, find the probability that a given oscillator among them is in a quantum state n. (*Hint:* use W_M of example 3.)

(ii) When an ideal gas of N monatomic molecules with total energy E is in thermal equilibrium, show that the probability of a given particle having an energy $\varepsilon = p^2/2m$ is proportional to $\exp(-\varepsilon/kT)$. [*Hint:* use $\Omega_0(E, N)$ of example 2.]

9. A vessel of volume V contains N gas molecules. Let n be the number of molecules in a part of the vessel of volume v. Considering that the probability of finding a certain molecule in v is equal to v/V in the thermal equilibrium state of this system,

(i) find the probability distribution $f(n)$ of the number n, and

(ii) calculate $\bar{n}$ and $\overline{(n - \bar{n})^2}$.

(iii) Making use of Stirling's formula, show that when N and n are both large $f(n)$ is approximately Gaussian.

(iv) Show that in the limit of $v/V \to 0$ and $V \to \infty$, with N/V = constant, $f(n)$ approaches the Poisson distribution $f(n) = \mathrm{e}^{-\bar{n}}(\bar{n})^n/n!$.

10. As illustrated in the figure, a string with a lead ball of mass m is slowly pulled upward through a small hole. Consider the work done on the system during this process and find the change in energy and the frequency of this pendulum during this "adiabatic process", assuming the amplitude of the pendulum to be small.

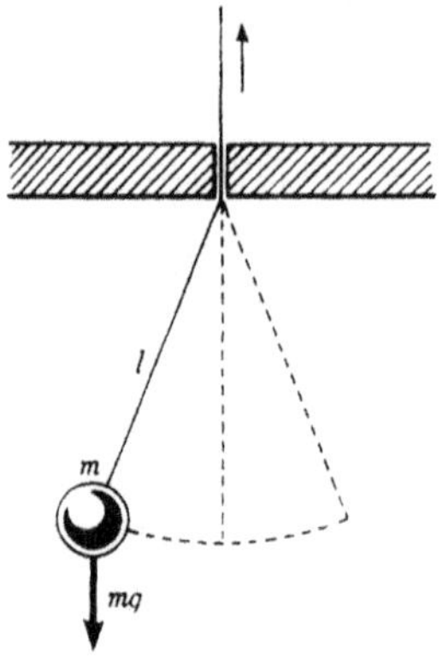

11. Prove on the basis of the principle of entropy increase that, when a coordinate x varies extremely slowly through a quasi-static adiabatic process (1.35), the entropy $S(x)$ does not change. (*Hint:* consider $\mathrm{d}S/\mathrm{d}t$ as a function of $\mathrm{d}x/\mathrm{d}t$.)

12. As in the figure, a system under consideration is contained in a box made of insulating walls and a movable piston. A weight w is placed on the piston. Regarding the total system including the weight as an isolated system, derive from the microcanonical ensemble the relation

$$p = -\left(\frac{\partial E}{\partial V}\right)_S .$$

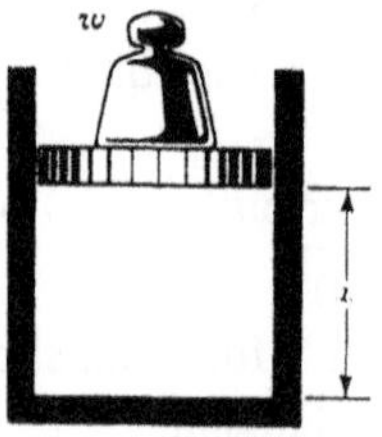

Apply this formula to an ideal gas of monatomic molecules and prove the equation of state $p = \frac{2}{3} E/V$.

13. There is a one-dimensional chain consisting of $n(\gg 1)$ elements, as is seen in the figure. Let the length of each element be a and the distance between the end points x. Find the entropy of this chain as a function of x and obtain the relation between the temperature T of the chain and the force (tension) which is necessary to maintain the distance x, assuming the joints to turn freely.

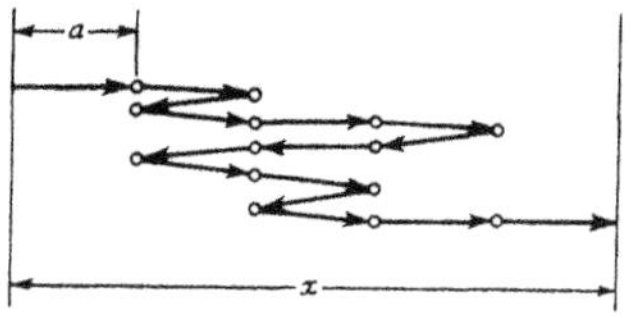

14. There is a system consisting of N oscillators of frequency ν. Discussing this system classically,

(i) find the number of states, and

(ii) using this result, derive the relation between the energy and the temperature of this system.

15. For an oscillator with mass m and angular frequency ω, calculate the partition function (i) classically and (ii) quantum mechanically. (iii) Find also the internal energy, the entropy and the heat capacity of a system consisting of N such oscillators as a function of temperature.

16. Show that, if the temperature is uniform, the pressure of a gas in a uniform gravitational field decreases with height according to $p(z) = p(0) \times \exp(-mgz/kT)$, where m is the mass of a molecule.

17. Consider an ideal gas consisting of N particles obeying classical statistics. Suppose that the energy of one particle ε is proportional to the magnitude of momentum p, $\varepsilon = cp$. Find the thermodynamic functions of this ideal gas without considering the internal structure of the particles.

18. We are given an ideal gas consisting of N monatomic molecules and a system consisting of N oscillators. Supposing that they have a canonical distribution at temperature T, find the most probable value E^* of the total energy E of the respective systems and confirm that it agrees with the mean value $\bar{E}$ in the canonical distribution.

19. Show that the grand canonical distribution function of a classical ideal gas of monatomic molecules is given by

$$\Xi = e^{\lambda f}.$$

What is the significance of f and λ?

20. Consider a monatomic ideal gas of N molecules in a volume V. Show, with the help of the T-μ distribution, that the number n of molecules contained in a small element of volume v is given by the Poisson distribution,

$$P_n = e^{-\bar{n}}(\bar{n})^n/n!$$

21. Show that in the T-p distribution the quantity

$$G(T, p, N) = -kT \log Y$$

is equal to the Gibbs free energy where Y is the partition function.

22. Classify the following particles according to Fermi or Bose statistics:

α-particle, ^3He, H_2 molecule, positron, $^6Li^+$ ion and $^7Li^+$ ion.

23. Show that, when the density of a gas consisting of particles with mass m is sufficiently low and its temperature is sufficiently high so that the condition,

mean de Broglie wavelength $\ll$ mean distance between particles

is satisfied, one can use Boltzmann statistics as a good approximation irrespective of whether the particles obey Fermi or Bose statistics.

24. Let p_s be the probability that a system is in a state s with energy E_s. Show that if the entropy is defined by

$$S = -k \sum_s p_s \log p_s$$

the values of the p_s which make S a maximum under the condition that the mean energy of the system is E, follows the canonical distribution.

[B]

25. Derive from an elementary molecular kinetic theory Poisson's equation, $pV^\gamma =$ constant, for a quasi-static adiabatic process of an ideal gas. (*Hint*: note especially the role played by intermolecular collisions.)

26. Discuss the number of states $\Omega_0(E)$ of a system consisting of N oscillators and show that this system is normal in the sense discussed in § 1.6.

27. N atoms are arranged regularly so as to form a perfect crystal. If one replaces n atoms among them ($1 \ll n \ll N$) from the lattice sites to interstices of the lattice, this becomes an imperfect crystal with n defects of the Frenkel type. The number N' of interstitial sites into which an atom can enter is of the same order of magnitude as N. Let w be the energy necessary to remove an atom from a lattice site to an interstitial site. Show that, in the equilibrium state at temperature T such that $w \gg kT$, the following relation is valid:

$$\frac{n^2}{(N-n)(N'-n)} = \mathrm{e}^{-w/kT}, \qquad \text{or} \qquad n \fallingdotseq \sqrt{NN'}\,\mathrm{e}^{-w/2kT}.$$

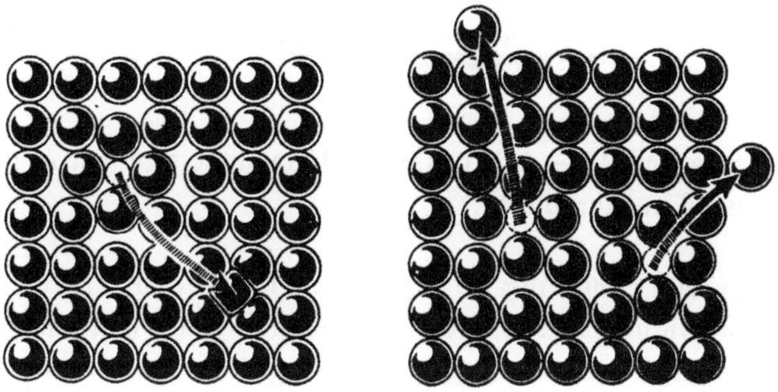

28. If n atoms in a perfect crystal formed by N atoms ($1 \ll n \ll N$) are displaced from lattice sites inside the crystal to lattice sites on the surface, it becomes imperfect, having defects of the Schottky type. Let w be the energy necessary to displace an atom from the inside to the surface. Show that in the equilibrium state at temperature T satisfying $w \gg kT$ one has

$$\frac{n}{N+n} = \mathrm{e}^{-w/kT}, \qquad \text{or} \qquad n \fallingdotseq N\mathrm{e}^{-w/kT}$$

if one can neglect any effect due to the change in the volume of the crystal.

29. Consider an adsorbent surface having N sites each of which can adsorb one gas molecule. Suppose that it is in contact with an ideal gas with the chemical potential μ (determined by the pressure p and the temperature T). Assuming that an adsorbed molecule has energy $-\varepsilon_0$ compared to one in a free state, determine in this case the covering ratio θ (the ratio of adsorbed molecules to adsorbing sites). (Use the grand canonical ensemble.) Find, in particular, the relation between θ and p in the case of monatomic molecules, utilizing the result for μ in example 8.

30. Find the fluctuation $\overline{(M - \bar{M})^2}$ of the total magnetic moment M of the spin system discussed in example 10.

31. Consider a system consisting of N particles. Let us divide all the quantum states of an individual particle into groups each of which contains states with nearly equal energy. Let the energy of the j-th group be ε_j and the number of states contained in it be C_j. A state of the whole system can then be specified by the set of the number of particles N_j in each group. (i) Show that the thermodynamic weight of the state specified by the set $\{N_j\}$ is given by

$$W\{N_j\} = \prod_j \frac{(N_j + C_j - 1)!}{N_j!(C_j - 1)!}, \qquad \text{(B. E.)}$$

$$W\{N_j\} = \prod \frac{C_j!}{N_j!(C_j - N_j)!}, \qquad \text{(F. D.)}$$

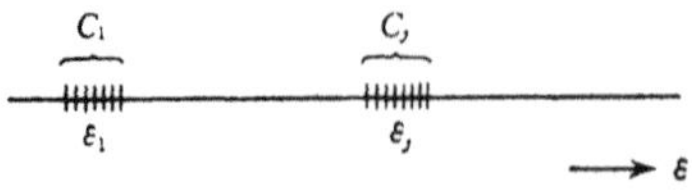

according to the statistics of the particle system. (ii) Supposing that the whole system is in contact with a heat bath at temperature T, find the most probable set among $\{N_j\}$ and derive from this result the B.E. (1.90) or F.D. distribution (1.91) as the probability of each state of an individual particle being occupied. (iii) Assuming that the energy of the whole system E is constant (the micro-canonical ensemble), derive the same result as in (ii).

32. Show that for an ideal gas the relation

$$p = \frac{2}{3}\frac{E_{\text{kin}}}{V}$$

holds irrespective of its statistics, where E_{kin} is the total kinetic energy. [*Hint:* Use (i) equation (1.94c) or (ii) alternatively, the relation $p = (\partial E/\partial V)_S$.]

33. Prove for an ideal gas the following equation concerning the grand canonical distribution:

(i) $\overline{(N - \bar{N})^2} = kT \dfrac{\partial}{\partial \mu} \bar{N}$,

(ii) in classical statistics $\quad \overline{(N - \bar{N})^2} = \bar{N}$,

(iii) in quantum statistics $\quad \overline{(n_\tau - \bar{n}_\tau)^2} = \bar{n}_\tau(1 \pm \bar{n}_\tau)$,

(+ : B. E., − : F. D.).

[C]

34. Show that, if one grants the equality of the time average of $X = \partial \mathscr{H}(p,q,x)/\partial x$ to its phase average (1.13) (the ergodic theorem), $\Omega_0(E, x)$ is invariant under a quasistatic adiabatic process in which x varies very slowly.

35. Suppose that a system of N particles, each of which can be in only two quantum states with $\pm\,\varepsilon_0$ (for example, a system of spins), is brought by some method into a state in which the total energy E is positive. What result does one get if one brings an ideal gas thermometer in contact with this system?

36. Let the spatial distribution of particles with charge e be given by the number density $n(\boldsymbol{r})$. If the potential of an external field is ϕ_{ext}, the total potential energy is

$$U = \tfrac{1}{2}e^2 \iint \frac{n(\boldsymbol{r})n(\boldsymbol{r}')}{|\boldsymbol{r} - \boldsymbol{r}'|}\,\mathrm{d}\boldsymbol{r}\,\mathrm{d}\boldsymbol{r}' + e\int n(\boldsymbol{r})\phi_{\text{ext}}\,\mathrm{d}\boldsymbol{r}\,.$$

Assume that the entropy of this system is

$$S = -k \int n(\mathbf{r})\{\log n(\mathbf{r}) - 1\}\, d\mathbf{r}$$

and find the equations which $n(\mathbf{r})$ and the static potential $\varphi(\mathbf{r})$ satisfy in the equilibrium state.

Solutions

1. Equation (1) derived in the solution of example 1 holds even if there are several kinds of molecules:

$$p = \frac{2}{3}\frac{1}{V}\overline{\sum_r \sum_i \frac{p_{ri}^2}{2m_r}} \tag{1}$$

where m_r is the mass of the r-th kind of molecule and p_{ri} the momentum of the i-th molecule of the r-th kind. Now,

$$p_r = \frac{2}{3}\frac{1}{V}\overline{\sum_i \frac{p_{ri}^2}{2m_r}}$$

is just the pressure which molecules of the r-th kind in volume V exert on the wall and has the same value as in the absence of molecules of other kinds: p_r is the partial pressure of the gas of the r-th kind. The total pressure is, according to (1), the sum of the partial pressures: $p = \sum_r p_r$.

2. Let the surface element in the gas be $\delta\sigma$ and its normal $\mathbf{n}$. If we denote by δP_n the normal component of the momentum transferred through the surface during the time δt, the pressure p is

$$p = \frac{\delta P_n}{\delta\sigma\delta t}. \tag{1}$$

In the case of an ideal gas, this momentum transfer is caused by the movement of molecules.

A molecule moving from the side A to B ($\mathbf{n}$ is directed from A to B) carries its momentum to the side B, while a molecule moving from B to A gives its

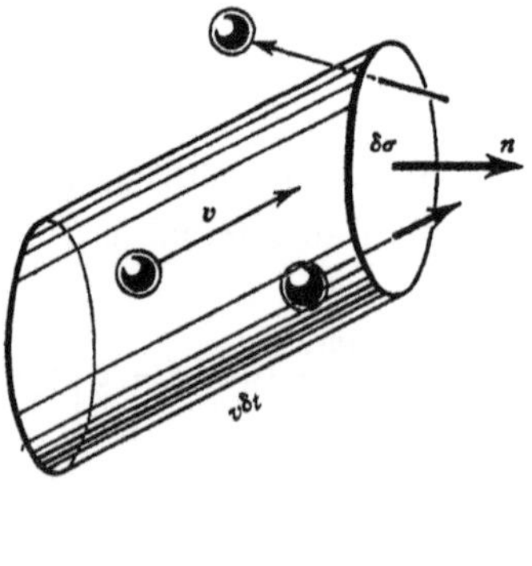

momentum $\boldsymbol{p}$ to the side A, in other words, it transfers momentum $-\boldsymbol{p}$ to the side B.

Among the molecules with velocity $\boldsymbol{v}$, those which pass through the surface element during the time δt are the ones in the oblique cylinder having base $\delta\sigma$ and slant height $\boldsymbol{v}\delta t$, as shown in the figure. The number of these molecules is given by $n(\boldsymbol{v})\ \delta\sigma v_x \delta t$ in terms of the number density of molecules $n(\boldsymbol{v})$. (We have chosen the x-axis along the normal of $\delta\sigma$.) Since such a molecule has the normal component of momentum mv_x, the contribution $\delta_1 P_n$ of the molecules moving from the side A to B is

$$\delta_1 P_n = \sum_{v_x > 0} n(\boldsymbol{v})\delta\sigma v_x \delta t m v_x .$$

One can treat the molecules moving from B to A with momentum $m\boldsymbol{v}(v_x < 0)$ in a similar way. Their number is $n(\boldsymbol{v})\delta\sigma \mid v_x \mid \delta t = -\ n(\boldsymbol{v})\delta\sigma v_x \delta t$ and each of them transfers momentum $-m\boldsymbol{v}$ to the side B. Hence the contribution $\delta_2 P_n$ of molecules passing from B to A is

$$\delta_2 P_n = \sum_{v_x < 0} n(\boldsymbol{v})\delta\sigma v_x \delta t m v_x .$$

Therefore,

$$\delta P_n = \delta_1 P_n + \delta_2 P_n = \sum_{\text{all } v} n(\boldsymbol{v}) m v_x^2 \delta\sigma\delta t .$$

Consequently, according to (1) the pressure p is given by

$$p = \sum_{v} n(\boldsymbol{v}) m v_x^2 = nm\overline{v_x^2} = \tfrac{1}{3} nm\,(\overline{v_x^2} + \overline{v_y^2} + \overline{v_z^2}) = \tfrac{2}{3} n\bar{\varepsilon}$$

in agreement with Bernoulli's formula. Here we have assumed the velocity distribution to be isotropic so that $\overline{v_x} = \overline{v_y} = \overline{v_z}$. In the case of the Maxwellian distribution given by (1.100), we get

$$p = nm \int_{-\infty}^{\infty} v_x^2 \mathrm{d}v_x \int_{-\infty}^{\infty} \mathrm{d}v_y \int_{-\infty}^{\infty} \mathrm{d}v_z \left(\frac{m}{2\pi kT}\right)^{\frac{3}{2}} \exp\left[-\frac{m}{2kT}(v_x^2 + v_y^2 + v_z^2)\right] = nkT \tag{2}$$

where use has been made of the well-known formulae,

$$\sqrt{\frac{a}{2\pi}} \int_{-\infty}^{\infty} \mathrm{e}^{-\frac{1}{2}ax^2} \mathrm{d}x = 1 \quad \text{(3a)}, \qquad \sqrt{\frac{a}{2\pi}} \int_{-\infty}^{\infty} x^2 \mathrm{e}^{-\frac{1}{2}ax^2} \mathrm{d}x = \frac{1}{a} . \quad \text{(3b)}$$

3. We proceed in the same manner as in the preceding problem. Since in the present case there is no molecule moving from the outside to the inside, the number of molecules leaking out during the time δt is

$$-\delta N = \sum_{v_x > 0} n(\boldsymbol{v}) A v_x \delta t .$$

Consequently, the rate of leaking is

$$-\frac{dN}{dt} = A\int_0^\infty dv_x \int_{-\infty}^\infty dv_y \int_{-\infty}^\infty dv_z\, v_x n f(v) = An\left(\frac{m}{2\pi kT}\right)^{\frac{1}{2}} \int_0^\infty v_x e^{-(m/2kT)v_x^2} dv_x \quad (1)$$

$$= An\left(\frac{m}{2\pi kT}\right)^{\frac{1}{2}} \cdot \int_0^\infty e^{-(m/2kT)v_x^2} d\left(\frac{v_x^2}{2}\right) = An\sqrt{\frac{kT}{2\pi m}} = A\sqrt{\frac{np}{2\pi m}} \quad (2)$$

(because of $p = nkT$). The velocity distribution of molecules leaking out is given by the integrand of (1), so that the probability of the x-component of the velocity being in the range between v_x and $v_x + dv_x$ is proportional to

$$v_x \exp\left(-\frac{m}{2kT} v_x^2\right) dv_x .$$

Normalizing this we have

$$f(v_x)\, dv_x = \frac{m}{kT} v_x e^{-mv_x^2/2kT} dv_x . \quad (3)$$

4. When an observer views, through the window of the furnace the light of wavelength λ_0 emitted by a molecule whose velocity has the velocity component v_x along the direction of the light ray, it appears to him to be light of wavelength $\lambda = \lambda_0(1 + v_x/c)$, due to the Doppler effect. Solving this for v_x, one gets

$$v_x = c\,\frac{\lambda - \lambda_0}{\lambda_0} . \quad (1)$$

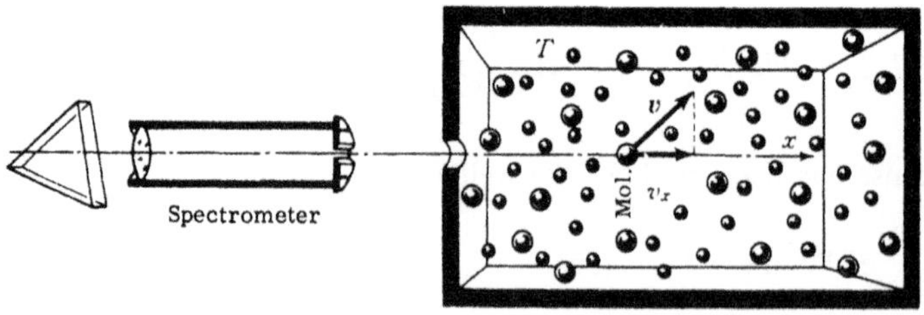

If one measures the intensity of light $I(\lambda)$ as a function of wavelength λ with a spectrometer, it is proportional to the number of molecules which can direct light of wavelength λ toward the spectrometer. Such molecules must have the velocity component v_x. Hence, using the Maxwellian distribution and letting the number density of molecules in the furnace be n, we get

$$I(\lambda) \propto \frac{n}{(2\pi kT/m)^{\frac{3}{2}}} \iint_{-\infty}^{\infty} \exp\left\{-\frac{m}{2kT}(v_x^2 + v_y^2 + v_z^2)\right\} dv_y\, dv_z \tag{2}$$

$$= n\sqrt{\frac{m}{2\pi kT}} \exp\left(-\frac{mv_x^2}{2kT}\right) \propto \exp\left(-\frac{mv_x^2}{2kT}\right).$$

Substituting this into (1) we obtain the desired equation. Note that it is Gaussian with respect to the wavelength λ and its width is of the order of

$$\lambda_0 \sqrt{kT/mc^2}.$$

5. (i) If we denote the momentum of the mass when it is moving in the positive direction by p, then it is $-p$ when the mass is moving in the opposite direction, since elastic collision with the wall at $x = l$ is assumed. The momentum does not depend on x in the interval $0 < x < l$. Hence, the trajectory of the phase point is as shown in left-hand figure.

(ii) Its energy is $E = p^2/2m$ and monotonically increasing with p. Consequently,

$$\Gamma_0(E) = l \times 2p = 2l\sqrt{2mE}.$$

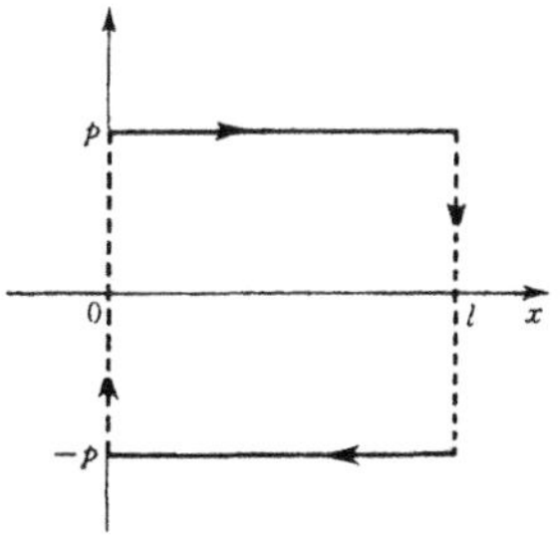

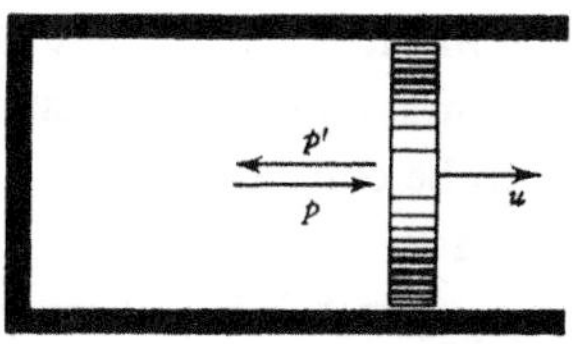

(iii) When the wall at $x = l$ is moving with velocity u, we have by virtue of the elastic collision, $(p/m) - u = (p'/m) + u$.

Hence, at each collision the magnitude of the momentum changes from p to $p' = p - 2mu$. During the time $\delta l/u$ for the position of the wall to change by δl, the particle collides $\delta l/u \div 2l/(p/m) = p\delta l(2mul)$ times. As a result the magnitude of its momentum changes by $\delta p = -2mu \times p\delta l/2mul =$

$-p\delta l/l$. Hence, $l\delta p + p\delta l = \delta(pl) = 0$. Therefore, from the result of (ii) we have $\delta\Gamma_0(E) = 0$.

(iv). The Schrödinger equation is

$$-\frac{\hbar^2}{2m}\frac{d^2}{dx^2}\psi(x) = E\psi(x).$$

Its solution is, in general, given by $\psi(x) = A\sin(\sqrt{2mE}\,x/\hbar) + B\cos(\sqrt{2mE}\,x/\hbar)$. Firstly, because $\psi(0)=0$, $B=0$. Secondly, because $\psi(l)=0$, we have $\sqrt{2mE}\,l/\hbar = \pi n$, $n = 1, 2, \ldots$ Hence, the eigenvalues of the energy are

$$E_n = (h^2/8ml^2)n^2 \qquad (n = 1, 2, \ldots).$$

From (5.20a) one gets

$$\Omega_0(E) = \sum_{E_n \le E} 1 = \left[\sqrt{\frac{8ml^2}{h^2}E}\right],$$

where the square brackets mean the largest integer smaller than the number inside. For large E, $\Omega_0(E) \sim 2l\sqrt{2mE}/h = \Gamma_0(E)/h$.

6. The Schrödinger equation is

$$-\frac{\hbar^2}{2m}\left(\frac{\partial^2}{\partial x^2} + \frac{\partial^2}{\partial y^2} + \frac{\partial^2}{\partial z^2}\right)\psi(x, y, z) = E\psi. \tag{1}$$

If one assumes the form, $\psi(x, y, z) \equiv X(x)Y(y)Z(z)$, one gets

$$-\left(\frac{\hbar^2}{2m}\frac{d^2X}{dx^2}\Big/X\right) - \left(\frac{\hbar^2}{2m}\frac{d^2Y}{dy^2}\Big/Y\right) - \left(\frac{\hbar^2}{2m}\frac{d^2Z}{dz^2}\Big/Z\right) = E.$$

Each term in the left hand side is a function, respectively of x, y, or z only and their sum is equal to a constant E at any point (x, y, z) in V. Hence each term has to be constant:

$$-\frac{\hbar^2}{2m}\frac{d^2X}{dx^2} = E^{(1)}X, \qquad -\frac{\hbar^2}{2m}\frac{d^2Y}{dy^2} = E^{(2)}Y, \qquad -\frac{\hbar^2}{2m}\frac{d^2Z}{dz^2} = E^{(3)}Z$$

and $E^{(1)} + E^{(2)} + E^{(3)} = E$.

Applying the result of the preceding problem to each term, one obtains

$$E_{n_1, n_2, n_3} = \frac{h^2}{8ml^2}(n_1^2 + n_2^2 + n_3^2), \qquad (n_i = 1, 2, \ldots). \tag{2}$$

The corresponding wave function is

$$\psi_{n_1, n_2, n_3}(x, y, z) = A\sin(n_1\pi x/l)\sin(n_2\pi y/l)\sin(n_3\pi z/l).$$

One must also prove that these exhaust all the possible quantum states, but we shall omit this here. Therefore,

$$\Omega_0(E) = \sum_{E_n = E} 1 = \sum_{n_1^2 + n_2^2 + n_2^2 \leq \frac{8ml^2}{h^2} E} 1 . \tag{3}$$

In a Cartesian coordinate system (n_1, n_2, n_3) each lattice point whose coordinates are all positive integers corresponds to one of the quantum states. $\Omega_0(E)$ is then equal to the number of lattice points inside a sphere of radius $\sqrt{8ml^2E/h^2}$. It is, therefore, approximately equal to $\frac{1}{8}$ of the volume of the sphere for large E. Consequently,

$$\Omega_0(E) \sim \frac{\frac{4}{3}\pi[8ml^2E/h^2]^{\frac{3}{2}}}{8} = \frac{4}{3}\pi \frac{V}{h^3}(2mE)^{\frac{3}{2}} . \tag{4}$$

The volume of the classical phase space is simply

$$\Gamma_0(E) = \int_{p^2 \leq 2mE} \mathrm{d}\boldsymbol{p}\,\mathrm{d}\boldsymbol{q} = V \int \mathrm{d}\boldsymbol{p} = V \cdot \frac{4}{3}\pi (2mE)^{\frac{3}{2}} . \tag{5}$$

If E is large, one indeed has $\Omega_0(E) \sim \Gamma_0(E)/h^3$. The density of states is, from (4),

$$\Omega(E) = \frac{\mathrm{d}\Omega_0(E)}{\mathrm{d}E} = 2\pi \frac{V}{h^2}(2m)^{\frac{3}{2}} E^{\frac{1}{2}} . \tag{6}$$

NOTE: † In this problem we have assumed that the walls of the box are rigid and do not allow particles to penetrate, so that we required the wave function to vanish at the walls. Instead of this boundary condition, one often uses the *periodic boundary condition*. In this condition the wave function is to repeat the same value with the period l in each direction of x, y, z. In this case the solution ψ of (1) is

$$\psi = A \exp\left[2\pi \mathrm{i}\left(\frac{n_1 x}{l} + \frac{n_2 y}{l} + \frac{n_3 z}{l}\right)\right]$$

$$\equiv A \exp\left[\mathrm{i}(k_x x + k_y y + k_z z)\right], \qquad k_x = 2\pi n_1/l, \text{ etc.}$$

and the energy is given by

$$E = \frac{h^2}{2m}(k_x^2 + k_y^2 + k_z^2) = \frac{h^2}{8ml^2}(n_1^2 + n_2^2 + n_3^2), \tag{2'}$$

where n_1, n_2, n_3, are positive or negative integers. To obtain $\Omega_0(E)$, one has, as before, only to count the number of lattice points inside a sphere of

† L. I. Schiff, *Quantum Mechanics* (McGraw-Hill, New York, 1949) Chapter II.

radius $\sqrt{8ml^2E/h^2}$. In the present case, however, the lattice constant is $2\pi/l$, twice as large as before. But, since one considers the whole volume of the sphere, the result for $\Omega_0(E)$ comes out the same.

7. If one denotes the mass by m, the coordinate by q and the momentum by p, the Hamiltonian is

$$\mathscr{H}(q,p) = \frac{p^2}{2m} + \tfrac{1}{2}m(2\pi\nu)^2q^2 .$$

By setting up the canonical equations according to (1.1) one can see this gives a harmonic oscillation with frequency ν. The surface of constant energy, $\mathscr{H}(q,p) = E$, is therefore given by

$$\frac{p^2}{2mE} + \frac{m(2\pi\nu)^2}{2E}q^2 = 1 .$$

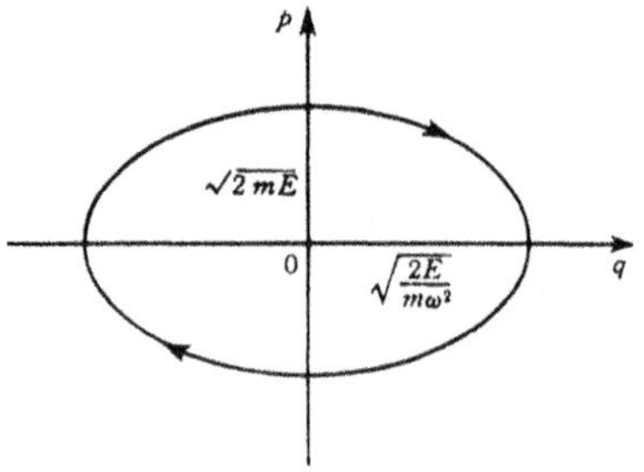

This is an ellipse whose axes are the q and the p axis, with intercepts $\sqrt{2E/m(2\pi\nu)^2}$ and $\sqrt{2mE}$, respectively. Hence, the volume of the phase space with energy below E is

$$\Gamma_0(E) = \pi\sqrt{\frac{2E}{m(2\pi\nu)^2}}\sqrt{2mE} = \frac{E}{\nu} . \tag{1}$$

Note that this result does not contain the mass m. In fact, by the canonical transformation,

$$p/\sqrt{m} = P , \qquad q\sqrt{m} = Q$$
$$(p\delta q - P\delta Q = 0)$$

one can get the Hamiltonian which does not contain m explicitly. Although the shape of the energy surface changes by this transformation (though still an ellipse), $\Gamma_0(E)$ is invariant.

The energy of a quantum state for an oscillator is given by

$$E_n = \tfrac{1}{2} h\nu + nh\nu, \qquad (n = 0, 1, 2, \dots).$$

Hence,

$$\Omega_0(E) = \sum_{n \leqq E} 1 = \left[\frac{E + \tfrac{1}{2} h\nu}{h\nu}\right]$$

where the square bracket means the largest integer smaller than the number inside. Since one can neglect $\tfrac{1}{2}h\nu$ when E is large, one gets $\Omega_0(E) \sim E/h\nu$. Since $\Gamma_0(E) = E/\nu$, one has

$$\frac{\Gamma_0(E)}{h} \sim \Omega_0(E).$$

8. (i) According to equation (1) in the solution to example 3, when the total energy of a system of N oscillators is $E = \tfrac{1}{2}Nh\nu + Mh\nu$ its thermodynamic weight is equal to

$$W_M = \frac{(M + N - 1)!}{M!(N - 1)!}.$$

When, furthermore, a given oscillator has an energy $\varepsilon_n = \tfrac{1}{2} h\nu + nh\nu$ the subsystem consisting of the rest of the $(N - 1)$ oscillators has an energy $E' = E - \varepsilon_n = \tfrac{1}{2}(N - 1)h\nu + (M - n)h\nu$. Hence, its thermodynamic weight is given by

$$W'_{M-n} = \frac{(M + N - n - 2)!}{(M - n)!(N - 2)!}.$$

By the principle of equal weight the desired probability is given by

$$Pr(n) = \frac{W'_{M-n}}{W_M} = \frac{M(M - 1)\dots(M - n + 1)(N - 1)}{(M + N - 1)\dots(M + N - n - 1)}.$$

Now, assuming that $N \gg 1$ and $M \gg n$, we approximate it by

$$Pr(n) \approx \frac{M^n N}{(M + N)^{n+1}} = \frac{N}{M + N}\left(\frac{M}{M + N}\right)^n.$$

If we write $M/N \equiv m$, the mean energy per oscillator in the whole system is $\tfrac{1}{2} h\nu + mh\nu$. In terms of m the above equation is rewritten as

$$Pr(n) = \frac{1}{1 + m}\left(\frac{m}{1 + m}\right)^n, \tag{1}$$

which can also be written in the form,

$$Pr(n) = \frac{e^{-\beta nh\nu}}{e^{\beta h\nu} - 1} \tag{1'}$$

if we introduce β by setting

$$\frac{m}{1+m} \equiv \mathrm{e}^{-\beta h\nu} \qquad \text{namely} \qquad m = \frac{1}{\mathrm{e}^{\beta h\nu}-1}. \tag{2}$$

We note that, although we have made an approximation in deriving equation (1), it satisfies the normalization condition, namely,

$$\sum_{n=0}^{\infty} Pr(n) = \frac{1}{1+m}\sum_{n=0}^{\infty}\left(\frac{m}{1+m}\right)^n = 1.$$

(ii) When the total energy of a system of N particles is approximately E, the thermodynamic weight is, according to (3) in the solution to example 2, given by

$$W(E, N) = \Omega_0'(E, N)\delta E = \frac{V^N}{h^{3N}N!}\tfrac{3}{2}N\cdot\frac{(2\pi mE)^{\frac{3}{2}N}}{\Gamma(\frac{3}{2}N+1)}\frac{\delta E}{E}.$$

If one specifies the energy ε of one particle, the thermodynamical weight of the remaining $N-1$ particles is then $W(E-\varepsilon, N-1)$. The probability of finding a particle in the microscopic state with energy ε is, therefore,

$$p(\varepsilon) = \frac{W(E-\varepsilon, N-1)}{W(E, N)} \propto \left[\frac{E-\varepsilon}{E}\right]^{\frac{3}{2}N} = \exp\left\{\tfrac{3}{2}N\log\left(1-\frac{\varepsilon}{E}\right)\right\}.$$

Here we take the limit of $E \to \infty$, $N \to \infty$, with $E/N \to \frac{3}{2}kT$. Expanding the logarithm and keeping the term which is linear in ε/E, we get $p(\varepsilon) \to \exp(-\varepsilon/kT)$. (One can easily get the normalized probability if one keeps the other factors in $W(E-\varepsilon, N-1)/W(E, N)$ which were omitted above.)

9. (i) The probability of finding a molecule in v is $p = v/V$ and that of finding it in $V - v$ is obviously $1 - p$. The probability of finding n molecules among N in v and the remaining $N - n$ molecules in $(V - v)$ is given by

$$f(n) = \frac{N!}{n!(N-n)!}p^n(1-p)^{N-n}. \tag{1}$$

(Since the distribution of the position of each molecule is independent and we do not specify which molecule is in v.)

(ii) It is convenient to use the generating function in order to calculate the mean value. If we define

$$F(x) = \sum_{n=0}^{N} f(n)x^n \tag{2}$$

and substitute (1), we obtain, with the help of the binomial theorem,

$$F(x) = \{(1-p) + px\}^N. \tag{3}$$

Since $F(1) = \sum f(n)$ and $F'(1) = \sum nf(n)$, we have

$$\bar{n} = \sum_{n=1}^{N} f(n)n = \frac{\mathrm{d}}{\mathrm{d}x}F(x)\Big|_{x=1} = Np\{1-p+px\}^{N-1}_{x=1} = Np = \frac{Nv}{V}. \tag{4}$$

$$\overline{n(n-1)} = \sum_{n=2}^{N} f(n)n(n-1) = \frac{\mathrm{d}^2}{\mathrm{d}x^2}F(x)\Big|_{x=1} = N(N-1)p^2\{1-p+px\}^{N-2}_{x=1}$$
$$= N(N-1)p^2. \tag{5}$$

Since $\overline{(n-\bar{n}^2)} = \overline{n^2} - \bar{n}^2 = \overline{n(n-1)} + \bar{n} - \bar{n}^2$, we obtain from (4) and (5)

$$\overline{(n-\bar{n})^2} = N(N-1)p^2 + Np - N^2p^2 = Np(1-p) = \frac{N}{V}v\left(1-\frac{v}{V}\right). \tag{6}$$

(iii) Assuming N and n to be large and using Stirling's formula, we obtain

$$\log f(n) \sim N\log N - n\log n - (N-n)\log(N-n) + n\log p +$$
$$+ (N-n)\log(1-p) \equiv \phi(n). \tag{7}$$

Hence, the value n^* of n which makes $f(n)$ maximum is given by

$$\phi'(n^*) = 0,$$

or

$$-\log\frac{n^*}{N-n^*} + \log\frac{p}{1-p} = 0, \qquad \text{whence } n^* = Np. \tag{8}$$

If we expand $\phi(n)$ around n^*, we have

$$\phi(n) = \phi(n^*) + \tfrac{1}{2}\phi''(n^*)(n-n^*)^2 + \dots$$
$$= \phi(n^*) - \frac{1}{2}\left(\frac{1}{n^*} + \frac{1}{N-n^*}\right)(n-n^*)^2 + \dots. \tag{9}$$

The higher order terms in the expansion are proportional to $(n-n^*)^m \times O(N^{-m+1})$, so that they can be neglected in the range where $(n-n^*) = O(N^{\frac{1}{2}})$. Consequently, within this range one can put

$$f(n) \propto \exp\left[-\frac{1}{2\Delta^2}(n-n^*)^2\right] \tag{10}$$

(Gaussian distribution), where the dispersion Δ is found from (9) to be

$$\Delta^2 = \frac{n^*(N - n^*)}{N} = Np(1 - p).$$

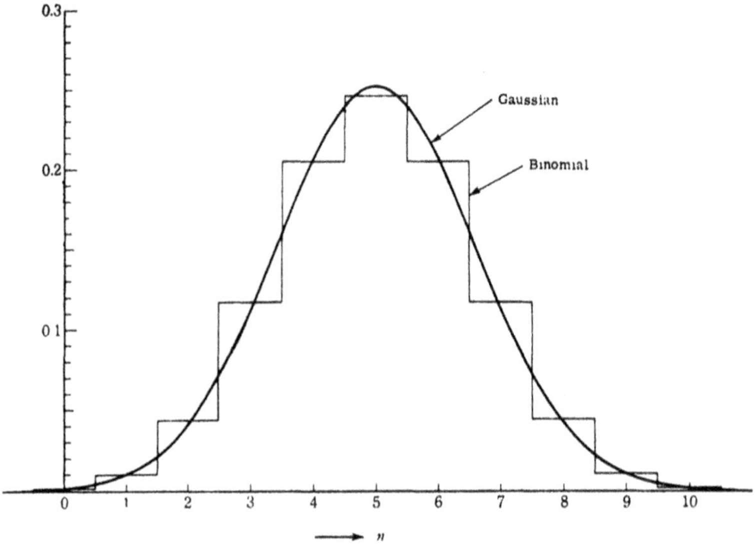

Both n^* and Δ^2 agree with (4) and (6).

(iv) In order to study the limit, it is most convenient to take this limit in the generating function (3) and then determine the limiting form of $f(n)$ as the expansion coefficient, occurring in (2). We thus write

$$\begin{aligned} \lim F(x) \equiv \bar{F}(x) &= \lim \exp\{N \log [(1 - p) + px]\} = \\ &= \exp [\lim N \log \{1 - p(1 - x)\}] \\ &= \exp [\lim \{- Np(1 - x) + NO(p^2)\}]. \end{aligned}$$

Since $\quad Np = Nv/V \to \bar{n}, \quad$ and $\quad N \times O(p^2) = NO(v^2/V^2) \to 0,$

$$\bar{F}(x) \equiv \sum_{n=0}^{\infty} \bar{f}(n) x^n = \exp[-\bar{n}(1 - x)]. \qquad (11)$$

Hence

$$\bar{f}(n) = \frac{(\bar{n})^n}{n!} e^{-\bar{n}}.$$

This is the Poisson distribution.

NOTE: One can derive (4) and (5) in a more elementary way. For instance,

$$\bar{n} = \sum_{n=1}^{N} nf(n) = Np \sum_{n=1}^{N} \frac{(N-1)!}{(n-1)!(N-n)!} p^{n-1}(1-p)^{N-n}$$
$$= Np\{p + (1-p)\}^{N-1} = Np\,.$$

10. Let us take the zero of potential energy at the height of the suspension point. The energy E_t is, when the angle of oscillation θ is assumed to be small, given by

$$E_t \equiv E - mgl = \tfrac{1}{2}ml^2\dot{\theta}^2 + \tfrac{1}{2}mgl\theta^2 - mgl\,, \tag{1}$$

where E is the part relevant to the pendulum motion ($-mgl$ is the potential energy at the equilibrium position). For simple harmonic oscillations, the time averages of the kinetic and the potential energy (over one period of time) are equal to each other:

$$l\overline{\dot{\theta}^2} = g\overline{\theta^2}$$

Hence,

$$E \equiv E_t + mgl = mgl\overline{\theta^2}\,. \tag{2}$$

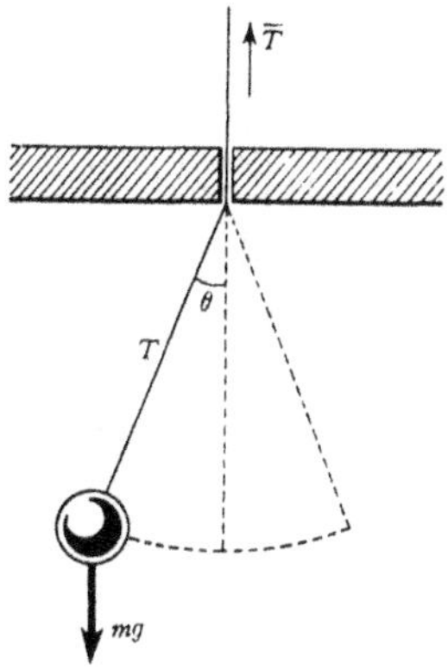

From a consideration of the forces along the direction of the string, one gets the tension T of the string,

$$T = ml\dot{\theta}^2 + mg\cos\theta = ml\dot{\theta}^2 + mg - \tfrac{1}{2}mg\theta^2\,.$$

Its time average is, therefore, $\bar{T} = mg + \frac{1}{2}\,mg\overline{\theta^2}$. When one pulls the string upward by $-\delta l$, the work done on this system is $-\bar{T}\delta l$ and the energy of the system increases by this amount. Therefore,

$$\delta E_t = -\bar{T}\delta l = -mg\delta l - \tfrac{1}{2} mg\overline{\theta^2}\delta l .$$

Hence

$$\delta E = \delta(E_t + mgl) = -\tfrac{1}{2} mg\overline{\theta^2}\delta l . \tag{3}$$

From (2) and (3) one gets

$$\frac{\delta E}{E} = -\frac{\delta l}{2l} = \frac{\delta v}{v} \tag{4}$$

(because $v = \sqrt{(g/l)}$). Integrating this, one obtains E/v = constant (adiabatic invariant). Since $\Gamma_0(E) = E/v$ as given in (1) of problem 7, this result is in agreement with the adiabatic theorem (1.36).

11. Since $\mathrm{d}S/\mathrm{d}t$ is a function of the time rate of change of x, $\mathrm{d}x/\mathrm{d}t$, it is obviously zero when $\mathrm{d}x/\mathrm{d}t = 0$. If one expands this in powers of $\mathrm{d}x/\mathrm{d}t$, one will not get the term linear in $\mathrm{d}x/\mathrm{d}t$ (and in general, the terms of odd powers). For, otherwise, one would have $\mathrm{d}S/\mathrm{d}t < 0$ in the process of varying x in the opposite direction, in contradiction with the principle of increase of entropy. Therefore, it has the form,

$$\frac{\mathrm{d}S}{\mathrm{d}t} = A\left(\frac{\mathrm{d}x}{\mathrm{d}t}\right)^2 + \dots, \qquad (A > 0)$$

from which one gets $\mathrm{d}S/\mathrm{d}x = A(\mathrm{d}x/\mathrm{d}t) \cdot \mathrm{d}x + \dots$ This shows that during the change $\mathrm{d}x$, if $\mathrm{d}x/\mathrm{d}t \to 0$, $\mathrm{d}S = 0$.

12. If the vertical coordinate of the weight is x, its potential energy is wx. If we denote the cross-section of the piston by A, the volume of the box is $V = xA$. Let $\Omega(E, V)$ be the density of states of the system under consideration. When the whole system has a total energy in the range $(E', E' + \delta E')$ and the weight is at the position x, the number of microscopic states of the whole system is given by

$$\Omega(E' - wx, xA)\delta E' .$$

The most probable value x^* of x is determined as the value which makes this a maximum:

$$-w\frac{\partial \log \Omega}{\partial E} + A\frac{\partial \log \Omega}{\partial V} = 0 .$$

With the help of (1.18) and (1.22) this can be written as

$$\frac{w}{A}\frac{\partial S(E, V)}{\partial E} = \frac{\partial S}{\partial V} .$$

The quantity $w/A = p$ is just the pressure on the piston. Hence one has

$$p = \left(\frac{\partial S}{\partial V}\right)_E \Big/ \left(\frac{\partial S}{\partial E}\right)_V = -\left(\frac{\partial E}{\partial V}\right)_S .$$

As given by equation (4) in the solution to example 2, S has the following form:

$$S = Nk\left\{\log\frac{V}{N} + \tfrac{3}{2}\log\frac{2E}{3N} + \ldots\right\} = Nk[\log(V E^{\frac{3}{2}}) + \ldots],$$

thus

$$\delta S = 0 \rightarrow \frac{\delta V}{V} + \frac{3}{2}\frac{\delta E}{E} = 0, \qquad \text{or} \qquad \left(\frac{\partial E}{\partial V}\right)_S = -\frac{2}{3}\frac{E}{V}.$$

Consequently one gets

$$p = \frac{2}{3}\frac{E}{V}. \tag{1}$$

Substituting $E = \frac{3}{2} NkT$, one obtains $p = NkT/V$.

13. In order to specify a possible configuration of the chain, we indicate successively, starting from the left end, whether each consecutive element is directed to the right (+) or to the left (−). In the case shown in the figure we gave in the problem we have (+ + − + + + − − − + + − + + +). The number of elements n_+ directed to the right and the number n_- of those directed to the left together determine the distance between the ends of the chain x:

$$x = (n_+ - n_-)a, \qquad n = n_+ + n_-,$$

hence

$$n_+ = \frac{na + x}{2a}, \qquad n_- = \frac{na - x}{2a}. \tag{1}$$

The number of configurations having the same x and hence the same n_+, n_- is given by

$$W(x) = \frac{n!}{n_+!\,n_-!}. \tag{2}$$

From (1.18) the entropy becomes, with the help of Stirling's formula,

$$S(x) = k\log W(x) = k(n\log n - n_+\log n_+ - n_-\log n_-) \tag{3}$$

$$= nk\left\{\log 2 - \frac{1}{2}\left(1 + \frac{x}{na}\right)\log\left(1 + \frac{x}{na}\right) - \frac{1}{2}\left(1 - \frac{x}{na}\right)\log\left(1 - \frac{x}{na}\right)\right\}.$$

Since the joints of the chain can turn freely, the internal energy does not depend on x. Only the entropy makes a contribution to the tension X, which can be obtained from the Helmholtz free energy

$$X = \left(\frac{\partial F}{\partial x}\right)_T = -T\left(\frac{\partial S}{\partial x}\right)_T = \frac{kT}{2a}\log\frac{1 + x/(na)}{1 - x/(na)} = \frac{kT}{na^2}x + \dots . \tag{4}$$

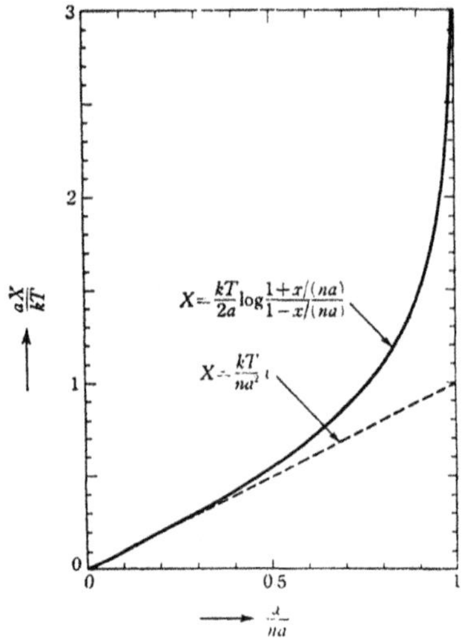

The last equality is the expansion formula for $x \ll na$, and its first term corresponds to Hooke's law.

NOTE 1: This chain is the simplest model embodying the essential property of rubber elasticity. See example 6 in Chapter 2 for the 3-dimensional case.

NOTE 2: If one lets X correspond to a magnetic field H and x to magnetization M, this problem becomes identical to example 10, although the method used is different.

14. (i) The Hamiltonian of this system is

$$\mathscr{H} = \sum_i \{\tfrac{1}{2} p_i^2 + \tfrac{1}{2}(2\pi\nu)^2 q_i^2\}. \tag{1}$$

Hence the phase integral $\Omega_0(E)$ is

$$\Omega_0(E) = \frac{\Gamma_0(E)}{h^N} = \frac{1}{h^N}\int_{\mathscr{H}\leq E} \prod_i \mathrm{d}q_i \mathrm{d}p_i .$$

By the transformation of variables, $p_i/\sqrt{2} \equiv x_i$, $2\pi\nu q_i/\sqrt{2} \equiv x_{N+i}$, (1) reduced to

$$\mathcal{H} = \sum_{i=1}^{2N} x_i^2 ,$$

so that

$$\Omega_0(E) = \frac{1}{(\pi h\nu)^N} \int_{\Sigma x_i^2 < E} \prod_1^{2N} \mathrm{d}x_i .$$

This is the volume of a $2N$ dimensional sphere with radius E. Using the result given in example 2, one gets

$$\Omega_0(E) = \frac{\pi^N}{(\pi h\nu)^N \Gamma(N+1)} E^N = \frac{1}{N!}\left(\frac{E}{h\nu}\right)^N .$$

(ii) Using Stirling's formula, one gets from (1.27)

$$S \simeq k \log \Omega_0 = Nk \log \frac{E}{h\nu} - N \log \frac{N}{\mathrm{e}} .$$

Hence, with the help of the relation (1.28) one obtains

$$\frac{1}{T} = \frac{\partial S}{\partial E} = \frac{Nk}{E} \qquad \text{or} \quad E = NkT .$$

15. (i) Since the Hamiltonian is $\mathcal{H} = p^2/2m + \frac{1}{2} m\omega^2 q^2$, the classical partition function is given by

$$Z_\mu = \int_{-\infty}^{\infty}\int_{-\infty}^{\infty} \mathrm{e}^{-\mathcal{H}/kT} \frac{\mathrm{d}p\,\mathrm{d}q}{h} = \frac{1}{h}\int_{-\infty}^{\infty} \mathrm{e}^{-p^2/2mkT}\mathrm{d}p \int_{-\infty}^{\infty} \mathrm{e}^{-mw^2q^2/2kT}\,\mathrm{d}q$$

$$= \frac{1}{h}\sqrt{2\pi mkT}\sqrt{\frac{2\pi kT}{m\omega^2}} = \frac{2\pi kT}{h\omega} = \frac{kT}{\hbar\omega} . \tag{1}$$

(ii) In quantum mechanics, since the energy level is $\varepsilon_n = (n + \frac{1}{2})h\nu$ the partition function is given by

$$Z_\mu = \sum_n \mathrm{e}^{-\varepsilon_n/kT} = \sum_{n=0}^{\infty} \mathrm{e}^{-(n+1/2)\hbar\omega/kT} = \mathrm{e}^{-\hbar\omega/2kT} \sum_{n=0}^{\infty} \{\mathrm{e}^{-\hbar\omega/kT}\}^n$$

$$= \frac{\mathrm{e}^{-\hbar\omega/2kT}}{1 - \mathrm{e}^{-\hbar\omega/kT}} = \frac{1}{\mathrm{e}^{\hbar\omega/2kT} - \mathrm{e}^{-\hbar\omega/2kT}} = \frac{1}{2\sinh\{\hbar\omega/2kT\}} . \tag{2}$$

We shall, in the following, consider only case (2), since the classical partition function (1) is the limit of (2) when $\hbar\omega/kT \to 0$.

(iii) Since the partition function Z of a system consisting of N independent oscillators is equal to the product of those for individual oscillators (example 7), we get

$$Z = Z_\mu^N = \left[2\sinh\frac{\hbar\omega}{2kT}\right]^{-N}. \tag{3}$$

Writing $\beta = 1/kT$, we obtain the internal energy of the system U,

$$U = -\frac{\partial}{\partial\beta}\log Z = N\frac{\partial}{\partial\beta}\log\sinh\frac{\beta\hbar\omega}{2} = N\frac{\hbar\omega}{2}\coth\frac{\hbar\omega}{2kT} \tag{4}$$

$$= N\left\{\frac{\hbar\omega}{2} + \frac{\hbar\omega}{\mathrm{e}^{\hbar\omega/kT}-1}\right\}.$$

The heat capacity C is given by differentiating the above equation with respect to temperature T:

$$C = \left(\frac{\partial U}{\partial T}\right)_{N,\omega} = Nk\frac{\mathrm{e}^{\hbar\omega/kT}}{\{\mathrm{e}^{\hbar\omega/kT}-1\}^2}\left(\frac{\hbar\omega}{kT}\right)^2. \tag{5}$$

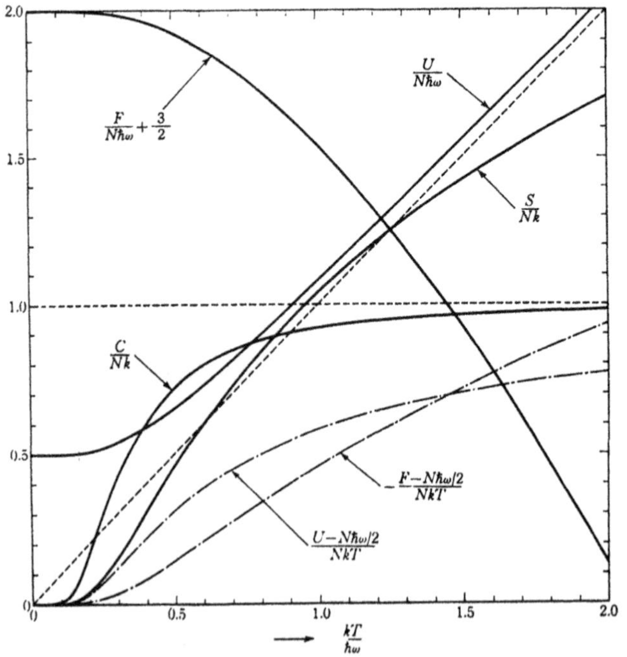

Fig. 1.14. Thermodynamic functions of harmonic oscillators.

To obtain the entropy S, we use the Helmholtz free energy

$$F = -kT\log Z = NkT\log\left(2\sinh\frac{\hbar\omega}{2kT}\right) = \tag{6}$$
$$= N\left[\tfrac{1}{2}\hbar\omega + kT\log\{1 - e^{-\hbar\omega/kT}\}\right]$$

in the expression $(U - F)/T$; then

$$S = Nk\left\{\frac{\hbar\omega}{2kT}\coth\frac{\hbar\omega}{2kT} - \log\left(2\sinh\frac{\hbar\omega}{2kT}\right)\right\}. \tag{7}$$

16. Since the temperature is constant the gas obeys the canonical distribution (1.70a), in which we use

$$\mathscr{H}_N = \sum_{i=1}^{N}\frac{p_i^2}{2m} + \sum_{i=1}^{N}u(z_i), \qquad u(z) = mgz. \tag{1}$$

The distribution of each of N molecules being independent, the probability of finding any one molecule in a volume element ΔV at height z is proportional to $\exp(-mgz/kT)\Delta V$. Therefore, the density of the gas at z is given by

$$n(z) = n(0)\exp(-mgz/kT). \tag{2}$$

Since the pressure p is given by $p(z) = n(z)kT$, we get $p(z) = p(0)\exp(-mgz/kT)$.

NOTE: Although equation (2) is obvious in this case, in order to obtain the density in a general case one must calculate the mean value $\overline{n(\boldsymbol{r})}$ of

$$n(\boldsymbol{r}) = \sum_{i=1}^{N}\delta(\boldsymbol{r}_i - \boldsymbol{r}) = \lim_{\Delta V\to 0}\frac{1}{\Delta V}\sum_{i=1}^{N}\phi_{\Delta V}(\boldsymbol{r}_i - \boldsymbol{r}) \tag{3}$$

where $\delta(\boldsymbol{r}_i - \boldsymbol{r}) = \delta(x_i - x)\delta(y_i - y)\delta(z_i - z)$, the Dirac delta function, and $\phi_{\Delta V}(\boldsymbol{r}_i - \boldsymbol{r})$ is a function which is equal to 1 when $\boldsymbol{r}_i$ is in the volume element ΔV around $\boldsymbol{r}$ and is 0 otherwise. Calculating this mean value by means of (1.70a), we find

$$\overline{n(\boldsymbol{r})} = \frac{N\int e^{-\beta u(\boldsymbol{r}')}\delta(\boldsymbol{r}' - \boldsymbol{r})\,d\boldsymbol{r}'}{\int e^{-\beta u(\boldsymbol{r})}\,d\boldsymbol{r}} = Ce^{-\beta u(\boldsymbol{r})} = Ce^{-\beta mgz}.$$

17. As has been shown in example 7, the partition function for an aggregate of almost independent bodies is equal to the product of those for the individual bodies. In the case of an ideal gas, however, one must divide the product of the partition functions for individual particles by $N!$ because of the indistinguishability of particles: (1.71a). Hence,

$$Z(V,T,N) = \frac{1}{N!}\left\{\frac{V}{h^3}\iiint_{-\infty}^{\infty} e^{-\varepsilon/kT} dp_x dp_y dp_z\right\}^N. \tag{1}$$

Because $\varepsilon = cp$, the integral over the momenta of a particle (p_x, p_y, p_z) becomes

$$\iiint e^{-\varepsilon/kT} dp_x dp_y dp_z = 4\pi \int_0^\infty e^{-cp/kT} p^2 dp = 8\pi (kT/c)^3. \tag{2}$$

Using Stirling's formula, $\log N! = N \log(N/e)$, one gets the Helmholtz free energy,

$$F(V,T,N) = -kT \log Z = -NkT \log\left(\frac{8\pi e k^3}{h^3 c^3}\frac{V}{N} T^3\right). \tag{3}$$

Therefore, the pressure p, the internal energy U and the enthalpy H are given by

$$p = -\left(\frac{\partial F}{\partial V}\right)_{T,N} = NkT\frac{d\log V}{dV} = \frac{NkT}{V}, \qquad \text{or} \quad pV = NkT, \tag{4}$$

$$U = -T^2\left[\frac{\partial}{\partial T}\left(\frac{F}{T}\right)\right]_{V,N} = T^2 Nk\frac{d\log T^3}{dT} = 3NkT, \tag{5}$$

and

$$H = U + pV = 4NkT \tag{6}$$

respectively. One can write down the entropy $S = (U - F)/T$ as well as the Gibbs free energy $G = F + pV$. The specific heats at constant volume C_V and at constant pressure C_p can be found from (5) and (6):

$$C_V = \left(\frac{\partial U}{\partial T}\right)_{V,N} = 3Nk, \qquad C_p = \left(\frac{\partial H}{\partial T}\right)_{p,N} = 4Nk. \tag{7}$$

NOTE: Since in relativistic mechanics the energy of a free particle is equal to $\mathscr{H} = c\sqrt{m^2c^2 + p^2}$ (c being the velocity of light), one may put $\varepsilon = cp$ for a particle with extremely large energy (an extreme relativistic particle) as was done in this problem. Note in this case the pressure is $p = \frac{1}{3} n\,\bar{\varepsilon}$ and the specific heat at constant volume is twice the value $\frac{3}{2}Nk$ for the non-relativistic energy, $\varepsilon = p^2/2m$.

18. According to the result of example 2 the density of states of an ideal gas is proportional to $E^{\frac{3}{2}N-1}$. Therefore, E^* giving

$$e^{-E/kT}E^{\frac{3}{2}N-1} = \max$$

($N \gg 1$) is determined by $-(E/kT) + \frac{3}{2}N \log E = \max$, from which one readily gets $E^* = \frac{3}{2}NkT$. This is equal to the mean value in the canonical distribution

$$\bar{E} = \frac{\int_0^\infty e^{-E/kT}E^{\frac{3}{2}N}\,dE}{\int_0^\infty e^{-E/kT}E^{\frac{3}{2}N-1}\,dE} = kT\frac{\Gamma(\frac{3}{2}N+1)}{\Gamma(\frac{3}{2}N)} = \frac{3}{2}NkT.$$

For a system of oscillators with energy $E = \frac{1}{2}Nh\nu + Mh\nu$ the canonical distribution with respect to M is given in terms of W_M of example 3 by

$$\frac{h\nu}{kT} = \frac{\partial}{\partial M}\log W_M, \qquad (M = M^*).$$

This can be calculated in the same way as in example 3 and the result is equation (3) in the solution. This result agrees with (4) obtained in solution 15.

19. From (1.74a) and (1.74c) one has

$$\Xi = \sum \lambda^N Z_N, \tag{1}$$

$$Z_N = \frac{1}{N!h^{3N}}\int e^{-\beta\mathscr{H}_N}\,d\Gamma = \frac{1}{N!}\left[\frac{1}{h^3}\int\!\!\int e^{-\beta p^2/2m}\,d\boldsymbol{x}\,d\boldsymbol{p}\right]^N = \frac{1}{N!}f^N \tag{2}$$

where

$$f = \frac{1}{h^3}\int\!\!\int e^{-\beta p^2/2m}\,d\boldsymbol{x}\,d\boldsymbol{p} = V\left(\frac{2\pi mkT}{h^2}\right)^{\frac{3}{2}} \tag{3}$$

is the partition function for a particle in a volume V (see example 8 for the calculation). Substituting (2) into (1) one obtains

$$\Xi = \sum_{N=0}^{\infty}\frac{\lambda^N}{N!}f^N = e^{\lambda f}. \tag{4}$$

By the definition of Ξ the quantity λ is the absolute activity. In fact,

$$\bar{N} = \frac{\sum N\lambda^N Z_N}{\sum \lambda^N Z_N} = \lambda\frac{\partial}{\partial\lambda}\log\Xi = \lambda f. \tag{5}$$

Hence

$$\lambda = \frac{\bar{N}}{f} = \frac{\bar{N}}{V}\bigg/\left(\frac{2\pi mkT}{h^2}\right)^{\frac{3}{2}}, \tag{6}$$

and

$$kT\log\lambda = kT\log\frac{\bar{N}}{V} - \tfrac{3}{2}kT\log\frac{2\pi mkT}{h^2}. \tag{7}$$

Equation (7) gives $kT\log\lambda = \mu$ = the chemical potential. This equation agrees with (6) in example 8.

NOTE: $h/\sqrt{mkT}$ is the de Broglie wavelength λ_B of a particle having an energy $p^2/2m = \frac{1}{2}kT$. λ gives the mean number of particles in a volume $\lambda_B^3/(2\pi)^{\frac{3}{2}}$.

20. If one considers the volume element v as a system, the remaining volume $V - v$ will play the role of a particle (as well as heat) source for this system. Therefore, one can regard v as a system obeying the T-μ distribution. In particular, the distribution of the number n of particles in v is, according to (1.75), given by

$$P_n = \lambda^n Z_n / \Xi\,. \tag{1}$$

Borrowing the results of the preceding problem, one has $Z_n = f^n/n!$ and $\Xi = e^{\lambda f}$. Hence,

$$P_n = \lambda^n f^n e^{-\lambda f}/n!\,. \tag{2}$$

With the help of (5) in the solution to the preceding problem, $\bar{n} = \lambda f$, (2) can be rewritten as $P_n = e^{-\bar{n}}(\bar{n})^n/n!$.

NOTE: This result is in accord with (iv) of example 9.

21. Following example 11, one can first prove that $\log Y(T,p,N) \propto N$ and then derive the equation in question from the relation, $-kT\partial\log Y/\partial N = \mu(T,p)$, where $\mu(T,p)$ is the mean of $\mu(T, N/V)$ (p = constant). Here we shall give another proof.

By definition (1.77c),

$$Y = \int_0^\infty Z_N(V)e^{-\beta pV}\,dV \qquad (\beta = 1/kT)\,.$$

Approximating the integrand by the largest term in the expansion, we have

$$Y = \int_0^\infty e^{-\beta pV^*}Z_N(V^*)\exp\left[-\frac{\frac{1}{2}\partial^2\log Z_N(V^*)}{\partial V^{*2}}\cdot(V-V^*)^2 + \ldots\right]dV\,,$$

higher order terms in [ ] being negligible. To the integral over V only the neighborhood of V^* makes a contribution. Since, however, the integral contributes to $\log Y$ a term of the order of $\log N$ which is negligible compared to N, we may put

$$-kT\log Y(T,p,N) = -kT\log Z_N(T,V^*) + pV^* = F_N(T,V^*) + pV^*\,. \tag{1}$$

In the above equations V^* is determined by the condition of maximum,

$$\frac{\partial \log Z_N(V^*)}{\partial V^*} = -\frac{p}{kT}, \quad \text{or} \quad \frac{\partial F_N(T, V^*)}{\partial V^*} = -p. \tag{2}$$

The equations (1) and (2) are just the procedure of obtaining the Gibbs free energy G from F by the Legendre transformation. Therefore, $G(T, p, N) = -kT \log Y(T, p, N)$.

22. The proton = P, neutron = N, electron = e^-, and positron = e^+ obey Fermi statistics. One can obtain the answer by seeing how many Fermi particles a given particle is made of:

α-particle = 2P + 2N = (Bose particle),
^{3}He = 2P + N + 2e^- = (Fermi particle),
H_2 molecule = 2(P + e^-) = (Bose particle), positron = e^+ = (Fermi particle),
$^6Li^+$ ion = 3P + 3N + 2e^- = (Bose particle),
$^7Li^+$ ion = 3P + 4N + 2e^- = (Fermi particle).

23. If $e^{(\varepsilon_\tau - \mu)/kT} \gg 1$ in (1.90) and (1.91), one can make the approximation $\bar{n}_\tau = e^{-(\varepsilon_\tau - \mu)/kT}$ in both the F. D. and B. E. cases. In other words one can apply Boltzmann statistics. Condition (1) may be expressed by $\bar{n}_\tau \ll 1$. (1)

Now, substituting (1) into (1.93), one has

$$N = e^{\mu/kT} \sum_\tau e^{-\varepsilon_\tau/kT}.$$

Calculating the sum classically, one gets

$$N = e^{\mu/kT} \frac{V}{h^3} \int e^{-p^2/2mkT} \mathrm{d}p = e^{\mu/kT} V \frac{(2\pi mkT)^{\frac{3}{2}}}{h^3},$$

so that

$$e^{\mu/kT} = \frac{N}{V} \frac{h^3}{(2\pi mkT)^{\frac{3}{2}}}.$$

Therefore, in order that (1) is satisfied by all $\varepsilon_\tau \geqq 0$, one must have

$$\left(\frac{V}{N}\right)^{\frac{1}{3}} \gg \frac{h}{\sqrt{2\pi mkT}} = \frac{\lambda_B}{\sqrt{2\pi}}, \tag{2}$$

i.e., it is necessary that the average de Broglie wavelength be much smaller than the mean distance between particles.

As is stated in the note to the solution of problem 19, λ_B is the de Broglie wavelength of a particle having a kinetic energy, and (2) requires that λ_B be much smaller that the mean distance between particles (or that the number of particles in a volume $\lambda_B^3 / (2\pi)^{\frac{3}{2}}$ be much smaller than 1).

24. The problem is to maximize S under the condition

$$\sum_s p_s = 1, \qquad (1) \qquad \sum_s E_s p_s = E = \text{constant}. \qquad (2)$$

We shall use Lagrange's method of undetermined multipliers (see NOTE to § 1.11). We multiply (1) and (2) by constants λ and κ, respectively, and add them to S. Then, supposing all p_s to be independent, we maximize $S + \kappa E + \lambda$. In the expression

$$\delta(S + \kappa E + \lambda) = \delta \sum_s \{-kp_s \log p_s + \kappa E_s p_s + \lambda p_s\}$$

$$= \sum_s (-k \log p_s - k + \kappa E_s + \lambda)\delta p_s + \frac{1}{2}\sum_s \left(-\frac{k}{p_s}\right)(\delta p_s)^2 + \dots$$

we put the coefficient of δp_s equal to 0 and get

$$p_s = \exp\left\{\frac{\kappa}{k} E_s + \frac{\lambda}{k} - 1\right\}.$$

Since this is positive, the coefficient of $(\delta p_s)^2$ is negative and we always have a maximum. We determine λ and κ such that they satisfy (1) and (2).
Since E_s has, in general, no upper limit, we must have $\kappa < 0$ in order for $\sum p_s$ and $\sum E_s p_s$ to converge. Therefore, writing $\kappa/k \equiv -\beta (\beta > 0)$, we get from (1)

$$1 = \sum_s p_s = e^{\lambda/k - 1} \sum_s e^{-\beta E_s} \equiv e^{\lambda/k - 1} Z(\beta).$$

Hence

$$p_s = \frac{e^{-\beta E_s}}{Z(\beta)}.$$

By (2) β is determined as a function of E.

DIVERTISSEMENT 4

Ergodic Theorem. A physical quantity A defined by a function $A(p, q)$ changes in time since the dynamical state $(p, q) = \mathrm{P}$ moves in the phase space (as indicated by P_t) according to the Hamilton equation of motion (1.1). Thus the time average of A may be defined by

$$\langle A \rangle_{\text{time}} = \lim_{T \to \infty} \frac{1}{T} \int_0^T A(t)\, dt, \quad A(t) \equiv A(\mathrm{P}_t).$$

The ergodic theorem states that $\langle A \rangle_{\text{time}} = \bar{A}$, $\bar{A}$ being the phase average defined by (1.13). The weight $|\operatorname{grad} \mathscr{H}|^{-1}$ appearing in (1.13) can be understood by recognizing that it represents the length of time for a phase point to pass the neighborhood of the point P, because $|\text{grad}\,\mathscr{H}| = \{\sum_i (p_i^2 + q_i^2)\}^{\frac{1}{2}}$ is the velocity of a phase point and $ds/|\text{grad}\,\mathscr{H}| = d\tau$ is the time for the phase point to traverse the length ds which is taken along the phase trajectory (the surface element $d\sigma$ is then written as $d\sigma = ds \cdot dl$, dl being taken perpendicular to ds). This remark only indicates that the phase average (1.13) is properly weighted, but provides no verification of the ergodic theorem.

Boltzmann assumed that a phase trajectory P_t passes in the course of time any point on the ergodic surface, which he supposed to support the equality (1.15). But this assumption cannot be true simply because a manifold ∞^1 can never cover a manifold ∞^{2f-1}. Therefore the original assumption of Boltzmann, which was called the *ergodic hypothesis*, was replaced by another assumption, i.e. the *quasi-ergodic hypothesis*, that the phase trajectory passes any neighborhood of any point on the ergodic surface.

The conjecture made by physicists in the last decades of the nineteenth century was proved only in the 1930's by mathematicians, namely by Birkhoff, Neumann and Koopman, and the theory was further refined by their followers like Hopf, Oxtoby and others. Physicists are now told by mathematicians that the ergodic theorem has been completely proved.

The mathematical theory is based upon the observation that the Hamiltonian equation of motion (1.1) defines a continuous point transformation (or Abbildung) in phase space which conserves a certain measure. This is a well-known theorem, called the Liouville theorem: Take an arbitrary subset of phase points or more intuitively just a domain with a certain volume. The phase points move by natural motion, if the boundary of the domain will move in time, but the measure of the subset, or the volume of the deformed domain, remains invariant. In other words, natural motion resembles a flow of an incompressible fluid in $2f$-dimensional phase space. If certain integrals of motion exist, the motion of phase points is restricted to limited parts of an ergodic surface. However, usual dynamical systems do not have integrals other than such simple integrals as the total energy, the total linear momentum or angular momentum. By redefining the phase average referred to such a limited part, if integrals are present, the ergodic theorem can be proved by the above mentioned "measure-conserving" property of dynamical motion, and the quasi-ergodic hypothesis is proved in such an *indecomposable* part of the ergodic surface. A compact and critical review of ergodic theories is available in Khinchin's booklet *Mathematical Foundation of Statistical Mechanics*, translated by Gamow (Dover, 1948). One volume of the proceedings of the International School of Physics at Varenna, Italy, is devoted to ergodic theories [*Ergodic Theories* (Academic Press, 1961)]. (See also Divertissement 7.)

25. Let us move the piston slowly with velocity u, as in (iii) of problem 5. A molecule has the component of its momentum perpendicular to the wall of the piston changed from p_x to $p'_x = p_x - 2mu$ at each collision. Hence its energy changes by

$$\delta\varepsilon(p_x) = \frac{1}{2m}(p_x - 2mu)^2 - \frac{1}{2m}p_x^2$$
$$\fallingdotseq -2up_x$$

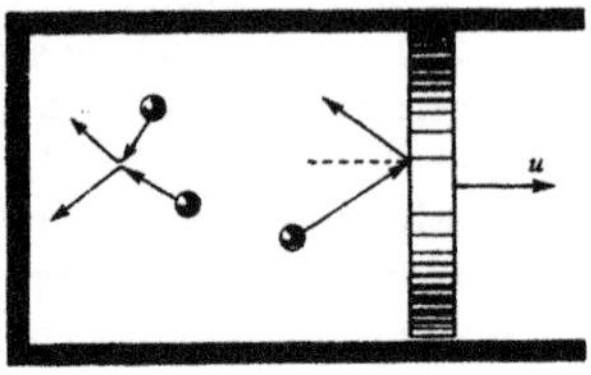

(where the term proportional to u^2 is neglected). Summing the changes for all the molecules colliding on the wall during the time δt, one can find the variation of the kinetic energy E_{kin} of the gas (letting the area of the piston be A):

$$\delta E = A\int_0^\infty \mathrm{d}p_x \int_{-\infty}^\infty \mathrm{d}p_y \int_{-\infty}^\infty \mathrm{d}p_z nf(\boldsymbol{p})\cdot\frac{p_x}{m}\cdot\delta\varepsilon(p_x)\delta t$$
$$= -A\int_0^\infty \mathrm{d}p_x \int_{-\infty}^\infty \mathrm{d}p_y \int_{-\infty}^\infty \mathrm{d}p_z nf(\boldsymbol{p})\cdot\frac{p_x^2}{m}\cdot 2u\delta t$$
$$= -(Au\delta t).\int_{-\infty}^\infty \mathrm{d}p_x \int_{-\infty}^\infty \mathrm{d}p_y \int_{-\infty}^\infty \mathrm{d}p_z nf(\boldsymbol{p})\cdot\frac{p_x^2}{m}$$
$$= -\delta V\cdot n\frac{2}{3}\bar{\varepsilon} = -\frac{2}{3}\frac{\delta V}{V}E_{\text{kin}}. \tag{1}$$

On the other hand, as obtained in example 1,

$$p = \tfrac{2}{3}n\bar{\varepsilon}, \qquad \text{or} \quad pV = \tfrac{2}{3}E_{\text{kin}}, \tag{2}$$

so that one has

$$\frac{\delta p}{p} + \frac{\delta V}{V} = \frac{\delta E_{\text{kin}}}{E_{\text{kin}}}. \tag{3}$$

(i) If, for simplicity, we consider a gas composed of monatomic molecules, we have $E = E_{\text{kin}}$. Substituting (1) into (3), one gets

$$\frac{\delta p}{p} + \frac{5}{3}\frac{\delta V}{V} = 0, \qquad \text{or} \quad pV^{\frac{5}{3}} = \text{const.} \tag{4}$$

(ii) If one assumes $E = aE_{\rm kin}$ (where a is a constant, $a > 1$), taking into account the internal degree of freedom, then (1) becomes $a\delta E_{\rm kin}/E_{\rm kin} = -\frac{2}{3}\delta V/V$. Substituting this into (3), one has

$$\frac{\delta p}{p} + \gamma\frac{\delta V}{V} = 0, \qquad \text{hence} \qquad pV^{\gamma} = \text{const}, \tag{5}$$

where

$$\gamma = \frac{(\frac{3}{2}a + 1)}{\frac{3}{2}a} = \frac{(\frac{3}{2}aR + R)}{\frac{3}{2}aR}. \tag{6}$$

On the other hand, since for one mole of the gas, $E_{\rm kin} = \frac{3}{2}R$, one has, taking into account the internal degree of freedom, $C_V = \frac{3}{2}aR$ and $C_p = C_V + R = \frac{3}{2}aR + R$. Therefore, (6) can be written as $\gamma = C_p/C_V$. This agrees with the result obtained by thermodynamics.

NOTE: Although the immediate effect of the moving wall is to increase the kinetic energy only of the direction perpendicular to the wall, we have assumed that this increase is evenly distributed among all the degrees of freedom due to intermolecular collisions.

26. Let us measure E without the energy of the zero point oscillation $\frac{1}{2}h\nu N$. According the example 3,

$$\Omega_0(E) = \sum_{M=0}^{E/h\nu} W_M = \sum_{M=0}^{E/h\nu} \frac{(M + N - 1)!}{M!(N-1)!} = \frac{(E/h\nu + N)!}{(E/h\nu)!N!}.$$

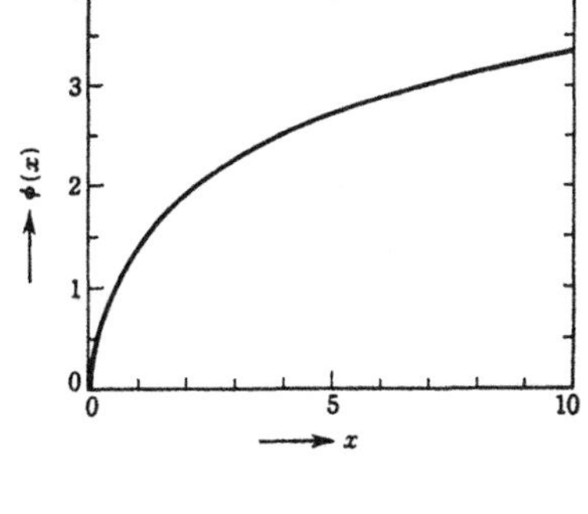

One can readily write down this result by noting that $\Omega_0(E)$ is equal to the number of ways of distributing the energy E among $(N + 1)$ oscillators. One can find the asymptotic properties for large N by studying

$$\log \Omega_0(E) \approx \left(\frac{E}{h\nu} + N\right)\log\left(\frac{E}{h\nu} + N\right) - \frac{E}{h\nu}\log\frac{E}{h\nu} - N\log N$$
$$= N\left\{\left(1 + \frac{E}{h\nu N}\right)\log\left(1 + \frac{E}{h\nu N}\right) - \frac{E}{h\nu N}\log\frac{E}{h\nu N}\right\}$$

obtained by means of Stirling's formula. Hence $\phi(x)$ of (1.24) is

$$\phi(x) = (1 + x)\log(1 + x) - x\log x\,.$$

For $x > 0$ one has

$$\phi'(x) = \log\left(\frac{1 + x}{x}\right) > 0\,, \qquad \lim_{x \to 0+} \phi'(x) = +\infty\,,$$
$$\phi''(x) = \frac{1}{1 + x} - \frac{1}{x} < 0\,.$$

Since, furthermore, $\lim_{x \to 0+} \phi(x) = 0$ and

$$\lim_{x \to +\infty} \phi(x) = \lim_{x \to +\infty} \left\{\log\left(1 + \frac{1}{x}\right)^x + \log(1 + x)\right\} = +\infty\,,$$
$$\phi(x) > 0.$$

Consequently, condition (1.25) is satisfied.

27. Because $n \ll N$ and $n \ll N'$, one may assume that interstitial atoms and vacancies are separated from each other by sufficiently large distances. In this case one can suppose that the energy of the imperfect crystal is higher than that of the perfect crystal by the amount

$$E(n) = nw\,. \tag{1}$$

The number of possible ways of removing n atoms from N lattice points and distributing them among N' interstitial sites is given by

$$W(n) = \frac{N!}{n!(N - n)!}\,\frac{N'!}{n!(N' - n)!}\,. \tag{2}$$

Therefore, that part of the partition function which is due to defects of the Frenkel type is given by the following equation:

$$Z = \sum W(n)\,\mathrm{e}^{-E(n)/kT}\,. \tag{3}$$

The energy $E(n)$ increases with n and $\exp\{-E(n)/kT\}$ decreases rapidly. But, since $W(n)$ increases rapidly with n, $W(n)\exp\{-E(n)/kT\}$ in (3) has a very sharp maximum. The value of n giving this maximum (the most probable value) corresponds to the value expected in thermal equilibrium. By Stirling's formula,

$$\log\{W(n)\mathrm{e}^{-E(n)/kT}\} = N\log N + N'\log N' - 2n\log n +$$

$$- (N-n)\log(N-n) - (N'-n)\log(N'-n) - \frac{w}{kT}n. \qquad (4)$$

Since the desired value n is to make this expression a maximum, one has

$$0 = \frac{\partial}{\partial n}\log\{W(n)\mathrm{e}^{-E(n)/kT}\} = -2\log n + \log(N-n) + \log(N'-n) - \frac{w}{kT},$$

or

$$\frac{n^2}{(N-n)(N'-n)} = \mathrm{e}^{-w/kT}. \qquad (5)$$

Because $n \ll N$ and $n \ll N'$, one may approximate the denominator by NN'.
NOTE 1: When n is of the same order of magnitude as N and N', approximations (1) and (2) become poor. When $w \gg kT$, however, we get $n \ll N, N'$ from result (5), so that there is no inconsistency in using (1) and (2).
NOTE 2: According to (1.18) the entropy of the state of an imperfect crystal specified by n is $S(n) = k\log W(n)$, so that $F(n) = E(n) - TS(n) = -kT\log[W(n)\exp\{-E(n)/kT\}]$ is the corresponding Helmholtz free energy. The largest term in the sum of (3) is the term with n which makes this free energy minimum. The value of $F(n)$ for this n can be approximated by $F = -kT\log Z$.

28. One can proceed as in the preceding problem. When n atoms are displaced to the surface, there are n holes and N atoms in $N+n$ lattice points. Hence the entropy is

$$S(n) = k\log\frac{(N+n)!}{n!N!} = k[(N+n)\log(N+n) - n\log n - N\log N]. \qquad (1)$$

($(N+n)!/n!N!$ is the number of ways of distributing holes and atoms.) Since the energy is $E(n) = nw$ as in the preceding problem, the free energy is

$$F(n) = E(n) - TS(n) = nw - kT\{(N+n)\log(N+n) - n\log n - N\log N\}. \qquad (2)$$

The n which minimizes this expression is the most probable value of n and can be regarded as the equilibrium value:

$$0 = \left(\frac{\partial F}{\partial n}\right)_{T,N} = w - kT \log \frac{N+n}{n} = 0.$$

Hence
$$\frac{n}{N+n} = e^{-w/kT}, \quad \text{or} \quad n/N = e^{-w/kT} \qquad (n \ll N).$$

29. When N_1 molecules are adsorbed, the energy is equal to $-N_1\varepsilon_0$. Since the number of states is equal to the number of possible ways of distributing N_1 identical molecules among N adsorbing points, which is $N!/N_1!(N-N_1)!$, one has

$$Z_{N_1} = \frac{N!}{N_1!(N-N_1)!} e^{\beta N_1 \varepsilon_0}.$$

Therefore, the partition function in the T-μ distribution is, according to (1.74c), given by

$$\Xi = \sum_{N_1=0}^{N} Z_{N_1} e^{\beta N_1 \mu} = \sum \frac{N!}{N_1!(N-N_1)!} [e^{\beta(\varepsilon_0+\mu)}]^{N_1} = [1 + e^{\beta(\varepsilon_0+\mu)}]^N. \tag{1}$$

The probability $Pr(N_1)$ of N_1 molecules being adsorbed is given by $e^{\beta N_1 \mu} Z_{N_1}/\Xi$. Hence the mean value of the number of adsorbed molecules is

$$\bar{N}_1 = \sum_{N_1=0}^{N} N_1 Pr(N_1) = \frac{1}{\Xi} \sum_{N_1=0}^{N} N_1 e^{\beta N_1 \mu} Z_{N_1} = \frac{\partial}{\beta \partial \mu} \log \Xi.$$

Substituting this into (1), one obtains

$$\bar{N}_1 = \frac{N}{\beta} \frac{\partial}{\partial \mu} \log [1 + e^{\beta(\varepsilon_0+\mu)}]. \tag{2}$$

Therefore, one obtains the covering ratio θ,

$$\theta = \frac{\bar{N}_1}{N} = \frac{1}{e^{-\beta(\varepsilon_0+\mu)} + 1}. \tag{3}$$

Since μ is the chemical potential of the particle source, that is, that of the ideal gas (p, T) which is in contact with the system, one obtains from equation (6) in the solution of example 8

$$e^{\mu/kT} = \frac{p}{kT}\left(\frac{h^2}{2\pi m kT}\right)^{\frac{3}{2}}, \tag{4}$$

if one can apply the Boltzmann statistics. Substituting this into (3), one finds

$$\theta = \frac{p}{p + p_0(T)}, \qquad \text{where} \qquad p_0(T) = \left(\frac{2\pi mkT}{h^2}\right)^{\frac{3}{2}} e^{-\varepsilon_0/kT}\, kT. \tag{5}$$

NOTE: This equation for θ is called Langmuir's isotherm. One can write down $p_0(T)$ in a slightly more general form. If one denotes the partition function of the gas per molecule (per unit volume) as $z_g(T)$ and that of a molecule on the adsorbing point as $z_{ad}(T)$, it follows that

$$p_0(T) = \frac{z_g(T)}{z_{ad}(T)} kT.$$

30. Each magnetic moment is either oriented in the + direction (parallel to the magnetic field) or − (antiparallel). The respective probabilities are

$$p_+ = e^{\beta\mu H}/Z_1, \qquad \text{and} \qquad p_- = e^{-\beta\mu H}/Z_1, \qquad (Z_1 = 2\cosh(\beta\mu H)). \tag{1}$$

We label each spin with a number 1, ... , N and denote the moments by $\mu_1, \ldots, \mu_N$. Each of the μ_i takes on the value $+\mu$ or $-\mu$ independently with probability (1). Hence we have

$$\bar{M} = \overline{\sum_{i=1}^{N} \mu_i} = \sum \bar{\mu}_i = N\mu(p_+ - p_-) = N\mu \tanh(\beta\mu H)$$

and

$$\overline{(M - \bar{M})^2} = \overline{(\sum_{i=1}^{N} (\mu_i - \bar{\mu}_i))^2} = \sum_{i=1}^{N} \overline{(\mu_i - \bar{\mu}_i)^2}$$

$$= \sum_{i=1}^{N} (\overline{\mu_i^2} - \bar{\mu}_i^2) = N\mu^2\{1 - \tanh^2(\beta\mu H)\} = N\mu^2/\cosh^2(\beta\mu H),$$

where we have used the fact that the distributions of individual μ_i are independent from each other so that $\overline{(\mu_i - \bar{\mu}_i)(\mu_j - \bar{\mu}_j)} = \overline{(\mu_i - \bar{\mu}_i)} \cdot \overline{(\mu_j - \bar{\mu}_j)} = 0 \quad (i \neq j)$.

ALTERNATIVE SOLUTION

If one denotes the number of +spins by N_+ and that − spins by N_-, the energy is $E = -(N_+ - N_-)\mu H$. Hence in the canonical distribution the probability of having the state (N_+, N_-) is given by

$$\frac{N!}{N_+!N_-!} e^{(N_+ - N_-)\mu H/kT} \Big/ Z_N = \frac{N!}{N_+!N_-!} p_+^{N_+} p_-^{N_-}$$

$$= \frac{N!}{N_+!(N - N_+)!} p_+^{N_+}(1 - p_+)^{N - N_+}.$$

This is the binomial distribution discussed in problem 9. According to equations (4) and (6) in its solution we have

$$\bar{N}_+ = Np_+ , \qquad \overline{(N_+ - \bar{N}_+)^2} = Np_+(1 - p_+) = Np_+p_- .$$

Substituting $M = (N_+ - N_-)\mu$, we get $\bar{M} = N\mu(p_+ - p_-)$ and

$$\begin{aligned}\overline{M^2} &= \mu^2\overline{(N_+ - N_-)^2} = \mu^2\overline{(2N_+ - N)^2} = \mu^2[\overline{4N_+^2} - 4N\bar{N}_+ + N^2] \\ &= \mu^2[4\overline{(N_+ - \bar{N}_+)^2} + 4\bar{N}_+^2 - 4N\bar{N}_+ + N^2] \\ &= \mu^2[4Np_+p_- + N^2(p_+ - p_-)^2] = 4N\mu^2 p_+p_- + \bar{M}^2 ,\end{aligned}$$

and hence

$$\overline{(M - \bar{M})^2} = \frac{N\mu^2}{\cosh^2(\beta\mu H)} .$$

31. (i) (a) B. E. statistics. The number of ways of distributing N_j particles among the C_j quantum states in the j-th group, allowing any number of particles to enter into each state, is equal to the number of ways of distributing N_j balls among C_j boxes, as was discussed in example 3. Hence, according to (1) in the solution of example 3, it is given by $(N_j + C_j - 1)!/N_j!(C_j - 1)!$. The desired answer is obtained by multiplying together these numbers for all j.

(b) Since in Fermi statistics only one particle can occur in each state, the number of ways of choosing N_j from among C_j objects is equal to $C_j!/N_j!\,(C_j - N_j)!$. Multiplying together for all j, we get the answer.

(ii) In the canonical distribution the probability of $\{N_j\}$ being realized is proportional to

$$W\{N_j\}\exp(-\sum_j \beta\varepsilon_j N_j) , \qquad (\beta = 1/kT) . \tag{1}$$

To obtain $\{N_j^*\}$ which makes this maximum, we take the logarithm of this and put the variation with respect to $\{N_j\}$ equal to 0. It amounts to minimizing the free energy F,

$$\frac{F\{N_j\}}{kT} = +\beta\sum\varepsilon_j N_j - \log W\{N_j\}$$

$$\begin{cases} = +\beta\sum_j\varepsilon_j N_j - \sum_j\{(C_j + N_j)\log(C_j + N_j) - N_j\log N_j - C_j\log C_j\} , & \text{(B. E.)} \\ = +\beta\sum_j\varepsilon_j N_j - \sum_j\{-(C_j - N_j)\log(C_j - N_j) - N_j\log N_j + C_j\log C_j\} , & \text{(F. D.)}\end{cases}$$

which we obtain with the help of Stirling's formula, assuming N_j, C_j and $C_j \pm N_j$ are all very large. Taking the variation, we have

$$\frac{\delta F}{kT} = \sum \left[\beta \varepsilon_j + \log \frac{N_j}{C_j \pm N_j} \right] \delta N_j , \tag{2}$$

(+ or − sign refers to the B. E. or F. D.).

Since $\sum N_j = N$ is given,

$$\delta N = \sum \delta N_j = 0 . \tag{3}$$

We use Lagrange's method of undetermined multipliers denoting the multiplier for (3) by $-\mu/kT$,

$$\frac{\delta(F - \mu N)}{kT} = \sum_j \left[\beta(\varepsilon_j - \mu) + \log \frac{N_j}{C_j \pm N_j} \right] = 0 . \tag{4}$$

Hence we get the most probable distribution

$$\frac{N_j^*}{C_j \pm N_j^*} = e^{-\beta(\varepsilon_j - \mu)}, \qquad \text{or} \qquad \frac{N_j^*}{C_j} = \frac{1}{e^{\beta(\varepsilon_j - \mu)} \mp 1} . \tag{5}$$

The coefficient μ is determined by

$$N = \sum_j N_j = \sum \frac{C_j}{\{e^{\beta(\varepsilon_j - \mu)} \mp 1\}} . \tag{6}$$

(iii) In the microcanonical distribution the most probable distribution is that which maximizes $W\{N_j\}$ under the conditions $E = \sum \varepsilon_j N_j$ and $N = \sum N_j$. Therefore, as in the case of (ii), one can find the desired condition from

$$\delta \log W = \sum_j \delta N_j \frac{\log (C_j \pm N_j)}{N_j} = 0 , \qquad \delta E = \sum \delta N_j \varepsilon_j = 0 ,$$

$$\delta N = \sum \delta N_j = 0 .$$

Multiplying equation (2) by $-\beta (= -1/kT)$ and equation (3) by $\beta\mu$ (Lagrange's undetermined multipliers) and adding them together, we again obtain (4) as the variational equation. In this case the undetermined multipliers β and μ are determined by requiring

$$E = \sum N_j \varepsilon_j = \sum \frac{C_j \varepsilon_j}{(e^{\beta(\varepsilon_j - \mu)} \mp 1)} \tag{7}$$

as well as (6).

32. From (1.94c) one finds

$$pV = \pm\frac{1}{\beta}\sum_{\tau}\log\{1 \pm e^{-\beta(\varepsilon_j-\mu)}\}.$$

If one denotes the density of states for one particle by $\Omega(\varepsilon) = \Omega_0'(\varepsilon)$, this can be written in integral form as

$$pV = \pm\frac{1}{\beta}\int_0^\infty \log\{1 \pm e^{\beta(\mu-\varepsilon)}\}\,\Omega(\varepsilon)\,d\varepsilon.$$

Integrating by parts one obtains

$$pV = \pm\frac{1}{\beta}\Big[\log\{1 \pm e^{\beta(\mu-\varepsilon)}\}\,\Omega_0(\varepsilon)\Big]_0^\infty + \int_0^\infty \frac{1}{e^{\beta(\varepsilon-\mu)} \pm 1}\,\Omega_0(\varepsilon)\,d\varepsilon. \qquad (1)$$

If $\Omega_0(\varepsilon)$ increases with ε like any power of ε, the upper limit of the first term vanishes. Also the lower limit vanishes because of the condition $\Omega_0(0) = 0$, so that only the integral of the second term remains.

When the energy is of the form $p^2/2m$, $\Omega_0(\varepsilon) \propto \varepsilon^{\frac{3}{2}}$ as we have seen in problem 6. Hence

$$\varepsilon\frac{d\Omega_0}{d\varepsilon} = \tfrac{3}{2}\Omega_0. \qquad (2)$$

Substituting this into (1), we find

$$pV = \tfrac{2}{3}\int_0^\infty \frac{\varepsilon}{e^{\beta(\varepsilon-\mu)} \pm 1}\frac{d\Omega_0}{d\varepsilon}\,d\varepsilon = \tfrac{2}{3}\int_0^\infty \frac{\varepsilon}{e^{\beta(\varepsilon-\mu)} \pm 1}\,\Omega(\varepsilon)\,d\varepsilon = \tfrac{2}{3}E_{\text{kin}}.$$

ALTERNATIVE SOLUTION

Since $p = T(\partial S/\partial V)_E = (\partial S/\partial V)_E/(\partial S/\partial E)_V = -(\partial E/\partial V)_S$ ((1.40): also see problem 12), one can find p by studying the change in the energy as V is changed with the entropy kept constant. As given by (2) in the solution to problem 6, the kinetic energy ε of a particle is inversely proportional to the second power of the length l of the edge of the box for given (n_1, n_2, n_3), namely, $\varepsilon \propto V^{-\frac{2}{3}}$. If one keeps the quantum numbers (n_1, n_2, n_3) fixed and varies only the volume of the box, then obviously the entropy does not change. Hence, under the condition S = constant,

$$\delta E_{\text{kin}}/E_{\text{kin}} = -\tfrac{2}{3}\delta V/V.$$

Therefore

$$(\partial E_{\text{kin}}/\partial V)_S = -\tfrac{2}{3}E_{\text{kin}}/V, \qquad \text{whence} \quad p = \tfrac{2}{3}E_{\text{kin}}/V.$$

33. (i) From (5.75) one has

$$\bar{N} = \frac{\sum_{N=0}^{\infty} N\,e^{\beta\mu N} Z_N}{\Xi}, \qquad \Xi = \sum e^{\beta\mu N} Z_N .$$

Hence,

$$\frac{\partial}{\partial\mu}\bar{N} = \beta\frac{\sum N^2 e^{\beta\mu N} Z_N}{\Xi} - \beta\frac{(\sum N e^{\beta\mu N} Z_N)^2}{\Xi^2}$$

$$= \beta(\overline{N^2} - \bar{N}^2) = \beta\overline{(N-\bar{N})^2} .$$

(ii) In classical statistics, one has $\bar{N} = e^{\beta\mu} f$, by (1.99),

and thus
$$\frac{\partial}{\partial\mu}\bar{N} = \beta e^{\beta\mu} f = \beta\bar{N} .$$

Therefore using the result in (i), one obtains

$$\overline{(N-\bar{N})^2} = \bar{N} .$$

(iii) In the case of quantum statistics, one has from equation (7) of the solution in example 12

$$Pr(n_\tau) = \frac{e^{\beta(\mu-\varepsilon_\tau)n_\tau}}{\sum e^{\beta(\mu-\varepsilon_\sigma)n_\sigma}} .$$

Therefore, if we write $\xi_\tau \equiv \sum_{n_\tau} e^{\beta(\mu-\varepsilon_\tau)n_\tau}$, we have, just as in (1), the relation

$$\frac{\partial}{\partial\mu}\overline{n_\tau} = \beta\overline{(n_\tau - \overline{n_\tau})^2}$$

between

$$\overline{n_\tau} = \frac{\sum n_\tau e^{\beta(\mu-\varepsilon_\tau)n_\tau}}{\xi_\tau}, \qquad \overline{n_\tau^2} = \frac{\sum n_\tau^2 e^{\beta(\mu-\varepsilon_\tau)n_\tau}}{\xi_\tau} .$$

Using (1.90, 91), we get

$$\frac{\partial}{\partial\mu}\frac{1}{e^{\beta(\varepsilon_\tau-\mu)} \mp 1} = \beta\frac{e^{\beta(\varepsilon_\tau-\mu)}}{(e^{\beta(\varepsilon_\tau-\mu)} \mp 1)^2} = \beta\overline{n_\tau}(1 \pm \overline{n_\tau}) .$$

Consequently we obtain the required relation.

34. When the parameter x is changed to $x + dx$, the change in $\Omega_0(E, x)$ is

$$d\Omega_0 = \Omega_0(E + dE, x + dx) - \Omega_0(E, x) = \left(\frac{\partial\Omega_0}{\partial E}\right)_x dE + \left(\frac{\partial\Omega_0}{\partial x}\right)_E dx . \quad (1)$$

The first term is the change in Ω_0 due to the shift of the surface $\mathscr{H}(x) = E$ to

$\mathcal{H}(x) = E + \mathrm{d}E$ resulting from the change in energy in this process, while the second term is due to the shift of the surface $\mathcal{H}(x) = E$ to $\mathcal{H}(x + \mathrm{d}x) = E$ caused by the change of the parameter x. If x is assumed to vary by the amount $\mathrm{d}x$ with uniform velocity during the time from 0 to τ, the change in the energy E in this process is

$$\mathrm{d}E = \int_0^\tau \frac{\mathrm{d}\mathcal{H}}{\mathrm{d}t}\,\mathrm{d}t = \int_0^\tau \left[\left\{\frac{\partial \mathcal{H}}{\partial p}\dot{p} + \frac{\partial \mathcal{H}}{\partial q}\dot{q}\right\} + \frac{\partial \mathcal{H}}{\partial x}\dot{x}\right]\mathrm{d}t = \int_0^\tau \frac{\partial \mathcal{H}}{\partial x}\dot{x}\,\mathrm{d}t$$

$$= \dot{x}\int_0^\tau X\,\mathrm{d}t = \langle X\rangle_{\text{time}}\cdot \mathrm{d}x\,. \tag{2}$$

(The quantity in { } in the second integral is 0 because of the equation of motion (1.1.).) Making the assumption of ergodicity, one obtains

$$\mathrm{d}E = \langle X\rangle_{\text{time}}\,\mathrm{d}x = \langle X\rangle_{\text{phase}}\,\mathrm{d}x = \mathrm{d}x \int_{\mathcal{H}=E} \frac{X(p,q,x)}{|\text{grad}\,\mathcal{H}|}\,\mathrm{d}\sigma \Big/ \int_{\mathcal{H}=E} \frac{\mathrm{d}\sigma}{|\text{grad}\,\mathcal{H}|}\,. \tag{3}$$

Let n be the normal at a point A on the surface S_1, $\mathcal{H}(p, q, x) = E$ and B be the point at which it intersects the surface S_2, $\mathcal{H}(p, q, x + \mathrm{d}x) = E + \mathrm{d}E$ and C the point at which it intersects the surface S_3, $\mathcal{H}(p, q, x + \mathrm{d}x) \equiv \mathcal{H}(p, q, x) + X\mathrm{d}x = E$. Then, since

$$\overrightarrow{\mathrm{AB}}\cdot \text{grad}\,\mathcal{H} = \mathrm{d}E,$$
$$\overrightarrow{\mathrm{AC}}\cdot \text{grad}\,\mathcal{H} = -X\mathrm{d}x\,,$$

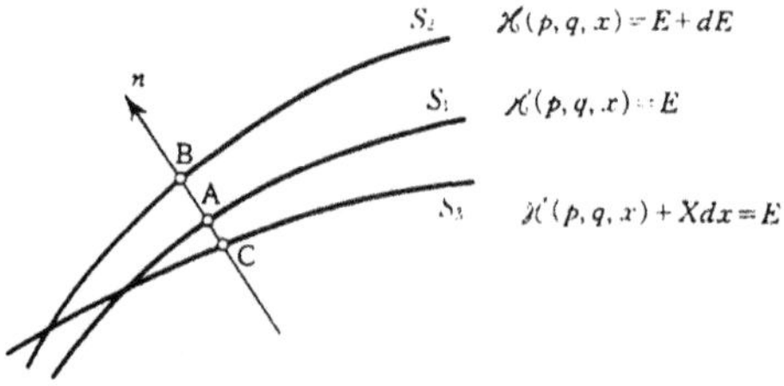

the volume enclosed by S_1 and S_2, $(\partial\Omega_0/\partial E)\mathrm{d}E$, and the volume enclosed by S_1 and S_3, $(\partial\Omega_0/\partial x)\mathrm{d}x$ are, respectively, equal to

$$\frac{\partial\Omega_0}{\partial E}\mathrm{d}E = \mathrm{d}E \int_{\mathcal{H}=E} \frac{\mathrm{d}\sigma}{|\text{grad}\,\mathcal{H}|}, \qquad \frac{\partial\Omega_0}{\partial x}\mathrm{d}x = -\,\mathrm{d}x \int_{\mathcal{H}=E} \frac{X\,\mathrm{d}\sigma}{|\text{grad}\,\mathcal{H}|}\,. \tag{4}$$

Substituting (4) into (1) one obtains

$$d\Omega_0 = dE \int_{\mathscr{H}=E} \frac{d\sigma}{|\mathrm{grad}\,\mathscr{H}|} - dx \int_{\mathscr{H}=E} \frac{X\, d\sigma}{|\mathrm{grad}\,\mathscr{H}|}$$

which vanishes because of (3).

35. If we denote the numbers of particles in the states $\pm\varepsilon_0$ by N_+ and N_-, $(N_+ + N_- = N)$, respectively, we have

$$E = (N_+ - N_-)\varepsilon, \qquad S = k \log \Omega = \log (N!/N_+!N_-!)$$

(see example 4). Rewriting S with the help of Stirling's formula, we find

$$S(E) = Nk\left[\log 2 - \tfrac{1}{2}\left\{\left(1 - \frac{E}{N\varepsilon_0}\right)\log\left(1 - \frac{E}{N\varepsilon_0}\right) + \left(1 + \frac{E}{N\varepsilon_0}\right)\log\left(1 + \frac{E}{N\varepsilon_0}\right)\right\}\right]. \tag{1}$$

Hence we have

$$\frac{1}{T} = \frac{k}{\varepsilon_0}\log\frac{N\varepsilon_0 - E}{N\varepsilon_0 + E}. \tag{2}$$

For $E > 0$, therefore, $T < 0$ $(\partial S/\partial E < 0)$, that is, it is in a state of negative temperature, and is not a normal system. Let us bring such a system in contact with an ideal gas thermometer consisting of n molecules, in order to measure its temperature. Denoting the energy of the gas by E_g and the number of states by $\Omega_g(E_g)$, we have

$$\Omega_g(E_g) \propto E_g^{\frac{3}{2}n} \tag{3}$$

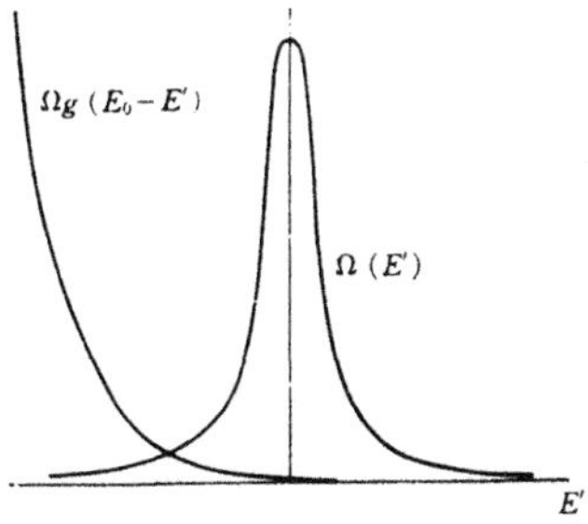

(as we saw in example 2). When the new equilibrium condition is established, the most probable partition of energy is determined by

$$\Omega(E')\Omega_g(E_0 - E') = \max, \qquad E_0 = E' + E_g, \tag{4}$$

where E_g^0 is the energy which is initially contained in the gas thermometer. Now $\Omega(E')$ has a sharp maximum at $E' = 0$, while $\Omega_g(E_0 - E')$ decreases rapidly with E' (see (3)) and hence its maximum lies in the region where $E' < 0$. Since the gas thermometer is small compared to the spin system ($n \ll N$), the maximum of the product occurs nearly at $E' = 0$. In other words, if one wants to measure the temperature of a system which is at negative temperature by means of a (small) thermometer, the original state of the latter is drastically disturbed and it is brought to a state with $T = \infty$, i.e., the thermometer will indicate $T = \infty$.

Let us, indeed, examine condition (4). It means that the temperatures of both systems are equal, namely

$$T(E') = T_g(E_0 - E'). \tag{5}$$

Therefore, using (2) and the relation

$$kT_g = \tfrac{2}{3} E_g'/n = \frac{2}{3n}(E_0 - E'),$$

we obtain

$$\frac{\varepsilon_0}{kT'} = \log \frac{N\varepsilon_0 - E'}{N\varepsilon_0 + E'} = \frac{3n\varepsilon_0}{2(E_0 - E')}. \tag{6}$$

This equation has the solution $E' < 0$. In order to obtain its value, we expand the logarithm, expecting it to satisfy $| E'/N\varepsilon_0 | \ll 1$:

$$\frac{-E'}{N\varepsilon_0} \doteqdot \frac{3}{4} \frac{n\varepsilon_0}{(E_0 - E')} \doteqdot \frac{3}{4} \frac{n\varepsilon_0}{E_0}. \tag{7}$$

If we recall that the initial energy is $E_0 = E' + E_g \sim E' = O(N\varepsilon_0)$, we have $E'/N\varepsilon_0 = O(n/N)$ which is practically zero. This means that almost all the excess energy $E(\gg E_g^0)$ has been transferred to the gas thermometer. The temperature value measured by this process thus depends on the size of the thermometer.

36. We shall obtain $n(\boldsymbol{r})$ which makes the Helmholtz free energy $U - TS$ a minimum under the condition that the total number of particles is fixed. Since we have

$$\delta U = e^2 \iint \frac{n(\boldsymbol{r}')\mathrm{d}\boldsymbol{r}'}{|\boldsymbol{r} - \boldsymbol{r}'|} \delta n(\boldsymbol{r})\,\mathrm{d}\boldsymbol{r} + e \int \phi_{\text{ext}} \delta n(\boldsymbol{r})\,\mathrm{d}\boldsymbol{r},$$

$$T\delta S = -\int kT \log n \cdot \delta n(\boldsymbol{r})\,\mathrm{d}\boldsymbol{r},$$

$$\delta N = \int \delta n(\boldsymbol{r})\,\mathrm{d}\boldsymbol{r}$$

we obtain the following equation for n:

$$kT \log n + e^2 \int \frac{n(\boldsymbol{r}')}{|\boldsymbol{r} - \boldsymbol{r}'|} \mathrm{d}\boldsymbol{r}' + e\phi_{\mathrm{ext}} - \mu = 0, \tag{1}$$

where we have introduced Lagrange's undetermined multiplier μ. Because the electrostatic potential ϕ is given by

$$\phi = \phi_{\mathrm{ext}} + e \int \frac{n(\boldsymbol{r}')}{|\boldsymbol{r} - \boldsymbol{r}'|} \mathrm{d}\boldsymbol{r}', \tag{2}$$

we obtain from (1)

$$n = \exp\left\{ -\frac{1}{kT}(e\phi - \mu) \right\}, \tag{3}$$

or writing the number density at the point $\phi = 0$ as n_0,

$$n = n_0 \mathrm{e}^{-e\phi/kT} \tag{4}$$

Substituting (3) or (4) into (1), we obtain the equation for ϕ. When ϕ_{ext} is given, we can determine the spatial distribution of particles by solving this equation.

CHAPTER 2

APPLICATIONS OF THE CANONICAL DISTRIBUTION

The canonical distribution is the most frequently used distribution in actual applications of statistical mechanics. The reasons for this are: first, because the canonical distribution represents the condition of constant temperature which is certainly most easily realized in physical experiments and second, because it is the most suited for mathematical manipulations. Some of the basic aspects of the canonical distribution have been stated in the previous chapter, but here they will again be summarized together with certain additional remarks, particularly those on asymptotic evaluations for large systems. These are important for a clear understanding of the relation between thermodynamics and statistical mechanics. Similar methods can be applied to other generalized canonical distributions. PROBLEMS A in this chapter require only the knowledge contained in the FUNDAMENTAL TOPICS of Chapter 1 and in the elementary parts of FUNDAMENTAL TOPICS of the present chapter (in particular, advanced topics such as the Laplace transformation and density matrices are not needed).

Fundamental Topics

§ 2.1. GENERAL PROPERTIES OF THE PARTITION FUNCTION $Z(\beta)$

By definitions (1.71b) and (1.71c), the partition function of a system is given by

$$Z(\beta) = \sum_{l} e^{-\beta E_l} = \int_{E_0}^{\infty} e^{-\beta E}\, \Omega(E)\, dE \tag{2.1}$$

where E_0 is the lowest energy level and $\Omega(E)$ is the state density of the system. The latter satisfies the conditions

$$\Omega \geqq 0, \qquad \lim_{E\to\infty} \Omega(E)\, e^{-\alpha E} = 0 \qquad (\alpha > 0) \tag{2.2}$$

(see § 1.6). From (2.1) and (2.2) one can see the following:

1. If β is a real number, then the integral (2.1) converges for $\beta > 0$ and $Z(\beta)$ is finite. If β is a complex number, then the same applies for $\Re\beta > 0$.†

† $\Re\beta$ means the real part of β.

2. For $\beta > 0$, one has

$$\partial^2 \log Z/\partial\beta^2 > 0 \tag{2.3}$$

(see eq. (2) in the solution to problem 1).

3. For a composite system I + II, it holds that

$$Z_{\mathrm{I}+\mathrm{II}}(\beta) = Z_{\mathrm{I}}(\beta) Z_{\mathrm{II}}(\beta) \tag{2.4}$$

whereas the state density satisfies (1.43), i.e.

$$\begin{aligned} \Omega_{\mathrm{I}+\mathrm{II}}(E) &= \int \Omega_{\mathrm{I}}(E')\, \Omega_{\mathrm{II}}(E - E')\, \mathrm{d}E' \\ &= \int \Omega_{\mathrm{I}}(E - E'')\, \Omega_{\mathrm{II}}(E'')\, \mathrm{d}E'' . \end{aligned} \tag{2.5}$$

It is important to remember that the product property of the partition function corresponds to the folding (convolution) property of the state density.

4. The inverse formula to obtain $\Omega(E)$ from the partition function $Z(\beta)$ is

$$\begin{aligned} \Omega(E) &= \frac{1}{2\pi \mathrm{i}} \int_{\beta' - \mathrm{i}\infty}^{\beta' + \mathrm{i}\infty} Z(\beta)\, \mathrm{e}^{\beta E}\, \mathrm{d}\beta \qquad (\beta' > 0) \\ &\equiv \frac{1}{2\pi} \int_{-\infty}^{\infty} Z(\beta' + \mathrm{i}\beta'')\, \mathrm{e}^{(\beta' + \mathrm{i}\beta'')E}\, \mathrm{d}\beta'' \end{aligned} \tag{2.6}$$

in which the integration path runs in the complex plane of β as shown in Fig. 2.1.

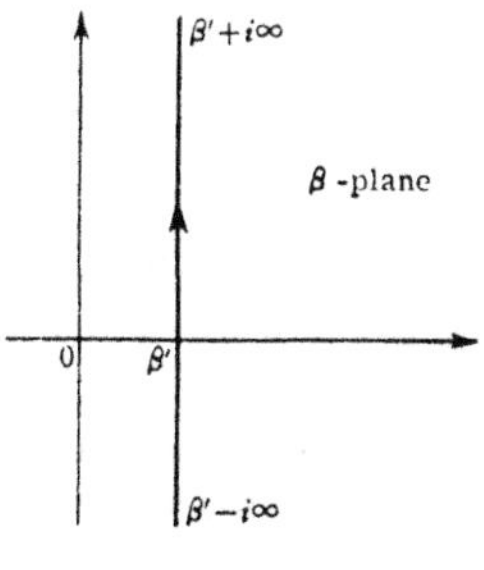

Fig. 2.1.

NOTE: Generally, for a real function $f(x)$ of x the function $g(s)$ defined by

$$\mathfrak{L}(f) \equiv g(s) = \int_0^{\infty} \mathrm{e}^{-sx} f(x)\, \mathrm{d}x$$

is called its Laplace transform. Here, s is a complex number. This integral is convergent over the half plane of s, satisfying $\Re s > s_0$. The convergence limit s_0 is called the convergence abscissa of the Laplace transform. For $\Re s > s_0$, $g(s)$ is an analytic function of s. It has at least one singular point on the line $\Re s = s_0$. The inverse transform of a Laplace transform is given by

$$f(x) = \frac{1}{2\pi i} \int_{s'-i\infty}^{s'+i\infty} g(s)\, e^{sx}\, ds$$

where the integration path is a straight line parallel to the imaginary axis, but it can be continuously deformed as long as the integral converges. $Z(\beta)$ is the Laplace transform of $\Omega(E)$. Therefore general theorems concerning Laplace transforms can be applied to these function. [Cf. Ian N. Sneddon, *Fourier Transforms* (McGraw-Hill, New York, 1951).]

§ 2.2+. ASYMPTOTIC EVALUATIONS FOR LARGE SYSTEMS

If the system under consideration is a large system which has a particle number N or a volume V of macroscopic size, its partition function $Z(\beta)$ will generally behave as

$$Z(\beta) = e^{N\phi(\beta)} \qquad \text{or} \qquad e^{V\psi(\beta)} \tag{2.7}$$

(as $N \to \infty$ or $V \to \infty$). Then an asymptotic evaluation of the state density $\Omega(E)$ (or an approximation which is valid for large N or V) is given as

$$\log \Omega(E) \simeq \log Z(\beta^*) + \beta^* E \tag{2.8}$$

where β^* is determined by the equation

$$\frac{\partial \log Z(\beta^*)}{\partial \beta^*} = -E \tag{2.9}$$

and so is a function of E. This formula is derived as follows:

Method of the steepest descent (Saddle point method):† In calculating $\Omega(E)$ by (2.6) one writes the integrand in the form

$$Z(\beta)e^{\beta E} = \exp\{\log Z(\beta) + \beta E\}$$

and chooses as integration path a straight line ($\beta^* - i\infty \to \beta^* + i\infty$) with β^* determined by (2.9). The exponent of the above expression is expanded

† Fowler and Darwin made the first use of the steepest descent method in statistical mechanics. Hence this is sometimes called the Fowler-Darwin method.

in the neighborhood of β^*. Then the integral (2.6) is approximated as follows:

$$\Omega(E) \simeq \frac{1}{2\pi} \int_{-\infty}^{\infty} \exp\left[\log Z(\beta^*) + \beta^* E - \frac{1}{2}\frac{\partial^2 \log Z(\beta^*)}{\partial \beta^{*2}} \beta''^2 + \dots\right] d\beta''$$

$$= \exp\{\log Z(\beta^*) + \beta^* E\} \times \left[2\pi \frac{\partial^2 \log Z(\beta^*)}{\partial \beta^{*2}}\right]^{-\frac{1}{2}} \{1 + \dots\}.$$

Since $\partial^2 \log Z(\beta^*)/\partial\beta^{*2} > 0$ by (2.3), the exponent has a maximum at $\beta'' = 0$. This maximum is extremely steep because N or V in (2.7) is very large. Therefore the most important contribution to the integral (2.6) comes from the neighborhood of $\beta'' \sim 0$. By an approximate integration over β'' one easily obtains the above result. The logarithm of this expression gives

$$\log \Omega \sim \{\log Z(\beta^*) + \beta^* E\} + \left[\log\left(2\pi \frac{\partial^2 \log Z}{\partial \beta^{*2}}\right)^{-\frac{1}{2}} + \dots\right].$$

The first term { } of this expression is O(N) or O(V), but the second term [ ] is only O(log N) or O(log V). Therefore the second term may be ignored in comparison with the first. Thus one gets (2.8).

Central limit theorem: Let $x_1, x_2, \dots x_n$ be random variables and consider the probability distribution of their sum

$$X = x_1 + x_2 + \dots + x_n .$$

For simplicity $x_1, x_2, \dots$ and x_n are assumed to obey the same distribution law

$$Pr(x, x + dx) = f(x)\,dx .$$

Then in the limit of large n the variable X obeys the normal distribution

$$Pr(X, X + dX) = \frac{1}{\sqrt{2\pi\Delta}} \exp\left\{-\frac{1}{2\Delta}(X - \bar{X})^2\right\} dX \qquad (2.10)$$

where

$$\bar{X} = n\bar{x} = n \int x f(x)\,dx ,$$
$$\Delta = \overline{(X - \bar{X})^2} = n\overline{(x - \bar{x})^2} = n \int (x - \bar{x})^2 f(x)\,dx .$$

This theorem is generalized to variables $x_1, x_2, \dots x_n$ which have different laws. The condition for the theorem is, qualitatively stated, that none of these variables is very much superior to the rest. For rigorous statements of the theorem, see: Cramer, *Mathematical Methods of Statistics* (Princeton Univ. Press, 1946).

Asymptotic evaluation by means of the central limit theorem:† One may assume a large system of macroscopic size to be composed of n ($\gg 1$) subsystems each of which has the same structure. Then the partition function has the form

$$Z(\beta) = \{z(\beta)\}^n, \qquad \text{where} \qquad z(\beta) = \int e^{-\beta\varepsilon}\omega(\varepsilon)\, d\varepsilon\,, \tag{2.11}$$

ω being the state density of the subsystem.

Let β be an arbitrary real number and consider the distribution law of ε,

$$f(\varepsilon)\, d\varepsilon = \frac{e^{-\beta\varepsilon}\omega(\varepsilon)\, d\varepsilon}{z(\beta)} \tag{2.12}$$

for each subsystem. The energy E of the total system, which is the sum

$$E = \varepsilon_1 + \varepsilon_2 + \dots + \varepsilon_n\,,$$

then follows, by the central limit theorem (2.10) if n is large, the normal distribution

$$F(E)\, dE = \frac{1}{(2\pi\Delta)^{\frac{1}{2}}} \exp\left\{-\frac{1}{2\Delta}(E - \bar{E})^2\right\} dE \tag{2.13}$$

where

$$\left.\begin{aligned} \bar{E} &= n\bar{\varepsilon} = -n\frac{\partial \log z(\beta)}{\partial\beta} = -\frac{\partial \log Z(\beta)}{\partial\beta}\,, \\ \Delta &= n\overline{(\varepsilon - \bar{\varepsilon})^2} = n\frac{\partial^2 \log z(\beta)}{\partial\beta^2} = \frac{\partial^2 \log Z(\beta)}{\partial\beta^2}\,. \end{aligned}\right\} \tag{2.14}$$

Since (2.12) is canonical, E must also obey a canonical law. Therefore (2.13) must be an asymptotic form of such a canonical distribution function, i.e.

$$\frac{e^{-\beta E}\Omega(E)}{Z(\beta)} \simeq \frac{1}{(2\pi\Delta)^{\frac{1}{2}}} \exp\left\{-\frac{1}{2\Delta}(E - \bar{E})^2\right\}. \tag{2.15}$$

If β here is chosen equal to β^* given by (2.9), then (2.15) yields

$$\log\left[\frac{e^{-\beta^* E}\Omega(E)}{Z(\beta^*)}\right] \sim 0\,.$$

This is equivalent to (2.8).

The maximum term in the partition function (see solution to problem 21, Chapter 1): Let E^* be the value of E which maximizes the integrand of (2.1)

† This method is first used by Khinchine, *Mathematical Foundation of Statistical Mechanics*, translated by Gamow (Dover). This method has the advantage over the steepest descent method that it is more suitable for a rigorous formulation.

of the partition function. If the integral is approximated by retaining only the contribution from its neighborhood, then the logarithm of the partition function is approximated by

$$\log Z(\beta) = -\beta E^* + \log \Omega(E^*), \tag{2.16}$$

$$\beta = \frac{\partial \log \Omega(E^*)}{\partial E^*}, \tag{2.17}$$

where $\log Z$ and $\log \Omega$ are assumed to be $O(N)$ or $O(V)$ so that corrections in $O(\log N)$ or $O(\log V)$ are omitted from (2.16).

§ 2.3+. ASYMPTOTIC EVALUATIONS AND LEGENDRE TRANSFORMATIONS OF THERMODYNAMIC FUNCTIONS

By putting $\beta^* = 1/kT$, one sees immediately that equations (2.8) and (2.9) coincide with the Legendre transformation

$$S(E) = \frac{1}{T}(E - F), \qquad \frac{\partial(F/T)}{\partial(1/T)} = E \tag{2.18}$$

which yields $S(E)$ from $F(T)$. Equations (2.8) and (2.9) also give

$$\frac{\partial \log \Omega(E)}{\partial E} = \left\{\frac{\partial \log Z(\beta^*)}{\partial \beta^*} + E\right\}\frac{\partial \beta^*}{\partial E} + \beta^* = \beta^*, \log Z(\beta^*) = \log \Omega - \beta^* E. \tag{2.19}$$

This coincides with the inverse Legendre transformation

$$\frac{\partial S}{\partial E} = \frac{1}{T}, \qquad \frac{F}{T} = \frac{E}{T} - S \tag{2.20}$$

which yields $F(T)$ from $S(E)$. Equations (2.16) and (2.17) are also identical with this transformation.

§ 2.4+. GRAND PARTITION FUNCTION $\Xi(\lambda)$

The grand partition function

$$\Xi(\lambda) = \sum_{N=0}^{\infty} \lambda^N Z_N \tag{2.21}$$

for a grand canonical distribution is not a Laplace transformation but is simply a power series constructed from Z_N. However, its properties are similar to the partition function $Z(\beta)$ as described in § 2.1:

1. $\Xi(\lambda)$ converges for $|\lambda| < \lambda_0$.

2. $$\lambda \frac{\partial}{\partial \lambda}\left(\lambda \frac{\partial \log \Xi}{\partial \lambda}\right) > 0. \tag{2.22}$$

3. For a composite system,

$$\Xi_{\mathrm{I+II}} = \Xi_{\mathrm{I}}\Xi_{\mathrm{II}}. \tag{2.23}$$

4.

$$Z_N = \frac{1}{2\pi \mathrm{i}} \oint \Xi(\lambda) \frac{\mathrm{d}\lambda}{\lambda^{N+1}} \tag{2.24}$$

where the integration is along a contour surrounding the origin.

5. The steepest descent method may be applied to the integral (2.24) by writing the integrand and integration variable λ as

$$\Xi(\lambda)\lambda^{-N} = \exp\{\log \Xi - N \log \lambda\}, \qquad \lambda = \lambda^* \mathrm{e}^{\mathrm{i}\phi}$$

and by carrying out the integration over ϕ in the neighborhood of $\phi \sim 0$. Then one obtains

$$\log Z_N = \log \Xi(\lambda^*) - N \log \lambda^*, \tag{2.25}$$

$$\lambda^* \frac{\partial \log \Xi(\lambda^*)}{\partial \lambda^*} = N \tag{2.26}$$

for very large N. Note that (2.25) gives $\log Z_N$ as a function of N, since $\lambda^*(N)$ in it is a function of N defined implicitly by (2.26).

6. If Ξ is asymptotically approximated by the maximum term in the summand (2.21), it becomes

$$\log \Xi(\lambda) \simeq \log Z_{N^*} + N^* \log \lambda \tag{2.27}$$

where $N^*(\lambda)$ is a function of λ defined by the equation

$$\log \lambda = -\frac{\partial \log Z_{N^*}}{\partial N^*}. \tag{2.28}$$

Note that (2.28) is equivalent to the relation which is obtained by differentiating (2.25) by N, i.e.,

$$\frac{\partial \log Z_N}{\partial N} = -\log \lambda^* + \left(\frac{\partial \log \Xi(\lambda^*)}{\partial \lambda^*} - \frac{N}{\lambda^*} \right) \frac{\partial \lambda^*}{\partial N} = -\log \lambda^*.$$

7. Equations (2.25) and (2.26) correspond to the thermodynamic transformation

$$F = N\mu + J, \qquad -\frac{\partial J}{\partial \mu} = N, \qquad (J = -pV), \tag{2.29}$$

and equations (2.27) and (2.28) to

$$-J \equiv pV = N\mu - F, \qquad \frac{\partial F}{\partial N} = \mu. \tag{2.30}$$

§ 2.5+. PARTITION FUNCTIONS FOR GENERALIZED CANONICAL DISTRIBUTIONS

What has been explained concerning the relations between Z and Ω and between Ξ and Z can be generalized to partition functions defined for other generalized canonical distributions. To a canonical distribution for a given x, there exists a corresponding canonical distribution in which the conjugate force X is prescribed instead of x. The latter partition function Z_X is a Laplace transform (or a generating function) of the partition function Z_x. Z_X and Z_x can be transformed into one another by mathematical transformations. If these transformations are evaluated asymptotically for a large (macroscopic) system, then the transformations coincide with the thermodynamic (Legendre) transformations between thermodynamic functions.

§ 2.6. CLASSICAL CONFIGURATIONAL PARTITION FUNCTIONS

Let a system consisting of N_A, N_B, ... particles of each species A, B, ... have the potential interaction energy $U(\boldsymbol{r}_1, \boldsymbol{r}_2, \ldots)$. The classical partition function of the system can be written in the form

$$Z_N = \left\{ \prod_A \left(\frac{2\pi m_A kT}{h^2} \right)^{\frac{3}{2}N_A} \right\} \frac{1}{N_A! N_B! \ldots} \int \ldots \int e^{-U/kT} \, d\Gamma_Q \tag{2.31}$$

in which the integration is already carried out for the momenta and $d\Gamma_Q$ represents the volume element of the configurational space for the coordinates of the particles having the dimension $3(N_A + N_B + \ldots)$. Then the integral

$$Q_{N_A, N_B \ldots} = \frac{1}{N_N! N_B! \ldots} \int \ldots \int e^{-U/kT} \, d\Gamma_Q \tag{2.32}$$

is called the *configurational partition function*. The probability that the particles realize a certain configuration of their positions is given by

$$Pr(d\Gamma_Q) = \frac{1}{N_A! N_B! \ldots} \frac{e^{-U/kT} \, d\Gamma_Q}{Q_N} . \tag{2.33}$$

Let $\{N\}$ represent all of the coordinates $\boldsymbol{r}_1, \boldsymbol{r}_2, \ldots \boldsymbol{r}_N$ of a system consisting of N identical particles. Then the configurational probability distribution function for these N particles can be written as

$$F_N\{N\} = \frac{1}{N!} \frac{e^{-U(N)/kT}}{Q_N} . \tag{2.34}$$

In some cases, it is necessary to find out probability distribution functions for one particle, two particles, ... and n particles picked out from the total of N particles. Such *reduced* distribution functions can be derived, in principle, by integrating (2.34).

§ 2.7+. DENSITY MATRICES

Definition: Let $\psi_{\alpha'}, \psi_{\alpha''}, \ldots$ be the quantum states of the system under consideration and let $\psi^*_{\alpha'}, \psi^*_{\alpha''}, \ldots$ be their Hermitean conjugates. They are mutually orthogonal and are normalized, namely,

$$(\psi^*_{\alpha'}, \psi_{\alpha''}) = \delta_{\alpha'\alpha''} \qquad (\langle\alpha' \mid \alpha''\rangle = \delta_{\alpha'\alpha''}). \tag{2.35}$$

If w_α gives the probability that the system is found to be in the state α, then the statistical properties of the system are represented by the operator †

$$\rho = \sum_\alpha w_\alpha \psi_\alpha \psi^*_\alpha = \sum_\alpha |\alpha\rangle w_\alpha \langle\alpha| . \tag{2.36}$$

The probability w_α is normalized as

$$\sum_\alpha w_\alpha = 1 . \tag{2.36'}$$

The operator defined by (2.36) is called a density matrix (sometimes, the normalization condition may not be fulfilled and yet the operator ρ is called the density matrix).

Let $\varphi_n \equiv |n\rangle$ ($n = 1, 2, \ldots$) be an arbitrary ortho-normal system in Hilbert space. Then the density matrix (2.36) is represented in this basis by

$$\langle n|\rho|m\rangle = \sum_\alpha \langle n|\alpha\rangle w_\alpha \langle\alpha|m\rangle . \tag{2.37}$$

By (2.37) one may write

$$\operatorname{Tr}\rho \equiv \sum_\alpha w_\alpha = 1 (= \sum \langle n|\rho|n\rangle), \tag{2.38}$$

where Tr means the *diagonal sum* or the *trace* of a matrix or an operator.

---

† An operator of the form $\psi_\alpha\psi_\alpha{}^*$ operates on a vector ψ in Hilbert space in such a way that

$$\psi_\alpha\psi_\alpha{}^* \cdot \psi = \psi_\alpha(\psi_\alpha{}^*\psi) .$$

Similarly it operates on a vector ψ^* in the conjugate space as

$$\psi^* \cdot \psi_\alpha\psi_\alpha{}^* = (\psi^*\psi_\alpha)\psi_\alpha{}^* .$$

In Dirac's notation we may write $\psi_\alpha \equiv |\alpha\rangle$, $\quad \psi_\alpha{}^* \equiv \langle\alpha^*|$, and so

$$(\psi_{\alpha'}{}^*\psi_{\alpha''}) = \langle\alpha' \mid \alpha''\rangle$$
$$\psi_\alpha\psi_\alpha{}^* \cdot \psi \equiv |\alpha\rangle\langle\alpha|\psi\rangle , \qquad \psi^*\psi_{\alpha\alpha}{}^* \psi \equiv \langle\psi^*|\alpha\rangle\langle\alpha| .$$

This way of writing the equations of operators and vectors is very convenient once one gets used to it.

Remember that such a diagonal sum is independent of the choice of the basic orthonormal system used to represent an operator as a matrix.

Expectation value of a dynamical quantity: For a statistical ensemble described by a given density operator ρ, the expectation value or the average of a dynamical quantity A is given by

$$\bar{A} = \sum w_\alpha \langle\alpha | A | \alpha\rangle = \operatorname{Tr} \rho A \tag{2.39}$$

where Tr ρA is the trace of the product of ρ and A, namely the diagonal sum of the product matrix of the matrices representing ρ and A in a certain representation.

The density matrix to represent a micro-canonical distribution has the form

$$\begin{aligned} \rho_E &= \sum_{E<E'<E+\delta E} \frac{\phi_{E'}\phi^*_{E'}}{W(E,\delta E)} \\ &\equiv \sum_{E<E'<E+\delta E} \frac{|E'\rangle\cdot\langle E'|}{W(E,\delta E)}, \end{aligned} \tag{2.40}$$

where $\phi_{E'} = |E'\rangle$ is the eigenvector corresponding to the eigenvalue E' of the system.

The density matrix to represent a canonical distribution has the form

$$\begin{aligned} \rho &= Z^{-1}\cdot \mathrm{e}^{-\beta\mathscr{H}} \qquad\qquad (\beta = 1/kT) \\ &\equiv Z^{-1}\cdot\sum_{E'} \mathrm{e}^{-\beta E'}\phi_{E'}\phi^*_{E'} \\ &\equiv Z^{-1}\cdot\sum_{E'} \mathrm{e}^{-\beta E'}| E'\rangle\langle E'| \end{aligned} \tag{2.41}$$

where $\phi_{E'} = |E'\rangle$ is an eigenvector of the Hamiltonian $\mathscr{H}$, i.e.,

$$\mathscr{H}\phi_{E'} = E'\phi_{E'} \qquad (\mathscr{H}\,|\,E'\rangle = E'\,|\,E'\rangle),$$

and

$$Z = \operatorname{Tr} \mathrm{e}^{-\beta\mathscr{H}} \equiv \sum_{E'} \mathrm{e}^{-\beta E'} \tag{2.42}$$

is the partition function of the system.

NOTE: In a canonical distribution, the average value of a dynamical variable A may be written as

$$\bar{A} = \frac{\operatorname{Tr} A \exp(-\beta\mathscr{H})}{\operatorname{Tr} \exp(-\beta\mathscr{H})}. \tag{2.43}$$

This corresponds to

$$\bar{A} = \frac{\int \mathrm{e}^{-\beta\mathscr{H}} A(p,q)\,\mathrm{d}\Gamma}{\int \mathrm{e}^{-\beta\mathscr{H}}\,\mathrm{d}\Gamma} \tag{2.44}$$

in classical statistical mechanics. In fact, one can prove that the trace operation goes into an integral in phase space over in the limit of $h \to 0$ (see: problem 33).

The expression (2.43) is very important in applications of quantum statistical mechanics. The traces may be calculated in some cases using explicit matrix representation by choosing a certain basis in Hilbert space of the system, or may be, under certain circumstances, more conveniently calculated by direct operational calculation without referring to a particular representation.

Examples

1. A classical system is in contact with a heat bath at $T°$ K. Show (i) that the mean kinetic energy per degree of freedom is equal to $\frac{1}{2}\,kT$ and (ii) that the relation

$$\overline{q_i \frac{\partial V}{\partial q_j}} = kT\delta_{ij} \qquad \left(\delta_{ij} = \begin{cases} 1 & (i=j) \\ 0 & (i \neq j) \end{cases}\right)$$

holds (law of equipartition of energy), provided that the potential energy V is $+\infty$ at the ends of the domains of the coordinates q_j.

SOLUTION

Both (i) and (ii) can be proved in a similar way. Let the momentum conjugate to a coordinate q_i be denoted by p_i. For the canonical distribution (1.70a), one calculates the mean value:

$$\begin{aligned}
\overline{p_i \frac{\partial \mathscr{H}}{\partial p_j}} &= C\int p_i \frac{\partial \mathscr{H}}{\partial p_j}\,\mathrm{e}^{-\beta\mathscr{H}}\,\mathrm{d}\Gamma \\
&= C\int p_i\left(-kT\frac{\partial\,\mathrm{e}^{-\beta\mathscr{H}}}{\partial p_j}\right)\mathrm{d}\Gamma \\
&= CkT\int\ldots\int\left(\Big[-p_i\mathrm{e}^{-\beta\mathscr{H}}\Big]_{p_j=-\infty}^{p_j=+\infty} + \int_{-\infty}^{+\infty}\mathrm{d}p_j\cdot\frac{\partial p_i}{\partial p_j}\mathrm{e}^{-\beta\mathscr{H}}\right)\mathrm{d}\Gamma_{(j)} \\
&= CkT\delta_{ij}\int \mathrm{e}^{-\beta\mathscr{H}}\,\mathrm{d}\Gamma = kT\delta_{ij}, \qquad (1)
\end{aligned}$$

where $C^{-1} = \int \mathrm{e}^{-\beta\mathscr{H}}\,\mathrm{d}\Gamma\left(= Z\prod N_A!h^{3N_A}\right)$.

The second equation (right-hand side) is transformed into the third one by partial integration over p_j, where $\mathrm{d}\Gamma_{(j)}$ is $\mathrm{d}\Gamma$ devoid of $\mathrm{d}p_j$. The relation (ii) can be proved by using q in place of p. In deriving (i) from equation (1), note that usually the momenta appear only in the kinetic energy term K of the Hamiltonian $\mathscr{H}$ and that K is a homogeneous quadratic function of momenta. Accordingly, one has, by Euler's theorem,

$$\sum_{i=1}^{f} p_i \frac{\partial K}{\partial p_i} = 2K, \qquad \text{and} \qquad \frac{\partial K}{\partial p_i} = \frac{\partial \mathscr{H}}{\partial p_i}. \tag{2}$$

Therefore,

$$\bar{K} = \tfrac{1}{2} \sum_{i=1}^{f} \overline{p_i \frac{\partial K}{\partial p_i}} = \tfrac{1}{2} \sum_{i=1}^{f} \overline{p_i \frac{\partial \mathscr{H}}{\partial p_i}} = \tfrac{1}{2} fkT, \qquad \text{or} \qquad \frac{1}{f} \bar{K} = \tfrac{1}{2} kT. \tag{3}$$

Specially, if $\mathscr{H} = \sum_i \frac{1}{2} a_i(q) p_i^2 + U(q)$ we have

$$\tfrac{1}{2} \overline{a_i p_i^2} = \tfrac{1}{2} \overline{p_i \frac{\partial \mathscr{H}}{\partial p_i}} = \tfrac{1}{2} kT.$$

2. Show that the electric polarization P of an ideal gas consisting of N diatomic molecules having a constant electric dipole moment μ is given by

$$P = \frac{N}{V} \mu \left\{ \coth\left(\frac{\mu E}{kT}\right) - \frac{kT}{\mu E} \right\},$$

where V is the volume of the gas and E is the external electric field. Prove that if $|\mu E| \ll kT$, the dielectric constant of the gas is given by

$$\varepsilon = 1 + 4\pi \frac{N}{V} \frac{\mu^2}{3\,kT}.$$

(The induced polarization in the molecules is disregarded, and the electric field acting on the molecules is assumed to be equal to the external field $\boldsymbol{E}$.)

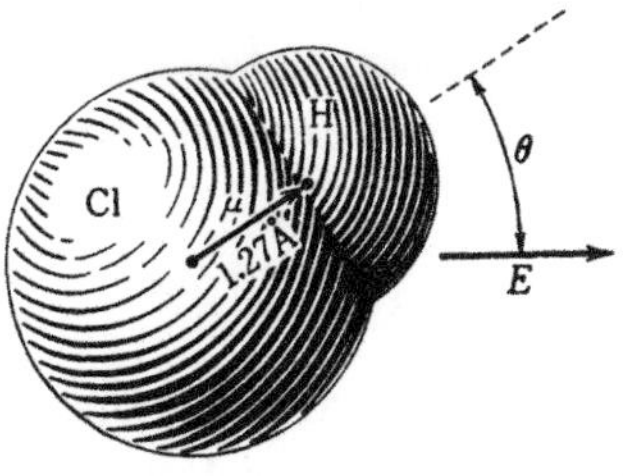

Fig. 2.2.

SOLUTION

Let the angle between the electric moment $\boldsymbol{\mu}$ and the external field $\boldsymbol{E}$ be denoted by θ. The potential energy is given by

$$u = -E\mu\cos\theta\,. \tag{1}$$

For a canonical distribution, the probability that the direction of $\boldsymbol{\mu}$ is in a solid angle element $\mathrm{d}\omega$ is given by

$$f(\theta)\,\mathrm{d}\omega = C\,\mathrm{e}^{-u/kT}\,\mathrm{d}\omega = C\exp\{E\mu\cos\theta/kT\}\,\mathrm{d}\omega\,, \tag{2}$$

where C is the normalization constant. Accordingly, the mean value of $\mu\cos\theta$ along the direction of $\boldsymbol{E}$ is given by

$$\overline{\mu\cos\theta} = \mu\int f(\theta)\cos\theta\,\mathrm{d}\omega = \mu\int\!\!\int\cos\theta\,\mathrm{e}^{E\mu\cos\theta/kT}\,\mathrm{d}\omega\Big/\int\!\!\int\mathrm{e}^{E\mu\cos\theta/kT}\,\mathrm{d}\omega\,. \tag{3}$$

Let $\mathrm{d}\omega = \sin\theta\,\mathrm{d}\theta\,\mathrm{d}\phi$, where ϕ is an azimuthal angle about the axis. The integrations over ϕ cancel each other and one has

$$\begin{aligned}\overline{\mu\cos\theta} &= \mu\int_0^\pi\cos\theta\,\mathrm{e}^{E\mu\cos\theta/kT}\sin\theta\,\mathrm{d}\theta\Big/\int_0^\pi\mathrm{e}^{E\mu\cos\theta/kT}\sin\theta\,\mathrm{d}\theta\\ &= \mu\int_{-1}^{1}\zeta\,\mathrm{e}^{(E\mu/kT)\zeta}\,\mathrm{d}\zeta\Big/\int_{-1}^{1}\mathrm{e}^{(E\mu/kT)\zeta}\,\mathrm{d}\zeta = \mu L(E\mu/kT)\,.\end{aligned} \tag{4}$$

Calculations are carried out as follows:

$$\tfrac{1}{2}\int_{-1}^{1}\mathrm{e}^{a\zeta}\,\mathrm{d}\zeta = \frac{1}{a}\sinh a\,, \qquad \tfrac{1}{2}\int_{-1}^{1}\zeta\,\mathrm{e}^{a\zeta}\,\mathrm{d}\zeta = \frac{\mathrm{d}}{\mathrm{d}a}\,\tfrac{1}{2}\int_{-1}^{1}\mathrm{e}^{a\zeta}\,\mathrm{d}\zeta\,,$$

or

$$\int_{-1}^{1}\zeta\,\mathrm{e}^{a\zeta}\,\mathrm{d}\zeta\Big/\int_{-1}^{1}\mathrm{e}^{a\zeta}\,\mathrm{d}\zeta = \frac{\mathrm{d}}{\mathrm{d}a}\log\left(\frac{1}{a}\sinh a\right) = \coth a - \frac{1}{a} \equiv L(a)\,. \tag{5}$$

Since the electric polarization P per unit volume is equal to the electric dipole moment of N/V molecules,

$$P = \frac{N}{V}\mu L\left(\frac{E\mu}{kT}\right). \tag{6}$$

In particular, if $E\mu \ll kT$, $L(a)$ can be expanded for $a \ll 1$:

$$L(a) = \frac{a}{3} - \frac{a^3}{45} + \ldots. \tag{7}$$

Taking only the first term proportional to E, one has $P = N\mu^2E/(3kTV)$. Since the electric displacement D and the dielectric constant ε are related by

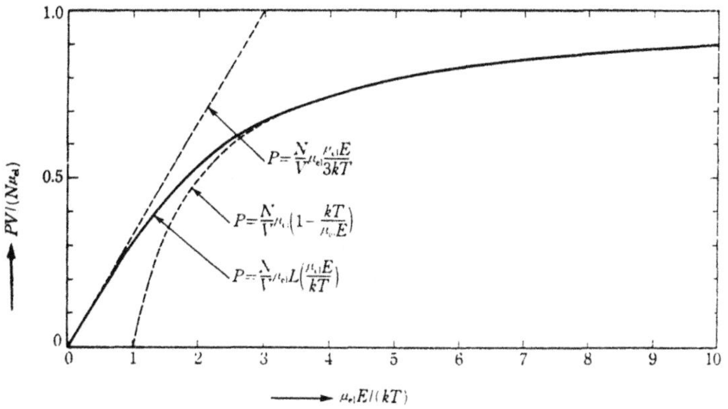

Fig. 2.3.

$$D = \varepsilon E = E + 4\pi P$$

$$\varepsilon = 1 + 4\pi \frac{N}{V} \cdot \frac{\mu^2}{3kT}.$$

ALTERNATIVE SOLUTION

In the orthodox method, the partition function Z is evaluated as

$$Z = Z_1^N / N! \tag{8}$$

where Z_1 is the partition function for a single molecule,

$$Z_1 = Z_{\text{trans}} \cdot Z_{\text{rot}} . \tag{9}$$

$Z_{\text{trans}} = V(2\pi mkT/h^2)^{\frac{3}{2}}$ is the contribution from translational motions while Z_{rot} is that from rotational motions. Since the Hamiltonian is given by

$$\mathscr{H}_{\text{rot}} = \frac{p_\theta^2}{2I} + \frac{p_\phi^2}{2I\sin^2\theta} + u(\theta) \tag{10}$$

(where θ and ϕ, the polar coordinates about the direction of $\boldsymbol{E}$ as an axis, specify the direction of the molecular axis or $\boldsymbol{\mu}$), one has

$$Z_{\text{rot}} = \int_0^{2\pi} \mathrm{d}\phi \int_0^{\pi} \mathrm{d}\theta \int_{-\infty}^{\infty} \int_{-\infty}^{\infty} \frac{\mathrm{d}p_\theta \, \mathrm{d}p_\phi}{h^2} \mathrm{e}^{-\mathscr{H}_{\text{rot}}/kT}$$

$$= \frac{IkT}{\hbar^2} \int_0^{\pi} \mathrm{e}^{\mu E \cos\theta / kT} \sin\theta \, \mathrm{d}\theta = \frac{2IkT}{\hbar^2} \sinh \frac{\mu E}{kT} \bigg/ \frac{\mu E}{kT} \tag{11}$$

($\hbar = h/2\pi$). On introducing this relation into (8), calculating $F = -kT \log Z$, and evaluating P by means of $VP = -(\partial F/\partial E)_{T,V,N}$, one is led to (6). The rest of the solution is omitted [P. Debye: *Polare Molekulen* (Hirzel, Leipzig, 1929)].

NOTE 1: Note that in this orthodox calculation, the correct weight $\sin\theta$ for θ, ϕ comes from the integral over p_θ and p_ϕ involving the rotational kinetic energy. This is general.

NOTE 2: For an HCl molecule, the internuclear distance $a = 1.27$ Å and the electric dipole moment $\mu = 1.03 \times 10^{-18}$ e.s.u. If the molecule has the structure H^+Cl^-, the electronic charge $e = 4.802 \times 10^{-10}$ e.s.u. leads to $\mu = ea = 6 \times 10^{-18}$ e.s.u. The observed moment is much smaller, because the HCl bond is largely covalent in character. In general, the electric dipole moments of molecules are of the order of 10^{-18} e.s.u. or 1 Debye unit. In an electric field of 1 e.s.u. = 300 V/cm at $T = 300°$ K, $\mu E/kT$ is of the order of 10^{-4}.

3. Atoms in a solid vibrate about their respective equilibrium positions with small amplitudes. Debye approximated the normal vibrations with the elastic vibrations of an isotropic continuous body and assumed that the number of vibrational mode $g(\omega)\,d\omega$ having angular frequencies between ω and $\omega + d\omega$ is given by

$$g(\omega) = \frac{V}{2\pi^2}\left(\frac{1}{c_l^3} + \frac{2}{c_t^3}\right)\omega^2 \equiv \frac{9N}{\omega_D^3}\omega^2 \qquad (\omega < \omega_D)$$
$$= 0 \qquad (\omega > \omega_D).$$

Here, c_l and c_t denote the velocities of longitudinal and transverse waves, respectively. The Debye frequency ω_D is determined by

$$\int_0^{\omega_D} g(\omega)\,d\omega = 3N$$

where N is the number of atoms and hence $3N$ is the number of degrees of freedom. Calculate the specific heat of a solid at constant volume with this model. Examine its temperature dependence at high as well as at low temperatures.

SOLUTION

The partition function for a harmonic oscillator having an angular frequency ω_j is given by

$$Z_j = \sum_{n=0}^{\infty} e^{-(n+\frac{1}{2})\hbar\omega_j/kT} = \left[2\sinh\frac{\hbar\omega_j}{2kT}\right]^{-1} \tag{1}$$

(see Chapter 1, problem 15). The partition function for a system of independent harmonic oscillators having various characteristic frequencies is given by $Z = \prod Z_j$, where the oscillators are numbered by j. Accordingly, the Helmholtz free energy is given by

$$F = -kT\log Z = kT\sum_j \log\left(2\sinh\frac{\hbar\omega_j}{2kT}\right). \tag{2}$$

Let the number of oscillators whose angular frequencies lie between ω and $\omega + \mathrm{d}\omega$ be denoted by $g(\omega)\,\mathrm{d}\omega$. Then,

$$F(T, V, N) = kT\int_0^\infty \log\left(2\sinh\frac{\hbar\omega}{2kT}\right)g(\omega)\,\mathrm{d}\omega \tag{3}$$

where V is the volume of a crystal: N is the number of atoms in the crystal. The internal energy is given by

$$U = -T^2\left[\frac{\partial}{\partial T}\left(\frac{F}{T}\right)\right]_{V,N} = \int_0^\infty \varepsilon(\omega, T)g(\omega)\,\mathrm{d}\omega \tag{4}$$

where $\varepsilon(\omega, T)$ is the mean energy of harmonic oscillators having an angular frequency ω:

$$\varepsilon(\omega, T) = \frac{\hbar\omega}{2}\coth\frac{\hbar\omega}{2kT} = \frac{\hbar\omega}{2} + \frac{\hbar\omega}{e^{\hbar\omega/kT} - 1}. \tag{5}$$

Accordingly, the heat capacity at constant volume is given by

$$C_V = \left(\frac{\partial U}{\partial T}\right)_{V,N} = k\int_0^\infty \frac{e^{\hbar\omega/kT}}{\{e^{\hbar\omega/kT} - 1\}^2}\left(\frac{\hbar\omega}{kT}\right)^2 g(\omega)\,\mathrm{d}\omega. \tag{6}$$

When this result is divided by the mass of the crystal, it gives the specific heat.

Introducing into this equation the previous relation for $g(\omega)$ on the basis of Debye's approximation, one has

$$C_V = 3Nk\cdot\frac{3}{\omega_D^3}\cdot\int_0^{\omega_D} \frac{e^{\hbar\omega/kT}}{\{e^{\hbar\omega/kT} - 1\}^2}\left(\frac{\hbar\omega}{kT}\right)^2\cdot\omega^2\,\mathrm{d}\omega$$

$$= 3Nk\cdot 3\left(\frac{T}{\Theta_D}\right)^3\int_0^{\Theta_D/T}\frac{\xi^4 e^\xi\,\mathrm{d}\xi}{(e^\xi - 1)^2} = 3Nkf\left(\frac{\Theta_D}{T}\right). \tag{7}$$

Here, $\hbar\omega/kT = \xi$. Furthermore,

$$\Theta_D = \hbar\omega_D/k \tag{8}$$

is called Debye's characteristic temperature or the Debye temperature. The integral (7) can be evaluated as a function of Θ_D/T.

In particular, if $T \gg \Theta_D$, $\Theta_D/T \ll 1$ and $\xi \ll 1$. Therefore, the integrand in (7), $\xi^4 e^\xi/(e^\xi - 1)^2 \sim \xi^2$, can be approximated by ξ^2. Integration yields

$$C_V \simeq 3Nk \cdot 3\left(\frac{T}{\Theta_D}\right)^3 \int_0^{\Theta_D/T} \xi^2 \, d\xi = 3Nk. \tag{9}$$

This is the classical value: the relation is nothing but Dulong-Petit's law. At low temperatures, $T \ll \Theta_D$. The upper limit of the integral (7) is very large and may be replaced by infinity:

$$C_V \simeq 3Nk \cdot 3\left(\frac{T}{\Theta_D}\right)^3 \int_0^\infty \frac{\xi^4 e^\xi}{(e^\xi - 1)^2} d\xi = 3Nk \cdot A\left(\frac{T}{\Theta_D}\right)^3. \tag{10}$$

In other words, C_V is proportional to T^3. The constant A is given by

$$A = 3\int_0^\infty \frac{\xi^4 e^\xi}{(e^\xi - 1)^2} d\xi = \tfrac{4}{5}\pi^4. \tag{11}$$

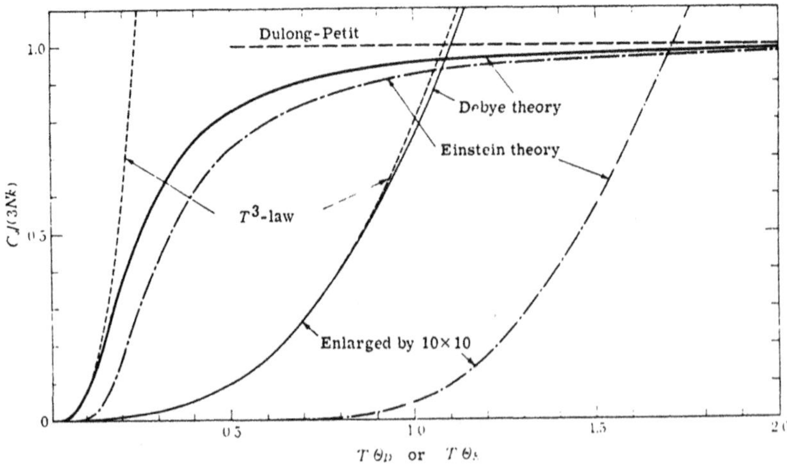

Fig. 2.4.

NOTE 1: Numerical tables are available for the function $f(x)$ in (7). [Fowler-Guggenheim: *Statistical Thermodynamics* (Cambridge 1939, p. 144).] The curve is shown in Fig. 2.4. At temperatures below 0.083 Θ_D the errors involved in equation (10) amount to less than 1 %.

NOTE 2: In calculating (11), $e^{\xi}/(e^{\xi}-1)^2$ is expanded in a power series of $e^{-\xi}$ and each term is integrated. A formula for ζ-functions is applied to the resulting series:

$$\int_0^\infty \frac{\xi^s e^\xi}{(e^\xi - 1)^2} d\xi = s! \sum_{n=1}^\infty \frac{1}{n^s} = s!\zeta(s).$$

$\zeta(2) = \frac{1}{6}\pi^2$, $\zeta(3) = 1.202$, $\zeta(4) = \pi^4/90$ (Jahnke-Emde, *Funktionentafeln*, p. 319).

NOTE 3: In the Debye approximation, $g(\omega)$ is derived in the following way. For longitudinal waves, a relation $\omega = c_l q$ holds between the frequency ω and the wave number q. For a crystal of cubic shape and size L^3, the number of vibrational modes having their wave number vector $\boldsymbol{q}$ in a volume element $d\boldsymbol{q} = dq_x dq_y dq_z$ is equal to $(L^3/8\pi^3)\, dq_x dq_y dq_z$ (see Chapter 1, problem 6). Therefore, the number of modes lying between $q = |\boldsymbol{q}|$ and $q + dq$ is given by

$$\iint_{q<\sqrt{q_x^2+q_y^2+q_z^2}<q+dq} \frac{L^3}{8\pi^3} dq_x dq_y dq_z = \frac{L^3}{8\pi^3} \cdot 4\pi q^2 dq = \frac{V}{2\pi^2} q^2 dq. \qquad (12)$$

Rewriting this in terms of ω, one has $(V/2\pi^2 c_l^3)\,\omega^2 d\omega$ for the number of vibrational modes lying between ω and $\omega + d\omega$. A similar relation holds for transverse vibrations, except that for each $\boldsymbol{q}$ two planes of polarization (directions of vibration) exist. Consequently, the number of modes of transverse vibrations lying between ω and $\omega + d\omega$ is given by $2(V/2\pi^2 c_t^3)\omega^2\, d\omega$.

In total, one has $g(\omega)$ given above. For N atoms, the number of mechanical degrees of freedom is $3N$ and the number of vibrational modes of a crystal lattice is $3N - 6 \sim 3N$ (6 for translational and rotational motions of a crystal as a whole). In order to have agreement between these numbers, the distribution $g(\omega)$ is assumed to terminate with ω_D. For rigorous calculations, the equations for vibration must be solved by taking into account the specific structure of the crystal lattice. The results show that for small ω the curve resembles that resulting from Debye's approximation and that for large ω a discrepancy appears (Fig. 2.5). However, the discrepancy is not appreciable for averaged quantities such as specific heat: for this, Debye's approximation agrees well with observations. Θ_D is constant for a given substance.

NOTE 4: Einstein did not consider the distribution $g(\omega)$ but assumed that all vibrational modes have one and the same frequency ω_E. Therefore, he obtained

$$C_V = 3Nk \frac{e^{\Theta_E/T}}{(e^{\Theta_E/T} - 1)^2} \left(\frac{\Theta_E}{T}\right)$$

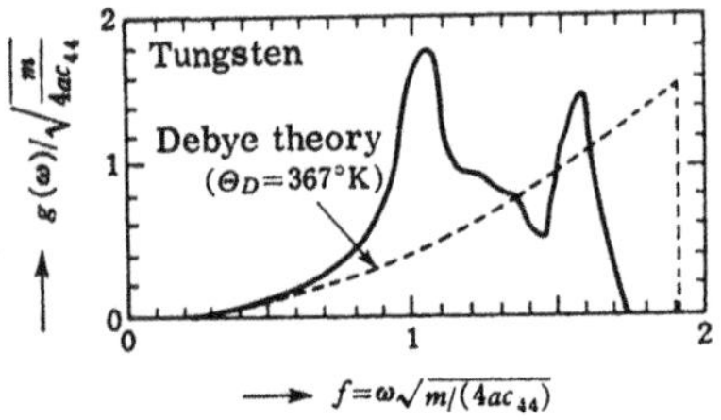

Fig. 2.5. (P.C. Fine, Phys. Rev. **56** (1939) 355.)

($\Theta_E = \hbar\omega_E/k$). This approximation gives poor agreement with experiments especially at low temperatures (Fig. 2.4).

4. Electromagnetic waves in a box of volume V are in thermal equilibrium with their environment at a temperature T. Derive the relation between the intensity of radiation leaking from a small hole bored through the wall of the box and the wavelength (Planck's radiation formula).

SOLUTION

Like elastic waves in a crystal (example 3), electromagnetic waves in a box constitute a system of independent harmonic oscillators. However, in contrast to the case of elastic waves, longitudinal electromagnetic waves do not exist. Let the number of oscillators having their angular frequencies between ω and $\omega + d\omega$ be denoted by $Vg(\omega)\, d\omega$. In the same manner as used in obtaining equation (12) of example 3, one finds

$$g(\omega) = \frac{\omega^2}{\pi^2 c^3} \qquad \text{(per unit volume)} \tag{1}$$

(the two directions of polarization introduce a factor 2), where c is the velocity of light. Disregarding the zero-point energy, the mean energy of oscillators in thermal equilibrium at T is given by

$$\varepsilon(\omega, T) = \frac{\hbar\omega}{e^{\hbar\omega/kT} - 1}. \tag{2}$$

Hence, the energy (per unit volume) of electromagnetic waves having their angular frequencies between ω and $\omega + d\omega$ is given by

$$u(\omega, T)\,d\omega = \varepsilon(\omega, T)g(\omega)\,d\omega = \frac{\hbar\omega^3}{\pi^2c^3}\frac{d\omega}{e^{\hbar\omega/kT} - 1}. \tag{3}$$

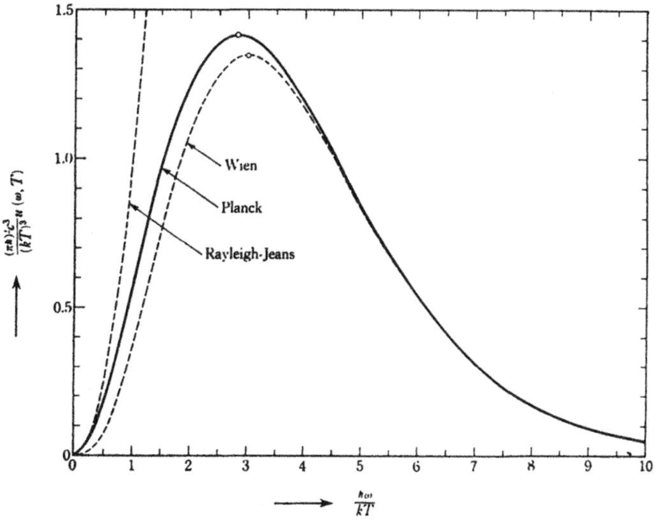

Fig. 2.6.

Rewriting this in terms of wavelength $\lambda = 2\pi c/\omega$, the energy of electromagnetic waves having their wavelengths between λ and $\lambda + d\lambda$ is given by

$$u(\lambda, T)\,d\lambda = u(\omega, T)\,d\omega = \frac{8\pi hc}{\lambda^5}\frac{d\lambda}{e^{hc/(\lambda kT)} - 1}. \tag{4}$$

Equation (3) or (4) is Planck's radiation formula.

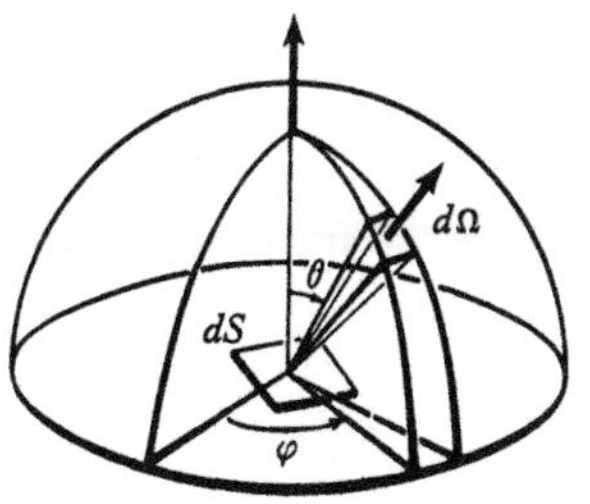

Fig. 2.7.

It is assumed that the thermal equilibrium of the electromagnetic waves is not disturbed even when a small hole is bored through the wall of the box.

The area of the hole is $\mathrm{d}S$. The energy which passes in unit time through a solid angle $\mathrm{d}\Omega$ making an angle θ with the normal to $\mathrm{d}S$ is

$$J(\lambda, T, \theta)\,\mathrm{d}\lambda\,\mathrm{d}\Omega\,\mathrm{d}S = c\mathrm{u}(\lambda, T)\,\mathrm{d}\lambda\cdot\cos\theta\,\frac{\mathrm{d}\Omega}{4\pi}\,\mathrm{d}S\,, \tag{5}$$

provided that only the contribution from electromagnetic waves having their wavelengths between λ and $\lambda + \mathrm{d}\lambda$ is taken into account.
The right hand side is divided by 4π, because the energy density u comprises all waves propagating along different directions. $J(\lambda, T, \theta)$ is a measure for the intensity of radiation. Introducing equation (4) into (5), one finds

$$J(\lambda, T, \theta) = 2\,\frac{hc^2}{\lambda^5}\,\frac{1}{\mathrm{e}^{hc/(\lambda kT)} - 1}\cos\theta\,. \tag{6}$$

NOTE 1: At high temperatures and for long wavelengths ($\lambda T \gg ch/k = 1.437$ cm deg) the exponential function in the denominator of Planck's radiation formula (4) can be expanded and the following approximate equations are obtained:

$$\mathrm{u}(\lambda, T) \cong \frac{8\pi}{\lambda^4}kT, \qquad \text{or} \qquad u(\omega, T) \cong g(\omega)kT. \tag{7}$$

Equation (7) is called Rayleigh-Jeans' radiation formula, which can be obtained if electromagnetic waves are treated as classical waves and the mean energy kT is allotted to each degree of freedom in accordance with the equipartition law. At low temperatures and for short wavelengths ($\lambda T \ll ch/k$), the 1 in the denominator of (4) can be ignored and one has

$$\mathrm{u}(\lambda, T) \cong \frac{8\pi hc}{\lambda^5}\,\mathrm{e}^{-hc/(\lambda kT)}, \qquad \text{or} \qquad u(\omega, T) \cong \hbar\omega\,\mathrm{e}^{-\hbar\omega/kT}\,g(\omega)\,. \tag{8}$$

This is Wien's radiation formula, which can be derived if electromagnetic waves are represented by an ideal gas consisting of classical particles each having an energy $\hbar\omega$.
NOTE 2: For the internal energy U of electromagnetic waves enclosed in a box, equation (3) gives

$$U = V\int_0^\infty u(\omega, T)\,\mathrm{d}\omega = V\,\frac{(kT)^4}{\pi^2c^3\hbar^3}\int_0^\infty \frac{x^3\,\mathrm{d}x}{\mathrm{e}^x - 1} = VaT^4 \tag{9}$$

where $a = \pi^2k^4/(15\,c^3\hbar^3) = 7.569 \times 10^{-15}\,\mathrm{erg\,cm^{-3}\,deg^{-4}}$. Equation (9) is

Stefan-Boltzmann's law. The integral in this equation is evaluated by means of

$$\int_0^\infty \frac{x^s\,dx}{e^x - 1} = \sum_{n=1}^\infty \int_0^\infty x^s e^{-nx}\,dx = s!\sum_{n=1}^\infty \frac{1}{n^{s+1}} = s!\zeta(s+1), \qquad (s = 1,2,3,\ldots),$$

as in NOTE 2 of the preceding example.

DIVERTISSEMENT 5

Heat radiation: The history of the theory of heat radiation has a very important and unique position in the whole history of science. Towards the end of the nineteenth century, three major fields of classical physics: mechanics, electromagnetism and thermodynamics had already been established. The problem of heat radiation was, so to say, the frontier or a touchstone where all of these fields of classical physics were focused, and in fact it disclosed all of their strengths and weaknesses. It also marked the most brilliant rise of German physics headed by such eminent physicists as Stefan, Boltzmann, Wien, Lummer and, finally, Planck.

For those of us who grew up after the birth of quantum mechanics, it may be difficult to realize fully the importance of the problem. It is still interesting to note that the great discovery of h by Max Planck was made through his study of such a macroscopic phenomena as heat radiation, rather than through more direct study of atomic phenomena which were already known at that time. In a sense, this is understandable. Statistical mechanics had to suffer from serious difficulties which were, as we very well know today, bound to classical mechanics. Classical statistical mechanics had to face serious contradictions with physical reality, because it aimed to explain nature by atomic pictures for which it had no exact idea; but classical mechanics was doomed to be overthrown by quantum theory. Readers are recommended to refer to: Max Planck, *Vorlesungen über die Theorie der Wärmestrahlung* (Barth, Leipzig); Max Planck, *Einführung zur theoretischen Physik*, Bd V., *Theorie der Wärme* (Hirzel, Leipzig).

5⁺. Derive the number of states $W(E)$ for a given total energy from the partition function $Z_N(\beta)$ of a system of N oscillators having a characteristic angular frequency ω. Calculate the entropy $S(E)$ by asymptotic calculation for large N.

SOLUTION

The partition function for a single harmonic oscillator is

$$Z_1(\beta) = \sum_{n=0}^\infty e^{-\beta(n+\frac{1}{2})\hbar\omega} = \frac{e^{-\frac{1}{2}\beta\hbar\omega}}{1 - e^{-\beta\hbar\omega}}$$

(see Chapter 1, problem 15). Accordingly, for a system of N oscillators,

$$Z_N(\beta) = Z_1(\beta)^N = e^{-\frac{1}{2}N\beta\hbar\omega}(1 - e^{-\beta\hbar\omega})^{-N}. \tag{1}$$

In order to evaluate $W(E)$ accurately, one finds from (1)

$$Z_N(\beta) = e^{-\frac{1}{2}N\beta\hbar\omega} \sum_{M=0}^{\infty} \binom{-N}{M} (-1)^M e^{-M\beta\hbar\omega}$$

$$= \sum_{M=0}^{\infty} \frac{(N+M-1)!}{(N-1)!M!} e^{-(M+\frac{1}{2}N)\beta\hbar\omega}. \tag{2}$$

Compare this with Z defined as

$$Z_N(\beta) = \int e^{-\beta E} \Omega(E)\, dE, \tag{3}$$

then

$$\Omega(E) = \frac{(N+M-1)!}{(N-1)!M!} \delta(E - [M + \tfrac{1}{2}N]\,\hbar\omega), \tag{4}$$

where δ is Dirac's delta function. Equation (4) implies that there are $W(M) = (N + M - 1)!/(N - 1)!M!$ states for each discrete energy level $E = (M + \frac{1}{2}N)\hbar\omega$ ($M = 0, 1, \ldots$). The result is identical with that obtained in Chapter 1, example 3.

From (2.6), one has

$$\Omega(E) = \frac{1}{2\pi i} \int Z_N(\beta) e^{\beta E}\, d\beta$$

$$= \frac{1}{2\pi i} \int (1 - e^{-\beta\hbar\omega})^{-N} e^{\beta(E - \frac{1}{2}N\hbar\omega)}\, d\beta. \tag{5}$$

Setting $E' = E - \frac{1}{2}N\hbar\omega$, and applying the asymptotic formulae (2.8) and (2.9) to this equation, one has

$$\log \Omega = -N \log(1 - e^{-\beta\hbar\omega}) + \beta E', \tag{6}$$

$$N \frac{\hbar\omega\, e^{-\beta\hbar\omega}}{1 - e^{-\beta\hbar\omega}} = E'. \tag{7}$$

Equation (7) yields

$$e^{-\beta\hbar\omega} = \frac{E'}{N\hbar\omega} \Big/ \left(1 + \frac{E'}{N\hbar\omega}\right), \qquad 1 - e^{-\beta\hbar\omega} = \left(1 + \frac{E'}{N\hbar\omega}\right)^{-1}$$

Substituting this into (6), one has

$$S = k\log\Omega(E) = k\left[N\log\left(1+\frac{E'}{N\hbar\omega}\right) - \frac{E'}{\hbar\omega}\log\left\{\frac{E'}{N\hbar\omega}\Big/\left(1+\frac{E'}{N\hbar\omega}\right)\right\}\right]$$

$$= kN\left[\left(1+\frac{E'}{N\hbar\omega}\right)\log\left(1+\frac{E'}{N\hbar\omega}\right) - \frac{E'}{N\hbar\omega}\log\frac{E'}{N\hbar\omega}\right]$$

$$= kN\left[\left(\frac{E}{N\hbar\omega}+\frac{1}{2}\right)\log\left(\frac{E}{N\hbar\omega}+\frac{1}{2}\right) - \left(\frac{E}{N\hbar\omega}-\frac{1}{2}\right)\log\left(\frac{E}{N\hbar\omega}-\frac{1}{2}\right)\right].$$

The result is identical with that obtained by applying Stirling's formula to $k\log W(M)$ (Chapter 1, example 3).

NOTE 1: For $s' > 0$ the following relation holds:

$$\frac{1}{2\pi i}\int_{s'-i\infty}^{s'+i\infty} \frac{e^{sx}}{s}\,dx = \begin{cases} 1 & (x > 0) \\ 0 & (x < 0). \end{cases}$$

Therefore, at least formally,

$$\frac{1}{2\pi i}\int_{s'-i\infty}^{s'+i\infty} e^{sx}\,dx = \delta(x). \tag{8}$$

On introducing the expansion formula (2) into (5) and applying the relation (8) to each term, one is led to (4) (although mathematically this is not rigorous).

Putting $e^{-\beta\hbar\omega} = x$, one has from (2)

$$W_M \equiv \frac{(N+M-1)!}{(N-1)!M!}$$

$$= \text{the coefficient of } x^M \text{ in the expansion of } (1-x)^{-N}.$$

NOTE 2: *Fowler-Darwin's method*: In general, if the energies of a system are integral multiples of a unit ε, the partition function can be defined as

$$f(x) = \sum W_M e^{-\beta M\varepsilon} \equiv \sum W_M x^M \qquad (x = e^{-\beta\varepsilon}). \tag{9}$$

Integration in the complex plane around the origin $x = 0$ in conformance with Cauchy's theorem,

$$W_M = \frac{1}{2\pi i}\int_C f(x)\frac{dx}{x^{M+1}} = \left(\frac{d}{dx}\right)^M f(x)\Big|_{x=0}, \tag{10}$$

gives the number of states W_M for $E = M\varepsilon$. For $\log f(x) = O(N)$ and $N \gg 1$, the asymptotic evaluation is carried out as follows: x^* (usually real and positive) is determined by

$$g(x) = \log f(x) - M \log x, \qquad g'(x^*) = \frac{\mathrm{d}}{\mathrm{d}x^*} \log f(x^*) - \frac{M}{x^*} = 0. \quad (11)$$

Equation (10) is integrated along a circle passing through x^*. On this path, the integrand has a sharp maximum at x^* ($N \gg 1$). Accordingly, as in § 1.2, the integration is approximated near this point. Then,

$$W_M = \frac{1}{2\pi} \int \exp\left[g(x^*) + \tfrac{1}{2} g''(x^*)(x^* \mathrm{e}^{\mathrm{i}\varphi} - x^*)^2 + \ldots\right] \mathrm{d}\varphi$$

$$\sim \frac{1}{2\pi} \int \exp\left[g(x^*) - \tfrac{1}{2} x^{*2} g''(x^*) \varphi^2\right] \mathrm{d}\varphi = \frac{1}{x^*[2\pi g''(x^*)]^{\frac{1}{2}}} \mathrm{e}^{g(x^*)},$$

or

$$\log W_M \sim g(x^*) + \mathrm{O}(\log N).$$

x^* is fixed by (11). Considering that $x^* = \mathrm{e}^{-\beta^* \varepsilon}$, the results are identical with (2.8) and (2.9). Instead of using Stirling's formula, Fowler and Darwin carried out asymptotic evaluations along this line for actual calculations. At the same time, they attempted to systematize statistical mechanics.

6⁺. For the system of N electric dipoles considered in example 2, the component of the resultant moment $\boldsymbol{M} = \sum_{i=1}^{N} \boldsymbol{\mu}_i$ along an arbitrary direction z is specified between M_z and $M_z + \mathrm{d}M_z$. The corresponding phase volume (number of states) is $\Omega(M_z)\,\mathrm{d}M_z$. Show that the following asymptotic estimation holds for large N:

$$\log \Omega(M_z) \simeq N\left[\text{const.} + \log(\sinh x/x) - \xi x\right],$$

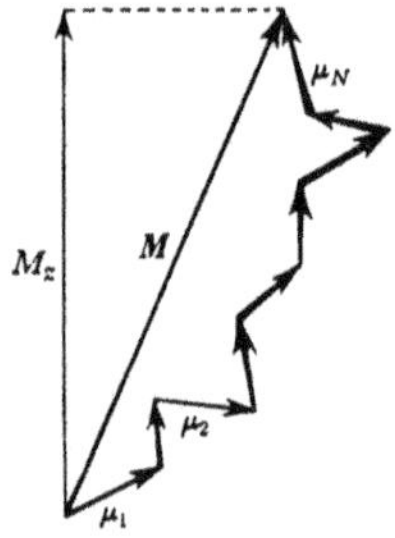

Fig. 2.8.

where $M_z/N\mu = \xi$ and $x = L^{-1}(\xi)$. (L is Langevin's functions; L^{-1} is its inverse function.) In particular, in the range $M_z/N\mu \ll 1$, $\Omega(M_z) \propto$ const. $\exp\left[-3M_z^2/(2N\mu^2)\right]$. Derive this approximate relation.

SOLUTION

The partition function Z evaluated in example 2 (2nd solution) is

$$Z = \int \exp(\mathscr{E}M_z/kT)\,\Omega(M_z)\,\mathrm{d}M_z, \tag{1}$$

where $\mathscr{E}$ is the field strength and M_z is the component of $\boldsymbol{M}$ along the field direction, because M_z appears only in the potential energy term $-\mathscr{E}M_z$ in the total energy. The asymptotic estimation valid for large N is given by the Legendre transformation of the thermodynamic function $\log Z(\mathscr{E})$ having $\mathscr{E}$ as its independent variable into another thermodynamic function $\log \Omega(M_z)$ having M_z as its independent variable.

$$\log \Omega(M_z) = \log Z(\mathscr{E}) - M_z\mathscr{E}/kT, \tag{2}$$

$$\partial \log Z(\mathscr{E})/\partial \mathscr{E} = M_z/kT\,; \tag{3}$$

here $\mathscr{E}$ is evaluated from (3) as a function of M_z and is introduced into the right-hand side of (2).

As given by (8), (9) and (11) in the alternative solution to example 2, the partition function has the following form:

$$Z = \frac{\alpha^N}{N!}\left(\sinh\frac{\mu\mathscr{E}}{kT}\Big/\frac{\mu\mathscr{E}}{kT}\right)^N \tag{4}$$

(α is independent of $\mathscr{E}$). Like (4) in example 2, (3) is given by

$$M_z/(N\mu) = L(\mathscr{E}\mu/kT) \tag{5}$$

On the other hand, (2) can be written as

$$\log\Omega(M_z) = \text{const.} + N\log\left(\sinh\frac{\mu\mathscr{E}}{kT}\Big/\frac{\mu\mathscr{E}}{kT}\right) - N\frac{M_z}{N\mu}\cdot\frac{\mathscr{E}\mu}{kT}\,. \tag{6}$$

Setting $\mu\mathscr{E}/kT = x$ and $M_z/N\mu = \xi$, one has (5) and (6) in the form already given.

For $\xi \ll 1$, $L(x) \fallingdotseq \frac{1}{3}x$, $x = L^{-1}(\xi) \fallingdotseq 3\xi$, and therefore

$$\sinh x/x = 1 + \tfrac{1}{6}x^2 + \ldots = 1 + \tfrac{3}{2}\xi^2 + \ldots\,.$$

From this, one has

$$\begin{aligned}\log\Omega(M_z) &= \text{const.} + N\left[\log(1 + \tfrac{3}{2}\xi^2 + \ldots) - 3\xi^2\right] \fallingdotseq \text{const.} - \tfrac{3}{2}N\xi^2 \\ &= \text{const.} - 3M_z^2/(2N\mu^2)\,.\end{aligned}$$

NOTE: The asymptotic estimation by the saddle point method is strictly carried out in the following way: By setting $M_z/\mu = X$ and $\mu\mathscr{E}/kT = x$, (1) is transformed into

$$Z(x) = \int e^{xX}\Omega(X)\,dX\,. \tag{7}$$

By inverse transformation we have

$$\Omega(X) = \frac{1}{2\pi i}\int_{x^*-i\infty}^{x^*+i\infty} Z(x)e^{-xX}\,dx = \frac{1}{2\pi i}\int_{x^*-i\infty}^{x^*+i\infty} \exp\left[\log Z(x) - xX\right]dx\,. \tag{8}$$

Now x^*, which is real in the convergence region of (7), is chosen as a saddle point for $\log Z(x) - xX$. The condition is

$$\frac{\partial \log Z(x^*)}{\partial x^*} = X\,, \qquad \text{or} \qquad L(x^*) = X/N \equiv \xi\,.$$

The exponent in (8) is written as $\log Z(x^*) - x^*X + \frac{1}{2}(\partial^2 \log Z/\partial x^2)_{x=x^*}\cdot (x-x^*)^2 + \ldots$ and $x = x^* + iy$. With this approximation, (8) is integrated over y from $-\infty$ to $+\infty$:

$$\Omega(X) \sim Z(x^*)e^{-x^*X}\left[2\pi\frac{\partial^2 \log Z(x^*)}{\partial x^{*2}}\right]^{-\frac{1}{2}};$$

therefore,

$$\log\Omega = \log Z(x^*) - x^*X + o(N) = \text{const.} + N\log(\sinh x^*/x^*) - x^*X + o(N)\,.$$

7$^+$. Give the coordinate representation of the density matrix $\rho = \exp(-\beta\mathscr{H})$ ($\mathscr{H} = p^2/2m$) for a single free particle of mass m. Calculate its limiting value for $L \to \infty$, assuming that the wave function of the particle is periodic in a cube of volume L^3. (The wave function $\psi(x, y, z)$ satisfies the boundary condition that $\psi(x+L, y, z) = \psi(x, y, z)$ and similar conditions for y and z.)

SOLUTION

For a free particle, the wave equation $\mathscr{H}\varphi_E = E\varphi_E$ takes the following form:

$$-\frac{\hbar^2}{2m}\left(\frac{\partial^2}{\partial x^2} + \frac{\partial^2}{\partial y^2} + \frac{\partial^2}{\partial z^2}\right)\varphi_E = E\varphi_E\,. \tag{1}$$

Hence the solution has the form $\varphi_E(x, y, z) = C\exp\left[i\,(k_x x + k_y y + k_z z)\right]$. In order to satisfy the boundary conditions, one must have

$$k_x L = 2\pi n_x \qquad (n_x \text{ is an integer. Similar relations hold for } y \text{ and } z). \tag{2}$$

The energy is given by

$$E(n_x, n_y, n_z) \equiv E_k = \frac{\hbar^2 k^2}{2m} = \frac{h^2}{2mL^2}(n_x^2 + n_y^2 + n_z^2). \tag{3}$$

By definition, the coordinate representation of the density matrix ρ (2.41) is

$$\langle \boldsymbol{r} | \rho | \boldsymbol{r}' \rangle = \sum_E \langle \boldsymbol{r} | E \rangle \mathrm{e}^{-\beta E} \langle E | \boldsymbol{r}' \rangle = \sum_E \varphi_E(\boldsymbol{r}) \mathrm{e}^{-\beta E} \varphi_E^*(\boldsymbol{r}')$$

(here ρ is not normalized). Substituting the aforementioned eigenvalues and eigenfunctions into this equation, one is led to

$$\langle \boldsymbol{r} | \rho | \boldsymbol{r}' \rangle = \sum_{n_x=-\infty}^{\infty} \sum_{n_y=-\infty}^{\infty} \sum_{n_z=-\infty}^{\infty} \frac{1}{L^3} \exp\left[-\frac{\beta h^2}{2mL^2}(n_x^2 + n_y^2 + n_z^2) + \frac{2\pi \mathrm{i}}{L}\{n_x(x - x') + n_y(y - y') + n_z(z - z')\} \right]. \tag{4}$$

The coefficient $1/L^3$ is introduced in order to normalize the wave function for volume L^3. Equation (4) can now be factored into three parts involving x, y and z, respectively.

$$\langle \boldsymbol{r} | \rho | \boldsymbol{r}' \rangle = \langle x | \rho_x | x' \rangle \langle y | \rho_y | y' \rangle \langle z | \rho_z | z' \rangle. \tag{5}$$

Here

$$\langle x | \rho_x | x' \rangle = \sum_{n_x=-\infty}^{\infty} \frac{1}{L} \exp\left[-\frac{\beta h^2}{2mL^2} n_x^2 + \frac{2\pi \mathrm{i}}{L} n_x(x - x') \right]. \tag{6}$$

In the limit of $L \to \infty$,

$$\sum_{n_x=-\infty}^{\infty} \frac{1}{L} \to \frac{1}{2\pi} \int_{-\infty}^{\infty} \mathrm{d}k_x, \qquad \left(k_x = \frac{2\pi}{L} n_x \right). \tag{7}$$

With this substitution, (6) can be transformed into

$$\langle x | \rho_x | x' \rangle = \frac{1}{2\pi} \int_{-\infty}^{\infty} \mathrm{d}k_x \exp\left[-\frac{\beta \hbar^2}{2m} k_x^2 + \mathrm{i}k_x(x - x') \right]$$

$$= \left(\frac{m}{2\pi\beta\hbar^2} \right)^{\frac{1}{2}} \exp\left\{ -\frac{m}{2\beta\hbar^2}(x - x')^2 \right\}. \tag{8}$$

Since similar relations hold for y and z also, (4) yields

$$\langle \boldsymbol{r} | \rho | \boldsymbol{r}' \rangle = \left(\frac{m}{2\pi\beta\hbar^2} \right)^{\frac{3}{2}} \exp\left[-\frac{m}{2\beta\hbar^2}(\boldsymbol{r} - \boldsymbol{r}')^2 \right]. \tag{9}$$

NOTE: The integral (8) can be evaluated by

$$\int_{-\infty}^{\infty} \exp[-\tfrac{1}{2}ax^2 + ixy]\,dx =$$

$$= \int_{-\infty}^{\infty} \exp\left[-\tfrac{1}{2}a\left(x - \frac{iy}{a}\right)^2 - \frac{y^2}{2a}\right]dx = \sqrt{\frac{2\pi}{a}}\,e^{-y^2/2\omega}. \tag{10}$$

The relation:

$$\int_{-\infty}^{\infty} e^{-\frac{1}{2}ax^2}\,dx = \int_{-\infty}^{\infty} e^{-\frac{1}{2}a(x-i\alpha)^2}\,dx$$
$$= \sqrt{2\pi/a} \tag{11}$$

can be proved by integrating $\exp(-\frac{1}{2}az^2)$ along a closed path shown in Figure 2.9 (the integral vanishes) and letting $X \to \infty$. The portions of the integral over $-X - ia \to -X$ and $X - ia \to X$ vanish.

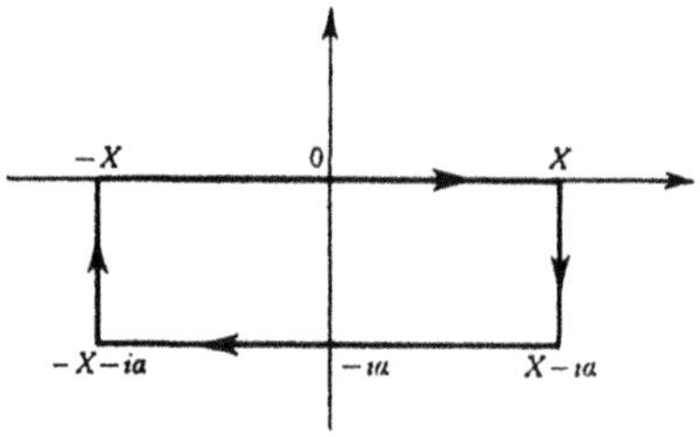

Fig. 2.9.

Problems

[A]

1. Show that the energy fluctuation in a canonical distribution is given by

$$\overline{(E - \bar{E})^2} = kT^2 C_V$$

where T is the absolute temperature and C_V is the heat capacity at constant volume. Prove the following relation in a similar manner:

$$\overline{(E - \bar{E})^3} = k^2\left\{T^4\left(\frac{\partial C_V}{\partial T}\right)_V + 2T^3 C_V\right\}.$$

Show that, in particular, for an ideal gas consisting of N monatomic molecules (disregard the internal structure) these equations can be reduced to

$$\frac{\overline{(E-\bar{E})^2}}{\bar{E}^2}=\frac{2}{3N}, \qquad \frac{\overline{(E-\bar{E})^3}}{\bar{E}^3}=\frac{8}{9N^2}.$$

2. A weight of mass m is fixed to the middle point of a string of length l as shown in the figure and rotates about an axis joining the ends of the string. The system is in contact with its environment at a temperature T. Calculate the tension X acting between the ends of the string in terms of its dependence upon the distance x between the ends.

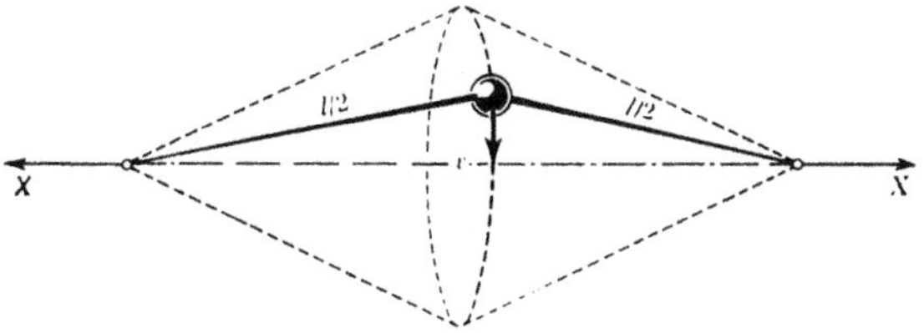

3. An ideal gas consisting of N particles of mass m (classical statistics being obeyed) is enclosed in an infinitely tall cylindrical container placed in a uniform gravitational field, and is in thermal equilibrium. Calculate the classical partition function, Helmholtz free energy, mean energy and heat capacity of this system (cf. Chapter 1, problem 15).

4. A cylinder of radius R and length L rotates about its axis with a constant angular velocity ω. Evaluate the density distribution of an ideal gas enclosed in the cylinder. Ignore the effect of gravitation. Carry out classical calculations assuming that thermal equilibrium is established at T. (*Hint:* The Hamiltonian that describes the motion in a rotating coordinate system is $H^* = H - \omega L$, where H is the Hamiltonian in the coordinate system at rest and L denotes the angular momentum. Apply the canonical distribution for H^*.)

5. In special relativity, the momentum components and the energy of a mass point m are given by

$$p_i=\frac{mv_i}{\sqrt{1-(v/c)^2}} \qquad (i=x,y,z),$$

and

$$\varepsilon=\frac{mc^2}{\sqrt{1-(v/c)^2}}$$

respectively, where c is the velocity of light; $v = \sqrt{(v_x^2 + v_y^2 + v_z^2)}$ is the velocity; and v_x, v_y and v_z are the velocity components of the mass point. Show that the Maxwell-Boltzmann distribution yields

$$\overline{\tfrac{1}{2} m v_i^2 / \sqrt{1-(v/c)^2}} = \tfrac{1}{2} kT.$$

(*Hint:* Rewrite ε in terms of momenta and apply the results of example 1.)

6. Generalized coordinates specifying the state of a system having $3N$ degrees of freedom are denoted by $q_1, q_2, \ldots, q_{3N}$. The force corresponding to a coordinate q_j is X_j. (When the Hamiltonian of the system is expressed by H, $X_j = -\partial H/\partial q_j$.) Show that

$$\overline{\sum_{j=1}^{3N} q_j X_j} = -3NkT$$

where T is the absolute temperature. This is the virial theorem.

In particular, when a gas made up of N molecules having an intermolecular interaction potential $U(q_1, q_2, \ldots, q_{3N})$ is enclosed in a vessel of volume V, the virial theorem takes the following form:

$$pV = NkT - \tfrac{1}{3} \overline{\sum_{j=1}^{3N} q_j \frac{\partial U}{\partial q_j}}$$

where p is the pressure exerted by the gas molecules on the wall of the vessel. Here, $q_1, q_2, \ldots, q_{3N}$ are Cartesian coordinates specifying the coordinates of these N molecules.

7. An evacuated box of volume V is in contact with a heat bath at T. Show that the pressure exerted on the vessel wall by thermal radiation prevailing in this empty box is equal to $\frac{1}{3}$ of the energy density u of thermal radiation.

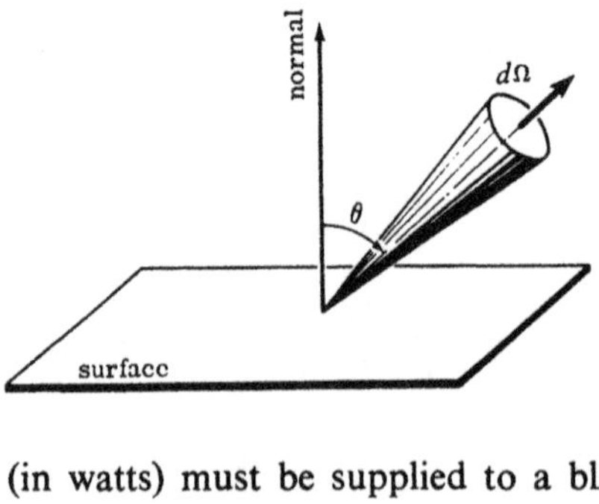

8. How much energy (in watts) must be supplied to a black body of 1 cm^3 placed in an evacuated vessel to maintain it at 500, 800 and 1000° K, respectively? It is assumed that the wall of the vessel is made of a black body at 300° K.

NOTE: A black body (perfect black body) absorbs all radiation incident on it. The energy emitted from a unit area of its surface in unit time through a solid angle $d\Omega$ making an angle θ with the normal to the surface is given by

$$J(\omega, T, \theta)\, d\omega\, d\Omega = c \cdot u(\omega, T)\, d\omega \cdot \cos\theta\, d\Omega / 4\pi .$$

Here, c is the velocity of light; T is the absolute temperature of the black body; and $u(\omega, T)$ is the energy density defined in the solution to example 4.

9. Suppose a magnetic moment is capable of assuming any one of the discrete values $g\mu_B m$ (m = magnetic quantum number $= J, J-1, J-2, \dots -J+1, -J$) as its component along the direction of a magnetic field H. Calculate the magnetization M of a body containing n such magnetic moments per unit volume. Evaluate the magnetic susceptibility χ for a weak field at high temperature ($g\mu_B JH \ll kT$). Examine the cases when $J = \frac{1}{2}$ and $J \to \infty$ ($\mu_B \to 0$ and $g\mu_B J \to \mu_0$). Assume that the interactions between the magnetic moments are negligible.

10. A surface having N_0 adsorption centers has $N (\leqslant N_0)$ gas molecules adsorbed on it. Show that the chemical potential μ of the molecules is given by

$$\mu = kT\left(\log \frac{N}{N_0 - N} - \log a(T)\right).$$

Disregard interactions between the adsorbed molecules. $a(T)$ is the partition function of a single adsorbed molecule.

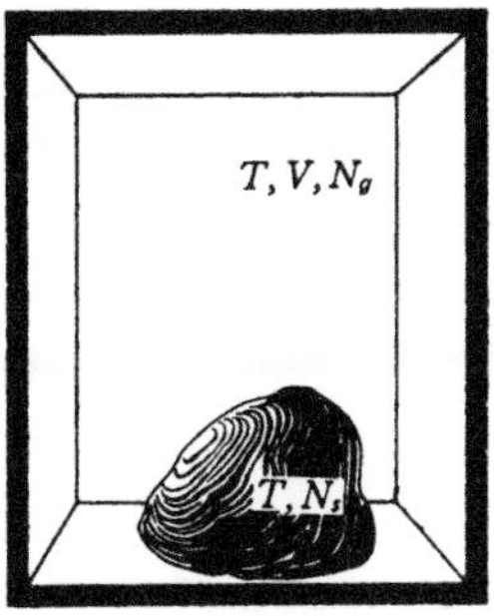

11. A solid and a vapor made up of the same kind of atoms are in equilibrium in a closed vessel of volume V at T° K. Let it be assumed that the partition function of the solid consisting of N_s atoms is given in the form $Z_s(T, N_s) = z_s(T)^{N_s}$ and that the vapor is an ideal gas having N_g molecules. Show that the equilibrium condition for $N_s \gg 1$ and $N_g \gg 1$ is given approximately by

$$N_g = \frac{z_g(T, V)}{z_s(T)}.$$

Here, $z_g(T, V)$ is the partition function of a gas molecule, which is assumed to be monatomic for simplicity. Assume that the volume occupied by the solide is negligible in comparison with V.

Plot the Helmholtz free energy as a function of N_s. Discuss the condition under which equilibrium is established.

12. Consider equilibrium between a solid and a vapor made up of monatomic molecules. It is assumed that the energy ϕ is required per atom for transforming the solid into separate atoms. For simplicity, take the Einstein model for the vibration of atoms in the solid, i.e., assume that each atom is represented by a three-dimensional harmonic oscillator performing vibration with a frequency ω about its equilibrium position independently of others. Evaluate the vapor pressure as a function of temperature.

13. N monomeric units are arranged along a straight line to form a chain molecule. Each monomeric unit is assumed to be capable of being either in an α state or in a β state. In the former state, the length is a and the energy

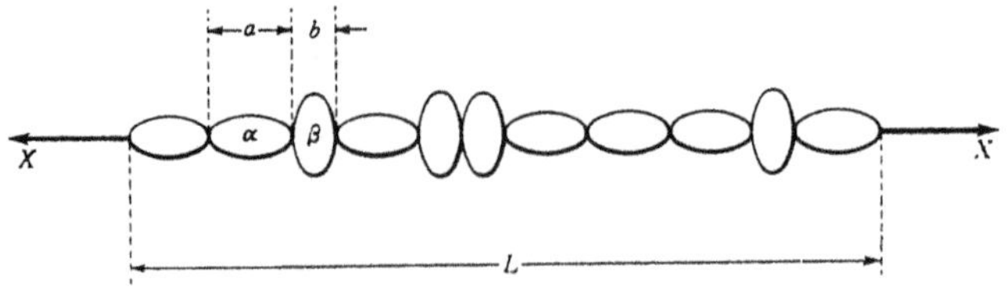

is E_α. The corresponding values in the latter state are b and E_β. Derive the relation between the length L of the chain molecule and the tension X applied between the both ends of the molecule. Use the canonical ensemble at constant tension.

14. Show that the partition function $Z(N, V, T)$ of a canonical ensemble satisfies the following relation:

$$N\left(\frac{\partial \log Z}{\partial N}\right)_{V,T} + V\left(\frac{\partial \log Z}{\partial V}\right)_{N,T} = \log Z .$$

It is assumed for simplicity that the system is a one-component system made up of N particles and that only the volume V is involved as an external variable. T denotes the absolute temperature.

15. The grand partition functions of the phases I and II are given by $\Xi_{\mathrm{I}}(\lambda_{\mathrm{A}}, \lambda_{\mathrm{B}}, \ldots, T, V_{\mathrm{I}})$, and $\Xi_{\mathrm{II}}(\lambda_{\mathrm{A}}, \lambda_{\mathrm{B}}, \ldots, T, V_{\mathrm{II}})$ respectively. The two phases are in mutual equilibrium with particles A, B, ... interchangeable between the phases. Show that the number of particles, N'_{A}, N''_{A}, N'_{B}, N''_{B}, ... in phase I or II are determined by

$$N'_{\mathrm{A}} : N''_{\mathrm{A}} = \frac{\partial \log \Xi_{\mathrm{I}}}{\partial \lambda_{\mathrm{A}}} : \frac{\partial \log \Xi_{\mathrm{II}}}{\partial \lambda_{\mathrm{A}}},$$

$$N'_{\mathrm{A}} + N''_{\mathrm{A}} = N_{\mathrm{A}} \text{ (total number of particles A), etc.}$$

[B]

16. Evaluate the contribution of a one-dimensional anharmonic oscillator having a potential $V(q) = cq^2 - gq^3 - fq^4$ to the heat capacity. Discuss the dependence of the mean value of the position q of the oscillator on the temperature T. Here, c, g and f are positive constants. Usually, $g \ll c^{\frac{3}{2}}/\sqrt{(kT)}$ and $f \ll c^2/kT$.

17. As shown in the figure, a chain molecule consists of N units, each having

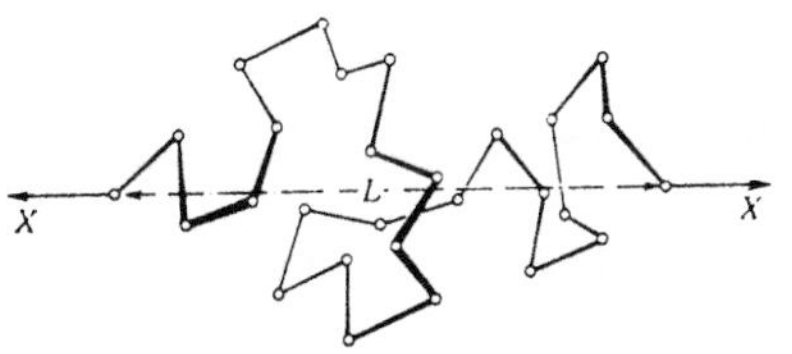

a length a. The units are joined so as to permit free rotation about the joints. Derive the relation between the tension X acting between both ends of the three-dimensional chain molecule and the distance L between the ends.

(Assume that the vibrational and other forms of energy are independent of the shape of the molecule.)

18. Derive Planck's law for heat radiation in a dispersive medium, the refractive index n of which depends on the frequency ν of radiation.

19. Wave motions, having the dispersion formula (relation between the frequency ω and the wave number q) $\omega = Aq^n$, exist in a solid. They yield a specific heat because they are excited as thermal motion at high temperature. Show that the specific heat is proportional to $T^{3/n}$ at low temperature. (*Hint:* Follow the derivation of Debye's specific heat of solids or of the heat radiation.)

20. The figure shows the dependence of the heat capacity C_V of a solid at constant volume on the temperature. C_∞ at high temperature is the classical

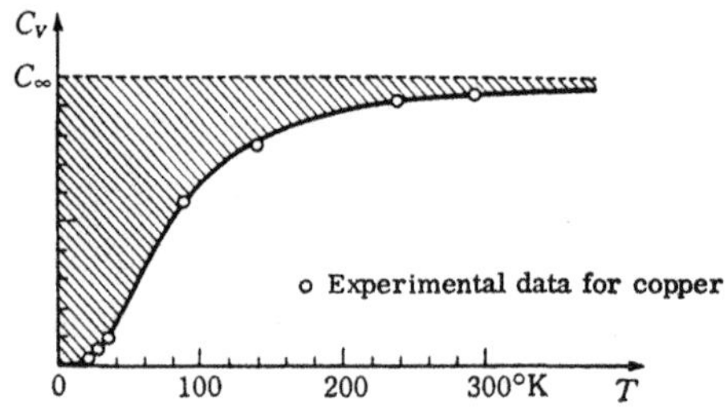

value (Dulong-Petit's law). Show that the shaded area above the heat capacity curve in the figure corresponds to the zero-point energy.

21. Discuss the heat capacities of one-dimensional and two-dimensional crystals by following the three-dimensional Debye model discussed in example 3.

22. The potential energy of interatomic bonds in a solid of volume V (the energy assumed by the solid when all atoms are at rest in their respective equilibrium positions) is denoted by $\Phi(V)$. The normal frequencies of vibration performed by atoms near their equilibrium positions are expressed by $\nu_j(V)$ ($j = 1, 2, \ldots, 3N - 6$, where N is the number of atoms composing the solid). It is assumed that the change of the normal frequencies accompanying the change in the volume V can be expressed by

$$\frac{\partial \log \nu_j}{\partial \log V} = -\gamma \qquad (j = 1, 2, \dots, 3N - 6)$$

for all normal frequencies with a common constant $\gamma(> 0)$ (Grüneisen's assumption). Show that the pressure of the solid is given by

$$p = -\frac{\mathrm{d}\Phi}{\mathrm{d}V} + \gamma\frac{U}{V}.$$

Explain the physical meaning of the first and the second term. (This equation of state is known as Mie-Grüneisen's equation.) U is the part of internal energy depending on atomic vibrations.

Let it be assumed that

$$\Phi(V) = \frac{(V - V_0)^2}{2\kappa_0 V_0} \qquad (\kappa_0 \text{ and } V_0 \text{ are constants}).$$

Usually, γ is of the order of $1 \sim 3$ and $\kappa_0 C_V T/V_0 \ll 1$. (C_V is the heat capacity at constant volume: T is the absolute temperature.) Considering these relations, discuss the thermal expansion of solids at constant pressure $p = 0$.

23. Using expressions for the partition function of classical ideal gases, evaluate the state density $\Omega(E)$ by the transformation (2.6). For simplicity, the gas molecules are assumed to be of one kind. Ignore the internal degrees of freedom.

24. Evaluate the state density $\Omega(E)$ for a system composed of N ($N \gg 1$) classical harmonic oscillators having a frequency ω by the inverse transformation (2.6). Determine its asymptotic expression by the saddle-point method. Derive Stirling's formula from it.

25. Two rigid dipoles are in thermal equilibrium, with the distance between their centers equal to R. Calculate the mean force acting between these dipoles. Examine this case at high temperatures.

[C]

26. A system S is in contact with a heat source R. By means of an appropriate device W (work source), an energy nE (n is a positive or negative constant) is supplied from W to R whenever an energy E is given from R to S. Show that the canonical distribution for this system having such a special heat source is given by

$$\rho(E) \propto \exp\{-\beta(1-n)E\}.$$

27. Prove that the magnetic susceptibility of a system obeying classical mechanics and classical statistics is strictly equal to zero (Bohr-van Leeuwen's theorem). [*Hint:* Let the vector potential, from which the magnetic field can be derived be $\boldsymbol{A}$. The Hamiltonian for a system of charged particles in a magnetic field can be written as

$$\mathscr{H} = \sum_{j=1}^{N} \frac{1}{2m_j}\left\{\boldsymbol{p}_j + \frac{e_j}{c}\boldsymbol{A}(\boldsymbol{r}_j)\right\}^2 + U(\boldsymbol{r}_1, \ldots, \boldsymbol{r}_N).]$$

28. N particles on a line are subjected to an interaction potential expressible as a function of mutual separation alone. Treat this system classically. Prove that if only interactions between adjacent particles are taken into account, the relation between the pressure and the volume (distance L between the end particles) can be expressed by a simple single-valued function, and hence no peculiar phenomenon corresponding to a phase transition takes place. The order of arrangement of particles along the straight line is assumed to be unaltered.

29. The Hamiltonian of an electron in a magnetic field $\boldsymbol{H}$ is given by $\mathscr{H} = -\mu_B \boldsymbol{\sigma}\cdot\boldsymbol{H}$, where $\boldsymbol{\sigma}$ is Pauli's spin operator and μ_B stands for the Bohr magneton. Evaluate (i) the density matrix in the diagonalized representation of σ_z, (ii) the density matrix in the diagonalized representation of σ_x, and (iii) the average values of σ_z in these representations, the z-axis being taken along the field direction.

30. Evaluate the matrix elements of the density matrix, $\rho = \exp(-\beta\mathscr{H}), \mathscr{H} = p^2/2m + \frac{1}{2}m\omega^2 q^2$ of a one-dimensional harmonic oscillator in the q-representation. Discuss, in particular, the limiting case of $\hbar\omega/kT = \beta\hbar\omega \ll 1$. [*Hint:* The eigenfunction $\psi_n(q)$ for the eigenvalue

$$E_n = (n + \tfrac{1}{2})\,\hbar\omega$$

is given by

$$\psi_n(q) = \left(\frac{m\omega}{\pi\hbar}\right)^{\frac{1}{4}} \frac{H_n(\xi)}{\sqrt{2^n n!}} e^{-\frac{1}{2}\xi^2}, \qquad \xi = \sqrt{\frac{m\omega}{\hbar}}\,q.$$

The Hermite polynomials $H_n(\xi)$ are defined as

$$H_n(\xi) = (-1)^n e^{\xi^2}\left(\frac{d}{d\xi}\right)^n e^{-\xi^2} = \frac{e^{\xi^2}}{\sqrt{\pi}}\int_{-\infty}^{\infty}(-2iu)^n e^{-u^2+2i\xi u}\,du.$$

Use the last expression in the integral form.]

31. Evaluate $\overline{q^2}$ and $\overline{p^2}$ from the density matrix in the q-representation obtained in the foregoing problem for a one-dimensional harmonic oscillator.

32. Show that the q-representation of an (unnormalized) density matrix exp $(-\beta\mathscr{H})$ for a Hamiltonian $\mathscr{H}(p, q)$ is given by

$$\langle q'|e^{-\beta\mathscr{H}}|q''\rangle = \exp\left[-\beta\mathscr{H}\left(\frac{\hbar}{i}\frac{\partial}{\partial q'}, q'\right)\right]\delta(q'-q'').$$

Here,

$$\mathscr{H}\left(\frac{\hbar}{i}\frac{\partial}{\partial q'}, q'\right)$$

is the q-representation of $\mathscr{H}$ and $\delta(q'-q'')$ is Dirac's delta function.

Apply the results to a free particle $\mathscr{H} = p^2/2m$ and evaluate the density matrix in the q-representation.

33. A system is made up of N particles, the Hamiltonian being given by

$$\mathscr{H} = \sum \frac{p_i^2}{2m} + V(r_1, \ldots, r_N).$$

The particles are assumed to be distinguishable. Calculate the r-representation $\langle r_1, \ldots, r_N|e^{-\beta\mathscr{H}}|r_1'', \ldots, r_N''\rangle$ of the density matrix $\exp(-\beta\mathscr{H})$ for the system. Show that in the limit of $h \to 0$, the partition function $\mathrm{Tr}(\exp(-\beta\mathscr{H}))$ agrees with the classical value:

$$Z_{cl} = \frac{1}{h^{3N}}\int \cdots \int \exp[-\beta\mathscr{H}(p_1, \ldots, r_1, \ldots)]\,dp_1 \ldots dr_1 \ldots .$$

(*Hint:* Use the results of problem 32.)

Solutions

1. From the definition (2.1) of the partition function, one has

$$\bar{E} = \frac{1}{Z}\int e^{-\beta E} E\Omega(E)\,dE = -\frac{Z'}{Z} = -\frac{\partial}{\partial\beta}\log Z, \left(Z' = \frac{\partial Z}{\partial\beta}, \beta = \frac{1}{kT}\right). \quad (1)$$

Differentiating both sides with respect to β, one obtains

$$\frac{\partial \bar{E}}{\partial \beta} = - \frac{\partial^2}{\partial \beta^2} \log Z = - \frac{Z''}{Z} + \left(\frac{Z'}{Z}\right)^2 = - \overline{E^2} + \bar{E}^2 = - \overline{(E - \bar{E})^2}, \quad (2)$$

$$\frac{\partial^2 \bar{E}}{\partial \beta^2} = - \frac{Z'''}{Z} + 3\frac{Z''}{Z}\frac{Z'}{Z} - 2\left(\frac{Z'}{Z}\right)^3 = \overline{E^3} - 3\overline{E^2}\bar{E} + 2\bar{E}^3 = \overline{(E - \bar{E})^3}. \quad (3)$$

On the other hand,

$$\frac{\partial \bar{E}}{\partial \beta} = - kT^2 \frac{\partial \bar{E}}{\partial T} = - kT^2 C_V,$$

$$\frac{\partial^2 \bar{E}}{\partial \beta^2} = - kT^2 \frac{\partial}{\partial T}(- kT^2 C_V) = k^2 \left\{T^4 \frac{\partial C_V}{\partial T} + 2T^3 C_V\right\}.$$

Therefore, one is led to the equations given in the problem. In particular, for an ideal gas consisting of N monatomic molecules,

$$\bar{E} = \tfrac{3}{2} NkT, \qquad C_V = \tfrac{3}{2} Nk.$$

Accordingly,

$$\frac{\overline{(E - \bar{E})^2}}{\bar{E}^2} = \frac{kT^2 C_V}{\bar{E}^2} = \frac{2}{3N},$$

$$\frac{\overline{(E - \bar{E})^3}}{\bar{E}^3} = \frac{2k^2 T^3 C_V}{\bar{E}^3} = \frac{8}{9N^2}.$$

2. For a given distance x between the ends of the string, the position of the weight is specified by an azimuthal angle about the axis joining the ends of the string. The weight is $r = \frac{1}{2}\sqrt{l^2 - x^2}$ distant from the axis. The canonical momentum is $p_\varphi = mr^2 \,\mathrm{d}\varphi/\mathrm{d}t$. When the weight rotates with an angular velocity $\omega = \mathrm{d}\phi/\mathrm{d}t$, one must balance the centrifugal force $mr\omega^2$ acting on it with the tension

$$X = \frac{mr\omega^2}{4r/x} = \frac{xp_\varphi^2}{4mr^4} = \frac{2x}{l^2 - x^2} \cdot \frac{p_\varphi^2}{2mr^2} \quad (1)$$

applied to the ends of the string. Since the moment of inertia is mr^2, the kinetic energy is $p_\varphi/(2mr^2)$, the mean value of which is $\frac{1}{2} kT$ in accordance to the equipartition law (cf. example 1). Accordingly, the mean value of the tension to be applied to the ends of the string is caculated from (1) as

$$\bar{X} = \frac{x}{l^2 - x^2} kT. \tag{2}$$

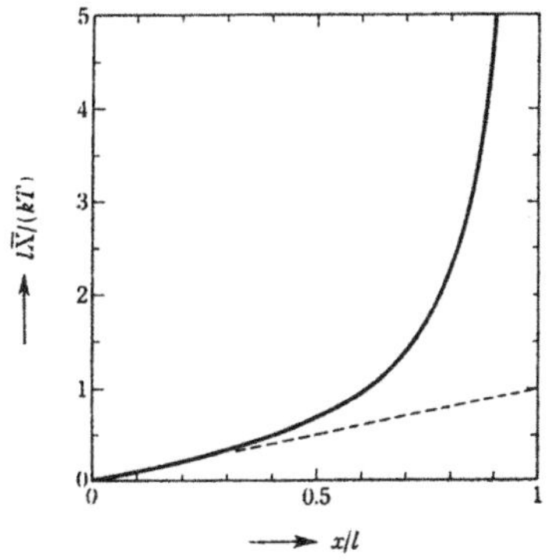

ALTERNATIVE SOLUTION

The Hamiltonian $\mathscr{H}$ of this system is $\mathscr{H} = p_\varphi^2/(2mr^2) = 2p_\varphi^2/m(l^2 - x^2)$. Therefore, the partition function $Z = \int\int e^{-\mathscr{H}/kT}\, d\varphi\, dp_\varphi/h$ has a factor involving x, i.e., $Z \propto (l^2 - x^2)^{\frac{1}{2}}$. From $\bar{X} = -kT\partial \log Z/\partial x$ one can readily derive (2).

3. The z-axis used to measure the vertical position of particles is taken along the axis of the vessel. The translational Hamiltonian per particle is given by

$$\varepsilon_{\text{trans}} = \frac{p_x^2 + p_y^2 + p_z^2}{2m} + mgz, \tag{1}$$

where g is the gravitational constant; p_x, p_y and p_z are momenta conjugate to the coordinates, x, y, and z, respectively. The partition function for a single particle is given by

$$Z_{\text{trans}} = \frac{1}{h^3} \int\int_\sigma dx\, dy \int_0^\infty dz \int\int\int_{-\infty}^{\infty} dp_x\, dp_y\, dp_z\, e^{-\varepsilon_{\text{trans}}/kT}$$

$$= \frac{\sigma}{h^3} \int_0^\infty e^{-zmg/kT} dz \cdot (2\pi mkT)^{\frac{3}{2}} = \sigma \frac{kT}{mg} \left(\frac{2\pi mkT}{h^2}\right)^{\frac{3}{2}} \tag{2}$$

($z = 0$ at the bottom of the vessel; σ = cross section of the vessel). Therefore, the partition function and the Helmholtz free energy for N particles are

$$Z = \frac{1}{N!}(Z_{\text{trans}})^N = \frac{1}{N!}\left(\sigma\frac{kT}{mg}\right)^N \left(\frac{2\pi mkT}{h^2}\right)^{\frac{3}{2}N}, \tag{3}$$

$$F = -kT\log Z = -NkT\log\left[\frac{\sigma kTe}{Nmg}\left(\frac{2\pi mkT}{h^2}\right)^{\frac{3}{2}}\right]. \tag{4}$$

(The latter is equal to the free energy of an ideal gas in a cylindrical vessel of a cross section σ and a height kT/mg in the absence of the gravitational field.) The energy E and the heat capacity C_V are given by

$$E = kT^2\frac{\partial \log Z}{\partial T} = \tfrac{5}{2}NkT, \qquad C_V = \tfrac{5}{2}Nk.$$

The latter is greater than the heat capacity $C_V = \frac{3}{2}Nk$ of an ideal gas at constant volume, because with increasing temperature, molecules rise and increase the potential energy.

4. The gas is in thermal equilibrium with regard to the coordinate system rotating with an angular velocity ω. For simplicity, the internal structure of the gas molecules is disregarded. The mass center of a gas molecule is at x_j, y_j and z_j in the rotating coordinate system, the conjugate momenta are denoted by p_{xj}, p_{yj} and p_{zj}. The Hamiltonian in the rotating coordinate system is

$$\mathscr{H} = \sum_{j=1}^{N}\left\{\frac{p_{xj}^2 + p_{yj}^2 + p_{zj}^2}{2m} + \phi(x_j, y_j, z_j)\right\} - \omega\sum_{j=1}^{N}(x_j p_{yj} - y_j p_{xj}) \tag{1}$$

where N stands for the total number of molecules, m is the mass of a molecule, and ϕ is the potential due to the vessel wall.

$$\phi(x, y, z) = \begin{cases} 0, & (x^2 + y^2 \leqq R^2, \quad 0 \leqq z \leqq L) \\ +\infty, & \text{otherwise} \end{cases} \tag{2}$$

where the z-axis is taken along the axis of the cylinder. The summation in the second term of the right-hand side of (1) is the z-component of the total angular momentum of gas molecules. The density of gas molecules at a point, x, y and z, or r, φ and z in cylindrical coordinates, is given by

$$\begin{aligned}\rho(x, y, z) &= m\sum_{j=1}^{N}\delta(x - x_j)\delta(y - y_j)\delta(z - z_j) \\ &= m\sum_{j=1}^{N}\delta(r - r_j)\delta(\varphi - \varphi_j)\frac{1}{r}\delta(z - z_j).\end{aligned} \tag{3}$$

Here, δ is Dirac's delta function. By use of the Hamiltonian (1), (3) is averaged over the canonical distribution (1.70a) to obtain the required density distribution. First, the partition function (1.71a) is calculated as

$$Z = \frac{1}{N!h^3}\prod_{j=1}^{N}\int_0^L dz_j\int_0^R r_j dr_j\int_0^{2\pi} d\varphi_j\iiint_{-\infty}^{\infty} dp_{xj}dp_{yj}dp_{zj}\times$$

$$\times\exp\left\{-\frac{p_{xj}^2+p_{yj}^2+p_{zj}^2}{2mkT}+\frac{\omega}{kT}(x_jp_{yj}-y_jp_{xj})\right\}$$

$$=\frac{1}{N!}\left[\int_0^L dz\int_0^R r\,dr\int_0^{2\pi} d\varphi\left(\frac{2\pi mkT}{h^2}\right)^{\frac{3}{2}}\exp\left(\frac{m\omega^2r^2}{2kT}\right)\right]^N$$

$$=\frac{1}{N!}\left[\pi R^2L\left(\frac{2\pi mkT}{h^2}\right)^{\frac{3}{2}}\frac{2kT}{m\omega^2R^2}\left\{\exp\left(\frac{m\omega^2R^2}{2kT}\right)-1\right\}\right]^N. \tag{4}$$

In calculating the average value of (3), the integral over momenta is the same as before. Therefore,

$$\bar{\rho}=\frac{1}{ZN!}\left(\frac{2\pi mkT}{h^2}\right)^{\frac{3}{2}N}\prod_{j=1}^{N}\int_0^L dz_j\int_0^R r_j dr_j\int_0^{2\pi} d\varphi_j\exp\left(\frac{m\omega^2r_j^2}{2kT}\right)\times$$

$$\times\sum_{i=1}^{N} m\frac{\delta(r-r_i)}{r_i}\delta(\varphi-\varphi_i)\delta(z-z_i)$$

$$=\frac{1}{ZN!}\left(\frac{2\pi mkT}{h^2}\right)^{\frac{3}{2}N} Nm\exp\left(\frac{m\omega^2r^2}{2kT}\right)\times$$

$$\times\left[\pi R^2L\frac{2kT}{m\omega^2R^2}\left\{\exp\left(\frac{m\omega^2R^2}{2kT}-1\right)\right\}\right]^{N-1}. \tag{5}$$

Here $\bar{\rho}=0$, when the point (r, φ, z) is outside the cylinder. Introducing (4) into (5), one has

$$\overline{\rho(x,y,z)}=\frac{Nm}{\pi R^2L}\frac{\exp\left\{\frac{m\omega^2}{2kT}(x^2+y^2)\right\}}{\frac{2kT}{m\omega^2R^2}\left\{\exp\left(\frac{m\omega^2R^2}{2kT}\right)-1\right\}},(x^2+y^2\leqq R^2, 0\leqq z\leqq L). \tag{6}$$

NOTE: Note that $-(\frac{1}{2}m\omega^2)(x^2+y^2)$ appearing in (4) and (6) is the potential of the centrifugal force. In the classical treatment, Coriolis' force does not affect the statistical properties.

5. This is an application of example 1. Let ε be rewritten as a function of momenta. Then

$$\varepsilon = c\sqrt{p_x^2 + p_y^2 + p_z^2 + (mc)^2}\,. \tag{1}$$

Therefore, $\partial\varepsilon/\partial p_i = c^2 p_i/\varepsilon$. According to example 1 (i) with ε in place of $\mathscr{H}$, one has

$$\overline{p_i\frac{\partial\varepsilon}{\partial p_j}} = \overline{\frac{c^2 p_i p_j}{\varepsilon}} = \delta_{ij}kT\,. \tag{2}$$

By expressing p_i in terms of ε and v_i, one has

$$\overline{mv_i v_j/\sqrt{1-(v/c)^2}} = \delta_{ij}kT\,. \tag{3}$$

The required relation is obtained if one puts $i = j$.

6. Equations already verified in example 1

$$\overline{q_i\frac{\partial\mathscr{H}}{\partial q_j}} = kT\delta_{ij}\,, \qquad X_j = -\frac{\partial\mathscr{H}}{\partial q_j}\,,$$

evidently lead to

$$\sum_{j=1}^{3N}\overline{q_j X_j} = -3NkT\,. \tag{1}$$

It is assumed that conditions are satisfied for the term integrated by partial integration to vanish.

In order to prove the second relation, note that X_j is made up of two parts: one derived from the intermolecular potential U and the other due to the potential of the vessel wall. The latter is denoted by F_j.

$$X_j = -\frac{\partial U}{\partial q_j} + F_j\,, \tag{2}$$

F_j is the component of force exerted by the wall on a molecule along the q_j direction. It is appreciable only when the molecule having the coordinate q_j, is close to the wall. N position vectors, $\boldsymbol{r}_1, \boldsymbol{r}_2, \ldots, \boldsymbol{r}_N$, are used in place of $q_1, q_2, \ldots, q_{3N}$. Correspondingly, $F_1, F_2, \ldots, F_{3N}$ are rewritten as $\boldsymbol{F}_1, \boldsymbol{F}_2, \ldots, \boldsymbol{F}_N$. By this change of notation, one has

$$\overline{\sum_{j=1}^{3N} q_j F_j} = \overline{\sum_{s=1}^{N} \boldsymbol{r}_s\cdot\boldsymbol{F}_s}\,. \tag{3}$$

When $\boldsymbol{r}_s$ is on the wall surface, $-\boldsymbol{F}_s$ gives the force exerted by molecule s

on the wall. When the force is summed up and averaged, it gives the pressure exerted on the wall surface by the gas. Therefore, one has

$$\overline{\sum_{s=1}^{N} \boldsymbol{r}_s \cdot \boldsymbol{F}_s} = \iint \boldsymbol{r} \cdot (-p\boldsymbol{n})\,\mathrm{d}f = -p \iiint \operatorname{div} \boldsymbol{r}\,\mathrm{d}V = -3pV \tag{4}$$

(cf. alternative solution of example 1 in Chapter 1)

$$\sum_{j=1}^{3N} \overline{q_j X_j} = \sum_{s=1}^{N} \overline{\boldsymbol{r}_s \cdot \boldsymbol{F}_s} - \sum_{j=1}^{3N} \overline{q_j \frac{\partial U}{\partial q_j}}.$$

Finally, expressions (1) and (2) are substituted into this last relation.

ALTERNATIVE SOLUTION

See the hint and solution for problem 12 in Chapter 4.

7. The number of modes of stationary electromagnetic waves (ranging over angular frequencies between ω and $\omega + \mathrm{d}\omega$) in a box of volume V can be calculated from (1) in the solution of example 4 as

$$V g(\omega)\,\mathrm{d}\omega = \frac{\omega^2 V}{\pi^2 c^3}\mathrm{d}\omega .$$

When the whole range of frequencies is considered, the free energy $F(V, T)$ may be calculated, in a similar manner to the derivation of (3) in the solution to example 3, as

$$F(V, T) = -kT \log Z = kT \int_0^\infty \frac{V\omega^2}{\pi^2 c^3}\,\mathrm{d}\omega \cdot \log\left(2\sinh\frac{\hbar\omega}{2kT}\right)$$

where Z = partition function. Therefore, one has

$$p = -\frac{\partial F}{\partial V} = -kT \int_0^\infty \frac{\omega^2\,\mathrm{d}\omega}{\pi^2 c^3} \log\left(2\sinh\frac{\hbar\omega}{2kT}\right).$$

By partial integration,

$$p = -kT\frac{\omega^3}{3\pi^2 c^3}\log\left(2\sinh\frac{\hbar\omega}{2kT}\right)\Bigg|_0^\infty + kT\int_0^\infty \frac{\omega^3\,\mathrm{d}\omega}{3\pi^2 c^3}\cdot\frac{\hbar}{2kT}\coth\frac{\hbar\omega}{2kT}$$

$$= \frac{1}{3}\int_0^\infty \frac{\omega^2\,\mathrm{d}\omega}{\pi^2 c^3}\left\{\tfrac{1}{2}\hbar\omega + \frac{\hbar\omega}{e^{\hbar\omega/kT}-1}\right\} = \frac{1}{3}\int_0^\infty \frac{\omega^2\,\mathrm{d}\omega}{\pi^2 c^3}\varepsilon(\omega, T) = \tfrac{1}{3}u .$$

Here, $\varepsilon(\omega, T)$ is the mean energy of a single oscillator of frequency ω, and hence u is the total energy of radiation per unit volume.

ALTERNATIVE SOLUTION

The derivation is simple by use of $p = -(\partial U/\partial V)_s$ (problem 12 in Chapter 1). The frequency and wave number of an electromagnetic wave are related by $\omega = cq$. For a given mode $(q_x, q_y, q_z) = (\pi/L)(n_x, n_y, n_z)$ (n_x, n_y, and n_z are integers), q is proportional to L^{-1}. During an adiabatic change of V, the frequency ω of each mode changes in proportion to L^{-1}. Accordingly, for each mode,

$$\varepsilon \propto \hbar\omega \propto L^{-1} \propto V^{-\frac{1}{3}}.$$

Therefore,

$$\frac{\partial \varepsilon}{\partial V} = -\frac{1}{3}\frac{\varepsilon}{V}, \qquad \text{thus} \quad p = \frac{1}{3}\frac{U}{V} = \tfrac{1}{3}u.$$

8. Let the total energy emitted per unit time from a surface element $\mathrm{d}S$ of a black body at T be denoted by $J\mathrm{d}S$. Then one has

$$\begin{aligned} J &= \int_0^\infty \mathrm{d}\omega \int_0^{2\pi} \int_0^{\frac{1}{2}\pi} J(\omega, T, \theta) \sin\theta \, \mathrm{d}\theta \, \mathrm{d}\varphi \\ &= \tfrac{1}{2}c \int_0^\infty u(\omega, T) \mathrm{d}\omega \int_0^{\frac{1}{2}\pi} \cos\theta \sin\theta \, \mathrm{d}\theta \\ &= \tfrac{1}{4}c \int_0^\infty u(\omega, T) \, \mathrm{d}\omega . \end{aligned} \tag{1}$$

Calculations performed in deriving (9) of the solution to example 4 yield

$$J = \sigma T^4, \qquad \sigma = \frac{\pi^2 k^4}{60 \cdot c^2 \hbar^3} = 5.672 \times 10^{-5} \,\mathrm{erg\,cm^{-2}\,sec^{-1}\,deg^{-4}}. \tag{2}$$

The constant σ is known as *Stefan-Boltzmann's constant.*

Since a cube of 1 cm edge length has a surface area of 6 cm^2, the energy emitted per unit time from a black body at temperature T to its environment is given by

$$W(T) = 6 \times 5.672 \times 10^{-5} \times T^4 = 3.403 \times 10^{-4} \times T^4 \,\mathrm{erg\,sec^{-1}}. \tag{3}$$

This amount of energy is completely absorbed by the surrounding wall, which in turn emits heat radiation as a black body at 300° K. The heat radiation emitted by the wall is partly absorbed by the central black body, the amount being equal to the energy the central black body would emit if it were at 300° K. This is evident from considerations on thermal equilibrium. Accordingly, the net energy loss of the central black body per unit time is given by

$$W(T) - W(300^\circ K) = 3.403 \times 10^{-4} \times (T^4 - 300^4)\,\text{erg sec}^{-1}. \quad (4)$$

Unless this amount of energy is supplied to the black body, one cannot maintain the black body at T. Put $T = 500$, 800 and 1000° K to obtain the answer:

$$W(500^\circ K) - W(300^\circ K) =$$
$$= 3.40 \times 10^{-4} \times (500^4 - 300^4) = 1.850\,\text{joule sec}^{-1}\,(= \text{watt}),$$
$$W(800^\circ K) - W(300^\circ K) =$$
$$= 3.40 \times 10^{-4} \times (800^4 - 300^4) = 13.64\,\text{joule sec}^{-1}\,(= \text{watt}),$$
$$W(1000^\circ K) - W(300^\circ K) =$$
$$= 3.40 \times 10^{-4} \times (1000^4 - 300^4) = 33.8\,\text{joule sec}^{-1}\,(= \text{watt}).$$

Divertissement 6

Negative temperature: What is the reason for the requirement that the temperature must be positive? If the temperature T is negative, a canonical distribution function will be infinite for $E \to \infty$, which looks surely absurd. Physical systems in the real world have those structures which assure positive values of temperature. Thermodynamics indeed is based upon this fact. The logical construction of statistical mechanics by means of probability theory is general, but it yields thermodynamics only when it is applied to normal systems in the sense mentioned in § 1.6.

Apart from reality, one may imagine systems which are not normal. A system can be abnormal if its energy levels E are bounded from above as well as from below. An example of this has been given in the spin system as treated in example 5 and problem 35, Chapter 1. However, a real physical system consists of a number of particles which have kinetic energy of translational motion or of oscillation. As a result, E, the energy of such a system, has no upper bound but certainly has a lower bound. For example, the electron in a hydrogen atom feels the potential $-e^2/r$, but it can never be at rest at the origin $r = 0$. The reason for this is of course the uncertainty principle of quantum mechanics i.e. $\Delta p \Delta x \geqq h$, which would then make $\Delta p = 0$ and accordingly the kinetic energy infinite. This statement may be made quite generally: namely, there should exist a lowest value of energy for any real system because of the uncertainty principle. More rigorously speaking, this statement depends on the nature of the interaction potential of particles. But there seems to be no interaction present in the real world which violates the above assertion of the existence of a lowest level.

It must be observed, therefore, that the validity of thermodynamics implicitly rests upon the microscopic structure of matter and upon the quantum-mechanical laws which govern the microscopic world.

Thus, a negative temperature can never exist in equilibrium states. It is, however, possible to create it in certain transient processes. Suppose that spins are aligned by a magnetic field and then the magnetic field is suddenly reversed in its direction. If the spins cannot follow the magnetic field, there must be more spins in higher Zeeman levels with respect to the new field than in the lower levels. If this distribution is described by a canonical distribution, the temperature must be negative. Therefore, by this reversal of the magnetic field the spin system will be put in a state of negative temperature. Other degrees of freedom of the system must be at the original temperature which is positive, as usual. If the spins have no interaction with such degrees of freedom, the spin system remains in such a negative temperature state. In reality, a certain degree of interaction must be present, and this drives the whole system to a new equilibrium, so that the spin temperature must approach a final positive value. Temperature may more conveniently be measured by $1/T$. Then negative temperatures continue smoothly to change to positive temperatures. In this sense, negative temperatures are hotter than positive temperatures.

Purcell and Pound, Phys. Rev. **81** (1951) 279, see also N. F. Ramsey, Phys. Rev. **103** (1956) 20, performed an experiment to create a state of negative temperature for nuclear spins in LiF. The concept of negative temperature has become more popular in recent times, particularly in connection with masers.

9. Consider a unit volume. Its energy in a magnetic field H is given by

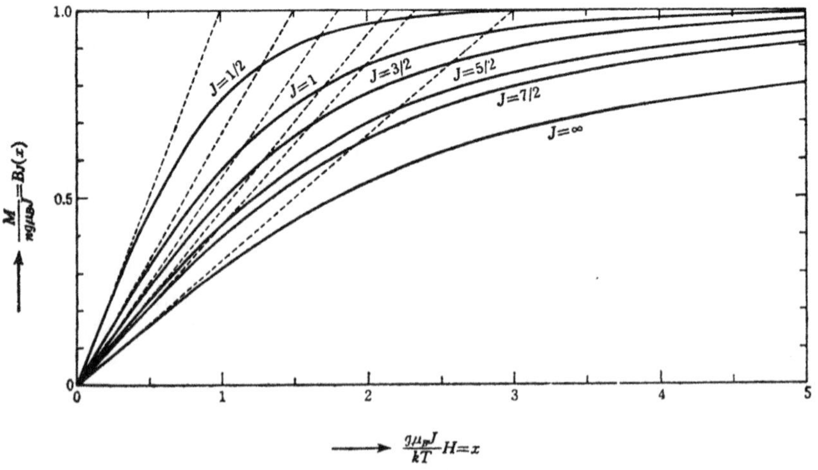

Fig. 2.10.

$$E = -MH = -g\mu_B H \sum_{i=1}^{n} m_i . \tag{1}$$

Here, m_i stands for the magnetic quantum number of the i-th particle and M is the total magnetic moment. The partition function Z is given by

$$\begin{aligned} Z &= \sum_{m_1=-J}^{J} \cdots \sum_{m_n=-J}^{J} \exp(\beta MH) \\ &= \prod_{j=1}^{n} \left\{ \sum_{m_j=-J}^{J} \exp(\beta g\mu_B H m_i) \right\} \qquad (\beta = 1/kT) \\ &= \left\{ \sinh\left(\beta g\mu_B H \frac{2J+1}{2}\right) \Big/ \sinh\left(\frac{1}{2}\beta g\mu_B H\right) \right\}^n , \end{aligned} \tag{2}$$

where use is made of the relation

$$\sum_{k=-n}^{n} x^k = x^{-n} \sum_{l=0}^{2n} x^l = \frac{(x^{2n+1}-1)}{x^n(x-1)} = \frac{x^{n+\frac{1}{2}} - x^{-n-\frac{1}{2}}}{x^{\frac{1}{2}} - x^{-\frac{1}{2}}} .$$

The mean magnetization is given by

$$\bar{M} = kT \frac{\partial}{\partial H} \log Z = n g\mu_B J \cdot B_J\left(\frac{g\mu_B J H}{kT}\right), \tag{3}$$

where

$$\begin{aligned} B_J(x) &= \frac{\mathrm{d}}{\mathrm{d}x} \log\left\{ \sinh\left(\frac{2J+1}{2J}x\right) \Big/ \sinh\frac{x}{2J} \right\} \\ &= \frac{2J+1}{2J} \coth\left(\frac{2J+1}{2J}x\right) - \frac{1}{2J}\coth\frac{x}{2J} \end{aligned} \tag{4}$$

is called the *Brillouin function*. Since for $x \ll 1$,

$$B_J(x) \cong \frac{J+1}{3J} x ,$$

one has

$$\bar{M} = n \frac{J(J+1) g^2 \mu_B^2}{3kT} H , \qquad \chi = \frac{\partial \bar{M}}{\partial H} = n \frac{J(J+1) g^2 \mu_B^2}{3kT} \tag{5}$$

for $g\mu_B JH/(kT) \ll 1$. For $J = \frac{1}{2}$, $B_{\frac{1}{2}}(x) = \tanh x$. Therefore, (5) leads to

$$\chi = n g^2 \mu_B^2 / 4kT . \tag{6}$$

At the limit of $J \to \infty$ and $g\mu_B J \to \mu_0$,

$$B_\infty(x) = \coth x - 1/x = L(x) . \tag{7}$$

Therefore, (3) and (5) lead to

$$\bar{M} = n\mu_0 L\frac{\mu_0 H}{kT}, \qquad \chi = \frac{n\mu_0^2}{3kT}. \tag{8}$$

NOTE 1: $\mu_B = e\hbar/(2mc)$ is referred to as the Bohr magneton and g is called Landé's g factor. For a single electron, $g \sim 2$ and $J = \frac{1}{2}$. In general, the magnetic moments in paramagnetic substances are attributable to electron spin and the orbital motion of electrons. J assumes integral or half-integral values. Also, g depends on the substance.
NOTE 2: Equation (5) expresses Curie's law ((1.21) in Chapter 1). Fig. 2.10 illustrates (3).
NOTE 3: Here, particles are assumed to be capable of taking various magnetic quantum states independently. This is valid when the particles exist separately (for instance, as separate ions in crystals) or when they obey Boltzmann statistics in gases (see the next problem). However, it does not apply to degenerate gases (cf. example 2 and problem 3 in Chapter 4).

10. An adsorption center having no adsorbed molecule on it is referred to as state 0, and that having a molecule of energy ε_i is denoted by i. Each of N_0 such centers can assume any of the states 0, 1, 2, ... independently of others. The absolute activity being expressed by λ, the grand partition function is given by

$$\Xi = [1 + \lambda \sum_{i=1}^{\infty} \exp(-\varepsilon_i/kT)]^{N_0} = (1 + \lambda a)^{N_0} \tag{1}$$

where

$$a = \sum \exp(-\varepsilon_i/kT).$$

For a given $\lambda = \exp(\mu/kT)$ (μ = chemical potential), the mean N is given by

$$N = \lambda\frac{\partial \log \Xi}{\partial \lambda} = N_0\frac{\lambda a}{1 + \lambda a}, \quad \text{thus} \quad \lambda a = \frac{N}{N_0 - N},$$

$$\text{and} \quad \mu = kT\left[\log\frac{N}{N_0 - N} - \log a\right].$$

NOTE: By expanding (1), one has

$$\Xi = \sum_N \frac{N_0! a^N}{N!(N_0 - N)!}\lambda^N = \sum_N Z_N \lambda^N.$$

Therefore,

$$Z_N = \frac{N_0!}{N!(N_0 - N)!}a^N.$$

This can be obtained directly. μ can now be calculated as $-\ kT\partial \log Z_N/\partial N = \mu$.

11. According to classical statistics, the partition function of an ideal gas consisting of N_g molecules each having a partition function $z_g(T, V)$ is given by

$$Z_g(T, V, N_g) = \frac{\{z_g(T, V)\}^{N_g}}{N_g!}.$$

Accordingly, by use of Stirling's formula the Helmholtz free energy of the gas phase is calculated as

$$F_g(T, V, N_g) = -\ kT \log Z_g = -\ kTN_g\{\log z_g(T, V) - \log N_g + 1\}. \quad (1)$$

On the other hand, the partition function of the solid phase is given by $Z_s(T, N_s) = [z_s(T)]^{N_s}$. Therefore, the Helmholtz free energy of the solid phase is given by

$$F_s(T, N_s) = -\ kT \log Z_s = -\ kTN_s \log z_s(T). \quad (2)$$

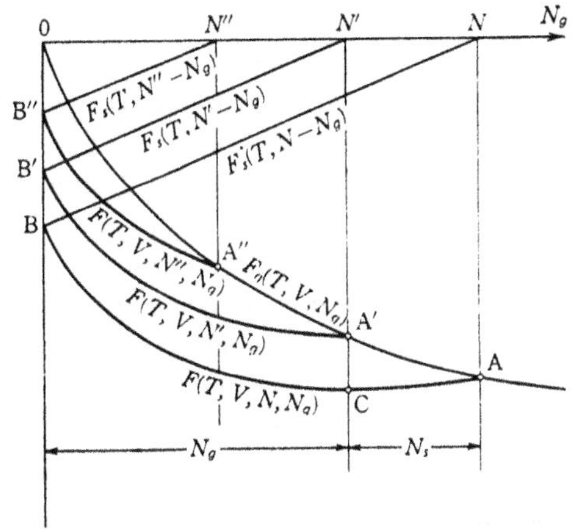

The most probable distribution (N_g, N_s) corresponding to thermal equilibrium is determined by

$$F(T, V, N, N_g) = F_g(T, V, N_g) + F_s(T, N_s) = \min. \quad (3)$$

as in the solution to example 9 in Chapter 1. The total number of atoms in the vessel is $N = N_g + N_s$, since the gas molecules are monatomic. This number is kept constant. Therefore, from (1), (2) and (3), one has

$$0 = \left(\frac{\partial F}{\partial N_g}\right)_{T,V,N} = -kT\left\{\log\frac{z_g(T,V)}{N_g} - \log z_s(T)\right\},$$

thus
$$N_g = \frac{z_g(T,V)}{z_s(T)}. \tag{4}$$

At a given temperature T, F_s is proportional to $N_s = N - N_g$ in accordance to (2). Hence, F_s plotted against N_g gives a straight line. Its inclination is determined by $z_s(T)$, which is constant at constant temperature. F_s is equal to zero at $N_g = N$. This point moves with N even at constant T. Because $z_g(T, V)/V$ is of the order of 10^{25} or more for common gases at room temperature, $z_g(T, V)/N_g \gg 1$. According to (1), the slope of the plot of F_g decreases with increasing N_g. The shape of the curve is determined by $z_g(T, V)$, i.e., by fixing both T and V. The free energy F of the entire system as given by (3) is the sum of F_s and F_g. The curve of F coincides with F_s at $N_g = 0$ and with F_g at $N_g = N$. The figure illustrates the general trend of the curves F_s, F_g and F for the total number of atoms equal to N, N' or N''. The slope of the curve F_g decreases with increasing N_g and hence becomes less than that of F_s for sufficiently large $N_g = N$. Accordingly, the curve AB of F has a minimum at C, the N_g value of which is given by (4). When the total number of atoms is as small as N'', the slope of F_g is steep. Indeed, it is always steeper than the slope of F_s for $N_g \leqslant N''$. In this case, no minimum appears on the curve F. Since F is smallest at $N_g = N''$, the corresponding state is realized as the state of thermal equilibrium in the vessel. In other words, only the gas phase exists in the volume V at T, because the number of atoms in the vessel is small. At the point A′, the slope of F_g is equal to that of F_s (*with the opposite sign*). The number of atoms corresponding to this point is denoted by N'. For $N \geq N'$, a minimum ($N_g \leqslant N$) appears and the solid is deposited. Accordingly, from (4) the condition for the formation of a solid phase is given by

$$N \geq \frac{z_g(T,V)}{z_s(T)}. \tag{5}$$

After the appearance of the solid phase, the increase of N leads only to the increase of N_s (growth of the solid), N_g being maintained at a value given by (4).

NOTE: The probability that, among the total of N atoms, N' atoms are in the gas phase and $N - N'$ atoms are in the solid phase is given by

$$Z_g(T, V, N')Z_s(T, N - N') = \frac{1}{N'!} z_g^{N'} z_s^{N-N'} = z_s^N \left(\frac{z_g}{z_s}\right)^{N'} \Big/ N'!\,.$$

This is proportional to $A^{N'}/N'!$ ($A = z_g/z_s$). The condition for its being a maximum is (3), while (4) is the corresponding value of N' or N_g. When $N < N_g$, this maximum is not realized. The largest value of the probability is at $N' = N$ (all the molecules are in the gas phase).

12. By assumption, the partition function of this solid composed of N_s atoms is given by

$$Z_s = e^{N_s \varphi/kT} \left\{2 \sinh \frac{\hbar\omega}{2kT}\right\}^{-3N_s} = \left[e^{\varphi/kT} \Big/ \left\{2 \sinh \frac{\hbar\omega}{2kT}\right\}^3 \right]^{N_s}. \qquad (1)$$

(An atom in the solid is stabilized by an energy φ compared with a free atom.) On the other hand, the partition function of the gas is

$$Z_g = \frac{1}{N_g!} \left(\frac{2\pi mkT}{h^2}\right)^{\frac{3}{2}N_g} V^{N_g}.$$

Therefore, as in the foregoing problem, the condition that $Z_s(N_s)\, Z_g(N_g) =$ = max. under condition that $N_s + N_g$ = constant (or that the total free energy is minimum) yields

$$N_g = \left(\frac{2\pi mkT}{h^2}\right)^{\frac{3}{2}} V \frac{e^{\varphi/kT}}{[2 \sinh(\hbar\omega/2kT)]^3}.$$

Introducing this into $p = N_g kT/V$, one has

$$p = kT \left(\frac{2\pi mkT}{h^2}\right)^{\frac{3}{2}} \left(2 \sinh \frac{\hbar\omega}{2kT}\right)^3 e^{-\varphi/kT}.$$

13. When N_α monomeric units of the chain molecule are in the α state and $N_\beta (= N - N_\alpha)$ units are in the β state, the energy and the length of the molecule are given by

$$E(N_\alpha, N_\beta) = N_\alpha E_\alpha + N_\beta E_\beta, \qquad L(N_\alpha, N_\beta) = N_\alpha a + N_\beta b. \qquad (1)$$

As in (1.76), the canonical ensemble for the molecule under tension X applied between both ends of the molecule gives the probability

$$P(N_\alpha, N_\beta) = \exp\left[-\frac{1}{kT}\{E(N_\alpha, N_\beta) - X \cdot L(N_\alpha, N_\beta)\}\right] \Big/ Y(X, T) \qquad (2)$$

for one of the states specified by N_α and N_β. The number of states having the same N_α and N_β is $N!/N_\alpha!\,N_\beta!$ or the number of ways in which the states α and β are allotted to N monomeric units. Accordingly, the partition function Y is given by

$$Y = \sum_{N_\alpha=0}^{N} \frac{N!}{N_\alpha! N_\beta!} \exp\left[-\frac{E(N_\alpha, N_\beta) - X \cdot L(N_\alpha, N_\beta)}{kT}\right]$$
$$= (e^{(Xa-E_\alpha)/kT} + e^{(Xb-E_\beta)/kT})^N. \tag{3}$$

Therefore, the mean of the molecular lengths L is calculated as

$$\bar{L} = kT\left(\frac{\partial \log Y}{\partial X}\right)_{T,N}$$
$$= N(a\,e^{(Xa-E_\alpha)/kT} + b\,e^{(Xb-E_\beta)/kT})/(e^{(Xa-E_\alpha)/kT} + e^{(Xb-E_\beta)/kT}). \tag{4}$$

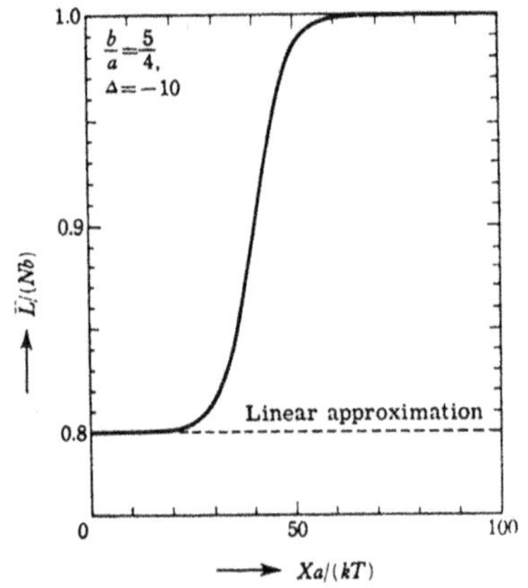

This gives the dependence of the molecular length on tension. In particular, when the tension is weak (Xa and Xb are sufficiently small compared with kT), one has

$$\bar{L} \cong \frac{a + b\,e^\Delta}{1 + e^\Delta} + \frac{X}{kT}\left\{\frac{a^2 + b^2 e^\Delta}{1 + e^\Delta} - \left(\frac{a + b\,e^\Delta}{1 + e^\Delta}\right)^2\right\} \tag{5}$$

by ignoring terms involving higher powers than X. Here $\Delta = (E_\alpha - E_\beta)/kT$. NOTE: The chain molecule treated in this problem presents a simplified model for keratin molecules in wool. It is presumed that this is the mechanism underlying the elasticity of wool.

NOTE: One may consider that the energy of a monomeric unit is $E_\alpha - Xa$ or $E_\beta - Xb$ depending on whether the unit is in the α or β state. By this reasoning, one can immediately write the last expression in (3) for the partition function of N independent units. In fact, it is unnecessary to be bothered with the number of states $N!/N_\alpha!N_\beta!$. This short cut is a merit of the use of the general canonical distribution. The readers are advised to become familiar with this way of reasoning.

14. The differential equation in question indicates that $\log Z$ is an extensive quantity. When x identical systems are assembled at a temperature T, the partition function of the resulting system is given by the partition function of a single system raised to the x-th power, provided that each component system is almost independent of others. Since the number of particles in the resultant is xN and the volume is xV, one has

$$Z(xN, xV, T) = [Z(N, V, T)]^x. \tag{1}$$

Taking the logarithms of both sides and differentiating them partially with x, one is led to

$$\left\{N \frac{\partial}{\partial(xN)} + V \frac{\partial}{\partial(xV)}\right\} \log Z(xN, xV, T) = \log Z(N, V, T).$$

The required equation can be obtained if x on the left-hand side is put equal to 1.

NOTE: Since $F = -kT \log Z$ is the Helmholtz free energy, the pressure p and the chemical potential μ are given by $p = -(\partial F/\partial V)_{N,T}$, and $\mu = (\partial F/\partial N)_{V,T}$, respectively. On the other hand, since the Gibbs free energy, $F + pV$, is $N\mu$ for a one-component system, the following relation holds:

$$F - \left(\frac{\partial F}{\partial V}\right)_{N,T} V = N\left(\frac{\partial F}{\partial N}\right)_{V,T} \tag{2}$$

Rewriting this, one gets the differential equation in question.

15. Let the partition functions for the phases I and II having N_A' and N_A'' particles of A, N_B' and N_B'' particles of B, and so on, respectively, be denoted by $Z_{\mathrm{I}}(N_A', N_B', \ldots)$ and $Z_{\mathrm{II}}(N_A'', N_B'', \ldots)$. Then the most probable distribution subject to the conditions that

$$N_A' + N_A'' = N_A = \text{const.}, \qquad N_B' + N_B'' = N_B = \text{const.}, \ldots$$

is determined by making $Z_{\mathrm{I}}(N'_A, \ldots)\, Z_{\mathrm{II}}(N''_A, \ldots)$ a maximum or by

$$\delta \log (Z_{\mathrm{I}} Z_{\mathrm{II}}) = \frac{\partial \log Z_{\mathrm{I}}}{\partial N'_A} \delta N'_A + \frac{\partial \log Z_{\mathrm{II}}}{\partial N''_A} \delta N''_A = 0, \qquad \delta N'_A + \delta N''_A = 0.$$

This condition is

$$\frac{\partial \log Z_{\mathrm{I}}}{\partial N'_A} = \frac{\partial \log Z_{\mathrm{II}}}{\partial N''_A}, \ldots . \tag{1}$$

On the other hand, since the relation between grand partition functions, $\Xi_{\mathrm{I}}(\lambda'_A, \ldots)$, $\Xi_{\mathrm{II}}(\lambda''_A, \ldots)$, and partition functions, Z_{I}, Z_{II}, is given by (2.27), (1) is nothing but $\lambda'_A = \lambda''_A = \lambda_A, \ldots$ Since (2.26) holds for each phase in addition, one has

$$N'_A : N''_A = \lambda_A \frac{\partial \log \Xi_{\mathrm{I}}}{\partial \lambda_A} : \lambda_A \frac{\partial \log \Xi_{\mathrm{II}}}{\partial \lambda_A} = \frac{\partial \log \Xi_{\mathrm{I}}}{\partial \lambda_A} : \frac{\partial \log \Xi_{\mathrm{II}}}{\partial \lambda_A}.$$

16. The part of the partition function of an oscillator originating from the potential function $V(q)$ is

$$z = \int_{-\infty}^{\infty} \mathrm{e}^{-V(q)/kT} \mathrm{d}q = \int_{-\infty}^{\infty} \exp\left(-\frac{cq^2}{kT}\right) \exp\left(\frac{gq^3 + fq^4}{kT}\right) \mathrm{d}q. \tag{1}$$

Since the coefficients g and f are small, the integral is largely determined by the first exponential function, which is appreciable only in the range $q^2 \lesssim kT/c$. In this range, $g|q|^3/kT \lesssim g\sqrt{kT}/(2c)^{\frac{3}{2}} \ll 1$, and $f\,|q|^4/kT \lesssim fkT/(2c)^2 \ll 1$. Therefore, the second exponential function in (1) can be expanded:

$$\begin{aligned} z &= \int_{-\infty}^{\infty} \exp\left(-\frac{cq^2}{kT}\right) \left\{1 + \frac{gq^3 + fq^4}{kT} + \frac{1}{2}\left(\frac{gq^3 + fq^4}{kT}\right)^2 + \ldots\right\} \mathrm{d}q \\ &= \int_{-\infty}^{\infty} \exp\left(-\frac{cq^2}{kT}\right) \left\{1 + \frac{fq^4}{kT} + \frac{1}{2}\frac{g^2 q^6}{(kT)^2} + \ldots\right\} \mathrm{d}q \\ &= \sqrt{\frac{\pi kT}{c}} \left\{1 + \frac{3}{4}\frac{f}{kT}\left(\frac{kT}{c}\right)^2 + \frac{1}{2}\frac{15}{8}\frac{g^2}{(kT)^2}\left(\frac{kT}{c}\right)^3 + \ldots\right\} \\ &= \sqrt{\frac{\pi kT}{c}} \left(1 + \frac{3}{4}\frac{fkT}{c^2} + \frac{15}{16}\frac{g^2 kT}{c^3} + \ldots\right). \end{aligned} \tag{2}$$

The contribution to the internal energy per oscillator is

$$u = kT^2 \frac{\partial \log z}{\partial T} = kT^2 \frac{\mathrm{d}}{\mathrm{d}T}\left[\frac{1}{2}\log\frac{\pi kT}{c} + \log\left\{1 + \frac{3}{4}\left(\frac{f}{c^2} + \frac{5}{4}\frac{g^2}{c^3}\right)kT + \ldots\right\}\right]$$

$$= \frac{1}{2}kT + kT^2 \frac{\mathrm{d}}{\mathrm{d}T}\left\{\frac{3}{4}\left(\frac{f}{c^2} + \frac{5}{4}\frac{g^2}{c^3}\right)kT + \ldots\right\}$$

$$= \frac{1}{2}kT + \frac{3}{4}\left(\frac{f}{c^2} + \frac{5}{4}\frac{g^2}{c^3}\right)(kT)^2 + \ldots. \tag{3}$$

Accordingly, the contribution to the heat capacity per oscillator is given by

$$C = \frac{\partial u}{\partial T} = \frac{k}{2} + \frac{3}{2}k\left(\frac{f}{c^2} + \frac{5}{4}\frac{g^2}{c^3}\right)kT + \ldots. \tag{4}$$

The first term in (3) as well as (4) is the value for a harmonic oscillator, while the second term is due to anharmonicity.

The mean value of q is finite because $V(q)$ is not an even function of q:

$$\bar{q} = \frac{1}{z}\int_{-\infty}^{\infty} q\,\mathrm{e}^{-V(q)/kT}\,\mathrm{d}q$$

$$= \frac{1}{z}\int_{-\infty}^{\infty} \exp\left(-\frac{cq^2}{kT}\right) q \left\{1 + \frac{gq^3 + fq^4}{kT} + \frac{1}{2}\frac{g^2q^6}{(kT)^2} + \ldots\right\}\mathrm{d}q$$

$$= \frac{1}{z}\int_{-\infty}^{\infty} \exp\left(-\frac{cq^2}{kT}\right)\left(\frac{gq^4}{kT} + \ldots\right)\mathrm{d}q$$

$$= \frac{1}{z}\sqrt{\frac{\pi kT}{c}}\left(\frac{3}{4}\frac{gkT}{c^2} + \ldots\right) = \frac{3}{4}\frac{gkT}{c^2} + \ldots. \tag{5}$$

The proportionality of q to T indicates that the position of the oscillator shifts as in thermal expansion.

17. Segments are numbered as 1, ..., N in sequence from one end. The state of the i-th segment is specified by its direction, θ_i and φ_i. (A line connecting the two ends of the molecule is chosen as the polar axis of polar coordinates.) For a given molecular shape, the number of microscopic states of the whole molecule is proportional to the product

$$\prod_{i=1}^{N} \mathrm{d}\omega_i$$

of elementary solid angles $d\omega_i = \sin\theta_i \, d\theta_i \, d\varphi_i$. The partition function of the canoniçal distribution under conditions of constant X is given by

$$Y = \int_{-\infty}^{\infty} e^{\beta X l} dl \quad \underset{l < \sum^N a\cos\theta_i < l + dl}{\int \cdots \int} \quad d\omega_1 \dots d\omega_N . \tag{1}$$

The second integral corresponds to the number of microscopic states falling in an infinitesimal length dl at a length l. Since the T-X canonical distribution is taken and integration is carried out over l, one has

$$Y = \int \cdots \int e^{\beta X \Sigma a \cos\theta_i} d\omega_1 \dots d\omega_N$$
$$= \prod_{i=1}^{N} \int e^{\beta X a \cos\theta_i} d\omega_i = \left[\iint e^{\beta X a \cos\theta} \sin\theta \, d\theta \, d\varphi \right]^N = \left[4\pi \frac{\sinh(\beta X a)}{\beta X a} \right]^N \tag{2}$$

(cf. (11) in the solution to example 2). Therefore, the relation between tension X and l is

$$l = \frac{1}{\beta} \frac{\partial \log Y}{\partial X} = NkT \frac{\partial}{\partial X} \log\left(\sinh\frac{Xa}{kT} \Big/ \frac{Xa}{kT} \right) = NaL\left(\frac{Xa}{kT} \right). \tag{3}$$

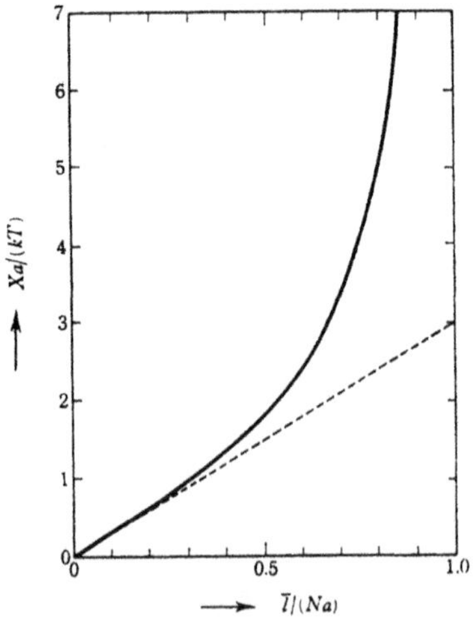

Here $L(x)$ is the Langevin function (cf. (5) in example 2).

NOTE 1: This problem involves the probability distribution of the sum of N independent vectors. The number of microscopic states $\Omega(l)$ for a given l

is equal to the second integral in (1). The integral has already been evaluated in example 6. (Let $M_z \to l$ and $\mu \to a$ into $\Omega(M_z)$.) Since

$$F(l) = -TS(l) = -kT\log\Omega(l)$$

one has

$$\frac{\partial F(l)}{\partial l} = X = -kT\frac{\partial \log\Omega(l)}{\partial l}.$$

If the tension X is calculated from the results obtained in example 6, one gets the same answer as (3). This is naturally to be expected. (Consider the correspondence between this problem and example 2.)

NOTE 2: This problem presents an idealized model for rubber-like chain molecules [cf. P. J. Flory, *Principles of Polymer Chemistry* (Cornell Univ. Press, Ithaca, 1953)].

18. The wave number q of radiation in a medium having a refractive index $n(\nu)$ is given by

$$q = n(\nu)2\pi\nu/c. \tag{1}$$

The number of modes in the wave number range from q to $q + \mathrm{d}q$ of electromagnetic vibration in a medium of volume V is given by $Vq^2\mathrm{d}q/\pi^2$, as has already been calculated in examples 3 and 4. By transforming this by means of (1) into the distribution over frequencies, one has

$$\mathrm{d}q = \frac{2\pi}{c}\left\{n(\nu) + \nu\frac{\mathrm{d}n(\nu)}{\mathrm{d}\nu}\right\}\mathrm{d}\nu = \frac{2\pi n(\nu)}{c}\frac{\mathrm{d}\log\{n(\nu)\nu\}}{\mathrm{d}\log\nu}\mathrm{d}\nu. \tag{2}$$

Therefore, the number of states having frequencies in the range from ν to $\nu + \mathrm{d}\nu$ is given by

$$Vg(\nu)\,\mathrm{d}\nu = \frac{V}{\pi^2}\left\{\frac{2\pi n(\nu)}{c}\right\}^3\nu^2\frac{\mathrm{d}\log\{n(\nu)\nu\}}{\mathrm{d}\log\nu}\mathrm{d}\nu. \tag{3}$$

Multiplying this by the mean energy of a harmonic oscillator,

$$\varepsilon(\nu, T) = \frac{h\nu}{e^{h\nu/kT} - 1}, \tag{4}$$

one is led to Planck's radiation formula:

$$u(\nu, T) = \frac{8\pi n^3(\nu)h\nu^3}{c^3}\frac{\mathrm{d}\log\{n(\nu)\nu\}}{\mathrm{d}\log\nu}\frac{1}{e^{h\nu/kT} - 1}. \tag{5}$$

19. As mentioned in example 3, the number of states of waves in the three-dimensional space having the wave number vector (q_x, q_y, q_z) in a volume $dq_x dq_y dq_z$ is given by $V dq_x dq_y dq_z/(2\pi)^3$. According to the dispersion formula, the distribution function (spectral function) $g(\omega)$ of vibrational modes is

$$g(\omega) = \frac{V}{(2\pi)^3} 4\pi q^2 \Big/ \left|\frac{d\omega}{dq}\right| = \frac{VA^{-3/n}}{2\pi^2 n} \omega^{(3-n)/n}, \qquad (\omega \leq \omega_c). \tag{1}$$

The spectrum is terminated at ω_c on the high-frequency side in conformity with the number of degrees of freedom. With this spectral function, the contribution to the specific heat is calculated by

$$C = k \int_0^{\omega_c} \frac{e^{\hbar\omega/kT}}{\{e^{\hbar\omega/kT} - 1\}^2} \left(\frac{\hbar\omega}{kT}\right)^2 \frac{VA^{-3/n}}{2\pi^2 n} \omega^{(3-n)/n} d\omega$$

$$= \frac{VA^{-3/n} k}{2\pi^2 n} \left(\frac{kT}{\hbar}\right)^{3/n} \int_0^{T_c/T} \frac{e^{\xi} \xi^{1+3/n}}{(e^{\xi} - 1)^2} d\xi \tag{2}$$

by use of (6) in the solution to example 3. Here $\hbar\omega_c$ is written as kT_c. At low temperature $T \ll T_c$, one may integrate up to $+\infty$. Therefore, $C \propto T^{3/n}$.
NOTE: $n = 1$ corresponds to elastic waves in solids (lattice vibrations) treated in example 3 while $n = 2$ applies to magnetization waves (spin waves) taking place in ferromagnetic substances.

20. As discussed in the solution to example 3, the heat capacity of a solid at constant volume can be expressed by means of the spectral distribution $g(\omega, V, N)$ of normal vibrations of a crystal lattice:

$$C_V(V, T, N) = \int_0^{\infty} \frac{\partial \varepsilon(\omega, T)}{\partial T} g(\omega, V, N) d\omega. \tag{1}$$

Here ε is the mean energy as a normal oscillator having an angular frequency ω:

$$\varepsilon(\omega, T) = \frac{\hbar\omega}{2} + \frac{\hbar\omega}{e^{\hbar\omega/kT} - 1}. \tag{2}$$

Since $\varepsilon(\omega, T) \to kT$ at the limit as $T \to \infty$, (1) gives the classical value:

$$C_\infty(V, T, N) = \int_0^{\infty} \frac{\partial(kT)}{\partial T} g(\omega, V, N) d\omega. \tag{3}$$

This is equal to k multiplied by the number ($3N-6$ in example 3) of degrees of freedom (Dulong-Petit's law), as is clear from the definition of $g(\omega, V, N)$:

$$\int_0^\infty g(\omega, V, N)\,\mathrm{d}\omega = \text{degrees of freedom of a crystal } -6. \qquad (4)$$

The shaded area in the figure is calculated from (1) and (3) as

$$\int_0^\infty [C_\infty(V,T,N) - C_V(V,T,N)]\,\mathrm{d}T$$

$$= \int\int_0^\infty \left[\frac{\partial}{\partial T}\{kT - \varepsilon(\omega, T)\}\right] g(\omega, V, N)\,\mathrm{d}T\,\mathrm{d}\omega$$

$$= \int_0^\infty \Big[kT - \varepsilon(\omega, T)\Big]_{T=0}^{T=\infty} g(\omega, V, N)\,\mathrm{d}\omega = \int_0^\infty \varepsilon(\omega, 0) g(\omega, V, N)\,\mathrm{d}\omega. \qquad (5)$$

As is clear from (2), $\varepsilon(\omega, 0) = \frac{1}{2}\hbar\omega$ is the zero-point energy of a harmonic oscillator. Accordingly, (5) is equal to the zero-point energy $U(V, 0, N)$ of the solid. The internal energy of the solid $U(V, T, N)$ is given by (4) in the solution to example 3.

21. The change in dimension does not affect (2) in the solution to example 3. Only the distribution function $g(\omega)$ of frequency is altered. Corresponding to (12) in the solution to example 3, the number of normal frequencies lying between q and $q + \mathrm{d}q$ of a one-dimensional crystal is $(L/2\pi)\mathrm{d}q$. Let the velocities of longitudinal and transverse waves be denoted by c_l and c_t, respectively. For longitudinal and transverse vibrations, the relation between the frequency ω and the wave number q is given, respectively, by

$$\omega_l = c_l q, \qquad \omega_t = c_t q.$$

Accordingly,

$$\left.\begin{aligned} g(\omega)\,\mathrm{d}\omega &= \frac{L}{2\pi}\left(\frac{1}{c_l} + \frac{2}{c_t}\right)\mathrm{d}\omega = \frac{3N}{\omega_\mathrm{D}}\,\mathrm{d}\omega \qquad && (\omega < \omega_\mathrm{D}), \\ &= 0 && (\omega > \omega_\mathrm{D}). \end{aligned}\right\} \qquad (1)$$

Here the terminating frequency ω_D in the Debye approximation is determined by

$$\int_0^{\omega_D} g(\omega)\,d\omega = 3N,$$

$$\omega_D = \frac{2\pi N}{L}c = 2\pi c/a \qquad (a = L/N,\ c^{-1} = \tfrac{1}{3}(c_l^{-1} + 2c_t^{-1})) \tag{2}$$

(which can be considered as the mean frequency corresponding to the wave length a). In the case of a two-dimensional crystal,

$$\iint_{q<(q_x^2+q_y^2)^{\frac{1}{2}}<q+dq} \frac{L^2}{(2\pi)^2}\,dq_x\,dq_y = 2\frac{L^2}{(2\pi)^2}\pi q\,dq. \tag{3}$$

Accordingly,

$$\left.\begin{aligned} g(\omega)\,d\omega &= \frac{L^2}{2\pi}\left(\frac{1}{c_l^2} + \frac{2}{c_t^2}\right)\omega\,d\omega = 6N\frac{\omega\,d\omega}{\omega_D^2} \qquad (\omega < \omega_D), \\ &= 0 \qquad (\omega > \omega_D). \end{aligned}\right\} \tag{4}$$

By use of (6) in the solution to example 3, the heat capacity can be written for one-dimensional and two-dimensional oscillators, respectively, as

$$C_1 = 3Nk\int_0^{\omega_D} \frac{e^{\hbar\omega/kT}}{\{e^{\hbar\omega/kT} - 1\}^2}\left(\frac{\hbar\omega}{kT}\right)^2\frac{d\omega}{\omega_D} = 3Nk\frac{T}{\Theta}\int_0^{\Theta/T}\frac{\xi^2 e^\xi}{(e^\xi - 1)^2}\,d\xi, \tag{5}$$

$$C_2 = 6Nk\int_0^{\omega_D} \frac{e^{\hbar\omega/kT}}{\{e^{\hbar\omega/kT} - 1\}^2}\left(\frac{\hbar\omega}{kT}\right)^2\frac{\omega\,d\omega}{\omega_D^2} = 6Nk\left(\frac{T}{\Theta}\right)^2\int_0^{\Theta/T}\frac{\xi^3 e^\xi}{(e^\xi - 1)^2}\,d\xi. \tag{6}$$

Here, $\Theta = \hbar\omega_D/k$. If $T \gg \Theta$, $C \to 3Nk$ in all cases (Dulong-Petit's classical value). If $T \ll \Theta$,

$$\begin{aligned} C_1 \sim 3Nk\frac{T}{\Theta}\int_0^\infty \frac{\xi^2 e^\xi\,d\xi}{(e^\xi - 1)^2} &= 3Nk\frac{T}{\Theta}\cdot 2\zeta(2) \\ &= \pi^2 Nk\frac{T}{\Theta}, \end{aligned} \tag{7}$$

$$\begin{aligned} C_2 \sim 6Nk\left(\frac{T}{\Theta}\right)^2\int_0^\infty \frac{\xi^3 e^\xi\,d\xi}{(e^\xi - 1)^2} &= Nk\cdot 36\zeta(3)\left(\frac{T}{\Theta}\right)^2 \\ &= 43.26Nk\left(\frac{T}{\Theta}\right)^2, \end{aligned} \tag{8}$$

cf. NOTE 2 to example 3 in calculations.

NOTE: Selenium and tellurium form crystals in which atomic chains are arranged in parallel as shown in Figure 2.21. For this reason, the T^1-law holds in a certain temperature range.

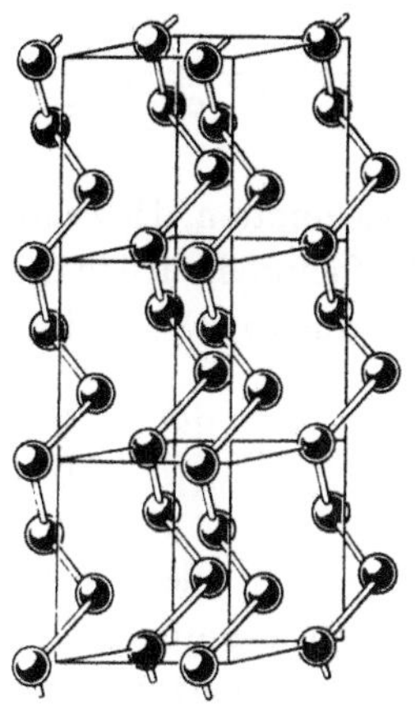

However, at very low temperatures, the T^3-law replaces it. Similarly, graphite obeys the T^2-law.

22. The state of the solid in question is specified by its volume V and the quantum numbers $n_j (j = 1, 2, \ldots, 3N - 6)$ of normal vibrations. The energy of this state is given by

$$E(V, n_1, n_2, \ldots) = \Phi(V) + \sum_{j=1}^{3N-6} (n_j + \tfrac{1}{2}) h\nu_j(V). \tag{1}$$

Accordingly, the partition function of the solid is

$$\begin{aligned} Z(T, V, N) &= \sum_{n_1=0}^{\infty} \sum_{n_2=0}^{\infty} \cdots \mathrm{e}^{-E(V, n_1, n_2, \cdots)/kT} \\ &= \mathrm{e}^{-\Phi(V)/kT} \prod_{j=1}^{3N-6} \frac{1}{2 \sinh \{h\nu_j(V)/2kT\}} . \end{aligned} \tag{2}$$

Therefore, the Helmholtz free energy is given by

$$F(T, V, N) = -kT \log Z = \Phi(V) + kT \sum_{j=1}^{3N-6} \log \left\{ 2 \sinh \frac{h\nu_j(V)}{2kT} \right\}. \tag{3}$$

The pressure is calculated as

$$p = -\left(\frac{\partial F}{\partial V}\right)_{T,N} = -\frac{\mathrm{d}\Phi}{\mathrm{d}V} - kT \sum_{j=1}^{3N-6} \coth\left(\frac{h\nu_j}{2kT}\right)\frac{h}{2kT}\frac{\mathrm{d}\nu_j}{\mathrm{d}V}$$

$$= -\frac{\mathrm{d}\Phi}{\mathrm{d}V} - \sum_{j=1}^{3N-6}\left\{\tfrac{1}{2}h\nu_j + \frac{h\nu_j}{\mathrm{e}^{h\nu_j/kT}-1}\right\}\frac{\mathrm{d}\log\nu_j}{\mathrm{d}V}. \tag{4}$$

According to Grüneisen's assumption, $\mathrm{d}\log\nu_j/\mathrm{d}V = -\gamma/V$ is independent of the numbering j and can be factored outside the summation, which gives the internal energy $U(V, T, N)$ due to atomic vibrations:

$$p = -\frac{\mathrm{d}\Phi}{\mathrm{d}V} + \gamma\frac{U}{V}. \tag{5}$$

The first term on the right-hand side is a pressure originating from the fact that the distances between the atoms at rest in the equilibrium state of the solid differ from their normal lengths (which corresponds to the volume $V = V_0$ leading to $\mathrm{d}\Phi(V)/\mathrm{d}V = 0$). The second term stands for a pressure which appears because, when the solid is compressed to decrease its volume V, the vibrations of the atoms about their equilibrium positions become more violent ($\nu_j \propto V^{-\gamma}$) and resist the compression.

If the given formula of parabolic type is assumed for $\Phi(V)$, the equation of state (5) can be transformed into

$$p = -\frac{V - V_0}{\kappa_0 V_0} + \gamma\frac{U(V, T, N)}{V}. \tag{6}$$

The volume expansion coefficient at constant pressure is defined as $\alpha = (\partial V/\partial T)_{p,N}/V$. Differentiating both sides of (6) partially with respect to T at fixed p and N (V is a function of p, N and T), one gets

$$0 = -\frac{\alpha V}{\kappa_0 V_0} + \gamma\left\{-\frac{U}{V^2}\alpha V + \frac{\alpha V}{V}\left(\frac{\partial U}{\partial V}\right)_{T,N} + \frac{1}{V}\left(\frac{\partial U}{\partial T}\right)_{V,N}\right\}. \tag{7}$$

Since the heat capacity at constant volume is $C_V = (\partial U/\partial T)_{V,N}$ and the vibrational part of the internal energy is

$$U(T, V, N) = \sum_{j=1}^{3N-6}\left\{\tfrac{1}{2}h\nu_j + \frac{h\nu_j}{\mathrm{e}^{h\nu_j/kT}-1}\right\}, \tag{8}$$

one has

$$C_V = k \sum_{j=1}^{3N-6} \frac{e^{h\nu_j/kT}}{\{e^{h\nu_j/kT} - 1\}^2} \left(\frac{h\nu_j}{kT}\right)^2, \tag{9}$$

$$\left(\frac{\partial U}{\partial V}\right)_{T,N} = \sum_{j=1}^{3N-6} \left[\frac{h\nu_j}{2} + \frac{h\nu_j}{e^{h\nu_j/kT} - 1} - kT \frac{e^{h\nu_j/kT}}{\{e^{h\nu_j/kT} - 1\}^2} \left(\frac{h\nu_j}{kT}\right)^2\right] \frac{d \log \nu_j}{dV}$$

$$= \gamma \frac{C_V T}{V} - \gamma \frac{U}{V}. \tag{10}$$

Rewriting (7) by use of (10), one obtains

$$\alpha \left\{\frac{V}{\kappa_0 V_0} + \gamma \frac{U}{V} + \gamma^2 \left(\frac{U}{V} - \frac{C_V T}{V}\right)\right\} = \gamma \frac{C_V}{V}.$$

Eliminating $\gamma U/V$ by use of (6), one has

$$\alpha = \frac{\gamma C_V}{V} \Bigg/ \left\{\frac{1}{\kappa_0} + (1 + \gamma) p + (2 + \gamma) \frac{V - V_0}{\kappa_0 V_0} - \gamma^2 \frac{C_V T}{V}\right\} \tag{11}$$

for the volume expansion coefficient at constant pressure. Clearly, the denominator is the reciprocal of the isothermal compressibility $\kappa = -(\partial V/\partial p)_{T,N}/V$. At temperatures satisfying $\kappa_0 C_V T/V_0 \ll 1$ at $p = 0$, the second term of the denominator in (11) vanishes and the fourth term is negligible compared with the first term. Moreover, since the approximation of $\Phi(V)$ by means of a parabolic function is feasible only for $|V - V_0| \ll V_0$, the third term may be disregarded in comparison with the first term. Then, (11) can be written as

$$\alpha \cong \frac{\gamma}{V_0} C_V \kappa_0. \tag{12}$$

In other words, α is proportional to C_V.

23. As shown by (1) in the solution to example 8 in Chapter 1, the classical translational partition function of an ideal gas is

$$Z(\beta) = \frac{V^N}{N!} \left(\frac{2\pi m}{\beta h^2}\right)^{\frac{3}{2}N}. \tag{1}$$

According to (2.6),

$$\Omega(E) = \frac{V^N}{N!} \left(\frac{2\pi m}{h^2}\right)^{\frac{3}{2}N} \times \frac{1}{2\pi i} \int_{\beta' - i\infty}^{\beta' + i\infty} \frac{e^{\beta E}}{\beta^{\frac{3}{2}N}} d\beta, \qquad (\beta' > 0). \tag{2}$$

(i) If $E < 0$, a semicircle of radius Y with β' as its center is conceived on the right side as shown in figure (a). Within this semicircle, no singular points exist for the integrand of (2). Therefore, the integral around the contour vanishes: $\int_C = 0$ (Cauchy's theorem).

At the limit of $Y \to \infty$, the integral along the semicircle is evidently zero, and thus

$$\int_{C^-} = \int_{\beta'-i\infty}^{\beta'+i\infty} = 0, \qquad \text{therefore} \qquad \Omega(E) = 0.$$

This is to be expected, since no states with $E < 0$ exist.

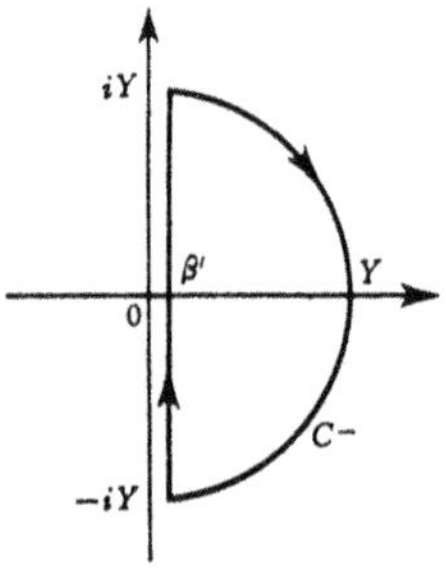

(a)

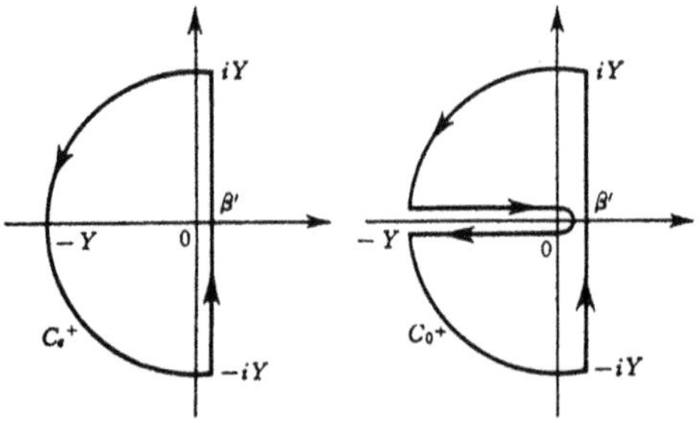

(b) (c)

(ii) If $E > 0$, one must treat the cases of even and odd N separately. For even N, a semicircle is conceived on the left side as shown in figure (b). The semicircle encloses the origin O, which is a pole in the β-plane. Therefore, Cauchy's theorem

$$\frac{1}{2\pi i}\oint f(z)\frac{dz}{(z-a)^{n+1}} = \frac{1}{n!}f^{(n)}(a)$$

leads to

$$\frac{1}{2\pi i}\int_{C_0^+}\frac{e^{\beta E}}{\beta^{\frac{1}{2}N}}d\beta = \frac{1}{(\frac{1}{2}N-1)!}\left(\frac{d}{d\beta}\right)^{\frac{1}{2}N-1} e^{\beta E}\bigg|_{\beta=0} = \frac{E^{\frac{1}{2}N-1}}{\Gamma(\frac{1}{2}N)}. \tag{3}$$

Now, the integral along the semi-circular arc vanishes at the limit of the radius tending to infinity. Therefore, the integral (2) is evaluated as (3). For odd N, the path of integration is chosen as shown in figure (c). In this case, $\beta = 0$ is a branch point of the second rank. Since this closed integration path encloses no singular points, the integral along the path vanishes. The integral around each of the quadrants vanishes at the limit of infinite radius. Therefore,

$$\frac{1}{2\pi i}\int_{\beta'-i\infty}^{\beta'+i\infty}\frac{e^{\beta E}}{\beta^{\frac{1}{2}N}}d\beta = \frac{1}{2\pi i}\int_{C}\frac{e^{\beta E}}{\beta^{\frac{1}{2}N}}d\beta.$$

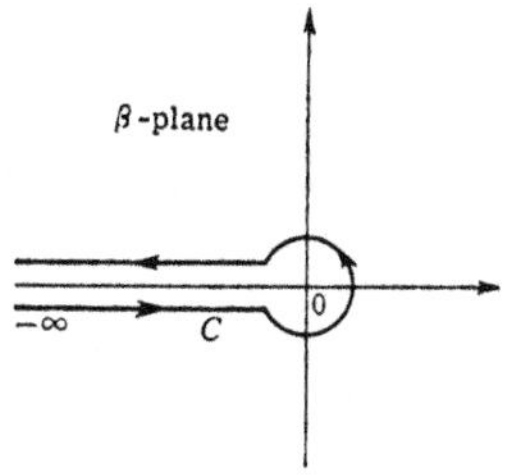

Here C stands for the path of integration shown in the figure above. In order to perform this integration, the power of β in the denominator of the integrand is reduced successively:

$$\frac{1}{2\pi i}\int_{C}\frac{e^{\beta E}}{\beta^{\frac{1}{2}N}}d\beta = \frac{E}{\frac{1}{2}N-1}\,\frac{1}{2\pi i}\int_{C}\frac{e^{\beta E}\,d\beta}{\beta^{\frac{1}{2}N-1}}$$

$$= \frac{E^{\frac{1}{2}N-\frac{1}{2}}}{(\frac{1}{2}N-1)(\frac{1}{2}N-2)\ldots\frac{3}{2}\cdot\frac{1}{2}}\,\frac{1}{2\pi i}\int_{C}\frac{e^{\beta E}}{\beta^{\frac{1}{2}}}d\beta. \tag{4}$$

The last integral can be transformed into a integral along the real axis in the γ-plane by means of $\gamma = \beta^{\frac{1}{2}}/i$:

$$\frac{1}{2\pi i}\int_{C}\frac{e^{\beta E}}{\beta^{\frac{1}{2}}}d\beta = \frac{1}{\pi}\int_{-\infty}^{\infty}e^{-E\gamma^2}d\gamma = \frac{1}{\sqrt{\pi E}}, \qquad (E>0). \tag{5}$$

By use of the relations for Γ-functions, $\Gamma(\frac{1}{2}) = \sqrt{\pi}$, and $\Gamma(z) = (z-1)\,\Gamma(z)$, (5) is introduced into (4). The resulting expression can be written in the same form as (3). Therefore,

$$\Omega(E) = \begin{cases} \dfrac{V^N}{N!}\left(\dfrac{2\pi m}{h^2}\right)^{\frac{3}{2}N} \dfrac{E^{\frac{3}{2}N-1}}{\Gamma(\frac{3}{2}N)}, & (E > 0) \\ 0, & (E \leqq 0). \end{cases} \tag{6}$$

For $E = 0$, one obtains the mean value of the aforementioned two expressions at the limit of $E \to 0$. Since both are equal to zero, the result for $E = 0$ is included in $E \leqq 0$ of (6).

NOTE: For any positive number n, the following relation holds:

$$\frac{1}{2\pi \mathrm{i}} \int_{s'-\mathrm{i}\infty}^{s'+\mathrm{i}\infty} \frac{\mathrm{e}^{sx}\,\mathrm{d}s}{s^{n+1}} = \frac{1}{2\pi \mathrm{i}} \int_C \frac{\mathrm{e}^{sx}}{s^{n+1}}\,\mathrm{d}s = \frac{x^n}{\Gamma(n+1)}. \tag{7}$$

This is rather a general definition of Γ- functions.

24. The partition function for a single oscillator is given by

$$Z_1 = \frac{1}{h}\iint \exp\left[-\beta\left(\frac{p^2}{2m} + \tfrac{1}{2}m\omega^2 q^2\right)\right]\mathrm{d}p\,\mathrm{d}q = \frac{1}{\beta\hbar\omega}.$$

Hence, for N oscillators,

$$Z_N = (\beta\hbar\omega)^{-N}. \tag{1}$$

The state density is given by

$$\Omega(E) = \frac{1}{2\pi\mathrm{i}} \int_{\beta'-\mathrm{i}\infty}^{\beta'+\mathrm{i}\infty} \mathrm{e}^{\beta E}(\beta\hbar\omega)^{-N}\,\mathrm{d}\beta \qquad (\Re\beta' > 0)$$

$$= \frac{1}{\hbar\omega}\frac{1}{2\pi\mathrm{i}} \int_{z'-\mathrm{i}\infty}^{z'+\mathrm{i}\infty} \mathrm{e}^{zE/\hbar\omega} z^{-N}\,\mathrm{d}z = \frac{1}{\hbar\omega}\left(\frac{E}{\hbar\omega}\right)^{N-1}\frac{1}{\Gamma(N)}. \tag{2}$$

(Calculations are similar to those for (3) in problem 23.) On the other hand, the second integral ($x \gg 1$, $N \gg 1$)

$$\phi(x) = \frac{1}{2\pi\mathrm{i}} \int_{z'-\mathrm{i}\infty}^{z'+\mathrm{i}\infty} \mathrm{e}^{zx} z^{-N}\,\mathrm{d}z = \frac{1}{2\pi\mathrm{i}} \int_{z'-\mathrm{i}\infty}^{z'+\mathrm{i}\infty} \exp(zx - N\log z)\,\mathrm{d}z$$

can be evaluated by the saddle-point method, the point being determined by

$$\frac{\mathrm{d}}{\mathrm{d}z}(zx - N\log z)\bigg|_{z=z^*} = x - \frac{N}{z^*} = 0, \qquad \text{or} \quad z^* = \frac{N}{x}.$$

Accordingly,

$$\log \phi(x) \sim N - N \log N + N \log x . \tag{3}$$

Therefore, corresponding to (2), one has

$$\log \Omega(E) \sim N \log \frac{E}{\hbar\omega} + N - N \log N . \tag{4}$$

Therefore,

$$\log \Gamma(N) \sim N \log N - N . \tag{5}$$

This is identical with

$$\log N! \sim N \log N - N \tag{6}$$

if unity is disregarded in comparison with N.

25. Let two dipole moments be denoted by the vectors $\boldsymbol{m}$ and $\boldsymbol{m}'$. The interaction energy Φ between them is given by

$$\Phi = -\boldsymbol{m} \cdot \frac{3\boldsymbol{R}\boldsymbol{R} - R^2 \mathbf{1}}{R^5} \cdot \boldsymbol{m}' \tag{1}$$

$$= -\frac{mm'}{R^3} \{2 \cos\theta \cos\theta' - \sin\theta \sin\theta' \cos(\varphi - \varphi')\} . \tag{2}$$

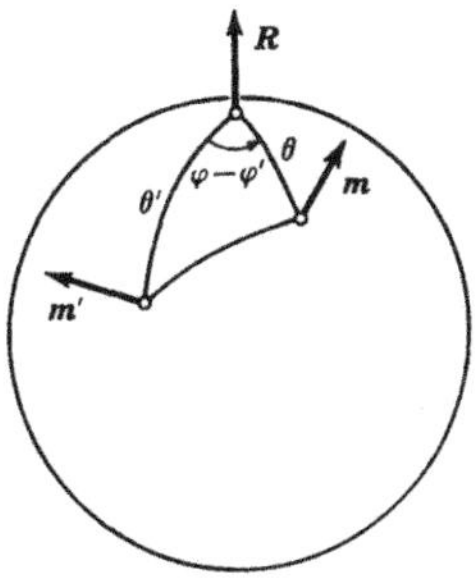

Here, (1) is expressed in the form of a dyadic, while (2) is written by means of angles shown in the figure. (Since as shown in example 2, the momenta involved in the integral give only the weights, $\sin\theta$ and $\sin\theta'$, for angular variables and do not appear in (2), they are not taken into account explicitly.) When these dipoles are distributed canonically at a temperature T, the partition function is given by

$$Z = \int d\omega \int d\omega' \exp\left(-\frac{\Phi}{kT}\right)$$
$$= \iint \sin\theta \, d\theta \, d\varphi \iint \sin\theta' \, d\theta' \, d\varphi' \times$$
$$\times \exp\left[-\frac{mm'}{kTR^3}\{2\cos\theta\cos\theta' - \sin\theta\sin\theta'\cos(\varphi-\varphi')\}\right].$$

At high temperatures ($mm'/kTR^3 \ll 1$), the exponential function can be expanded. The first term vanishes on integration, while the second term yields

$$\int\int\int\int d\cos\theta \, d\cos\theta' \, d\varphi \, d\varphi' \{2\cos\theta\cos\theta' - \sin\theta\sin\theta'\cos(\varphi-\varphi')\}^2 =$$
$$= (2\pi)^2 \int d\cos\theta \int d\cos\theta' \{4\cos^2\theta\cos^2\theta' + \tfrac{1}{2}\sin^2\theta\sin^2\theta'\} = (4\pi)^2 \times \tfrac{2}{3}.$$

Accordingly,

$$Z = (4\pi)^2 \left\{1 + \frac{1}{3}\frac{m^2 m'^2}{R^6 k^2 T^2} + \cdots\right\}. \tag{3}$$

On the other hand, the force acting between these dipoles is averaged as

$$\bar{\boldsymbol{F}} = -\overline{\frac{\partial\Phi}{\partial\boldsymbol{R}}} = -\iint d\omega \, d\omega' \frac{\partial\Phi}{\partial\boldsymbol{R}} \exp\left(-\frac{\Phi}{kT}\right) \Big/ \iint d\omega \, d\omega' \exp\left(-\frac{\Phi}{kT}\right)$$
$$= kT \frac{\partial}{\partial\boldsymbol{R}} \log Z.$$

Therefore, one has from (3)

$$\bar{\boldsymbol{F}} = -2\frac{m^2 m'^2}{kT}\frac{1}{R^7}\frac{\boldsymbol{R}}{R}. \tag{4}$$

Thus an attraction inversely proportional to R^7 acts between the two dipoles. NOTE: Dipoles have greater probability to orient themselves so as to attract each other. On the average, an attraction known as the Keesom force results.

26. Let W + R be regarded as a heat source for the system S. As shown in § 1.12, the distribution function for the system is given by

$$\rho(E) \propto \exp\left[\frac{1}{k}\{S_{W+R}(E) - S_{W+R}(0)\}\right], \tag{1}$$

$S_{W+R}(E)$ is the entropy of the system S having an energy E. It is the sum of the entropy S_W of the work source W and the entropy S_R of the heat source R. Since the former is invariant, (1) can be transformed into

$$\rho(E) \propto \exp\left[\frac{1}{k}\{S_R(E_R) - S_R(E_R^0)\}\right]. \tag{2}$$

Here, $S_R(E_R)$ is the entropy of the heat source R when the energy of the system S is E, and $S_R(E_R^0)$ is the corresponding value for $E = 0$.

When R gives energy E to S, it receives energy nE from W. Accordingly,

$$E_R = E_R^0 - E + nE. \tag{3}$$

Introducing this into (2), one gets

$$\rho(E) \propto \exp\left[\frac{1}{k}\{S_R(E_R^0 - (1-n)E) - S_R(E_R^0)\}\right]. \tag{4}$$

Further calculations can be made along the line shown in § 1.12.

27. In the presence of a magnetic field, the partition function for the given Hamiltonian is

$$\int \cdots \int \exp(-\mathscr{H}/kT)\,\mathrm{d}\boldsymbol{p}_1 \cdots \mathrm{d}\boldsymbol{p}_N \mathrm{d}\boldsymbol{r}_1 \cdots \mathrm{d}\boldsymbol{r}_N.$$

Integration over $\boldsymbol{p}_1, \ldots, \boldsymbol{p}_N$ yields only the factor $(2\pi mkT)^{\frac{3}{2}N}$, which does not depend on $\boldsymbol{A}$. Accordingly, the free energy is independent of the magnetic field $\boldsymbol{H}$. Hence, the magnetization $\boldsymbol{M} = -\partial F/\partial \boldsymbol{H}$ vanishes.

28. Let the coordinates and momenta of N particles be denoted by x_j and p_j ($j = 1, 2, \ldots, N$) and the interaction potential by $\varphi(|x_i - x_j|)$. The Hamiltonian of the system takes the following form:

$$\mathscr{H} = \sum_{j=1}^{N} \frac{p_j^2}{2m_j} + \sum_{j=1}^{N-1} \varphi(x_{j+1} - x_j). \tag{1}$$

Since the volume (length) of the system is $L = x_N - x_1 \geqq 0$, the partition function of the T-p distribution in (1.77) is

$$Y = \int_0^\infty \mathrm{d}L \int_{-\infty}^{\infty} \cdots \int \prod_{j=1}^{N} \frac{\mathrm{d}p_j}{h} \int \cdots \int \prod_{i=1}^{N} \mathrm{d}x_i\, \mathrm{e}^{-\beta(\mathscr{H}+pL)} - Q \cdot \prod_{j=1}^{N} \left(\frac{2\pi m_j kT}{h^2}\right)^{\frac{1}{2}}. \tag{2}$$

Here, integration over coordinates is performed under the conditions that the center of mass of the system of particles is at rest, the volume is L, and

$x_1 < x_2 < \ldots < x_N$. Since (2) contains an integral over L, the condition of constant volume is not obeyed.

$$Q = \int_{-\infty}^{\infty} \mathrm{d}x_N \int_{-\infty}^{x_N} \mathrm{d}x_{N-1} \cdots \int_{-\infty}^{x_2} \mathrm{d}x_1\, \delta\left(\frac{1}{M}\sum_{i=1}^{N} m_i x_i\right) \mathrm{e}^{-\beta(\Sigma\varphi + pL)}. \tag{3}$$

Here, δ is Dirac's delta function and $M = \sum_{i=1}^{N} m_i$ is the total mass of the system, the center of mass being chosen as the origin of coordinates. Introducing the new variables, $\xi_j = x_{j+1} - x_j (j = 1, 2, \ldots, N-1)$ and using the Fourier decomposition of the delta function,

$$\delta(x) = \frac{1}{2\pi} \int_{-\infty}^{\infty} \mathrm{d}\omega\, \mathrm{e}^{\mathrm{i}\omega x} \tag{4}$$

one can rewrite (3) in the following form:

$$\begin{aligned} Q &= \frac{1}{2\pi} \int_{-\infty}^{\infty} \mathrm{d}\omega \int_{-\infty}^{\infty} \mathrm{d}x_N\, \mathrm{e}^{\mathrm{i}\omega x_N} \prod_{j=1}^{N-1} \int_{0}^{\infty} \exp\left\{-\beta\varphi(\xi_j) - \left(\beta p + \frac{\mathrm{i}\omega}{M}\sum_{i=1}^{j} m_i\right)\xi_j\right\} \mathrm{d}\xi_j \\ &= \int_{-\infty}^{\infty} \mathrm{d}\omega\, \delta(\omega) \prod_{j=1}^{N-1} \int_{0}^{\infty} \exp\left\{-\beta\varphi(\xi) - \left(\beta p + \frac{\mathrm{i}\omega}{M}\sum_{i=1}^{j} m_i\right)\xi\right\} \mathrm{d}\xi \\ &= \left[\int_{0}^{\infty} \exp\{-\beta\varphi(\xi) - \beta p\xi\}\, \mathrm{d}\xi\right]^{N-1}. \end{aligned} \tag{5}$$

The mean value of the volume L is given by

$$\bar{L} = -\frac{1}{\beta}\left(\frac{\partial \log Y}{\partial p}\right)_\beta = -\frac{1}{\beta}\left(\frac{\partial \log Q}{\partial p}\right)_\beta = -\frac{N-1}{\beta}\frac{\partial \log I(\beta, p)}{\partial p}. \tag{6}$$

In (6), the following relations are used:

$$I(\beta, p) = \int_{0}^{\infty} f(x)\, \mathrm{e}^{-\beta p x}\, \mathrm{d}x, \qquad f(x) = \mathrm{e}^{-\beta\varphi(x)} \geqq 0. \tag{7}$$

Since this is the Laplace transformation of $f(x)$, $I(\beta, p)$ is normal and has no singular points provided that $p > 0$ and that the integral converges. In order to show that $\bar{L}$ is a monotonic function of the pressure p, it suffices to show that

$$\left(\frac{\partial \bar{L}}{\partial p}\right)_\beta = -\frac{N-1}{\beta}\frac{\partial^2 \log I}{\partial p^2} = -\frac{N-1}{\beta}\frac{II'' - I'^2}{I^2} \tag{8}$$

does not change its sign on writing $(\partial I/\partial p)_\beta = I'$. According to (7), along with $w(x) = f(x)\,e^{-\beta px} \geqq 0$, one has

$$I = \int_0^\infty w(x)\,dx\,, \qquad I' = \int_0^\infty (-\beta x) w(x)\,dx\,, \qquad I'' = \int_0^\infty (-\beta x)^2 w(x)\,dx\,. \tag{9}$$

Accordingly, I'/I is the mean value of $-\beta x$, while I''/I is that of $(-\beta x)^2$, both averaged with the weighting factor $w(x)$. Therefore,

$$\frac{I''}{I} - \left(\frac{I'}{I}\right)^2 = \frac{1}{I}\int_0^\infty \left(-\beta x - \frac{I'}{I}\right)^2 w(x)\,dx > 0 \tag{10}$$

gives the square of fluctuations in $-\beta x$ and hence is always positive. According to (8), this implies

$$\left(\frac{\partial \bar{L}}{\partial p}\right)_\beta < 0\,. \tag{11}$$

Divertissement 7

Foundation of Statistical Mechanics. Physics is the prime example of the exact sciences, and statistical mechanics constitutes one of the important bases of physics. Now, if one is told that the foundation of statistical mechanics is not entirely clear, he would feel rather astounded and puzzled. Being one of those engaged in this discipline, the writer feels a little embarrassed to admit this, but it is indeed the fact. There is, nevertheless, no doubt about the validity of statistical mechanics. When one is concerned with its application, it would be better if one uses it as much as possible without being too much disturbed.

The problem here is how to justify the principle of equal weight. There are rather many physicists who are not satisfied with the ergodic theorem, mentioned in § 1.3 and Divertissement 4. The mathematical proof of this theorem is too general and does not make use of the characteristic property of dynamical systems in physics such as we deal with in statistical mechanics. We tend to think, therefore, that in this theory we are not really grasping the physical circumstances which make statistical mechanics successful. One would feel that the basis for relating the value of actual observation to that which has to do with probability lies in the fact that an actual system consists of a large number of particles. Although this intuition is probably true, it has not been fully clarified.

Basic problems such as this are not raised just from philosophical interest alone. Indeed, we cannot avoid these problems when we want to advance the study of statistical mechanics to fields thus far unexplored, such as the problem of irreversible processes. Besides Khinchin's book (see Divertissement 4), the appendix to ter Haar's textbook† is also recommended as a concise introduction to this matter.

29. The components, σ_x, σ_y and σ_z, of Pauli's spin operator are operators satisfying the following relations:

$$\begin{aligned} \sigma_x^2 = \sigma_y^2 = \sigma_z^2 = 1\,, &\qquad \sigma_x\sigma_y = -\,\sigma_y\sigma_x = \mathrm{i}\sigma_z\,, \\ \sigma_y\sigma_z = -\,\sigma_z\sigma_y = \mathrm{i}\sigma_x\,, &\qquad \sigma_z\sigma_x = -\,\sigma_x\sigma_z = \mathrm{i}\sigma_y\,. \end{aligned} \tag{1}$$

Since $\boldsymbol{\sigma}\cdot\boldsymbol{H}$ is the scalar product of vectors, the Hamiltonian is

$$\mathscr{H} = -\,\mu_{\mathrm{B}}H\sigma_z\,, \tag{2}$$

the z-axis being taken along the direction of $\boldsymbol{H}$.

(i) Since $\sigma_z^2 = 1$, the eigenvalue of σ_z is $\pm\,1$. In the expression that diagonalizes σ_z,

$$\sigma_x = \begin{pmatrix} 0 & 1 \\ 1 & 0 \end{pmatrix}, \qquad \sigma_y = \begin{pmatrix} 0 & -\mathrm{i} \\ \mathrm{i} & 0 \end{pmatrix}, \qquad \sigma_z = \begin{pmatrix} 1 & 0 \\ 0 & -1 \end{pmatrix}. \tag{3}$$

In this expression, the density matrix $\rho = \mathrm{e}^{-\beta\mathscr{H}}/\mathrm{Tr}\,\mathrm{e}^{-\beta\mathscr{H}}$ can be written as

$$\rho = \frac{1}{\mathrm{e}^{\beta\mu_{\mathrm{B}}H} + \mathrm{e}^{-\beta\mu_{\mathrm{B}}H}} \begin{pmatrix} \mathrm{e}^{\beta\mu_{\mathrm{B}}H} & 0 \\ 0 & \mathrm{e}^{-\beta\mu_{\mathrm{B}}H} \end{pmatrix}. \tag{4}$$

The mean value of σ_z is evaluated from (2.43) as

$$\overline{\sigma_z} = \mathrm{Tr}(\rho\sigma_z) = \frac{\mathrm{e}^{\beta\mu_{\mathrm{B}}H} - \mathrm{e}^{-\beta\mu_{\mathrm{B}}H}}{\mathrm{e}^{\beta\mu_{\mathrm{B}}H} + \mathrm{e}^{-\beta\mu_{\mathrm{B}}H}} = \tanh(\beta\mu_{\mathrm{B}}H)\,. \tag{5}$$

(ii) From expression (3), the expression diagonalizing σ_x can be obtained by carrying out the transformation:

$$\begin{pmatrix} 1/\sqrt{2} & -1/\sqrt{2} \\ 1/\sqrt{2} & 1/\sqrt{2} \end{pmatrix}$$

and thus

$$\sigma_x = \begin{pmatrix} 1/\sqrt{2} & 1/\sqrt{2} \\ -1/\sqrt{2} & 1/\sqrt{2} \end{pmatrix} \begin{pmatrix} 0 & 1 \\ 1 & 0 \end{pmatrix} \begin{pmatrix} 1/\sqrt{2} & -1/\sqrt{2} \\ 1/\sqrt{2} & 1/\sqrt{2} \end{pmatrix} = \begin{pmatrix} 1 & 0 \\ 0 & -1 \end{pmatrix}. \tag{6}$$

Therefore, on applying this transformation to (4), one has

$$\begin{aligned} \rho &= \frac{1}{\mathrm{e}^{\beta\mu_{\mathrm{B}}H} + \mathrm{e}^{-\beta\mu_{\mathrm{B}}H}} \begin{pmatrix} 1/\sqrt{2} & 1/\sqrt{2} \\ -1/\sqrt{2} & 1/\sqrt{2} \end{pmatrix} \begin{pmatrix} \mathrm{e}^{\beta\mu_{\mathrm{B}}H} & 0 \\ 0 & \mathrm{e}^{-\beta\mu_{\mathrm{B}}H} \end{pmatrix} \begin{pmatrix} 1/\sqrt{2} & -1/\sqrt{2} \\ 1/\sqrt{2} & 1/\sqrt{2} \end{pmatrix} \\ &= \frac{1}{2}\begin{bmatrix} 1 & -\tanh(\beta\mu_{\mathrm{B}}H) \\ -\tanh(\beta\mu_{\mathrm{B}}H) & 1 \end{bmatrix}. \end{aligned} \tag{7}$$

† D. ter Haar: *Elements of Statistical Mechanics* (Reinhart and Co., New York, 1954).

In this representation, σ_z turns out to be

$$\sigma_z = \begin{pmatrix} 1/\sqrt{2} & 1/\sqrt{2} \\ -1/\sqrt{2} & 1/\sqrt{2} \end{pmatrix}\begin{pmatrix} 1 & 0 \\ 0 & -1 \end{pmatrix}\begin{pmatrix} 1/\sqrt{2} & -1/\sqrt{2} \\ 1/\sqrt{2} & 1/\sqrt{2} \end{pmatrix} = \begin{pmatrix} 0 & -1 \\ -1 & 0 \end{pmatrix}. \tag{8}$$

Therefore, its mean value evaluated by (2.43) is

$$\begin{aligned} \overline{\sigma_z} &= \mathrm{Tr}\frac{1}{2}\begin{pmatrix} 1 & -\tanh(\beta\mu_B H) \\ -\tanh(\beta\mu_B H) & 1 \end{pmatrix}\begin{pmatrix} 0 & -1 \\ -1 & 0 \end{pmatrix} \\ &= \frac{1}{2}\mathrm{Tr}\begin{pmatrix} \tanh(\beta\mu_B H) & -1 \\ -1 & \tanh(\beta\mu_B H) \end{pmatrix} = \tanh(\beta\mu_B H) \end{aligned} \tag{9}$$

in agreement with (5). This is naturally to be expected, because mean values (trace) are independent of the representation used (cf. note 1).

NOTE 1: When a unitary transformation ($UU^* = 1$, $U^* = U^{-1}$) is applied to a matrix A to yield $A' = UAU^{-1}$, $\mathrm{Tr}\, A = \mathrm{Tr}\, A'$. This is because

$$a'_{ik} = \sum_l \sum_m u_{il} a_{lm} u^*_{mk}, \quad \text{and thus} \quad \sum a'_{ii} = \sum_i \sum_l \sum_m u_{il} u^*_{mi} a_{lm} = \sum_l \sum_m \delta_{lm} a_{lm} = \sum_l a_{ll}.$$

NOTE 2: Since $\sigma_z^2 = 1$, one has

$$\begin{aligned} e^{\beta\mu_B H\sigma_z} &= \sum_{n=0}^{\infty} \frac{1}{n!}(\beta\mu_B H)^n \sigma_z^n = \sum_{n\ \text{even}} \frac{1}{n!}(\beta\mu_B H)^n + \sum_{n\ \text{odd}} \frac{1}{n!}(\beta\mu_B H)^n \sigma_z \\ &= \cosh(\beta\mu_B H) + \sigma_z \sinh(\beta\mu_B H). \end{aligned}$$

Since $\mathrm{Tr}\ 1 = 2$, $\mathrm{Tr}\ \sigma_z = 0$ and $\mathrm{Tr}\,\sigma_z^2 = 2$,

$$\mathrm{Tr}\, e^{\beta\mu_B H\sigma_z} = 2\cosh(\beta\mu_B H), \qquad \mathrm{Tr}\,\sigma_z e^{\beta\mu_B H\sigma_z} = 2\sinh(\beta\mu_B H),$$

$$\overline{\sigma_z} = \mathrm{Tr}\,\sigma_z e^{\beta\mu_B H\sigma_z}/\mathrm{Tr}\, e^{\beta\mu_B H\sigma_z} = \tanh(\beta\mu_B H).$$

Thus, it is often convenient to carry out calculations without resort to explicit expressions.

30. According to definition (2.41) (the normalization factor Z^{-1} being omitted),

$$\langle q' | e^{-\beta\mathscr{H}} | q'' \rangle = \sum_{n=0}^{\infty} e^{-\beta E_n} \psi_n(q')\psi_n(q'')$$

$$= \left(\frac{m\omega}{\pi\hbar}\right)^{\frac{1}{2}} \frac{e^{(\xi'^2+\xi''^2)/2}}{\pi} \int\int_{-\infty}^{\infty} \sum_{n=0}^{\infty} \frac{(-2uv)^n}{n!} e^{-(n+\frac{1}{2})\beta\hbar\omega} e^{-u^2+2i\xi' u} e^{-v^2+2i\xi'' v}\, du\, dv,$$

where the given integral expression is used. Summation over n yields, for the integrand,

$$e^{-\frac{1}{2}\beta\hbar\omega} \exp\{-u^2 + 2i\xi' u - v^2 + 2i\xi'' v - 2uv\, e^{-\beta\hbar\omega}\}.$$

By use of a general integral formula:

$$\int \cdots \int_{-\infty}^{\infty} \exp\left[-\frac{1}{2}\sum_j^n \sum_k^n a_{jk}x_jx_k + \mathrm{i}\sum_k^n b_kx_k\right]\mathrm{d}x_1 \cdots \mathrm{d}x_n$$

$$= \frac{(2\pi)^{\frac{1}{2}n}}{[\det(A)]^{\frac{1}{2}}}\exp(-\tfrac{1}{2}\textstyle\sum A_{jk}^{-1}b_jb_k) \qquad (1)$$

[where $\det(A)$ stands for the determinant corresponding to the matrix $(a_{jk}) = A$; A^{-1}_{jk} is an element of the inverse matrix A^{-1} of (a_{jk}) or $AA^{-1} = A^{-1}A = 1$. See note for its proof], one has

$$A = 2\begin{pmatrix} 1 & \mathrm{e}^{-\beta\hbar\omega} \\ \mathrm{e}^{-\beta\hbar\omega} & 1 \end{pmatrix}, \qquad \det(A) = 4(1 - \mathrm{e}^{-2\beta\hbar\omega}),$$

$$A^{-1} = \frac{1}{2(1-\mathrm{e}^{-2\beta\hbar\omega})}\begin{pmatrix} 1 & -\mathrm{e}^{-\beta\hbar\omega} \\ -\mathrm{e}^{-\beta\hbar\omega} & 1 \end{pmatrix}.$$

Accordingly,

$$\langle q' \,|\, \mathrm{e}^{-\beta\mathscr{H}} \,|\, q''\rangle = \left(\frac{m\omega}{\pi\hbar}\right)^{\frac{1}{2}} \frac{\mathrm{e}^{-\frac{1}{2}\beta\hbar\omega}}{(1-\mathrm{e}^{-2\beta\hbar\omega})^{\frac{1}{2}}} \times$$

$$\times \exp\left[\tfrac{1}{2}(\xi'^2 + \xi''^2) - \frac{1}{(1-\mathrm{e}^{-2\beta\hbar\omega})}(\xi'^2 - 2\xi'\xi''\mathrm{e}^{-\beta\hbar\omega} + \xi''^2)\right]$$

$$= \left[\frac{m\omega}{2\pi\hbar\sinh(\beta\hbar\omega)}\right]^{\frac{1}{2}} \exp\left[-\tfrac{1}{2}(\xi'^2 + \xi''^2)\coth(\beta\hbar\omega) + \frac{\xi'\xi''}{\sinh(\beta\hbar\omega)}\right]$$

$$= \left[\frac{m\omega}{2\pi\hbar\sinh(\beta\hbar\omega)}\right]^{\frac{1}{2}} \times$$

$$\times \exp\left[-\frac{m\omega}{4\hbar}\left\{(q'+q'')^2\tanh\tfrac{1}{2}\beta\hbar\omega + (q'-q'')^2\coth\tfrac{1}{2}\beta\hbar\omega\right\}\right]. \qquad (2)$$

Here, use is made of

$$\tanh(\tfrac{1}{2}\beta\hbar\omega) = \{\cosh(\beta\hbar\omega) - 1\}/\sinh(\beta\hbar\omega) = \frac{\sinh(\beta\hbar\omega)}{1+\cosh(\beta\hbar\omega)}.$$

When $\beta\hbar\omega \ll 1$, $\sinh(\beta\hbar\omega) \to \beta\hbar\omega$ and $\tanh(\frac{1}{2}\beta\hbar\omega) \to \frac{1}{2}\beta\hbar\omega$. Therefore, (2) leads to

$$\langle q' | \mathrm{e}^{-\beta H} | q''\rangle \to \left[\frac{m}{2\pi\hbar^2\beta}\right]^{\frac{1}{2}} \exp\left[-\frac{\beta m\omega^2}{8}(q'+q'')^2 - \frac{m}{2\beta\hbar^2}(q'-q'')^2\right]$$

$$= \exp\left(-\frac{\beta m\omega^2}{2}q'^2\right)\cdot\delta(q'-q'') \qquad (3)$$

where the relation:

$$\lim_{a\to\infty}\sqrt{\frac{a}{2\pi}}\,e^{-\frac{1}{2}a(x-y)^2}=\delta(x-y) \tag{4}$$

is used.

NOTE: Proof of formula (1): Vectors are written as $(x_1, \ldots, x_n) = \boldsymbol{x}$, $(b_1, \ldots, b_n) = \boldsymbol{b}$, etc. On introducing a new vector $\boldsymbol{y}$ defined by

$$A\boldsymbol{y}=\boldsymbol{b}\,,\qquad \boldsymbol{y}=A^{-1}\boldsymbol{b}$$

and putting $\boldsymbol{x} = \boldsymbol{z} + \mathrm{i}\boldsymbol{y}$, one can transform the exponent in (1) into

$$\begin{aligned}-\tfrac{1}{2}(\boldsymbol{x}A\boldsymbol{x})+\mathrm{i}(\boldsymbol{b}\boldsymbol{x}) &= -\tfrac{1}{2}((\boldsymbol{z}+\mathrm{i}\boldsymbol{y})A(\boldsymbol{z}+\mathrm{i}\boldsymbol{y}))+\mathrm{i}(\boldsymbol{b}(\boldsymbol{z}+\mathrm{i}\boldsymbol{y}))\\ &= -\tfrac{1}{2}(\boldsymbol{z}A\boldsymbol{z})-\mathrm{i}(\boldsymbol{z}A\boldsymbol{y})+\mathrm{i}(\boldsymbol{b}\boldsymbol{z})+\tfrac{1}{2}(\boldsymbol{y}A\boldsymbol{y})-(\boldsymbol{b}\boldsymbol{y})\\ &= -\tfrac{1}{2}(\boldsymbol{z}A\boldsymbol{z})-\tfrac{1}{2}(\boldsymbol{b}A^{-1}\boldsymbol{b}).\end{aligned}$$

Accordingly, by changing the integration variables from $x_1, \ldots, x_n$ to $z_1, \ldots, z_n$, the left-hand side of (1) is transformed into

$$\exp\Big[-\tfrac{1}{2}\sum_j\sum_k A_{jk}^{-1}b_jb_k\Big]\int\cdots\int\exp\Big(-\tfrac{1}{2}\sum_i\sum_k a_{ik}z_iz_k\Big)\mathrm{d}z_1\cdots\mathrm{d}z_n.$$

By an appropriate orthogonal transformation, $z_i = \sum_k t_{ik}\,\zeta_k$, it is possible to transform the second order expression $\sum_i\sum_k a_{ik}\,z_i\,z_k = \boldsymbol{z}A\boldsymbol{z}$ into a diagonal form $\sum\alpha_i\,\zeta_i^2$. Since the relation $\mathrm{d}z_1 \ldots \mathrm{d}z_n = \mathrm{d}\zeta_1 \ldots \mathrm{d}\zeta_n$ holds, the integral given above becomes equal to

$$\prod_{i=1}^{n}(2\pi/\alpha_i)^{\frac{1}{2}}.$$

Since, on the other hand,

$\alpha_1 \ldots \alpha_n$ = product of the eigenvalues of $A = \det(A)$, (1) is proved.

31. By the use of density matrices, one can write

$$\langle q^2\rangle = Z^{-1}\,\mathrm{Tr}\,(e^{-\beta\mathscr{H}}q^2) = Z^{-1}\int\mathrm{d}q\cdot q^2\langle q\,|\,e^{-\beta\mathscr{H}}\,|\,q\rangle, \tag{1}$$

$$\langle p^2\rangle = Z^{-1}\int\mathrm{d}q'\left[\left(\frac{\hbar}{\mathrm{i}}\frac{\partial}{\partial q'}\right)^2\langle q'\,|\,e^{-\beta\mathscr{H}}\,|\,q''\rangle\right]_{q''=q'}, \tag{2}$$

$$Z = \mathrm{Tr}\,e^{-\beta\mathscr{H}} = \int\mathrm{d}q\langle q\,|\,e^{-\beta\mathscr{H}}\,|\,q\rangle. \tag{3}$$

In integral (2), differentiation is first carried out with respect to q' and then integration is performed by putting q'' equal to q' (diagonal elements of $p^2 e^{-\beta H}$). From the results of the foregoing problem, one has

$$Z = \int_{-\infty}^{\infty} \left\{ \frac{m\omega}{2\pi\hbar \sinh(\beta\hbar\omega)} \right\}^{\frac{1}{2}} \exp\left[-\frac{m\omega}{\hbar} q^2 \tanh(\tfrac{1}{2}\beta\hbar\omega) \right] dq$$

$$= \left\{ \frac{m\omega}{2\pi\hbar \sinh(\beta\hbar\omega)} \right\}^{\frac{1}{2}} \left\{ \frac{\pi\hbar}{m\omega \tanh(\frac{1}{2}\beta\hbar\omega)} \right\}^{\frac{1}{2}} = \frac{1}{2\sinh(\frac{1}{2}\beta\hbar\omega)} = \frac{e^{-\frac{1}{2}\beta\hbar\omega}}{1 - e^{-\beta\hbar\omega}} \quad (4)$$

as expected. By means of

$$\int_{-\infty}^{\infty} x^2 e^{-\frac{1}{2}ax^2} dx \Big/ \int_{-\infty}^{\infty} e^{-\frac{1}{2}ax^2} dx =$$

$$= -2 \frac{d}{da} \log \int_{-\infty}^{\infty} e^{-\frac{1}{2}ax^2} dx = -2 \frac{d}{da} \log \sqrt{\frac{2\pi}{a}} = \frac{1}{a} \quad (5)$$

(1) is transformed into

$$\langle q^2 \rangle = \frac{\hbar}{2m\omega} \coth(\tfrac{1}{2}\beta\hbar\omega), \quad (6)$$

so that

$$\tfrac{1}{2}m\omega^2 \langle q^2 \rangle = \tfrac{1}{4}\hbar\omega \coth \tfrac{1}{2}\beta\hbar\omega = \tfrac{1}{2}\langle \mathscr{H} \rangle, \qquad \langle \mathscr{H} \rangle = \tfrac{1}{2}\hbar\omega \coth(\tfrac{1}{2}\beta\hbar\omega). \quad (7)$$

Now, (2) can be calculated in an elementary way by performing differentiation. The results are

$$\frac{1}{2m} \langle p^2 \rangle = \tfrac{1}{2} \langle \mathscr{H} \rangle, \qquad \langle p^2 \rangle = m \langle \mathscr{H} \rangle \quad (8)$$

in correspondence to (7). Equations (7) and (8) are naturally to be expected, because they imply that the energy is equally distributed between the kinetic energy and the potential energy.

32. In the q-representation of Schrödinger's equation,

$$\mathscr{H}\left(\frac{h}{i} \frac{\partial}{\partial q'}, q' \right) \varphi_l(q') = E_l \varphi_l(q') \quad (1)$$

$\varphi_l(q')$ is a normalized eigenfunction and E_l is an eigenvalue. According to definition (2.41), the q-representation of the density matrix is

$$\langle q' | e^{-\beta\mathscr{H}} | q'' \rangle = \sum_l e^{-\beta E_l} \varphi_l(q') \varphi_l^*(q''). \quad (2)$$

On applying (1) to

$$\exp(-\beta\mathscr{H}) = \sum_{n=0}^{\infty} (-\beta)^n \mathscr{H}^n/n!$$

repeatedly, one has

$$\exp\left\{-\beta\mathscr{H}\left(\frac{\hbar}{\mathrm{i}}\frac{\partial}{\partial q'}, q'\right)\right\}\cdot\varphi_l(q') = \mathrm{e}^{-\beta E_l}\varphi_l(q'). \tag{3}$$

Therefore, (2) can be written at least formally as

$$\begin{aligned}\langle q'|\,\mathrm{e}^{-\beta\mathscr{H}}\,|q''\rangle &= \exp\{-\beta\mathscr{H}(q')\}\sum_l \varphi_l(q')\varphi_l^*(q'') \qquad (4)\\ &= \exp\{-\beta\mathscr{H}(q')\}\,\delta(q'-q'').\end{aligned}$$

Although $\sum_l \varphi_l(q')\varphi_l^*(q'')$ does not converge in the true sense, it may be regarded as Dirac's δ function. This is because if φ_l $(l = 1, 2, \ldots)$ constitutes a complete set (for a functional space satisfying given boundary conditions), the relation

$$\int\sum_l \varphi_l(q')\varphi_l^*(q'')\,\mathrm{d}q''f(q'') = f(q') \tag{5}$$

holds for an arbitrary function, $f(q) = \sum_l c_l\varphi_l(q)$ (i.e., (5) is a definition of the δ function).

If for a free particle, the δ function is expressed by

$$\delta(\boldsymbol{x}'-\boldsymbol{x}'') = \left(\frac{1}{2\pi}\right)^3\int_{-\infty}^{\infty}\mathrm{e}^{\mathrm{i}\boldsymbol{k}(\boldsymbol{x}'-\boldsymbol{x}'')}\,\mathrm{d}\boldsymbol{k},$$

one has

$$\begin{aligned}\langle\boldsymbol{x}'|\,\mathrm{e}^{-\beta p^2/(2m)}\,|\boldsymbol{x}''\rangle &= \left(\frac{1}{2\pi}\right)^3\int_{-\infty}^{\infty}\exp\left(\frac{\beta\hbar^2}{2m}\frac{\partial^2}{\partial\boldsymbol{x}'^2}\right)\mathrm{e}^{\mathrm{i}\boldsymbol{k}(\boldsymbol{x}'-\boldsymbol{x}'')}\,\mathrm{d}\boldsymbol{k}\\ &= \left(\frac{1}{2\pi}\right)^3\int_{-\infty}^{\infty}\exp\left[-\frac{\beta\hbar^2}{2m}\boldsymbol{k}^2 + \mathrm{i}\boldsymbol{k}(\boldsymbol{x}'-\boldsymbol{x}'')\right]\mathrm{d}\boldsymbol{k}\\ &= \left(\frac{m}{2\pi\beta\hbar^2}\right)^{\frac{3}{2}}\exp\left[-\frac{m}{2\beta\hbar^2}(\boldsymbol{x}'-\boldsymbol{x}'')^2\right]\end{aligned}$$

(cf. example 7).

NOTE: In expression (4), an arbitrary functional system $\{\varphi_l\}$ can be used for $\delta(q'-q'') = \sum_l \varphi_l(q')\varphi_l^*(q'')$, provided that it is complete. They are not required to be eigenfunctions of (1). Evidently, this is because (5) holds in general. Therefore, in the representation (4) of the density matrix, any set of functions (for instance, plane waves) can be used as base.

33. As mentioned in the note to problem 32, any system of orthogonal functions can be used for evaluating the $\boldsymbol{r}$-representation of the density matrix, provided that it forms a complete set. If a particle exists in a box of volume L^3 and periodic conditions are imposed on it as boundary conditions, the plane waves:

$$L^{-\frac{3}{2}N}\exp\{\sum_{j=1}^{N} \mathrm{i}\boldsymbol{k}_j\boldsymbol{r}_j\}, \qquad (k_{jx} = 2\pi n_{jx}/L, \cdots) \tag{1}$$

constitute a complete set of orthogonal functions. Therefore,

$$\begin{aligned}\delta(\boldsymbol{r}'-\boldsymbol{r}'') &\equiv \delta(\boldsymbol{r}_1'-\boldsymbol{r}_1'')\delta(\boldsymbol{r}_2'-\boldsymbol{r}_2'')\ldots\delta(\boldsymbol{r}_N'-\boldsymbol{r}_N'')\\ &= L^{-3N}\sum_{\boldsymbol{k}_1}\ldots\sum_{\boldsymbol{k}_N}\exp\left\{\sum_{j=1}^{N}\mathrm{i}\boldsymbol{k}_j(\boldsymbol{r}_j'-\boldsymbol{r}_j'')\right\}\\ &\to \frac{1}{(2\pi)^{3N}}\int\ldots\int \mathrm{d}\boldsymbol{k}_1\ldots\mathrm{d}\boldsymbol{k}_N\exp\left\{\sum_j \mathrm{i}\boldsymbol{k}_j(\boldsymbol{r}_j'-\boldsymbol{r}_j'')\right\}.\end{aligned} \tag{2}$$

This is the limit at $L \to \infty$. By employing this, one has

$$\begin{aligned}\langle\boldsymbol{r}'|\mathrm{e}^{-\beta\mathscr{H}}|\boldsymbol{r}''\rangle &= \exp\left[-\beta\left\{\sum_j\frac{1}{2m}\left(\frac{\hbar}{\mathrm{i}}\frac{\partial}{\partial\boldsymbol{r}_j'}\right)^2 + V(\boldsymbol{r}_1',\ldots,\boldsymbol{r}_N')\right\}\right]\cdot\delta(\boldsymbol{r}'-\boldsymbol{r}'')\\ &= \frac{1}{(2\pi)^{3N}}\int\ldots\int\mathrm{d}\boldsymbol{k}_1\ldots\mathrm{d}\boldsymbol{k}_N\exp\left[-\sum_j\frac{\beta}{2m}\left(\frac{\hbar}{\mathrm{i}}\frac{\partial}{\partial\boldsymbol{r}_j'}\right)^2 - \beta V(\boldsymbol{r}')\right]\times\\ &\qquad\times\exp\left\{\sum_{j=1}^{N}\mathrm{i}\boldsymbol{k}_j(\boldsymbol{r}_j'-\boldsymbol{r}_j'')\right\}.\end{aligned} \tag{3}$$

Here

$$\frac{1}{\mathrm{i}}\frac{\partial}{\partial x}\mathrm{e}^{\mathrm{i}kx}\cdot\phi(x) = \mathrm{e}^{\mathrm{i}kx}\left(k+\frac{1}{\mathrm{i}}\frac{\partial}{\partial x}\right)\cdot\phi(x). \tag{4}$$

Hence

$$f\left(\frac{1}{\mathrm{i}}\frac{\partial}{\partial x}\right)\mathrm{e}^{\mathrm{i}kx}\phi(x) = \mathrm{e}^{\mathrm{i}kx}f\left(k+\frac{1}{\mathrm{i}}\frac{\partial}{\partial x}\right)\cdot\phi(x). \tag{5}$$

By repeated use of (4), (5) yields

$$\left(\frac{1}{\mathrm{i}}\frac{\partial}{\partial x}\right)^2 \mathrm{e}^{\mathrm{i}kx}\phi(x) = \frac{1}{\mathrm{i}}\frac{\partial}{\partial x}\mathrm{e}^{\mathrm{i}kx}\left(k + \frac{1}{\mathrm{i}}\frac{\partial}{\partial x}\right)\phi(x) = \mathrm{e}^{\mathrm{i}kx}\left(k + \frac{1}{\mathrm{i}}\frac{\partial}{\partial x}\right)^2 \phi(x),$$

$$\left(\frac{1}{\mathrm{i}}\frac{\partial}{\partial x}\right)^n \mathrm{e}^{\mathrm{i}kx}\phi(x) = \mathrm{e}^{\mathrm{i}kx}\left(k + \frac{1}{\mathrm{i}}\frac{\partial}{\partial x}\right)^n \phi(x).$$

On applying (5) to (3), one has

$$\langle \boldsymbol{r}' | \mathrm{e}^{-\beta\mathscr{H}} | \boldsymbol{r}'' \rangle = \frac{1}{(2\pi)^{3N}} \int \cdots \int \mathrm{d}\boldsymbol{k}_1 \ldots \mathrm{d}\boldsymbol{k}_N \exp\left\{\sum_{j=1}^{N} \mathrm{i}\boldsymbol{k}_j(\boldsymbol{r}'_j - \boldsymbol{r}''_j)\right\} \times$$
$$\times \exp\left[-\sum_j \frac{\beta}{2m}\left(\hbar\boldsymbol{k}_j + \frac{\hbar}{\mathrm{i}}\frac{\partial}{\partial \boldsymbol{r}'_j}\right)^2 - \beta V(\boldsymbol{r}')\right]. \tag{6}$$

On putting $\hbar\boldsymbol{k}_j = \boldsymbol{p}_j$ (this $\boldsymbol{p}_j$ is an ordinary number rather than an operator), (6) leads to

$$\langle \boldsymbol{r}' | \mathrm{e}^{-\beta\mathscr{H}} | \boldsymbol{r}'' \rangle = \frac{1}{h^{3N}} \int \cdots \int \mathrm{d}\boldsymbol{p}_1 \ldots \mathrm{d}\boldsymbol{p}_N \exp\left\{\sum \frac{\mathrm{i}\boldsymbol{p}_j}{\hbar}(\boldsymbol{r}'_j - \boldsymbol{r}''_j)\right\} \times$$
$$\times \exp\left[-\sum_j \frac{\beta}{2m}\left(\boldsymbol{p}_j + \frac{\hbar}{\mathrm{i}}\frac{\partial}{\partial \boldsymbol{r}'_j}\right)^2 - \beta V(\boldsymbol{r}')\right]. \tag{7}$$

In order to evaluate the partition function, one puts $\boldsymbol{r}'_j = \boldsymbol{r}''_j$ and integrates (7) over $\boldsymbol{r}'_1, \ldots, \boldsymbol{r}'_N$ in all space. In the limit of $\hbar \to 0$, the last factor in (7) becomes

$$\lim_{\hbar\to 0} \exp\left[-\sum_j \frac{\beta}{2m}\left(\boldsymbol{p}_j + \frac{\hbar}{\mathrm{i}}\frac{\partial}{\partial \boldsymbol{r}'_j}\right)^2 - \beta V(\boldsymbol{r}')\right] = \exp\left\{-\sum_j \frac{\beta}{2m}\boldsymbol{p}_j^2 - \beta V(\boldsymbol{r}')\right\}. \tag{8}$$

Therefore, the limiting value of the trace of (7) at $\hbar \to 0$ is given by

$$\begin{aligned}\lim_{\hbar\to 0} Z &= \lim_{\hbar\to 0} \int \cdots \int \mathrm{d}\boldsymbol{r}_1 \ldots \mathrm{d}\boldsymbol{r}_N \langle \boldsymbol{r} | \mathrm{e}^{-\beta\mathscr{H}} | \boldsymbol{r} \rangle \\ &= \int \cdots \int \mathrm{d}\boldsymbol{r}_1 \ldots \mathrm{d}\boldsymbol{r}_N \frac{1}{h^{3N}} \int \cdots \int \mathrm{d}\boldsymbol{p}_1 \ldots \mathrm{d}\boldsymbol{p}_N \exp\left[-\sum_j \frac{\beta}{2m}\boldsymbol{p}_j^2 - \beta V(\boldsymbol{r})\right] \\ &= Z_{\mathrm{cl}}\end{aligned} \tag{9}$$

in agreement with the classical partition function.

NOTE: This is a proof for the agreement between the quantum mechanical results and those of classical statistics in the limit of $\hbar \to 0$. This provides the basis for dividing the phase space into cells of h^{3N} in size and indicating a correspondence between the number of cells and the number of quantum mechanical states. In order to derive the factor $1/N!$ originating from the indistinguishability of identical particles, one must employ, in place of (1), a system of wave functions:

$$\varphi(\boldsymbol{k}_1, \dots, \boldsymbol{k}_N) = \frac{1}{\sqrt{N!n_1!n_2!\dots}} \sum \theta(P) P \prod_j (\mathrm{e}^{\mathrm{i}\boldsymbol{k}_j \boldsymbol{r}_j'}/L^{\frac{3}{2}}) \tag{10}$$

by taking into account the symmetry property of the particles. Here P stands for a permutation operator, the numbering of the particles being interchanged only among the set of $\boldsymbol{r}_1'$, ..., $\boldsymbol{r}_N'$. In Bose statistics, $\theta(P)$ is identically equal to 1, while in Fermi statistics, it is ± 1 depending on whether P is even or odd. Now, $n_{\boldsymbol{k}}$ is the number of particles in the $\boldsymbol{k}$-th state. It can be proved that when a δ function corresponding to (2) is formed from (10), the quantum mechanical δ function that replaces (2) on taking into account the symmetry property of the particles is

$$\delta_q(\boldsymbol{r}' - \boldsymbol{r}'') = \frac{1}{N!} \sum \theta(P) P \delta(\boldsymbol{r}_1' - \boldsymbol{r}_1'') \delta(\boldsymbol{r}_2' - \boldsymbol{r}_2'') \dots \delta(\boldsymbol{r}_N' - \boldsymbol{r}_N''). \tag{11}$$

For Fermi statistics, the proof is simple, because (10) is a determinant. For Bose statistics, however, it is fairly complicated. By use of this, calculations can be made in a similar manner as in (3) *et seq.* The result indicates that in the limit of $\hbar \to 0$, all terms in (11) vanish except the first one (term with $P = 1$). Accordingly, only $1/N!$ is introduced as a multiplication factor to (9). The details of the proof are omitted.

CHAPTER 3

STATISTICAL THERMODYNAMICS OF GASES

This chapter deals mainly with thermodynamic properties of ideal gases and nearly ideal gases. Real gases at ordinary temperatures and pressures may be in fact regarded as such, though not at low temperatures or at high pressures. In this approximation the translational motion of molecules is treated as classical, the quantum effects being disregarded. In most cases the effects of molecular interactions are considered only as corrections given by the second virial coefficient. Problems A and B are limited to such cases. Only a few examples are given in C which require more advanced treatment such as general cluster expansions of imperfect gases.

Fundamental Topics

§ 3.1. PARTITION FUNCTIONS OF IDEAL GASES

The partition function of an ideal gas is given by

$$Z_N(T, V) = \left(\frac{2\pi mkT}{h^2}\right)^{\frac{3}{2}N} \frac{V^N}{N!} j(T)^N = \frac{1}{N!} f^N. \tag{3.1}$$

This is valid when the temperature is so high and the density is so low that: (1) the non-degeneracy condition (1.95)

$$\frac{N}{V} \ll \frac{(2\pi mkT)^{\frac{3}{2}}}{h^3}$$

is fulfilled, and (2) the effects of molecular interaction can be neglected. Then Fermi- or Bose-statistics reduce simply to Boltzmann statistics (§ 1.15). In (3.1) f is the partition function of a single molecule and j is the internal partition function of a single particle. The thermodynamic functions are then given by

$$\bar{F} = -RT \log\left\{\frac{(2\pi mkT)^{\frac{3}{2}}}{h^3} \frac{V}{N_L}\right\} - RT - RT \log j(T), \tag{3.2a}$$

$$\bar{G} = RT\left[\log\left\{\frac{h^3}{(2\pi mkT)^{\frac{3}{2}}} \frac{p}{kT}\right\} - \log j\right], \tag{3.2b}$$

$$\bar{U} = \tfrac{3}{2}RT + \bar{U}_i, \qquad \bar{U}_i = -RT^2 \frac{\partial}{\partial T} \log j, \tag{3.2c}$$

$$\bar{C}_V = \tfrac{3}{2}R + \bar{C}_i, \qquad \bar{C}_i = R\frac{\partial}{\partial T}\left(-T^2\frac{\partial}{\partial T}\log j\right), \tag{3.2d}$$

where $\bar{F}$ is the molar Helmholtz free energy, $\bar{U}$ the molar internal energy, $\bar{G}$ the chemical potential or the molar Gibbs free energy, $\bar{C}_V$ the molar heat capacity at constant volume, $\bar{U}_i$ the molar energy of the internal degrees of freedom, and $\bar{C}_i$ is the corresponding internal molar heat capacity.

§ 3.2. INTERNAL DEGREES OF FREEDOM AND INTERNAL PARTITION FUNCTIONS

The internal partition function is written as

$$j(T) = \sum_l g_l e^{-\varepsilon_l/kT} \tag{3.3}$$

where ε_l is the energy of a single molecule, the translational part being eliminated, and l represents the quantum number of such an internal quantum state of a molecule, g_l being the degeneracy.

The internal motion of a molecule consists of (1) the electronic state, (2) the state of the nucleus, (3) the rotational state of the molecule as a whole and (4) the vibrational states within the molecule. Rigorously speaking, these four kinds of modes interact each other, but they may be approximately separated in most cases. In the following such approximate formulae are summarized for monatomic, diatomic, and polyatomic gases.

1. *Monatomic molecules:* Examples: He, Ne, A, ... In these monatomic molecules, the electrons usually form closed shells. Each atom is usually at the lowest level because the energy difference is very large between the ground state and the next excited state. The lowest electronic state has no degeneracy so that $g_e = 1$. (This may not be true at extremely high temperatures or when the atom is ionized.) The nucleus has the degeneracy $g_n = 2s_n + 1$ for possible orientations of nuclear spin if its magnitude is s_n. Therefore the internal partition function is

$$j = g_e \times g_n = 2s_n + 1\,. \tag{3.4}$$

2. *Diatomic molecules:* In many cases the lowest electronic state is non-degenerate and is separated by fairly large energy from the next excited state, so that the electronic internal partition function is simply $g_e = 1$. However, in some cases the lowest electronic state has a finite angular momentum and accordingly a finite degeneracy (example: O_2, $g_e = 3$). Also the excited states may be low enough that they have to be considered even at ordinary temperatures.

The vibrational motion of a molecule is influenced by the centrifugal force due to molecular rotation. However, provided that the temperature is not too high, this effect may be ignored, and the vibration and rotation may be separated as a first approximation. The nuclear state and the molecular rotation are independent in *heteronuclear* molecules, but they are closely coupled in *homonuclear* molecules through the restriction of Fermi- or Bose-statistics. Thus the internal partition function of a diatomic molecule is given by

$$j = g_e\, g_{nuc}\, r(T) v(T) \qquad \text{(heteronuclear molecule } AB) \tag{3.5a}$$

$$= g_e\, j_{nuc-rot}(T) v(T) \qquad \text{(homonuclear molecule } AA) \tag{3.5b}$$

$$\simeq \tfrac{1}{2} g_e\, g_{nuc}\, r_c(T) v(T) \qquad \text{(homonuclear molecule, high temperature)} \tag{3.5c}$$

where

$$\left.\begin{aligned} g_e &= \text{degeneracy of the lowest electronic state (ordinarily } g_e = 1) \\ g_{nuc} &= \text{degeneracy due to nuclear spin} \\ &= (2s_A + 1)(2s_B + 1) \qquad (AB \text{ molecule}) \\ &= (2s_A + 1)^2 \qquad (AA \text{ molecule}) \end{aligned}\right\} \tag{3.6}$$

$v(T)$ = vibrational partition function with frequency ν

$$= \frac{e^{-\Theta_v/2T}}{1 - e^{-\Theta_v/T}} \equiv \left[2 \sinh \frac{\Theta_v}{2T}\right]^{-1} \quad (\Theta_v = h\nu/k), \tag{3.7†}$$

$r(T)$ = rotational partition function with moment of inertia I

$$= \sum_{l=0}^{\infty} (2l + 1) \exp\left\{- l(l + 1) \frac{h^2}{8\pi^2 IkT}\right\} \tag{3.8††}$$

$$= \sum_{l=0}^{\infty} (2l + 1)\, e^{-l(l+1)\Theta_r/T} \qquad (\Theta_r = h^2/8\pi^2 Ik),$$

$r_c(T)$ = semi-classical rotational partition function (high temperature approximation $h^2/8\pi^2 IkT \ll 1$)

$$= \frac{8\pi^2 IkT}{h^2} = \frac{T}{\Theta_r} \qquad \text{(see example 1 and problem 3).} \tag{3.9}$$

---

† Vibrational quantum levels are given by

$$\varepsilon_n = (n + \tfrac{1}{2})h\nu, \qquad n = 0, 1, 2, \ldots$$

†† Rotational quantum levels of a linear rotator with the principal moments of inertia $(I, I, 0)$ are given by

$$\varepsilon_j = j(j + 1)h^2/8\pi^2 I.$$

The level j has the degeneracy $(2j + 1)$.

For a homonuclear molecule AA the nuclear-rotational partition function is given by the following:

1. When the nucleus A is a Fermi-particle with half integral spin ($s_n = \frac{1}{2}, \frac{3}{2}, \ldots$)

$$j_{\text{nuc-rot}} = s_n(2s_n + 1)r_{\text{even}} + (s_n + 1)(2s_n + 1)r_{\text{odd}}. \tag{3.10a}$$

EXAMPLE: H_2, $s_n = \frac{1}{2}$

2. When the nucleus A is a Bose-particle with an integral spin ($s_n = 0, 1, 2, \ldots$)

$$j_{\text{nuc-rot}} = (s_n + 1)(2s_n + 1)r_{\text{even}} + s_n(2s_n + 1)r_{\text{odd}}. \tag{3.10b}$$

EXAMPLE: D_2, $s_n = 1$.

In the above expressions, r_{even} and r_{odd} are:

$$r_{\text{even}} = \sum_{l=0,2,4,\ldots}^{\infty} (2l + 1)\exp\left\{-\frac{l(l + 1)h^2}{8\pi^2 IkT}\right\}, \tag{3.11a}$$

$$r_{\text{odd}} = \sum_{l=1,3,\ldots}^{\infty} (2l + 1)\exp\left\{-\frac{l(l + 1)h^2}{8\pi^2 IkT}\right\}. \tag{3.11b}$$

NOTE: The reason for the complexity of the coupling between the nuclear and rotational degrees of freedom in homonuclear molecules is due to the symmetry requirement of the nuclear-rotational wave function. According to a fundamental principle of quantum mechanics, the wave function must be symmetrical or anti-symmetrical with respect to a permutation of identical particles. This means that the wave function must remain unchanged for Bose particles when two Bose particles are interchanged, whereas it must change its sign when two Fermi particles are interchanged. The rotational wave function of a diatomic molecule is even or odd according to whether the quantum number j is even or odd with respect to interchange of the coordinates of its two atoms. For a given j, therefore, the nuclear spin wave function must be either symmetric or antisymmetric depending on the parity of j and the statistics of the nuclei. For two nuclei, each having the spin s_n, there are $(2s_n + 1)^2$ states in total, among which $(s_n + 1)(2s_n + 1)$ are symmetric and $s_n(2s_n + 1)$ are antisymmetric with respect to permutation of the nuclear spin variables.† Considering these, one can easily obtain (3.10a, b). In most cases, however, the high temperature approximation (3.5c) is quite satisfactory because the

† Let $f_1, \ldots f_m$ ($m = 2s_n + 1$) be the spin-functions representing different spin orientations of a nucleus with the spin s_n. Symmetric spin functions for two identical nuclei have the form $f_i(1)f_j(2) + f_j(1)f_i(2)$, and anti-symmetric spin functions have the form $f_i(1)f_j(2) - f_j(1)f_i(2)$. There are $\frac{1}{2}m(m + 1)$ of the former type and $\frac{1}{2}m(m - 1)$ of the latter type.

moment of inertia is so big that the rotational states may be treated in semi-classical approximation, and hence the rotational-nuclear coupling does not need to be considered in detail except for introduction of the *symmetry number*. However, the hydrogen molecule requires particular caution, because the rotational characteristic temperature Θ_r is fairly high. Thus we have ortho-hydrogen molecules (j = odd), para-hydrogen molecules (j = even), ortho-deuterium molecules (j = even) and para-deuterium molecules (j = odd).

3. *Polyatomic molecules.* In most cases, the lowest electronic state is far below the excited states. The moments of inertia are so large that the rotational motion can be treated classically (high temperature approximation). Therefore the symmetry requirement of the wave function, when the molecule contains identical atoms, need not be considered in such detail as (3.10). The rotational-nuclear partition function becomes simply the product of the classical rotational partition function and the degeneracy weight of nuclear spin states divided by the *symmetry number* γ. As an extension of (3.5c) the partition function of a polyatomic molecule is written as

$$j = g_e \frac{g_{\mathrm{nuc}} \cdot r(T)}{\gamma} v(T) \tag{3.12}$$

which usually holds as a satisfactory approximation. Here

g_e = degeneracy of the lowest electronic state,

g_{nuc} = degeneracy of the nuclear spin states $= \prod (2s_n + 1)$, (3.13)

$r(T)$ = rotational partition function (classical, high temperature approximation)

$$= \pi^{\frac{1}{2}} \left(\frac{8\pi^2 AkT}{h^2}\right)^{\frac{1}{2}} \left(\frac{8\pi^2 BkT}{h^2}\right)^{\frac{1}{2}} \left(\frac{8\pi^2 CkT}{h^2}\right)^{\frac{1}{2}}, \tag{3.14}$$

where A, B, C are the principal moments of inertia (see problem 4),

$v(T)$ = vibrational partition function

$$= \prod_{i=1}^{3n-6} \frac{e^{-\frac{1}{2}h\nu_i/kT}}{1 - e^{-h\nu_i/kT}} = \prod_{i=1}^{3n-6} \frac{e^{-\Theta_i/2T}}{1 - e^{-\Theta_i/T}} \quad (\Theta_i = h\nu_i/k), \tag{3.15}$$

where ν_i ($i = 1, 2, \ldots, 3n - 6$, n being the number of atoms in the molecule) are the normal frequencies,

γ = symmetry number = the number of identical configurations realized by rotation of a molecule when it contains identical nuclei.

Note that, however, *linear polyatomic molecules* (example: CO_2) must be treated by (3.9) rather than (3.14).

§ 3.3. MIXTURES OF IDEAL GASES

Except at low temperatures and at high densities, gas mixtures may also be treated as ideal gases. The partition function of a mixture of N_A molecules of type A, N_B of type B, ... confined in a box of volume V is given by

$$Z_{N_A, N_B, \ldots}(V, T) = \prod_{A, B, \ldots} \left(\frac{2\pi m_A kT}{h^2}\right)^{\frac{3}{2}N_A} \frac{V^{N_A}}{N_A!}(j_A)^{N_A} \equiv \prod_{A, B, \ldots} \frac{f_A^{N_A}}{N_A!} \tag{3.16}$$

where f_A, f_B, ... are the partition functions of molecules A, B, ..., and j_A, j_B, ... are the internal partition functions of each species of molecules. The Gibbs free energy then becomes

$$G = \sum N_A \mu_A = \sum n_A \bar{G}_A \qquad (n_A = \text{number of moles}),$$

$$\mu_A = -kT\left[\log\left\{\frac{(2\pi m_A kT)^{\frac{3}{2}}}{h^3}\frac{kT}{p}\right\} + \log j_A - \log\frac{N_A}{N}\right], \tag{3.17}$$

where $p = NkT/V$ = total pressure, $N = \sum_A N_A$ = total number of molecules. Other thermodynamic functions can be easily derived.

DIVERTISSEMENT 8

Gibbs' paradox: Imagine a gas mixture consisting of two kinds of molecules, one being white balls and the other red balls. When they are mixed at constant pressure and temperature the entropy increases by the amount

$$\Delta S_{\text{mix}} = K\left\{N_w \log\frac{N_w + N_r}{N_w} + N_r \log\frac{N_w + N_r}{N_r}\right\}$$

where N_w and N_r are the numbers of white and red balls respectively. Now suppose that the red color is caused to be bleached continuously from red to pink and then to white. Then how does the entropy of the mixture behave in such a process? When the molecules are all white, there cannot be any extra entropy of mixing. If it did exist, the entropy would no longer be extensive, which fact is of course contradictory to the basic property of entropy. But the mixing entropy must continue to exist in so far as the pink color can be distinguished from the white. Does it then suddenly jump down to zero at the moment when the pink color is completely bleached? Put differently, is the entropy such a strange, discontinuous function of the depth of color?

In the real physical world, a molecule can be identical with another, or it is distinctly differentiated from the other. The structure of nature is not continuous as with colored balls, but it is essentially discrete, because molecules consist of atoms which again consist of different numbers of protons, neutrons and electrons. It is interesting to point out that such macroscopic laws as thermodynamic laws reflect the discrete structure of the microscopic world, without which thermodynamics must cease to be valid.

§ 3.4. MOLECULAR INTERACTIONS

In real gases the molecules interact with each other and collide incessantly. This causes deviations from ideal gas behavior. Strictly speaking, the interaction between molecules may depend on the internal states of the molecules, but usually such complexities are ignored. Then the effect of molecular interaction is considered only for the translational part of the partition function, the internal partition functions simply being assumed unchanged, so that the total partition function takes the form

$$Z_N(T, V) = \left(\frac{2\pi mkT}{h^2}\right)^{\frac{3}{2}N} j(T)^N \cdot Q_N \tag{3.18}$$

for a gas consisting of N identical molecules. Q_N is the configurational partition function (2.32) in the classical approximation

$$Q_N = \frac{1}{N!}\int \cdots \int e^{-U(\boldsymbol{r}_1, \boldsymbol{r}_2, \dots)/kT} \mathrm{d}\boldsymbol{r}_1 \cdots \mathrm{d}\boldsymbol{r}_N \tag{3.19}$$

where

$$U = \sum_{(i,\,j)\,\text{pairs}} u(r_{ij}) \tag{3.20}$$

is the potential function used to represent the molecular interaction and is usually assumed to be the sum of the pair interaction potentials $u(r_{ij})$, r_{ij} being the distance between the i-th and j-th molecules. In so far as the internal motion is considered separately, $u(r_{ij})$ is a central force depending only on the distance r_{ij}. (The assumptions here made are not necessarily true. Except for the simple case of monatomic gases, the molecular interactions generally depend on the orientation of the molecules. The pair interaction assumption is not always rigorous. In spite of these complications, however, simplified models of molecular interactions often turn out to be fairly satisfactory in fitting the experimental data.)

The functional form of $u(r_{ij})$ can be, in principle, derived from theory. But there are a number of semi-empirical expressions commonly used. The most well-known example is the Lennard-Jones potential

$$u(r) = \frac{A}{r^n} - \frac{B}{r^m} \qquad (m < n) \tag{3.21}$$

where m is usually assumed to be 6 in accordance with the Van der Waals force and n is assumed to be 12. This choice of n has no fundamental physical basis but is made only for mathematical convenience.

§ 3.5. CLUSTER EXPANSION

Expansion of the configurational partition function: The distribution function of N molecules is given by

$$F\{N\} = Q_N^{-1} \exp(-U/kT) = Q_N^{-1} \prod_{(ij)} \exp\{-u(r_{ij})/kT\} \tag{3.22}$$

if (3.20) is assumed for the molecular interaction. Now define

$$e^{-u(r_{ij})/kT} = 1 + f_{ij}, \qquad f_{ij} = e^{-u(r_{ij})/kT} - 1. \tag{3.23}$$

Since the pair interaction potential $u(r_{ij})$ will behave like theupper curve of

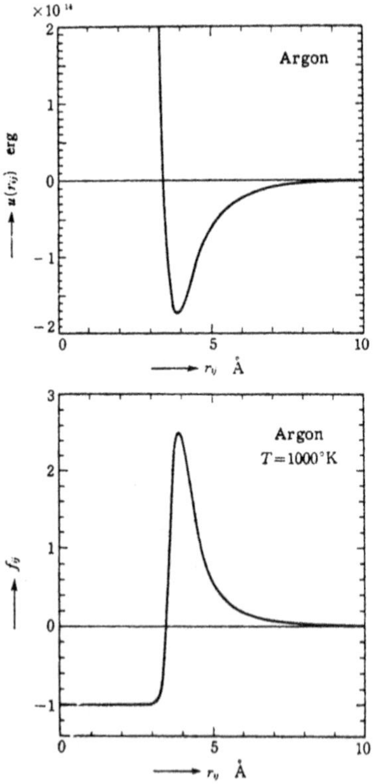

Fig. 3.1.

Figure 3.1, f_{ij} will vanish for r_{ij} greater than the range of the molecular force and will be -1 at small values of r_{ij} where the molecules strongly repel each other. In the intermediate range of f_{ij}, f_{ij} will take on positive values. From the definition of f_{ij}, one can write the expansion

$$\begin{aligned} e^{-U/kT} &= \prod_{ij}(1 + f_{ij}) \\ &= 1 + \sum_{i<j} f_{ij} + \sum\sum f_{ij} f_{i'j'} + \cdots \end{aligned} \tag{3.24}$$

Decomposition into clusters: The general term in the expansion of (3.24) is the product of a certain number of f-functions. This may be represented as a bond diagram which is constructed by connecting the molecules which appear in each f_{ij} function (see Fig. 3.2). Each group of moleculeus thus

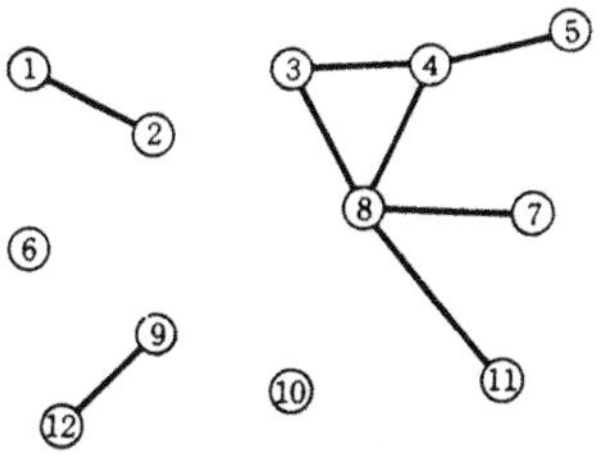

Fig. 3.2.

connected together is called a cluster. The size of a cluster is defined by the number of molecules in it.

Suppose N molecules are grouped into a number of clusters, m_l clusters of the size l $(l = 1, 2, \ldots)$, so that

$$N = \sum_{l=1}^{N} l m_l , \tag{3.25}$$

and in each cluster the molecules are connected in a certain pattern. The diagram thus obtained corresponds to a certain term of the expansion (3.24). Therefore, the sum of the expansion terms in (3.24) is carried out by (1) choosing $(m_1, m_2, \ldots, m_N)$ in all possible ways restricted by the condition (3.25), (2) grouping the molecules 1, 2, ... N into clusters (m_l clusters of the size l, $l = 1, 2, \ldots$) and (3) connecting molecules in each cluster in all possible ways as long as the cluster is not split up.

Cluster integral: For a cluster of size l, the cluster integral is defined by

$$b_l = \frac{1}{l!V} \int \cdots \int \sum_{\text{all possible bonding}} \left(\prod f_{ij}\right) d\boldsymbol{r}_1 \cdots d\boldsymbol{r}_l . \tag{3.26}$$

For a given grouping of N molecules into clusters, procedure (3) mentioned above and integration over all coordinates of the molecules give

$$\prod_l (l!Vb_l)^{m_l}.$$

Therefore integration of the expansion (3.24) over the molecular coordinates results in

$$\sum_{\substack{\{m_l\} \\ \Sigma l m_l = N}} \frac{N!}{\prod_l (l!)^{m_l} m_l!} \prod_l (l!Vb_l)^{m_l} = N! \sum_{\{m_l\}} \prod_l \frac{(Vb_l)^{m_l}}{m_l!} \tag{3.27}$$

where

$$N!/\prod_{l=1}^{N} (l!)^{m_l} m_l!$$

is the total number of ways of grouping N molecules into a set of clusters consisting of m_l clusters of size l ($l = 1, 2, \ldots N$). The summation is taken over all possible choices of $(m_1, m_2, \ldots)$ under the condition (3.25), i.e.

$$\sum_l l m_l = N.$$

Grand partition function: By (3.19) and (3.27) one has

$$Q_N = \sum_{\substack{\{m\} \\ \sum_{l=1}^{N} l m_l = N}} \prod_{l=1}^{N} \frac{(Vb_l)^{m_l}}{m_l!}. \tag{3.28}$$

Consider a grand-canonical ensemble for a given chemical potential μ (absolute activity $\lambda = \exp(\mu/kT)$) and calculate the partition function:

$$\Xi = \sum_{N=0}^{\infty} \lambda^N Z_N = \sum_{N=0}^{\infty} \left\{\lambda \left(\frac{2\pi mkT}{h^2}\right)^{\frac{3}{2}} j\right\}^N Q_N. \tag{3.29}$$

One inserts (3.28) into (3.29) and considers that the restriction for the summation over $\{m_l\}$ is now removed in the grand-canonical partition function because of the summation over N. Introducing the fugacity ζ by the definition

$$\lambda \left(\frac{2\pi mkT}{h^2}\right)^{\frac{3}{2}} j = \zeta \tag{3.30}$$

one obtains, from (3.29),

$$\begin{aligned} \Xi &= \sum_{N=0}^{\infty} \sum_{\substack{\{m_l\} \\ \Sigma l m_l = N}} \zeta^{\Sigma l m_l} \prod_{l=1}^{N} \frac{(Vb_l)^{m_l}}{m_l!} \\ &= \prod_{l=1}^{\infty} \sum_{m_l=0}^{\infty} \frac{1}{m_l!} (Vb_l \zeta^l)^{m_l} = \exp\left\{V \sum_{l=0}^{\infty} b_l \zeta^l\right\}. \end{aligned} \tag{3.31}$$

Equation of state: By (3.31) one has

$$\frac{p}{kT} = \sum_{l=1}^{\infty} b_l \zeta^l, \tag{3.32}$$

$$N = \lambda \frac{\partial \log \Xi}{\partial \lambda} = \zeta \frac{\partial \log \Xi}{\partial \zeta} = V \sum_{l=1}^{\infty} l b_l \zeta^l,$$

or

$$\frac{N}{V} \equiv \frac{1}{v} = \sum_{l=1}^{\infty} l b_l \zeta^l. \tag{3.33}$$

Virial expansion: By eliminating ζ from (3.32) and (3.33), one obtains the virial expansion formula

$$p = \frac{kT}{v}\left[1 - \sum_{s=1}^{\infty} \frac{s}{s+1} \beta_s v^{-s}\right]. \tag{3.34}$$

Here the constants β_s are called the irreducible cluster integrals. The general proof of this expansion is fairly complicated and so is omitted here. [See J. E. Mayer and M. G. Mayer, *Statistical Mechanics* (John Wiley and Sons) Chapters 13, 14.]

Examples

1. Give the high temperature expansion of the rotational partition function $r(T)$ (3.8) for hetero-nuclear diatomic molecules using Euler-MacLaurin's summation formula. Calculate the value of $r(T)$ at $T = 300.4°$ K for HCl ($\Theta_r/T = \hbar^2/2Ik = 15.02°$ K) and find the deviation from the classical value T/Θ_r.

SOLUTION

$r(T)$ is given by (3.8), i.e.,

$$r(T) = \sum_{l=0}^{\infty} (2l+1) \exp\left\{-\frac{\hbar^2 l(l+1)}{2IkT}\right\} \equiv \sum_{l=0}^{\infty} (2l+1) \exp\left\{-\frac{\Theta_r}{T} l(l+1)\right\}. \tag{1}$$

At high temperatures $T \gg \Theta_r$, one can utilize Euler-MacLaurins' summation formula:

$$\sum_{n=0}^{\infty} f(n) = \int_0^{\infty} f(x)\,dx + \tfrac{1}{2}f(0) - \tfrac{1}{12}f'(0) + \tfrac{1}{720}f'''(0) - \tfrac{1}{30240}f^{V}(0) + \cdots \tag{2†}$$

($f(x)$ being analytic for $0 < x < \infty$).

† Whittaker and Watson, *A Course in Modern Analysis*, p. 127.

Putting $f(x) = (2x+1)\exp\{-x(x+1)\sigma\}$, with $\sigma = \Theta_r/T$, one has

$$\int_0^\infty f(x)\,dx = \int_0^\infty (2x+1)e^{-x(x+1)\sigma}\,dx = \frac{1}{\sigma}\int_0^\infty e^{-\xi}\,d\xi = \frac{T}{\Theta_r}, \tag{3}$$

and

$$f(0) = 1, \quad f'(0) = 2 - \sigma, \quad f'''(0) = -12\sigma + 12\sigma^2 - \sigma^3,$$
$$f^{V} = 120\sigma^2 - 180\sigma^3 + 30\sigma^4 - \sigma^5, \cdots,$$

so that

$$r(T) = \frac{1}{\sigma} + \frac{1}{3} + \frac{\sigma}{15} + \frac{4\sigma^2}{315} + O(\sigma^3)$$
$$= \frac{T}{\Theta_r}\left\{1 + \frac{1}{3}\frac{\Theta_r}{T} + \frac{1}{15}\left(\frac{\Theta_r}{T}\right)^2 + \frac{4}{315}\left(\frac{\Theta_r}{T}\right)^3 + O\left(\frac{\Theta_r}{T}\right)^4\right\} \tag{4}$$

(Mulholland's formula). Equation (3) gives the classical partition function.

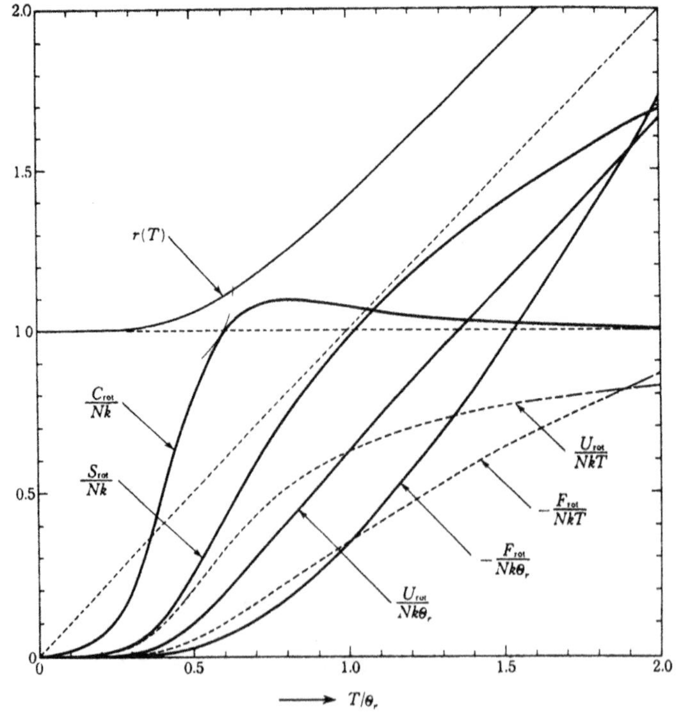

Fig. 3.3.

For HCl one has $\sigma = \frac{1}{20}$ at 300.4° K. Hence

$$r(T)_{\text{classical}} = 20$$
$$r(T) = 20.333\ldots \quad \text{(up to 2nd term in (4))}$$
$$r(T) = 20.33666\ldots \quad \text{(up to 3rd term in (4))}$$
$$r(T) = 20.336698\ldots \quad \text{(up to 4th term in (4))}.$$

NOTE: 1 At low temperatures $T \ll \Theta_r$, direct evaluation of the series (1) is possible. Figure 3.3 shows the behavior of $r(T)$ and related thermodynamic quantities.

NOTE 2: For HD, HT and DT, observations at low temperatures are possible. Figure 3.4 gives the molar heat capacities at constant pressure of these gases, Figure 3.5 gives those of H_2, D_2 and T_2.

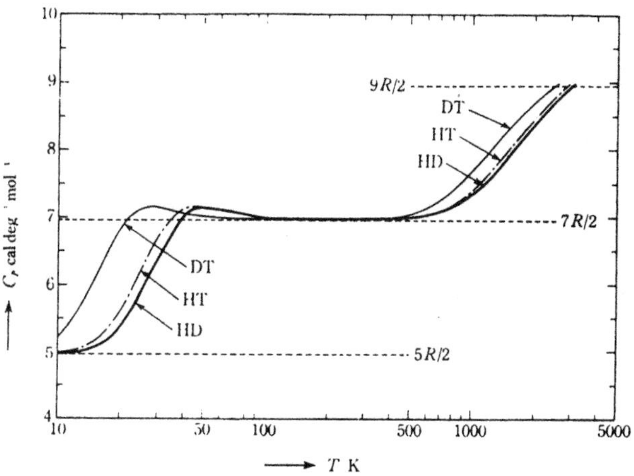

Fig. 3.4.

2. Evaluate the Helmholtz free energy, the entropy and the molar heat capacity at constant pressure of CO_2 gas at 0° C and 1 atm. Assume the ideal gas formula, and use the following data:

molecular weight $M = 44.010$

moment of inertia (as a linear molecule O–C–O) $I = 71.67 \times 10^{-40}$ g cm^2

normal modes of vibration: $\tilde{\nu}_1 = 667.3 = \tilde{\nu}_2, \tilde{\nu}_3 = 1383.3, \tilde{\nu}_4 = 2439.3$ cm^{-1}.

The electronic ground state of the CO_2 molecule is $^1\Sigma_g^+$ and hence is non-degenerate.

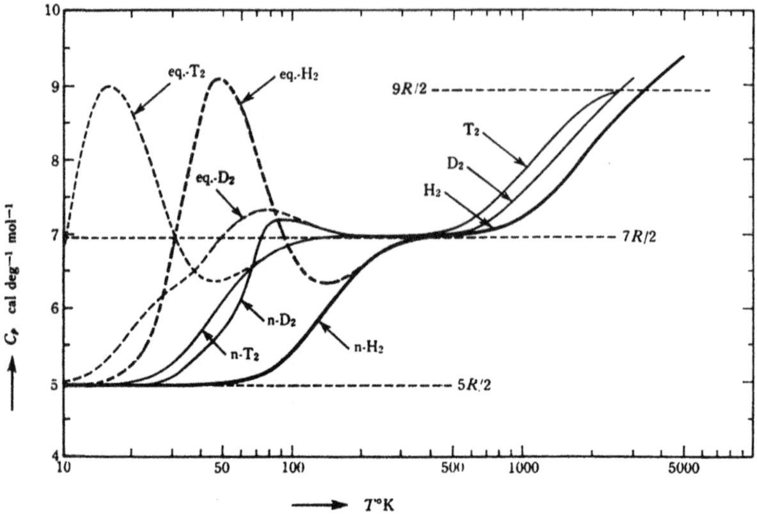

Fig. 3.5.

NOTE: $\tilde{\nu}_1$ and $\tilde{\nu}_2$ correspond to the bending modes, ν_3 to the mode in which the two C–O bonds oscillate out of, and $\tilde{\nu}_4$ to that in which they oscillate in phase (see Fig. 3.6).

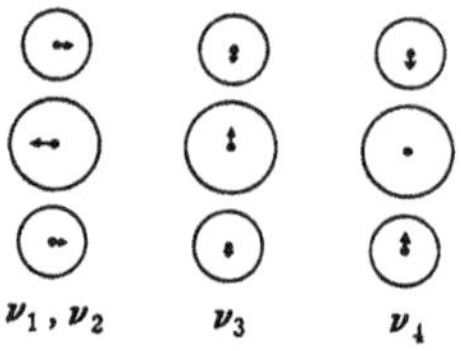

Fig. 3.6.

SOLUTION

The partition function for CO_2 gas is given by (3.1) and (3.12). Using the following values: the mass of CO_2 molecule, $m = 44.010 \times 1.65963 \times 10^{-24} = 0.7304 \times 10^{-22}$ g: $kT(T = 0°\text{C}) = 3.771 \times 10^{-14}$ erg, $kT/hc = 189.90\,\text{cm}^{-1}$, $RT = 2.271 \times 10^{-10}$ erg mol^{-1}, 1 atm ($= 1.01325 \times 10^6$ erg cm^{-3}); then one finds the contributions from the translational motion to be

$$F_{\text{trans}} = -RT\left[\log\left\{\left(\frac{2\pi mkT}{h^2}\right)^{\frac{3}{2}}\frac{kT}{p}\right\} + 1\right] = -RT \times (4.673 + 1)$$
$$= -1.0612 \times 10^{11}\,\text{erg mol}^{-1} = -2536\,\text{cal mol}^{-1}, \qquad (1)$$

$$\bar{S}_{\text{trans}} = R\left[\log\left\{\left(\frac{2\pi mkT}{h^2}\right)^{\frac{3}{2}}\frac{kT}{p}\right\}+\frac{5}{2}\right] = R \times (4.673 + 2.500)$$

$$= 59.64 \times 10^7\,\text{erg}\,\text{mol}^{-1}\,\text{deg}^{-1} = 14.254\,\text{cal}\,\text{mol}^{-1}\,\text{deg}^{-1}, \tag{2}$$

$$\bar{C}_{\text{trans}} = \tfrac{3}{2}R = 1.2471 \times 10^8\,\text{erg}\,\text{mol}^{-1}\,\text{deg}^{-1} = 2.981\,\text{cal}\,\text{mol}^{-1}\,\text{deg}^{-1}. \tag{3}$$

For the rotational motion, one can use the classical partition function (3.9) because $\Theta_r = \hbar^2/(2Ik) = 40.26 \times 10^{-40} \div (71.67 \times 10^{-40}) = 0.5617°\,\text{K}$ is small enough. The symmetry number γ in (3.2) is equal to 2. One has, therefore,

$$\bar{F}_{\text{rot}} = -RT\log\frac{r(T)}{\gamma} = -RT\log\frac{T}{\gamma\Theta_r} = -RT \times 5.494$$

$$= -1.2477 \times 10^{11}\,\text{erg}\,\text{mol}^{-1} = -2982\,\text{cal}\,\text{mol}^{-1}, \tag{4}$$

$$\bar{S}_{\text{rot}} = R\left(\log\frac{T}{\gamma\Theta_r}+1\right) = R \times (5.494 + 1)$$

$$= 53.99 \times 10^7\,\text{erg}\,\text{mol}^{-1}\,\text{deg}^{-1} = 12.905\,\text{cal}\,\text{mol}^{-1}\,\text{deg}^{-1}, \tag{5}$$

$$\bar{C}_{\text{rot}} = R = 8.314 \times 10^7\,\text{erg}\,\text{mol}^{-1}\,\text{deg}^{-1} = 1.9872\,\text{cal}\,\text{mol}^{-1}\,\text{deg}^{-1}. \tag{6}$$

The characteristic temperatures for the intra-molecular vibrations are found to be:

$$\Theta_1 = hc\tilde{\nu}_1/k = \Theta_2 = 1.4387 \times 667.3 = 960.0°\,\text{K} \quad (\Theta_1/T = 3.514),$$
$$\Theta_3 = hc\tilde{\nu}_3/k = 1990.1°\,\text{K} \quad (\Theta_3/T = 7.284),$$
$$\Theta_4 = hc\tilde{\nu}_4/k = 3509°\,\text{K} \quad (\Theta_4/T = 12.844).$$

Hence one obtains from (3.15)

$$\bar{F}_{\text{vib}} = -RT\log v(T) = -RT\sum_{i=1}^{4}\left\{-\frac{\Theta_i}{2T} - \log(1-e^{-\Theta_i/T})\right\}$$

$$= -RT(-13.578 + 0.061) = 3.070 \times 10^{11}\,\text{erg}\,\text{mol}^{-1}$$

$$= 7337\,\text{cal}\,\text{mol}^{-1}, \tag{7}$$

$$\bar{S}_{\text{vib}} = R\sum_{i=1}^{4}\left\{-\log(1-e^{-\Theta_i/T}) + \frac{\Theta_i/T}{e^{\Theta_i/T}-1}\right\}$$

$$= R(0.0612 + 0.2211) = 2.3470 \times 10^7\,\text{erg}\,\text{mol}^{-1}\,\text{deg}^{-1}$$

$$= 0.5610\,\text{cal}\,\text{mol}^{-1}\,\text{deg}^{-1}, \tag{8}$$

$$\bar{C}_{\text{vib}} = R\sum_{i=1}^{4}\frac{(\Theta_i/T)^2 e^{\Theta_i/T}}{(e^{\Theta_i/T}-1)^2} = R \times 0.8177$$

$$= 6.798 \times 10^7\,\text{erg}\,\text{mol}^{-1}\,\text{deg}^{-1}$$

$$= 1.6249\,\text{cal}\,\text{mol}^{-1}\,\text{deg}^{-1}. \tag{9}$$

There is no contribution from the electronic part of the partition function because of the non-degeneracy of the electronic ground state.

Summing up these contributions, one obtains

$$\bar{F} = 1819\,\text{cal mol}^{-1}, \qquad \bar{S} = 27.720\,\text{cal mol}^{-1}\,\text{deg}^{-1},$$
$$\bar{C}_V = 6.593\,\text{cal mol}^{-1}\,\text{deg}^{-1}.$$

NOTE: Figure 3.7 shows the temperature dependence of $\bar{C}_p$ for CO_2.

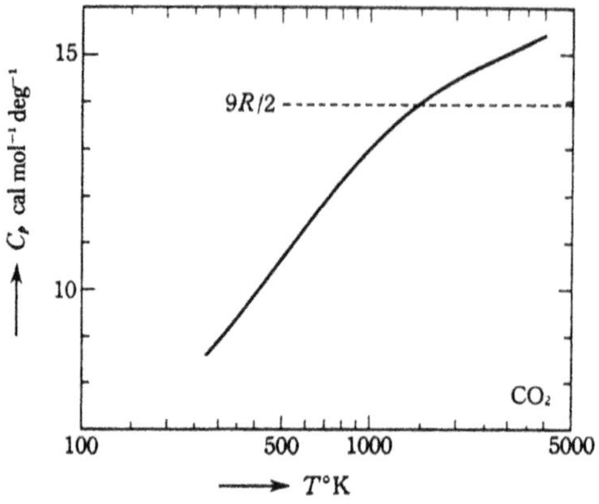

Fig. 3.7.

3. Consider an ideal gas molecule AB which undergoes the dissociation reaction $AB \rightleftarrows A + B$. If n_A, n_B and n_{AB} are the concentrations (the numbers of molecules per unit volume) of each molecule respectively, derive the law of mass action:

$$\frac{n_{AB}}{n_A n_B} = K(T) = \frac{V f_{AB}}{f_A f_B} e^{w_0/kT} = \left[\frac{(m_A + m_B)h^2}{2\pi m_A m_B kT}\right]^{\frac{3}{2}} \frac{j_{AB}^0}{j_A^0 j_B^0} e^{w_0/kT}$$

where f_A etc. are the partition functions per molecule, V is the volume of the container and j_A^0, ... are the partition functions for the internal degrees of freedom of each molecule. The zero of the energy for each molecule is chosen at the ground state (not including the zero point energy of the vibration) of the respective molecules so that $w_0 = \varepsilon_A^0 + \varepsilon_B^0 - \varepsilon_{AB}^0$ is the difference in the energy zeros.

SOLUTION

The partition function of the ideal gas occupying a volume V, which contains N_A, N_B and N_{AB} molecules, is given by (3.16);

$$Z(N_A, N_B, N_{AB}, V, T) = \frac{f_A^{N_A} f_B^{N_B} f_{AB}^{N_{AB}}}{N_A! N_B! N_{AB}!}. \tag{1}$$

This can also be considered to be proportional to the probability of finding N_A, N_B and N_{AB} molecules in V at temperature T. The equilibrium distribution is the most probable distribution and is determined by maximizing this Z:

$$\log Z(N_A, N_B, N_{AB}) = \sum_{\alpha = A, B, AB} \{N_\alpha \log f_\alpha - N_\alpha \log N_\alpha + N_\alpha\} = \max. \tag{2}$$

This is subject to the constraints

$$\begin{aligned}&\text{the total number of A} = N_A + N_{AB} = \text{const}\\ &\text{the total number of B} = N_B + N_{AB} = \text{const.}\end{aligned} \tag{3}$$

Therefore one obtains

$$\delta \log Z = \delta N_A \cdot \log(f_A/N_A) + \delta N_B \cdot \log(f_B/N_B) + \delta N_{AB} \cdot \log(f_{AB}/N_{AB}) = 0, \tag{4}$$

$$\delta N_A + \delta N_{AB} = 0, \qquad \delta N_B + \delta N_{AB} = 0, \tag{5}$$

so that

$$\delta \log Z = \{-\log(f_A/N_A) - \log(f_B/N_B) + \log(f_{AB}/N_{AB})\}\delta N_{AB} = 0,$$

and hence

$$\frac{N_{AB}}{N_A N_B} = \frac{f_{AB}}{f_A f_B}$$

or

$$\frac{n_{AB}}{n_A n_B} = \frac{V f_{AB}}{f_A f_B} \equiv K.$$

Substituting

$$f_A = \left(\frac{2\pi m_A kT}{h^2}\right)^{\frac{3}{2}} V j_A^0 e^{-\varepsilon^0{}_A/kT}$$

etc. and $m_{AB} = m_A + m_B$, one has the final result for K.

NOTE: If one uses the internal partition function $\prod (1 - e^{-\theta_\nu/T})$, which does not include the zero point energy of the vibration, instead of (3.7), one must put $w_0 = \varepsilon_A^0 + \varepsilon_B^0 - \varepsilon_{AB}^0 + \frac{1}{2}\sum (h\nu_A + h\nu_B - h\nu_{AB})$. This is equal to the energy change in this reaction at 0° K.

Problems

[A]

1. Assuming the ideal gas formula for argon, calculate the Helmholtz free energy, internal energy, entropy, and chemical potential per molecule. The atomic weight of argon is 39.94.

2. The energy difference between the lowest electronic state $^1S_0$ and the first excited state $^3S_1$ of the He atom is 159 843 cm^{-1}. Evaluate the fraction of excited atoms in He gas at 6000° K.
NOTE: $^1S_0$, $^3S_1$, etc. are symbols to indicate states of an atom. The superscript represents the spin multiplicity. In the present problem, 1 and 3 correspond to the degeneracies of the two states.

3. The rotational motion of a diatomic molecule is specified by two angular variables θ, φ and the corresponding canonical conjugate momenta p_θ, p_φ.

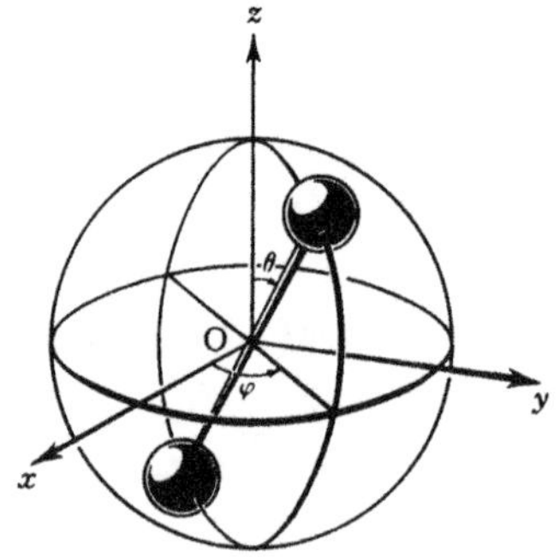

Assuming the form of the kinetic energy of the rotational motion to be

$$\varepsilon_{\text{rot}} = \frac{1}{2I} p_\theta^2 + \frac{1}{2I \sin^2\theta} p_\varphi^2 ,$$

derive the classical formula for the rotational partition function

$$r(T) = \frac{2IkT}{\hbar^2} \tag{3.9}$$

and calculate the corresponding entropy and specific heat.

4. Let the Hamiltonian of a rotator with principal moments of inertia (A, B, C) be

$$\mathscr{H} = \frac{1}{2A\sin^2\theta}\{(p_\phi - \cos\theta\, p_\psi)\cos\psi - \sin\theta \sin\psi\, p_\theta\}^2$$
$$+ \frac{1}{2B\sin^2\theta}\{(p_\phi - \cos\theta\, p_\psi)\sin\psi + \sin\theta\cos\psi\, p_\theta\}^2 + \frac{1}{2C} p_\psi^2 .$$

Here θ, ϕ and ψ are the Eulerian angles as shown in the figure. Derive the

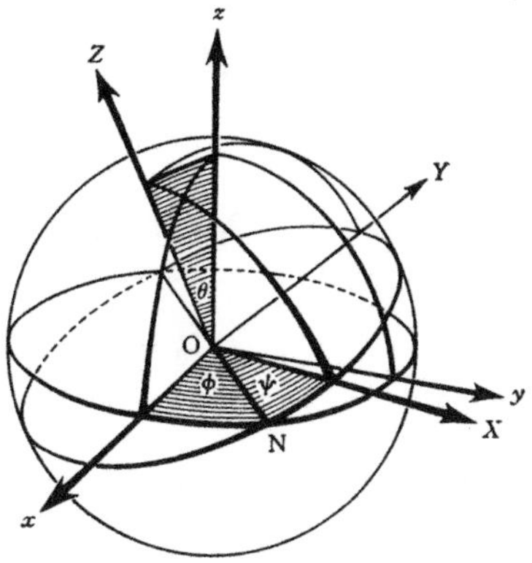

rotational partition function for the polyatomic molecule in the classical approximation. (*Hint:* Carry out the integration in the order p_θ, p_ϕ, p_ψ.)

5. The observed values of the rotational specific heat of a HD molecule are given in the figure. Explain the reason for the appearence of a maximum in this

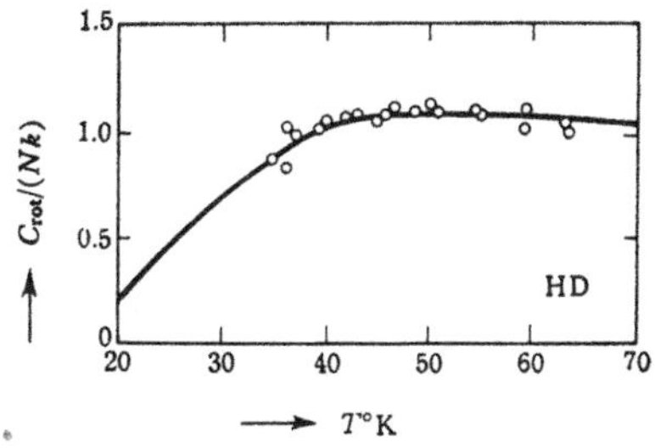

curve using the high temperature formula of the rotational partition function (cf. example 1).

6. Find the ratio of the number of ortho-hydrogen molecules to that of para-hydrogen molecules in a high temperature H_2 gas. How about the D_2 gas? NOTE: The ortho-hydrogen molecule has the greater nuclear spin weight. The nuclear spin of the H atom is $\frac{1}{2}$ and that of the D atom is 1.

7. The rate of ortho-para conversion of hydrogen molecules is so slow that ortho-hydrogen and para-hydrogen can be separated as though they are

different kinds of gases. Calculate the specific heat at low temperatures and show that para-hydrogen gas has the larger specific heat.

8. Calculate the molar heat content of NH_3 at 300° K. Assume the ideal gas formula and use the following data:

principal moments of inertia:

$$A = 4.44 \times 10^{-40}\,\mathrm{cm^2\,g}, \qquad B = C = 2.816 \times 10^{-40}\,\mathrm{cm^2\,g}$$

normal modes of vibration:

$$\bar{\nu}_1 = \bar{\nu}_2 = 3336\,\mathrm{cm}^{-1}, \qquad \bar{\nu}_3 = \bar{\nu}_4 = 950\,\mathrm{cm}^{-1},$$
$$\bar{\nu}_5 = 3414\,\mathrm{cm}^{-1}, \qquad \bar{\nu}_6 = 1627\,\mathrm{cm}^{-1}.$$

9. Calculate the contribution of the first excited electronic state of the O_2 molecule to the Helmholtz free energy of oxygen gas at 5000° K. Use:

$$\begin{pmatrix}\text{the first excited level } {}^1\Delta_g \\ \text{(degeneracy } g_e = 2)\end{pmatrix} - \begin{pmatrix}\text{the ground state } {}^3\Sigma_g \\ \text{(degeneracy } g_e = 3)\end{pmatrix} = 7824\ \mathrm{cm}^{-1}.$$

10. Evaluate the equilibrium constant $K_p = p_{\mathrm{N}}^2/p_{\mathrm{N_2}}$ of the dissociative reaction $N_2 \rightleftarrows 2N$ at 5000° K under the following assumptions: the characteristic temperature of rotation and that of vibration of the N_2 molecule are $\Theta_r = 2.84°$ K and $\Theta_v = 3.35 \times 10^3$ °K respectively. The dissociation energy is $D_0 = 169.3\ \mathrm{kcal\ mol^{-1}}$ (including the correction for the zero point energy of the vibrations). The electronic ground state of the N_2 molecule has no degeneracy, but that of the N atom has a degeneracy 4 due to electron spin.

11. Calculate, by statistical mechanics, the equilibrium constant for the reaction $D_2 + H_2 \rightleftarrows 2\,HD$ at temperatures high enough to allow the use of the classical approximation for the rotational motion.

12. Show that the equilibrium constant for a reaction such as $HCl + DBr \rightleftarrows DCl + HBr$ approaches unity at sufficiently high temperatures.

13. Derive the formula for the pressure p of a real gas using the virial theorem (problem 3.6)

$$p = \frac{kT}{v}\left[1 + \frac{1}{2v}\int_0^\infty \{1 - \mathrm{e}^{-u(r)/kT}\}\, 4\pi r^2\, \mathrm{d}r + \mathrm{o}\left(\frac{1}{v}\right)\right] \qquad (v = V/N),$$

where the inter-molecular potential $u(r)$ is assumed to be the function of the distance r only.

[B]

14. The energy levels of a symmetrical top molecule (which has two equal principal moments of inertia) are given by

$$\varepsilon_{l,\lambda,m} = \frac{h^2}{8\pi^2}\left\{\frac{l(l+1)}{A} + \lambda^2\left(\frac{1}{C} - \frac{1}{A}\right)\right\}$$

where A, A, and C are the principal moments of inertia and l, λ, m are the quantum numbers. Also, m and λ take integral values between l and $-l$. The energy is degenerate with respect to m. Calculate the rotational partition function and derive the classical limit of the formula.

15. The vibration of a diatomic molecule exhibits anharmonicity, when the amplitude becomes very large, owing to the characteristic form of the potential as shown in the figure. The energy levels can be given approximately

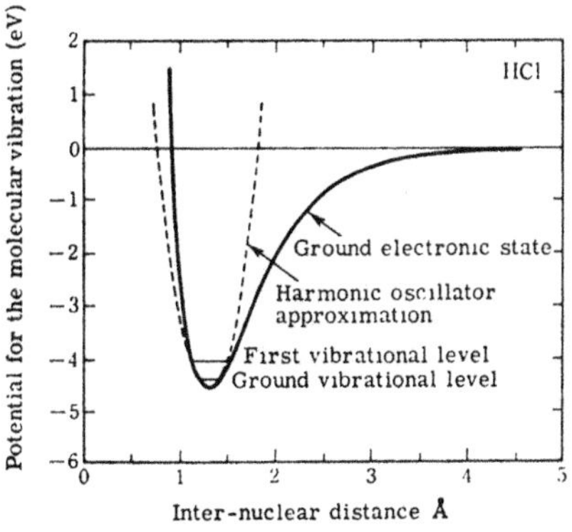

by

$$\varepsilon_n = (n + \tfrac{1}{2})h\nu - x_e(n + \tfrac{1}{2})^2 h\nu \qquad (n = 0, 1, 2, \cdots)$$

when x_e is a parameter used to represent the degree of anharmonicity. Calculate the effect of anharmonicity upon the vibrational specific heat up to the first order in x_e.

16. Answer the following questions concerning the CH_4 gas:

(i) How many degrees of freedom does the intra-molecular vibration have?

(ii) Give the nuclear-rotational partition function in high temperature approximation.

(iii) Give the chemical constant.

(iv) What are the important effects on the thermodynamic quantities when H is replaced by D in the series CH_4, CH_3D, CH_2D_2, CHD_3, CD_4?

17. One can assume that the inter-molecular forces are not affected by substituting atoms by their isotopes. Show that it is impossible to separate isotopes by use of gas reactions at high temperatures (where the internal motion of molecules can be treated classically).

18. Consider a gas composed of molecules interacting with the intermolecular potential

(a) $$u(r) = \alpha/r^n \qquad (n > 0)$$

(b) $$u(r) \begin{cases} = \infty & r < a \\ = -u_0 = \text{const } (< 0) & a < r < b \\ = 0 & r > b\,. \end{cases}$$

Calculate the second virial coefficient and the Joule-Thomson coefficient. Discuss the Joule-Thomson effect.

[C]

19. A H_2 molecule decomposes into H atoms when it is adsorbed upon a certain metallic surface. Derive the relationship between the quantity adsorbed and the pressure of the H_2 gas.

20. Consider a classical system of N identical molecules with no internal structure. Let $v^{-n}F_n(x_1, y_1, z_1, \ldots z_n)\,\mathrm{d}x_1 \ldots \mathrm{d}z_n$ ($v = V/N$, V being the volume) be the probability with which n particles fall into the region $(x_1, x_1 + \mathrm{d}x_1)$, ..., $(z_n, z_n + \mathrm{d}z_n)$. Derive the following relation assuming two-body interactions between particles:

$$\mathrm{e}^{-\mu/kT} = \left(\frac{2\pi mkT}{h^2}\right)^{\frac{3}{2}} v\left[1 + \sum_{n=1}^{N} \frac{1}{n!v^n}\int \ldots \int F_n(x_1, \ldots, z_n) \prod_{i=1}^{n} f_{0i}\,\mathrm{d}x_1 \ldots \mathrm{d}z_n\right]$$

where m = the mass of the particle, μ = the chemical potential, $f_{0i} = \exp(-u_{0i}/kT) - 1$ and u_{0i} is the potential of the inter-molecular force between the 0-th and i-th molecule.

21. Give the lowest order correction term which shows a deviation from the ideal gas value for each of the following quantities of a real gas:
Helmholtz free energy F, Gibbs free energy G, entropy S,
internal energy U, specific heat at constant volume C_V,
and specific heat at constant pressure C_p.
(Hint: Start from equation (3.31) and (3.30).)

22. Give an explicit formula for the cluster integral b_3 (3.26) and derive the relation

$$b_3 = \tfrac{1}{2}\beta_1^2 + \tfrac{1}{3}\beta_2$$

where β_1 and β_2 are the irreducible cluster integrals defined by

$$\beta_1 = \frac{1}{V}\iint f_{12}\,d\boldsymbol{r}_1\,d\boldsymbol{r}_2\,, \qquad \beta_2 = \frac{1}{2V}\iiint f_{12}f_{23}f_{31}\,d\boldsymbol{r}_1\,d\boldsymbol{r}_2\,d\boldsymbol{r}_3\,.$$

Solutions

1. From (3.1) and (3.2) one obtains the translational parts

$$\mu_{\text{trans}} = \frac{G_{\text{trans}}}{N} = kT\log\left\{\frac{p}{kT}\left(\frac{h^2}{2\pi mkT}\right)^{\frac{3}{2}}\right\} = kT\chi(T,p), \tag{1}$$

$$\frac{F_{\text{trans}}}{N} = kT(\chi - 1), \qquad \frac{U_{\text{trans}}}{N} = \tfrac{3}{2}kT, \qquad \frac{S_{\text{trans}}}{N} = k\left(\tfrac{5}{2} - \chi\right), \tag{2}$$

$$Z_{\text{trans}} = \exp\{-N(\chi - 1)\}\,. \tag{3}$$

At $T = 300°$ K and $p = 1$ atm $= 1.01325 \times 10^6$ erg cm^{-3} one has $kT = 4.1408 \times 10^{-14}$ erg, $\log(kT) = -30.81506$, $\log p = 13.82868$ and $\log\{p/(kT)^{\frac{5}{2}}\} = 90.86633$. Substituting these values and $\log \hbar = -62.11702$, $\log\{h^3/(2\pi)^{\frac{3}{2}}\} = -183.59424$†, one obtains

$$\chi(300°\,\text{K}, 1\,\text{atm}) = -92.72791 - \tfrac{3}{2}\log m = -10.59475 - \tfrac{3}{2}\log\frac{m}{m^*} \tag{4}$$

where m should be measured in grams and $m^* = 1.65963 \times 10^{-24}$g is the mass of the atom which has unit atomic weight. For argon (atomic weight 39.94), one has $\chi = -16.126$ from (4), $\mu_{\text{trans}} = -0.6677 \times 10^{-12}$ erg from equation (1) so that $F_{\text{trans}}/N = -0.7092 \times 10^{-12}$ erg, $U_{\text{trans}}/N =$

† Such a large number of significant figures is unnecessary in this case but is given for the sake of future reference.

0.6211×10^{-13} erg, $S_{\text{trans}}/N = 1.8808 \times 10^{-15}$ erg deg^{-1} from equation (2) and $Z_{\text{trans}} = \exp(17.126 \times N)$ from equation (3). The number of atoms per unit volume is $N = pV/kT = 2.447 \times 10^{19}$. Hence one has, finally,

$$F_{\text{trans}} = -1.7354 \times 10^7 \text{ erg},$$
$$S_{\text{trans}} = 4.6023 \times 10^4 \text{ erg deg}^{-1},$$
$$U_{\text{trans}} = 1.5598 \times 10^6 \text{ erg}$$
$$Z_{\text{trans}} = \exp(4.189 \times 10^{20}) = 10^{10^{20.26}}.$$

2. In a gas in thermal equilibrium at a temperature T the probability $P(l)$ of a gas molecule occupying an internal energy level ε_l is given by

$$P(l) = \frac{g_l \mathrm{e}^{-\varepsilon_l/kT}}{j(T)}. \tag{1}$$

Here g_l is the degeneracy of the l-th state and $j(T)$ is the partition function for the internal degree of freedom. Now only the lowest two states come into consideration since higher states have negligible probability. Considering the degeneracy, one has

$$P(^3S_1) = \frac{3\mathrm{e}^{-\varepsilon_1/kT}}{1 + 3\mathrm{e}^{-\varepsilon_1/kT}}. \tag{2}$$

For $\varepsilon_1 = 159\,843$ cm^{-1}, $T = 6000°$ K, one has $\mathrm{e}^{-\varepsilon_1/kT} = \mathrm{e}^{-2.3 \times 10^5/T} \fallingdotseq \mathrm{e}^{-38.3} \fallingdotseq 2.1 \times 10^{-11}$, an extremely small value. Hence the second term in the denominator can be neglected, and one finds that the fraction in question is only about 6×10^{-11}.

3. The rotational partition function in classical statistics is given by

$$r(T) = \frac{1}{h^2}\int_0^{\pi} \mathrm{d}\theta \int_0^{2\pi} \mathrm{d}\phi \int_{-\infty}^{\infty} \mathrm{d}p_\theta \int_{-\infty}^{\infty} \mathrm{d}p_\phi \exp\left[-\frac{1}{kT}\left(\frac{p_\theta^2}{2I} + \frac{1}{2I\sin^2\theta}p_\phi^2\right)\right].$$

The integration over ϕ is trivial and the integrations over p_θ and p_ϕ can be carried out using the formula

$$\int_{-\infty}^{\infty} \mathrm{e}^{-ax^2}\mathrm{d}x = \sqrt{\pi/a}.$$

After the final integration over θ, one obtains the required result. The Helmholtz free energy F_{rot} is given immediately by

$$F_{\text{rot}} = -kT\log r(T) = -kT\log(2IkT/\hbar^2),$$

and the corresponding entropy by

$$S_{\text{rot}} = -(\partial F_{\text{rot}}/\partial T)_V = k + k\log(2IkT/\hbar^2),$$

and hence the specific heat is given by

$$C_{V\text{rot}} = T(\partial S_{\text{rot}}/\partial T) = k \qquad \text{(these formulas are given per molecule)}.$$

NOTE: The rotational partition function $r(T)$ should always be greater than unity as may be seen from (3.8). This shows that the classical formula (3.9) becomes meaningless at temperatures $T < \Theta_{\text{r}} = \hbar^2/(2Ik)$.

4. The given Hamiltonian can be rewritten as

$$\mathscr{H} = \frac{1}{2}\left(\frac{\sin^2\psi}{A} + \frac{\cos^2\psi}{B}\right) \times$$

$$\times \left\{ p_\theta + \left(\frac{1}{B} - \frac{1}{A}\right) \frac{\sin\psi\cos\psi}{\sin\theta\left(\dfrac{\sin^2\psi}{A} + \dfrac{\cos^2\psi}{B}\right)} (p_\phi - \cos\theta\, p_\psi^2) \right\}^2 +$$

$$+ \frac{1}{2AB\sin^2\theta} \frac{1}{\left(\dfrac{\sin^2\psi}{A} + \dfrac{\cos^2\psi}{B}\right)} (p_\phi - \cos\theta\, p_\psi)^2 + \frac{1}{2C} p^2 . \qquad (1)$$

Substituting this into the classical formula

$$r(T) = \frac{1}{h^3} \int_0^\pi \int_0^{2\pi} \int_0^{2\pi} \int_{-\infty}^{\infty} \int_{-\infty}^{\infty} \int_{-\infty}^{\infty} e^{-\mathscr{H}/kT} \, d\theta\, d\phi\, d\psi\, dp_\theta\, dp_\phi\, dp_\psi , \qquad (2)$$

one can perform the integrations successively using the formula

$$\int_{-\infty}^{\infty} e^{-a(x+b)^2} dx = \int_{-\infty}^{\infty} e^{-ax^2} dx = (\pi/a)^{\frac{1}{2}} .$$

Integration over p_θ gives

$$(2\pi kT)^{\frac{1}{2}} \left\{\frac{\sin^2\psi}{A} + \frac{\cos^2\psi}{B}\right\}^{-\frac{1}{2}} \qquad (3)$$

over p_ϕ gives

$$(2\pi kTAB)^{\frac{1}{2}} \sin\theta \left(\frac{\sin^2\psi}{A} + \frac{\cos^2\psi}{B}\right)^{\frac{1}{2}} \qquad (4)$$

and over p_ψ gives

$$(2\pi kTC)^{\frac{1}{2}} . \qquad (5)$$

The integrations over θ, ϕ and ψ of the product (3), (4), (5) is trivial.

One obtains finally

$$r(T) = \pi^{\frac{1}{2}}\left(\frac{8\pi^2 AkT}{h^2}\right)^{\frac{1}{2}}\left(\frac{8\pi^2 BkT}{h^2}\right)^{\frac{1}{2}}\left(\frac{8\pi^2 CkT}{h^2}\right)^{\frac{1}{2}}. \tag{6}$$

5. The expansion formula for the rotational partition function of a heteronuclear diatomic molecule is given in (5) of example 1. Hence one has

$$\frac{F_{\text{rot}}}{N} = -kT\log r(T) = -kT\log\left[\frac{1}{\sigma}\left\{1+\frac{\sigma}{3}+\frac{\sigma^2}{15}+\frac{4\sigma^3}{315}+O(\sigma^4)\right\}\right]$$

$$= kT\log\sigma - kT\frac{\sigma}{3}\left\{1+\frac{\sigma}{30}+\frac{8\sigma^2}{945}+O(\sigma^3)\right\}$$

$$= kT\log\frac{\Theta_r}{T} - \frac{k\Theta_r}{3}\left\{1+\frac{1}{30}\frac{\Theta_r}{T}+\frac{8}{945}\left(\frac{\Theta_r}{T}\right)^2+O\left(\frac{\Theta_r}{T}\right)^3\right\}, \tag{1}$$

and

$$\frac{U_{\text{rot}}}{N} = -T^2\frac{\partial}{\partial T}\left(\frac{F_{\text{rot}}}{NT}\right) = kT - \frac{k\Theta_r}{3}\left\{1+\frac{1}{15}\frac{\Theta_r}{T}+\frac{8}{315}\left(\frac{\Theta_r}{T}\right)^2+O\left(\frac{\Theta_r}{T}\right)^3\right\}, \tag{2}$$

$$\frac{C_{\text{rot}}}{N} = \frac{\partial}{\partial T}\left(\frac{U_{\text{rot}}}{N}\right) = k\left\{1+\frac{1}{45}\left(\frac{\Theta_r}{T}\right)^2+\frac{16}{945}\left(\frac{\Theta_r}{T}\right)^3+O\left(\frac{\Theta_r}{T}\right)^4\right\}. \tag{3}$$

Equation (3) shows that the rotational molar heat content C_{rot}/N decreases toward a limiting value k when the temperature goes to infinity. On the other hand, C_{rot} must go to zero as $T \to 0$. Consequently, the C_{rot} versus T curve should have at least one maximum. (Detailed investigation shows that there is only one maximum.)

6. The nuclear spin of H is $\frac{1}{2}$. Therefore the nuclear spin weight of the ortho- or para-hydrogen molecule H_2 is 3 or 1 as given by (3.10a). The ratio of the number of ortho- and para-hydrogen molecules is given by

$$j_{\text{nuc-rot}}(\text{ortho}) : j_{\text{nuc-rot}}(\text{para}) = 3r_{\text{odd}} : r_{\text{even}}. \tag{1}$$

At high temperatures, $T \gg \Theta_r$, r_{even} and r_{odd} are both equal to $\frac{1}{2}\, r_c$ (3.9). Hence one obtains

$$n(\text{ortho-H}_2) : n(\text{para-H}_2) = 3 : 1 .$$

For the D_2 molecule, the ortho- para ratio of the nuclear spin weight, i.e. $(s_n+1)(2s_n+1) : s_n(2s_n+1)$ becomes $6:3$ ($s_n = 1$). Hence one has

$$n(\text{ortho-D}_2) : n(\text{para-D}_2) = 2 : 1 .$$

7. As given in the previous solution, one has from (3.10a)

$$j_{\text{nuc-rot}}(\text{ortho}) = 3r_{\text{odd}}, \qquad j_{\text{nuc-rot}}(\text{para}) = r_{\text{even}}, \tag{1}$$

where

$$r_{\text{odd}} = 3e^{-2\Theta/T} + 7e^{-12\Theta/T} + 11e^{-30\Theta/T} + \cdots, \tag{2}$$

$$r_{\text{even}} = 1 + 5e^{-6\Theta/T} + 9e^{-20\Theta/T} + \cdots. \tag{3}$$

The molar heat content is given by

$$C_{\text{nuc-rot}}(\text{ortho}) = R\frac{\partial}{\partial T}\left\{-\frac{\partial \log r_{\text{odd}}}{\partial(1/T)}\right\},$$

$$C_{\text{nuc-rot}}(\text{para}) = R\frac{\partial}{\partial T}\left\{-\frac{\partial \log r_{\text{even}}}{\partial(1/T)}\right\}.$$

At low temperatures one has, from (3),

$$C_{\text{nuc-rot}}(\text{ortho}) = \tfrac{700}{3} R(\Theta/T)^2 e^{-10\Theta/T} + \cdots, \tag{4}$$

$$C_{\text{nuc-rot}}(\text{para}) = 180\, R(\Theta/T)^2 e^{-6\Theta/T} + \cdots. \tag{5}$$

Thus, at low temperatures we should have $C_{\text{nuc-rot}}$ (para) > $C_{\text{nuc-rot}}$ (ortho) in this case.

NOTE: By comparing (2) and (3), one finds that r_{odd} decreases more rapidly than r_{even} as Θ/T increases. Correspondingly, the internal energy $U_{\text{rot, even}}$

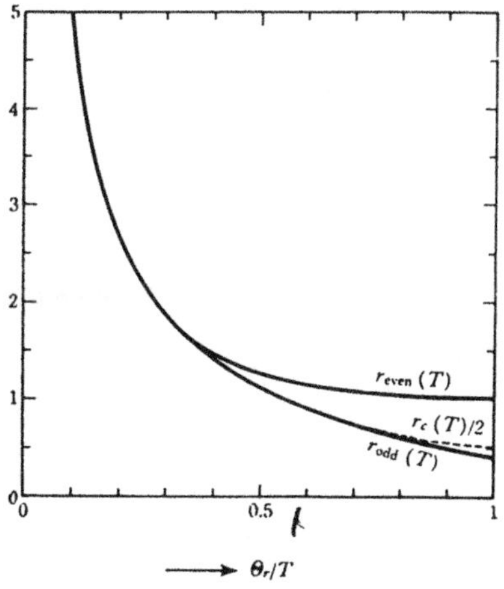

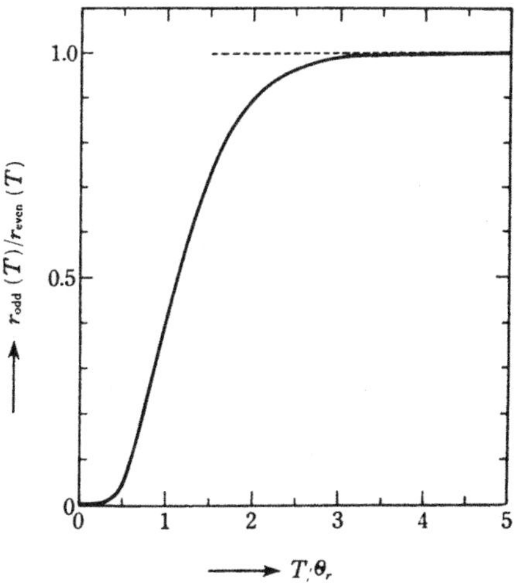

$(= -k\,\partial \log r_{\text{even}}/\partial(1/T))$ has a larger slope than $U_{\text{rot, odd}}$ and hence one knows that the specific heat of para-hydrogen (even) is larger than that of ortho-hydrogen (odd). Numerical results for these quantities are shown in the figures.

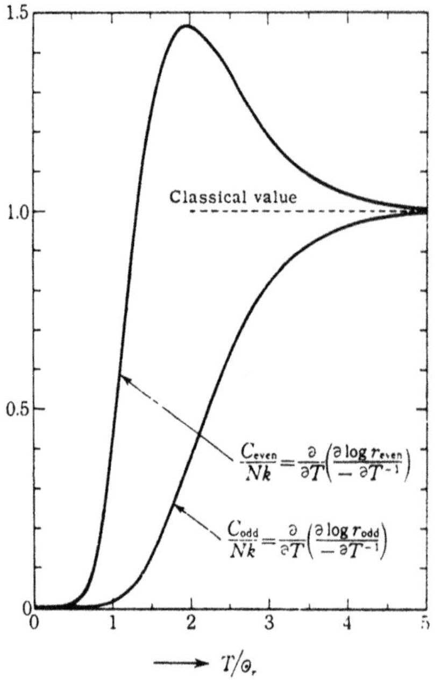

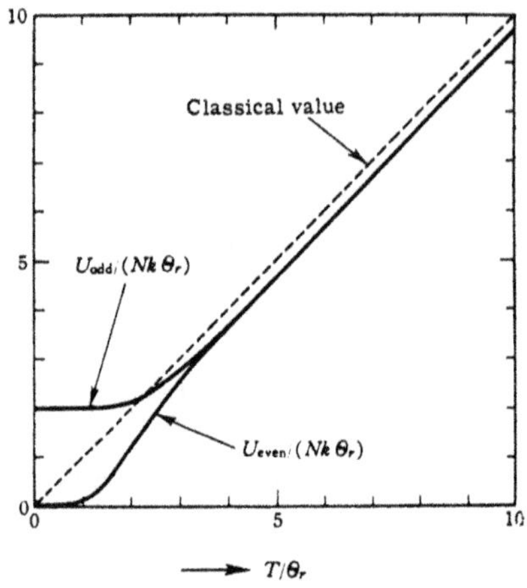

8. At 300° K the rotational motion can be treated classically, because $h^2/8\pi^2AK$ and $h^2/8\pi^2BK$ are of the order of 10° K. Consequently the molar heat content is given by

$$\bar{C}_V = 3R + R\sum_{i=1}^{6}\theta_i^2\frac{e^{\theta_i}}{(e^{\theta_i}-1)^2}, \qquad \theta_i = \frac{hc\bar{\nu}_i}{kT} = 1.4387\frac{\bar{\nu}_i}{T}. \tag{1}$$

Substituting the given numerical values, one obtains $\theta_1 = \theta_2 = 15.998$, $\theta_3 = \theta_4 = 4.56$, $\theta_5 = 16.372$, $\theta_6 = 7.80$. Hence one finds that only the θ_3 and θ_4 terms make appreciable contributions:

$$\theta_3^2 e^{\theta_3}/(e^{\theta_3}-1)^2 \fallingdotseq 0.22, \qquad \text{thus} \quad \bar{C}_V \fallingdotseq 3.44R\,.$$

9. One can treat the electronic partition function $e(T)$ as a separate factor in

$j(T)$ because the first excited electronic level $\varepsilon_1 = 7824\,\mathrm{cm}^{-1} = 11256°$ K is sufficiently high compared to the vibrational quantum ($\theta_v = 2229°$ K). Neglecting the contribution of the second excited state ε_2 which is about 2 times that of ε_1, one obtains

$$e(T) = 3 + 2\mathrm{e}^{-\varepsilon_1/kT}. \tag{1}$$

At $T = 5000°$ K, one has $\mathrm{e}^{-\varepsilon_1/kT} = \mathrm{e}^{-2.25} = 0.105$, and thus $e(T) = 3.21$. Hence the contribution to the Helmholtz free energy is given by

$$F_e = -RT \log e(T) = -4.16 \times 10^{11} \times 1.166\,\mathrm{erg\,mol^{-1}} = \\ = -4.85 \times 10^{11}\,\mathrm{erg\,mol^{-1}} = -11.59\,\mathrm{kcal\,mol^{-1}}. \tag{2}$$

So the correction term is seen to be

$$\Delta F_e = -RT\{\log(3 + 0.21) - \log 3\} = -RT \log 1.07 \simeq -0.07\,RT,$$

and

$$\Delta F_e \risingdotseq -0.7\,\mathrm{kcal\,mol^{-1}} \qquad \text{at } 5000°\,K.$$

NOTE 1: The value of the total Helmholtz free energy is 3.20×10^5 kcal mole^{-1} at 5000° K.

NOTE 2: Moreover, the contribution of the electronic part to the molar heat content cannot be neglected. The figure shows the experimental values

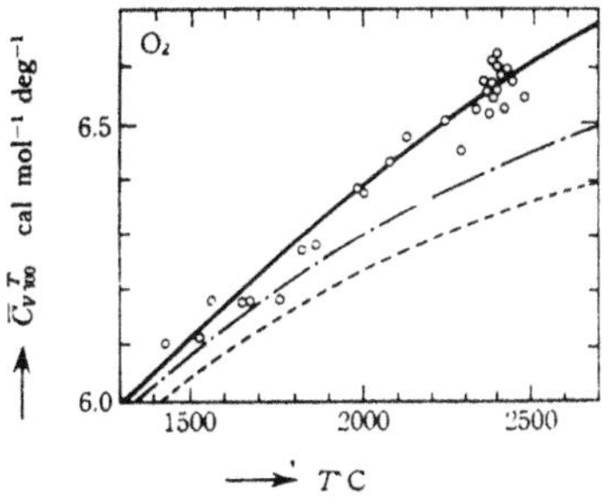

(indicated by open circles) of the mean molar heat content at constant volume $C_{V300}^{\;T}$ between 300° K and $T°$ K. The dashed line is the calculated value including only the vibrational term with $\Theta_v = 2229°$ K, the chain line includes the rotational term also and the solid line is calculated including the term $^1\Delta_g$.

10. The equilibrium constant K given in example (3) can be rewritten as

$$K_p(T) = \frac{p_\mathrm{N}^2}{p_{\mathrm{N}_2}} = \frac{n_\mathrm{N}^2}{n_{\mathrm{N}_2}} kT = kT\left[\frac{2\pi m_\mathrm{N}^2 kT}{2m_\mathrm{N}h^2}\right]^{\frac{3}{2}} \frac{\{j(\mathrm{N})\}^2}{j(\mathrm{N}_2)}\,\mathrm{e}^{-w/kT}, \tag{1}$$

where w is the dissociation energy including the zero-point energy correction. Hence the zero-point energy of the vibration can be omitted from $j(N_2)$. Treating the rotational motion classically, one has from (3.5c)

$$j(N_2) = \frac{1}{2}\frac{T}{\Theta_r}(1 - e^{-\Theta_v/T})^{-1} e(N_2) g_N^2 \tag{2}$$

and

$$j(N) = e(N) g_N . \tag{3}$$

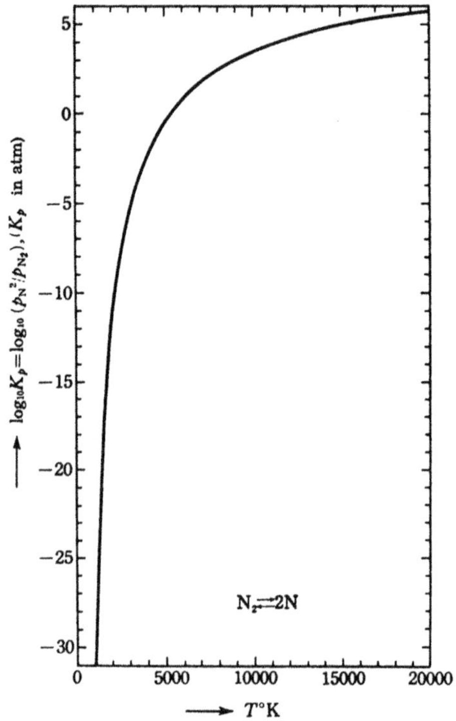

Substituting these into (1), one obtains

$$K_p(T) = kT\left[\frac{\pi m_N kT}{h^2}\right]^{\frac{3}{2}} \frac{2\Theta_r}{T}(1 - e^{-\Theta_v/T})\frac{\{e(N)\}^2}{e(N_2)} e^{-D_0/RT},$$

$$\log K_p = -\frac{D_0}{RT} + \frac{3}{2}\log\frac{\pi m_N kT}{h^2} + \log 2k\Theta_r + \log(1 - e^{-\Theta_v/T}) + \log\frac{\{e(N)\}^2}{e(N_2)} \tag{4}$$

in which the nuclear parts have disappeared. The following values are to be inserted into equation (4): $e(\mathrm{N}) = 4$, $e(\mathrm{N}_2) = 1$,

$$T = 5000°\,K, \qquad kT = 6.90 \times 10^{-13}\,\mathrm{erg},$$
$$RT = 9.94\,\mathrm{kcal\,mol^{-1}}, \qquad \Theta_v/T = 0.670,$$
$$\Theta_r = 2.84°\,K, \qquad D_0 = 169.3\,\mathrm{kcal\,mol^{-1}},$$
$$m_\mathrm{N} = 14.01 \times 1.66 \times 10^{-24}\,\mathrm{g} = 23.3 \times 10^{-24}\,\mathrm{g}.$$

It is to be noted that $(\pi m_\mathrm{N} kT/h^2)^{\frac{3}{2}} \cdot 2k\Theta_r$ on the right hand side, which is evaluated in c.g.s. units, has the dimension of a pressure. Hence the value of $K_p(T)$ calculated in atm is given by

$$\begin{aligned}\log K_p(\mathrm{atm}) &= -\log(1.01325 \times 10^6) + \log K_p(\mathrm{c.g.s.})\\ &= -13.83 + [-17.02 + 62.38 - 34.78 - 0.72 + \log 4]\\ &= -13.83 + 12.63 = -1.20,\end{aligned}$$

$$\log_{10} K_p(\mathrm{atm}) = -0.52.$$

NOTE: The figure shows the curve $\log_{10} K_p$ (atm) versus T.

11. As in example (3) one can obtain the numbers of molecules $N(\mathrm{D}_2)$ and $N(\mathrm{HD})$ from the conditions

$$Z(N(\mathrm{H}_2), N(\mathrm{D}_2), N(\mathrm{HD})) = \frac{f(\mathrm{H}_2)^{N(\mathrm{H}_2)}}{N(\mathrm{H}_2)!}\frac{f(\mathrm{D}_2)^{N(\mathrm{D}_2)}}{N(\mathrm{D}_2)!}\frac{f(\mathrm{HD})^{N(\mathrm{HD})}}{N(\mathrm{HD})!} = \max$$

$$2N(\mathrm{H}_2) + N(\mathrm{HD}) = \mathrm{const}, \qquad 2N(\mathrm{D}_2) + N(\mathrm{HD}) = \mathrm{const}$$

and hence

$$K_p = K_c = \frac{N(\mathrm{HD})^2}{N(\mathrm{H}_2)N(\mathrm{D}_2)} = \frac{f(\mathrm{HD})^2}{f(\mathrm{H}_2)f(\mathrm{D}_2)}$$
$$= \left\{\frac{m(\mathrm{HD})^2}{m(\mathrm{H}_2)m(\mathrm{D}_2)}\right\}^{\frac{3}{2}} \frac{j(\mathrm{HD})^2}{j(\mathrm{H}_2)j(\mathrm{D}_2)}. \tag{1}$$

Here one has assumed that the molecular binding in H_2, D_2 and HD is precisely the same and that there is no change in the ground state energy(the internal partition function includes the zero-point vibration). At high temperatures, where equation (3.5) and (3.6) can be used for the nuclear-rotational partition functions, one has

$$\frac{j_{\text{nuc-rot}}(\mathrm{HD})^2}{j_{\text{nuc-rot}}(\mathrm{H}_2)j_{\text{nuc-rot}}(\mathrm{D}_2)} = \frac{4I(\mathrm{HD})^2}{I(\mathrm{H}_2)I(\mathrm{D}_2)}$$
$$= 4\frac{\{m_\mathrm{H}m_\mathrm{D}/(m_\mathrm{H}+m_\mathrm{D})\}^2}{(\frac{1}{2}m_\mathrm{H})(\frac{1}{2}m_\mathrm{D})} = 16\frac{m_\mathrm{H}m_\mathrm{D}}{(m_\mathrm{H}+m_\mathrm{D})^2}$$

where the ratio of the moments of inertia has been replaced by the ratio of the reduced masses because all the nuclear distances are assumed to be equal. Now one obtains

$$K=4\frac{m_H+m_D}{2(m_H m_D)^{\frac{1}{2}}}\frac{(1-e^{-h\nu_{H_2}/kT})(1-e^{-h\nu_{D_2}/kT})}{(1-e^{-h\nu_{HD}/kT})^2}\exp(-\tfrac{1}{2}h(2\nu_{HD}-\nu_{H_2}-\nu_{D_2})/kT). \tag{2}$$

For $\Theta_{HD} \gg T$, $\quad \Theta_{HD} = h\nu_{HD}/k = 4324°\,K$ and

$$K = 4\frac{m_H+m_D}{2(m_H m_D)^{\frac{1}{2}}}\exp\{-\tfrac{1}{2}h(2\nu_{HD}-\nu_{H_2}-\nu_{D_2})/kT\}. \tag{3}$$

For $\Theta_{HD} \ll T$, then

$$K_\infty = 4\frac{m_H+m_D}{2(m_H m_D)^{\frac{1}{2}}}\frac{\nu_{H_2}\nu_{D_2}}{\nu_{HD}^2}. \tag{4}$$

Since the vibrational frequency is inversely proportional to the square root of the reduced mass, one has

$$\nu_{HD}:\nu_{H_2}:\nu_{D_2} = \left(\frac{m_H m_D}{m_H+m_D}\right)^{-\frac{1}{2}}:\left(\frac{m_H}{2}\right)^{-\frac{1}{2}}:\left(\frac{m_D}{2}\right)^{-\frac{1}{2}}. \tag{5}$$

Therefore, expression (4) becomes

$$K_\infty = 4.$$

Putting $m_D = 2m_H$, one has

$$\begin{aligned}\nu_{HD}:\nu_{H_2}:\nu_{D_2} &= 1:2/\sqrt{3}:\sqrt{2/3}\\ &= 1:1.155:0.817.\end{aligned}$$

Hence one obtains from (3)

$$\begin{aligned}K &= 3\sqrt{2}\exp(-0.0144\,\Theta_{HD}/T)\\ &= 4.24\exp(-62.3/T).\end{aligned}$$

The figure shows the temperature dependence of K_p.

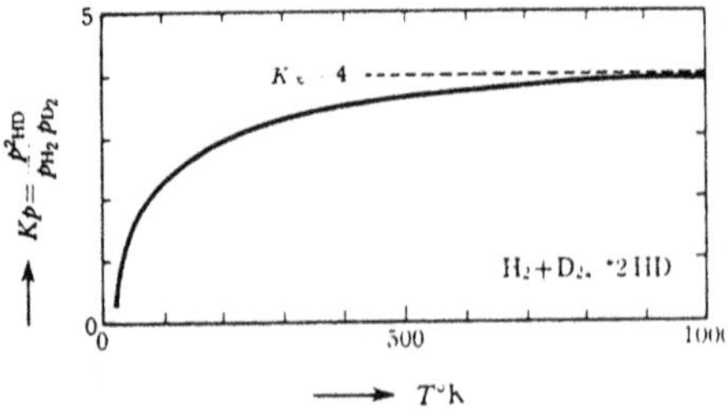

12. As in the previous problem, one has

$$K_p = K_c = \frac{N(\mathrm{DCl})N(\mathrm{HBr})}{N(\mathrm{HCl})N(\mathrm{DBr})} = \frac{f(\mathrm{DCl})f(\mathrm{HBr})}{f(\mathrm{HCl})f(\mathrm{DBr})}$$

$$= \left[\frac{m(\mathrm{DCl})m(\mathrm{HBr})}{m(\mathrm{HCl})m(\mathrm{DBr})}\right]^{\frac{3}{2}} \cdot \frac{I(\mathrm{DCl})I(\mathrm{HBr})}{I(\mathrm{HCl})I(\mathrm{DBr})} \cdot \frac{\nu(\mathrm{HCl})\nu(\mathrm{DBr})}{\nu(\mathrm{DCl})\nu(\mathrm{HBr})} \tag{1}$$

where the temperature has been assumed to be so high that the approximation (3.5a) for the rotation and the approximation $v(T) = kT/h\nu$, $h\nu \ll kT$ for the vibration can be used. Assuming that the inter-nuclear distance and also the binding force of the diatomic molecule do not change as a result of substitution of isotopes, one has

$$I(\mathrm{DCl}) : I(\mathrm{HCl}) = \mu(\mathrm{DCl}) : \mu(\mathrm{HCl}),$$
$$I(\mathrm{HBr}) : I(\mathrm{DBr}) = \mu(\mathrm{HBr}) : \mu(\mathrm{DBr}),$$
$$\nu(\mathrm{DCl}) : \nu(\mathrm{HCl}) = \mu(\mathrm{DCl})^{-\frac{1}{2}} : \mu(\mathrm{HCl})^{-\frac{1}{2}},$$
$$\nu(\mathrm{HBr}) : \nu(\mathrm{DBr}) = \mu(\mathrm{HBr})^{-\frac{1}{2}} : \mu(\mathrm{DBr})^{-\frac{1}{2}},$$

where μ is the reduced mass. Hence the right hand side of equation (1) becomes

$$\left[\frac{(m_\mathrm{D} + m_\mathrm{Cl})(m_\mathrm{H} + m_\mathrm{Br})}{(m_\mathrm{H} + m_\mathrm{Cl})(m_\mathrm{D} + m_\mathrm{Br})}\right]^{\frac{3}{2}} \cdot \frac{\mu(\mathrm{DCl})\mu(\mathrm{HBr})}{\mu(\mathrm{HCl})\mu(\mathrm{DBr})} \cdot \left[\frac{\mu(\mathrm{DCl})\mu(\mathrm{HBr})}{\mu(\mathrm{HCl})\mu(\mathrm{DBr})}\right]^{\frac{1}{2}}$$

$$= \left[\frac{(m_\mathrm{D} + m_\mathrm{Cl})(m_\mathrm{H} + m_\mathrm{Br})}{(m_\mathrm{H} + m_\mathrm{Cl})(m_\mathrm{D} + m_\mathrm{Br})}\right]^{\frac{3}{2}} \left[\frac{\mu(\mathrm{DCl})\mu(\mathrm{HBr})}{\mu(\mathrm{HCl})\mu(\mathrm{DBr})}\right]^{\frac{3}{2}} = 1\,.$$

DIVERTISSEMENT 9

H-theorem: As was mentioned in a previous Divertissement 3, Boltzmann introduced the concept of the H-function, which was shown never to increase in time as gas molecules continue their motion colliding with one another. Two serious objections were raised against this assertion. Loschmidt argued that this property of the H-function is contradictory to the reversible (symmetry in past and future) nature of mechanics. Zermelo, on the other hand, pointed out that it contradicts the well-known recurrence theorem of Poincaré. The latter theorem says that a phase point will come back again to any neighborhood of its initial position if one waits long enough (this is similar to the quasi-ergodic hypothesis), so that the H-function should return to its initial value some time, though one may have to wait for millions of years.

These objections caused very hot disputes at that time, but now they are not so difficult to resolve. We note here only that Boltzmann never proved his theorem by pure mechanics, but he introduced an assumption which is of a probabilistic nature and is usually called the Stosszahl-ansatz. An excellent review with criticisms will be found in Ehrenfests' article (Collected Scientific Papers, North-Holland Publ. Co., Amsterdam, 1959). We should also refer to ter Haar's book (see Divertissement 7).

It is still an important problem of statistical mechanics as to how to understand the irreversible nature of physical processes, or, more exactly, how to construct systematic theories of irreversible processes which allow quantitative calculations for any given physical processes. Extensive studies are still in progress in this direction throughout the world. Quantum-mechanical and classical perturbation theories have been formulated for this purpose by Van Hove, Prigogine and their collaborators. Different approaches by the Russian school headed by Bogolubov should be also mentioned. Useful review articles are found in: *Fundamental Problems in Statistical Mechanics*, compiled by E. G. D. Cohen (North-Holland Publ. Co., Amsterdam, 1962); *Studies in Statistical Mechanics*, Vol. 1, edited by J. de Boer and G. E. Uhlenbeck (North-Holland Publ. Co., Amsterdam, 1962).

13. Assuming the inter-molecular potential to be

$$U = \sum_{\langle ij \rangle} u(r_{ij}), \qquad u(\infty) = 0,$$

($\sum_{\langle ij \rangle}$ means summation over the pair ij), one obtains the virial theorem (Chapter 2, Problem (6))

$$pV = NkT - \frac{1}{3}\left\langle \sum_i \boldsymbol{r}_i \frac{\partial U}{\partial \boldsymbol{r}_i} \right\rangle$$

$$= NkT - \frac{1}{3}\sum_{\langle ij \rangle}\left\langle r_{ij}\frac{\mathrm{d}u}{\mathrm{d}r_{ij}}\right\rangle = NkT - \frac{N(N-1)}{6}\left\langle r\frac{\mathrm{d}u}{\mathrm{d}r}\right\rangle \tag{1}$$

where

$$\left\langle r\frac{\mathrm{d}u}{\mathrm{d}r}\right\rangle = \int \cdots \int r_{12}\frac{\mathrm{d}u}{\mathrm{d}r_{12}}\mathrm{e}^{-U/kT}\,\mathrm{d}\boldsymbol{r}_1 \cdots \mathrm{d}\boldsymbol{r}_N \Big/ \int \cdots \int \mathrm{e}^{-U/kT}\,\mathrm{d}\boldsymbol{r}_1 \cdots \mathrm{d}\boldsymbol{r}_N . \tag{2}$$

For large V, one knows that (2) is of the order V^{-1} to the first approximation. Consequently, p can be obtained correctly up to the order V^{-2} from (1). In the calculation, one can replace the denominator of (2) by V^N and retain only $u(r_{12})$ in the numerator of (2). Then one has

$$\left\langle r\frac{\mathrm{d}u}{\mathrm{d}r}\right\rangle \simeq V^{-2}\int \mathrm{d}\boldsymbol{r}_1 \int \mathrm{d}\boldsymbol{r}_2\, r_{12}\frac{\mathrm{d}u}{\mathrm{d}r_{12}}\mathrm{e}^{-u(r_{12})/kT} = V^{-1}\int \mathrm{d}\boldsymbol{r}\cdot r\frac{\mathrm{d}u}{\mathrm{d}r}\mathrm{e}^{-u(r)/kT}, \tag{3}$$

where $\boldsymbol{r} = \boldsymbol{r}_2 - \boldsymbol{r}_1$, $r = |\boldsymbol{r}|$.

After integration by parts, (3) becomes

$$V^{-1}\int_0^\infty 4\pi r^2\,\mathrm{d}r\cdot r\frac{\mathrm{d}u}{\mathrm{d}r}\mathrm{e}^{-u(r)/kT}$$

$$= V^{-1}kT\left\{4\pi r^3\{1-\mathrm{e}^{-u(r)/kT}\}\Big|_0^\infty - 12\pi\int_0^\infty r^2\,\mathrm{d}r\{1-\mathrm{e}^{-u(r)/kT}\}\right\}$$

$$= -\frac{12\pi}{V}kT\int_0^\infty r^2\,\mathrm{d}r\{1-\mathrm{e}^{-u(r)/kT}\}.$$

Hence one finally obtains

$$p = \frac{NkT}{V}\left\{1+\frac{2N}{V}\int_0^\infty \pi r^2\{1-\mathrm{e}^{-u(r)/kT}\}\,\mathrm{d}r + \mathrm{o}\left(\frac{N}{V}\right)\right\}.$$

14. Putting

$$\frac{h^2}{8\pi^2AkT}=\sigma,\qquad \frac{h^2}{8\pi^2kT}\left(\frac{1}{C}-\frac{1}{A}\right)=\tau \tag{1}$$

one has for the rotational partition function

$$r(T)=\sum_{l=0}^{\infty}\sum_{\lambda=-l}^{l}(2l+1)\exp\{-l(l+1)\sigma-\lambda^2\tau\}.$$

It will be difficult to express this in closed form. In the limit $\sigma\to 0$ and $\tau\to 0$, one has

$$r(T)\to\int_0^\infty (2l+1)\mathrm{e}^{-l(l+1)\sigma}\,\mathrm{d}l\int_{-l}^{l}\mathrm{e}^{-\lambda^2\tau}\,\mathrm{d}\lambda$$

$$= -2\int_0^\infty \frac{1}{\sigma}\,\mathrm{d}\{\mathrm{e}^{-l(l+1)\sigma}\}\cdot\int_0^l \mathrm{e}^{-\lambda^2\tau}\,\mathrm{d}\lambda$$

$$= \frac{2}{\sigma}\left[-\mathrm{e}^{-l(l+1)\sigma}\int_0^l \mathrm{e}^{-\lambda^2\tau}\,\mathrm{d}\lambda\Big|_0^\infty + \int_0^\infty \mathrm{e}^{-l(l+1)\sigma-l^2\tau}\,\mathrm{d}l\right]$$

(integration by parts)

$$= \frac{2}{\sigma}\int_0^\infty \mathrm{e}^{-l(l+1)\sigma-l^2\tau}\,\mathrm{d}l \doteq \frac{2}{\sigma}\int_0^\infty \mathrm{e}^{-l^2(\sigma+\tau)}\,\mathrm{d}l = \frac{\pi^{\frac{1}{2}}}{\sigma}\Big/(\sigma+\tau)^{\frac{1}{2}}. \tag{2}$$

(The use of the integration variable $x = l\sqrt{\sigma}$ will justify the approximation made in the next to last expression.) Substituting (1) into (2), one obtains

$$r(T) = \pi^{\frac{1}{2}} \cdot \frac{8\pi^2 AkT}{h^2} \cdot \left(\frac{8\pi^2 CkT}{h^2}\right)^{\frac{1}{2}}. \tag{3}$$

This is simply equation (3.14) for the case $B = A$.

15. The vibrational partition function takes the form

$$v(T) = \sum_{n=0}^{\infty} \exp\left[-\left(n+\frac{1}{2}\right)\frac{h\nu}{kT} + x_e\left(n+\frac{1}{2}\right)^2 \frac{h\nu}{kT}\right]$$

$$= \exp\left\{-\frac{u}{2} + x_e\frac{u}{4}\right\} \sum_{n=0}^{\infty} \exp\{-nu + x_e n(n+1)u\}, \quad u = h\nu/kT. \tag{1}$$

Calculation of the series is generally difficult, so that the calculation up to the first order in x_e will be given here. In the following the first factor in $v(T)$ will be omitted because it has nothing to do with specific heat.

$$v(T) = \sum_{n=0}^{\infty} e^{-nu}\{1 + x_e n(n+1)u + \cdots\}$$

$$= \frac{1}{1-e^{-u}} + x_e u\left\{\left(\frac{d}{du}\right)^2 - \frac{d}{du}\right\}\frac{1}{1-e^{-u}} + \cdots$$

$$= \frac{1}{1-e^{-u}}\left[1 + 2x_e u \frac{e^{-u}}{(1-e^{-u})^2} + \cdots\right]. \tag{2}$$

Using the last expansion, one has

$$\log v(T) = \log v_0(T) + 2x_e u\, e^{-u}/(1-e^{-u})^2 + \cdots \tag{3}$$

and hence, from

$$C_{vib} = R\frac{\partial}{\partial T}\left(T^2 \frac{\partial}{\partial T}\log v\right) = Ru^2 \frac{\partial^2}{\partial u^2}\log v$$

one obtains

$$C_{vib} = R\frac{u^2 e^u}{(e^u-1)^2}\left[1 + \frac{2x_e}{1-e^{-u}}\{-2 + u(1 + 3e^{-u} + 6e^{-2u})\}\right].$$

The second term in brackets is the correction term of order x_e. In the high temperature limit, $u \to 0$, one has

$$C_{\text{vib}} \doteqdot R\left\{\left(1 - \tfrac{1}{12}u^2 + \tfrac{1}{240}u^4\right) + 4x_e\left(\frac{1}{u} + \tfrac{1}{80}u^3\right)\right\}.$$

One sees that the correction term $4x_e R\,(kT/h\nu)$ increases with the temperature. NOTE: In a cruder approximation, one defines the corrected characteristic temperature $\Theta_v = (h\nu/k)(1 - 2x_e)$ from the energy difference between the ground level ($n = 0$) and the first excited level ($n = 1$), and then replaces the system with a harmonic oscillator of the frequency defined above. For the H_2 molecule, one has $x_e = 0.02685$ for the electronic ground state ${}^1\Sigma_g^+$, but the third order anharmonic term $y_e\,(n + \frac{1}{2})^3 h\nu$ will also be appreciable.

16. (i) The number of vibrational degrees of freedom $= 9$ (the total number of degrees of freedom $3 \times 5 = 15$) – (the translational, and rotational degrees of freedom $= 6$).

(ii) At high temperatures one can separate the nuclear spin and the rotational motion. Putting s_H (nuclear spin of H) $= \frac{1}{2}$, $s_C = 0$ (nuclear spin of $C^{12} = 0$) and the principal moments of inertia $A = B = C$, one obtains from (3.12) the partition function for this part $j_{\text{nuc-rot}}$

$$j_{\text{nuc-rot}} = \frac{1}{\gamma}(2s_C + 1)(2s_H + 1)^4 \pi^{\frac{1}{2}} \left(\frac{8\pi^2 AkT}{h^2}\right)^{\frac{1}{2}} \left(\frac{8\pi^2 BkT}{h^2}\right)^{\frac{1}{2}} \left(\frac{8\pi^2 CkT}{h^2}\right)^{\frac{1}{2}}. \tag{1}$$

A methane molecule has four threefold axes and hence the symmetry number is $\gamma = 3 \times 4 = 12$.

(iii) The partition function for methane gas will be given by (cf. (3.1))

$$Z_N(T, V) = \frac{V^N}{N!}\left(\frac{2\pi mkT}{h^2}\right)^{\frac{3}{2}N} (j_{\text{nuc-rot}})^N \left(\prod_{i=1}^{9} \frac{e^{-h\nu_i/2kT}}{1 - e^{-h\nu_i/kT}}\right)^N \tag{2}$$

where the classical approximation is used for the translational and the rotational motion, and the harmonic oscillator approximation is used for the vibration. One can evaluate the Gibbs free energy $\bar{G} = \bar{F} + pV$, $\bar{F} = -kT\log Z_N$ (per mole) as

$$\bar{G}(T, p) = \bar{U}_0 + RT\log p - 4RT\log T + RT\sum_{i=1}^{9}\log(1 - e^{-h\nu_i/kT}) - RTi_0, \tag{3}$$

$$i_0 = \log\left\{\frac{(2\pi m)^{\frac{3}{2}}k^{\frac{5}{2}}}{h^3}\,\frac{8\pi^2(2\pi k)^{\frac{3}{2}}(ABC)^{\frac{1}{2}}}{h^3}\,\frac{(2s_C + 1)(2s_H + 1)^4}{\gamma}\,g_0\right\} \tag{4}$$

where $g_0 = 1$ is the degeneracy of the lowest electronic state, and i_0 is the chemical constant required.

(iv) The changes in the thermodynamic functions for the substitution products are due to (a) the change in the fourth term in (3) caused by the changes in the frequencies corresponding to the difference in atomic mass, (b) the change in the total mass of the molecule, (c) the change of the factor $(ABC)^{\frac{1}{2}}$ in (4) due to the changes in the moments of inertia, (d) the change in the nuclear spins and (e) the change in the symmetry number. The discussion of (a) is cumbersome because it requires the theory of the normal mode and will be omitted here. The changes (b), (d) and (e) are tabulated below.

	m	γ	$\Pi(2s_n+1)$	$\Delta\bar{G}$
CH_4	16	12 (4 threefold axis)	$2^4 = 16$	0
CH_3D	17	3 (1 threefold axis)	$2^3 \cdot 3 = 24$	$-1.8827RT$
CH_2D_2	18	2 (1 twofold axis)	$2^2 \cdot 3^2 = 36$	$-2.7794RT$
CHD_3	19	3 (1 threefold axis)	$2 \cdot 3^3 = 54$	$-2.8605RT$
CD_4	20	12 (4 threefold axis)	$3^4 = 81$	$-1.9566RT$

In (c) one must calculate the principal moments of inertia. Considering that the mass of the C atom is large compared to that of the H and D atoms, and that the molecules have a high symmetry, one can assume that the center of mass lies at the position of the C atom even for molecules other than CH_4 and CD_4. Hence the moment of inertia tensor can be assumed to be symmetric and the corresponding factor in (4) will be put as log $A^{\frac{3}{2}}$. The ratio of A will be given approximately by

$$A(CH_4):A(CH_3D):A(CH_2D_2):A(CHD_3):A(CD_4) \fallingdotseq 4:5:6:7:8.$$

NOTE: A more rigorous calculation will not be very difficult. The structure of

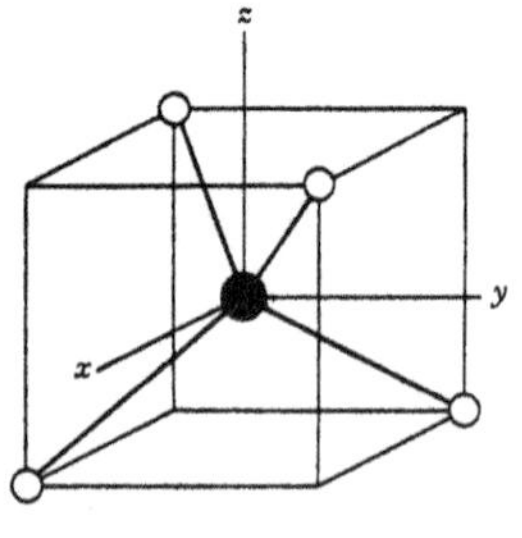

the CH_4 molecule can be expressed as in the figure, with the C atom at the (body) center of the cube and 4 H atoms in the corners. Taking the origin at the body center and considering the length of a side of the cube to be 2, one has the positions of the H atoms as (1, 1, 1), (1, 1, 1), ($\bar{1}, \bar{1}, \bar{1}$), $\bar{1}, \bar{1}, \bar{1}$). The matrix of the moment of inertia tensor T is given by

$$T = \sum_i m_i(r_i^2 \mathbf{1} - r_i r_i) + M(r_0^2 \mathbf{1} - r_0 r_0) \qquad (r_0 = \sum m_i r_i / M, \qquad M = \sum m_i)$$

where $\mathbf{1}$ is the unit matrix. Substituting values of the coordinates and the masses of H, D, and C atoms (1, 2 and 12), one gets the tensor for each molecule,

$$T(CH_4) = \begin{pmatrix} 8 & 0 & 0 \\ 0 & 8 & 0 \\ 0 & 0 & 8 \end{pmatrix},$$

$$T(CH_3D) = \begin{pmatrix} 10 + \frac{2}{17} & -\frac{18}{17} & -\frac{18}{17} \\ -\frac{18}{17} & 10 + \frac{2}{17} & -\frac{18}{17} \\ -\frac{18}{17} & -\frac{18}{17} & 10 + \frac{2}{17} \end{pmatrix},$$

$$T(CH_2D_2) = \begin{pmatrix} 12 + \frac{4}{18} & -2 & 0 \\ -2 & 12 + \frac{4}{18} & 0 \\ 0 & 0 & 12 + \frac{4}{18} \end{pmatrix},$$

$$T(CHD_3) = \begin{pmatrix} 14 + \frac{2}{19} & -\frac{1}{19} & -\frac{1}{19} \\ -\frac{1}{19} & 14 + \frac{2}{19} & -\frac{1}{19} \\ -\frac{1}{19} & -\frac{1}{19} & 14 + \frac{2}{19} \end{pmatrix},$$

$$T(CD_4) = \begin{pmatrix} 16 & 0 & 0 \\ 0 & 16 & 0 \\ 0 & 0 & 16 \end{pmatrix}.$$

The value of the product ABC of the principal values can be given by the value of the determinant of the matrix T.

17. Assuming the classical approximation for molecular motion including the internal degrees of freedom, one has

$$f = g_{\text{nuc}} \frac{V}{h^{3n}} \int \cdots \int e^{-\varepsilon_p/kT} \prod dp \cdot \int \cdots \int e^{-\varepsilon_q/kT} \prod' dq$$

for the partition function per molecule. Here g_{nuc} is the nuclear spin degeneracy, V is the volume of the container, n is the number of atoms in a molecule, ε_p is the sum of the kinetic energies of these atoms:

$$\varepsilon_p = \sum p_i^2 / 2m_i$$

and ε_q is the potential energy of the internal motion along the coordinate q. It is assumed that ε_q is invariant against substitution of isotopes, and hence the contribution to the partition function is also invariant. The contribution from the kinetic energy is given simply by $\prod_i (2\pi m_i kT/h^2)^{\frac{3}{2}}$. The equilibrium constant for the reaction

$$A + B + \cdots \rightleftarrows L + M + \cdots$$

is given by

$$K = \prod f_L / \prod f_A$$

(a generalization of example 3). The atoms which appear on both sides are exactly equal in total. Consequently, the atomic masses cancel out completely and K becomes independent of the isotopic mass. Hence chemical equilibrium cannot be shifted by isotopic substitution. Therefore chemical reactions cannot be used to separate isotopes under such conditions.

NOTE: At low temperatures, there will be some isotope effects in the chemical reaction and they can be utilized for isotope separation.

18. The second virial coefficient is given by $B(T) = -\frac{1}{2}\beta_1$ (3.24). From (3.32) and (3.33) one has $b_2 = \frac{1}{2}\beta_1$ and then

$$B(T) = \tfrac{1}{2}\int_0^\infty \{1 - e^{-u(r)/kT}\} 4\pi r^2 \, dr \tag{1}$$

(a) $u(r) = \alpha/r^n$ ($\alpha > 0$, $n > 3$). A formula which can be obtained from (1) by means of integration by parts will be convenient:

$$B(T) = -\frac{2\pi}{3kT}\int_0^\infty e^{-u(r)/kT}\frac{du(r)}{dr} r^3 \, dr = \frac{2\pi n\alpha}{3kT}\int_0^\infty \exp\left(-\frac{\alpha}{r^n kT}\right)\frac{dr}{r^{n-2}}.$$

Putting $x = \alpha/(r^n kT)$, one has the formula which contains the Γ-function ($n > 3$),

$$B(T) = \frac{2\pi}{3}\left(\frac{\alpha}{kT}\right)^{3/n}\int_0^\infty e^{-x}x^{-3/n}\,dx = \frac{2\pi}{3}\Gamma\left(\frac{n-3}{n}\right)\left(\frac{\alpha}{kT}\right)^{3/n}. \tag{2a}$$

(b) One has immediately, then,

$$B(T) = \frac{1}{2}\left\{\int_0^a 4\pi r^2\,dr + \int_a^b (1 - e^{u_0/kT})4\pi r^2\,dr\right\}$$

$$= \frac{2\pi}{3}\{a^3 - (e^{u_0/kT} - 1)(b^3 - a^3)\} = \frac{2\pi}{3}\{b^3 - e^{u_0/kT}(b^3 - a^3)\}. \tag{2b}$$

The Joule-Thomson coefficient is given by

$$\left(\frac{\partial T}{\partial p}\right)_H = \frac{1}{C_p}\left(T\frac{\mathrm{d}B}{\mathrm{d}T} - B\right). \tag{3}$$

Substituting (2a, b) into (3), one obtains

$$\left(\frac{\partial T}{\partial p}\right)_H = -\frac{2\pi}{3}\Gamma\left(\frac{n-3}{n}\right)\cdot\left(\frac{3}{n}+1\right)\cdot\left(\frac{\alpha}{kT}\right)^{3/n}\Big/ C_p, \tag{4a}$$

$$\left(\frac{\partial T}{\partial p}\right)_H = \frac{2\pi}{3}\left\{-b^3 + \left(1+\frac{u_0}{kT}\right)(b^3 - a^3)\mathrm{e}^{u_0/kT}\right\}\Big/ C_p. \tag{4b}$$

From (4a) one can see that $(\partial T/\partial p)_H < 0$ for case (a). This means the temperature increases by the Joule-Thomson expansion. In case (b), the potential will become repulsive for $u_0 = 0$. The Joule-Thomson coefficient becomes $(\partial T/\partial p)_H = -\frac{2}{3}\pi a^3$, which is negative as in case (a). For $u_0 > 0$, there is an attractive region of potential. At temperatures below a certain value, T_i, the second term in the bracket of (4b) surpasses the first term. Therefore one has $(\partial T/\partial p)_H > 0$, which means the temperature decreases due to Joule-Thomson expansion.

19. One has the equation H_2 (gas) $\rightarrow$ 2 H (adsorbed) + w, where w means the energy change in the process – dissociation and adsorption – not including the zero point vibration. The free energy of the adsorbed phase is given by

$$F(N) = -kT\log\frac{N_0!}{N!(N_0-N)!}\left[2\left(2\sinh\frac{h\nu_\parallel}{2kT}\right)^{-2}\left(2\sinh\frac{h\nu_\perp}{2kT}\right)^{-1}\mathrm{e}^{w/2kT}\right]^N,$$

and hence

$$\mu_\mathrm{H}(\text{adsorbed}) = \frac{\partial F}{\partial N} = -kT\log 2\left(2\sinh\frac{h\nu_\parallel}{2kT}\right)^{-2}\left(2\sinh\frac{h\nu_\perp}{2kT}\right)^{-1}\mathrm{e}^{w/2kT} + kT\log N/(N_0-N) \quad (1)$$

is the chemical potential of the adsorbed H atom. Here, 2 is the nuclear spin weight, $\nu_\perp$ and $\nu_\parallel$ are the frequencies of vibration of the H atom perpendicular to and parallel to the adsorbing surface, N_0 is the number of the adsorbing centers and N is the number of the adsorbed H atoms. The condition for equilibrium between the hydrogen gas and the adsorbed hydrogen is given by

$$2\mu_\mathrm{H}(\text{adsorbed}) = \mu_{\mathrm{H}_2}(\text{gas}).$$

Here μ_{H_2} is given by

$$\mu_{H_2} = -kT\left[\log\frac{(2\pi mkT)^{\frac{3}{2}}}{h^3}\frac{kT}{p} + \log 2^2\frac{8\pi^2 IkT}{2h^2}\left\{2\sinh\frac{h\nu}{2kT}\right\}^{-1}\right] \tag{2}$$

where I is the moment of inertia of the H_2 molecule and ν is the frequency of vibration.

From (1) and (2) one obtains

$$\frac{N}{N_0 - N} = Ap^{\frac{1}{2}}, \qquad \text{or} \quad \frac{N}{N_0} = \frac{Ap^{\frac{1}{2}}}{1 + Ap^{\frac{1}{2}}},$$

$$A = \frac{h^{\frac{3}{2}}}{(2\pi)^{\frac{3}{2}}m^{\frac{3}{2}}(kT)^{\frac{3}{2}}I^{\frac{1}{2}}}\frac{(2\sinh h\nu/2kT)^{\frac{1}{2}}\,e^{w/2kT}}{(2\sinh h\nu_{\perp}/2kT)(2\sinh h\nu_{\parallel}/2kT)^2}$$

$$\fallingdotseq \frac{h^{\frac{3}{2}}}{(2\pi)^{\frac{3}{2}}m^{\frac{3}{2}}(kT)^{\frac{3}{2}}I^{\frac{1}{2}}}\exp\{(\tfrac{1}{4}h\nu - \tfrac{1}{2}h\nu_{\perp} - h\nu_{\parallel} + \tfrac{1}{2}w)/kT\},$$

where the last expression gives the approximate form assuming such a high vibration frequency that only the zero point vibration may be considered.

20. From the Helmholtz free energy of the system $F(N, T, V)$, one has for $N \gg 1$

$$\mu = \left(\frac{\partial F}{\partial N}\right)_{T,V} = F(N+1, T, V) - F(N, T, V) = -kT\log\frac{Z(N+1, T, V)}{Z(N, T, V)},$$

or, by (3.18),

$$e^{-\mu/kT} = \left(\frac{2\pi mkT}{h^2}\right)^{\frac{3}{2}}\frac{Q(N+1, T, V)}{Q(N, T, V)}. \tag{1}$$

Let the particles which appear in $Q\,(N + 1, T, V)$ of the numerator be numbered as 0, 1, 2, ..., N. One has from (3.20)

$$U(0, 1, \cdots, N) = \sum_{i=1}^{N} u_{0i} + U(1, 2, \cdots, N), \tag{2}$$

thus

$$e^{-U(0,1,\cdots,N)/kT} = e^{-U(1,2,\cdots,N)/kT}\prod_{i=1}^{N}(1 + f_{0i}).$$

From (3.19)

$$Q(N+1, T, V) = \frac{1}{(N+1)!}\int\cdots\int e^{-U(0,1,\cdots,N)/kT}\,dx_0\,dy_0\,dz_0\,dx_1\cdots dz_N$$

$$= \frac{1}{N+1}\iiint dx_0\,dy_0\,dz_0\frac{1}{N!}\int\cdots\int e^{-U(1,2,\cdots,N)/kT}\times$$

$$\times\prod_{i=1}^{N}(1 + f_{0i})\,dx_1\cdots dz_N. \tag{3}$$

In the expansion of the product of the $1 + f_{0i}$, there will be $N!/(N-n)!n!$ terms of the product of nf_{0i}'s and these give the same contribution to Q because of the identity of the particles.

$$Q(N+1, T, V) = \frac{V}{N+1}\sum_{n=0}^{N}\frac{1}{(N-n)!n!}\int\cdots\int e^{-U(1,\cdots,N)/kT}\prod_{i=1}^{n} f_{0i}\,dx_1\cdots dz_N. \quad (4)$$

On the other hand, the probability $P_n(x_1, ..., z_n)\,dx_1 ... dz_n$ that a particular set of particles $(1, 2, ..., n)$ falls into the region $(x_1, x_1 + dx_1), ... (x_n, x_n + dx_n)$ is given by

$$P_n(x_1, \cdots, z_n) = \frac{1}{Q(N, T, V)}\frac{1}{N!}\int\cdots\int e^{-U(1,2,\cdots,N)/kT}\,dx_{n+1}\cdots dz_N. \quad (5)$$

A similar probability for unspecified n particles $v^{-n} F_n(x_1, ..., z_n)\,dx_1 ... dz_n$ is given by

$$\frac{1}{v^n}F_n(x_1, \cdots, z_n) = N(N-1)\cdots(N-n+1)P_n(x_1, \cdots, z_n)$$
$$= \frac{N!}{(N-n)!}P_n(x_1, \cdots, z_n). \quad (6)$$

Substituting (4), (5) and (6) into (1) one obtains the required result, using $V/(N+1) \fallingdotseq V/N = v$ for the first factor in (4).

21. Substituting (3.31) and (3.30) into (2.25) and (2.26)

$$\log Z_N = \log \Xi - N\log\lambda, \qquad N = \lambda\partial\log\Xi/\partial\lambda$$

one has

$$\log Z_N = V\sum_{l=1}^{\infty} b_l\zeta^l - N\log\zeta + N\log\left(\frac{2\pi mkT}{h^2}\right)^{\frac{3}{2}} j, \quad (1)$$

$$\sum_{l=1}^{\infty} lb_l\zeta^l = N/V \equiv v^{-1}. \quad (2)$$

Noting that $b_1 = 1$, one can rewrite (2) as

$$\zeta + 2b_2\zeta^2 + \cdots = v^{-1}, \quad \text{thus} \quad \zeta = v^{-1} - 2b_2\zeta^2 + \cdots. \quad (3)$$

The first term gives the result for the ideal gas. One can proceed with the next approximation in the following way. By successive approximation one obtains

$$\zeta = v^{-1} - 2b_2 v^{-2} + \cdots. \quad (4)$$

Substituting this into (1), one finds

$$F = -kT\log Z_N$$

$$= -NkT\left[v\{\zeta + b_2\zeta^2 + \cdots\} - \log v^{-1}(1 - 2b_2v^{-1} + \cdots) + \log\left(\frac{2\pi mkT}{h^2}\right)^{\frac{3}{2}}j\right]$$

$$= -NkT\left[(1 - b_2v^{-1} + \cdots) + (2b_2v^{-1} + \cdots) + \log\left(\frac{2\pi mkT}{h^2}\right)^{\frac{3}{2}}jv\right]$$

$$= -NkT\left[\log\left(\frac{2\pi mhT}{h^2}\right)^{\frac{3}{2}}jv + \{1 + b_2v^{-1} + \cdots\}\right]. \tag{5}$$

Hence

$$p = -\frac{\partial F}{N\partial v} = \frac{kT}{v}\{1 - b_2v^{-1} + \cdots\}, \tag{6}$$

$$G = F + pV = -NkT\left[\log\left(\frac{2\pi mkT}{h^2}\right)^{\frac{3}{2}}jv + 2b_2v^{-1} + \cdots\right] \tag{7}$$

and

$$S = -\left(\frac{\partial F}{\partial T}\right)_v$$

$$= Nk\left[\log\left(\frac{2\pi mkT}{h^2}\right)^{\frac{3}{2}}jv + \tfrac{3}{2} + T\frac{\partial}{\partial T}(\log j) + 1 + \frac{\partial}{\partial T}(Tb_2)\cdot v^{-1} + \cdots\right], \tag{8}$$

$$U = F + TS = NkT\left[\tfrac{3}{2} - T\frac{\partial}{\partial T}(\log j) + T\frac{\partial b_2}{\partial T}v^{-1} + \cdots\right], \tag{9}$$

$$H = G + TS = NkT\left[\tfrac{5}{2} - T\frac{\partial}{\partial T}(\log j) - \left(b_2 - T\frac{\partial b_2}{\partial T}\right)v^{-1} + \cdots\right], \tag{10}$$

$$C_V = \tfrac{3}{2}Nk + C_i + \frac{\partial}{\partial T}\left(T^2\frac{\partial b_2}{\partial T}\right)v^{-1} + \cdots, \tag{11}$$

$$C_p = \tfrac{5}{2}Nk + C_i + \frac{\partial}{\partial T}\left\{T^3\frac{\partial}{\partial T}\left(\frac{b_2}{T}\right)\right\}v^{-1} + \cdots. \tag{12}$$

NOTE: Consider these equations for the cases given in problem 18.

22. From definition (3.26), one knows that b_3 corresponds to a group of diagrams which can be obtained by connecting three molecules 1, 2, 3 with the bond f_{ij} in all possible ways. The number of proto-types of the diagram mentioned above is 2 as shown in the figure. The number of distributions of 1, 2, 3 in type (A) is three and in type (B) is one. Hence one has

$$b_3 = \frac{1}{3!V}\iiint\{f_{12}f_{13}+f_{23}f_{21}+f_{31}f_{32}+f_{12}f_{23}f_{31}\}\,d\boldsymbol{r}_1\,d\boldsymbol{r}_2\,d\boldsymbol{r}_3. \quad (1)$$

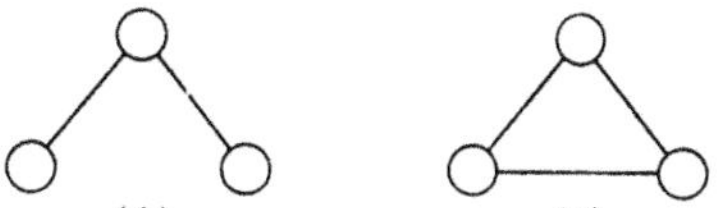

(A) (B)

The terms of type (A) contribute the same values to the integral. For instance, $f_{12}f_{13}$ can be integrated over $\boldsymbol{r}_2$ and $\boldsymbol{r}_3$ independently and yields the value $\beta_1{}^2$. Hence one obtains from (1)

$$b_3 = \tfrac{1}{6}\{3\beta_1^2 + 2\beta_2\} = \tfrac{1}{2}\beta_1^2 + \tfrac{1}{3}\beta_2,$$

where β_1 and β_2 are given in the problem.

NOTE: For cluster integrals higher than the third order, the analysis is made in a similar way. Each term in the cluster integral containing l molecules corresponds to a diagram which is obtained by connecting l molecules by means of the bond f_{ij}. A diagram will be called reducible if it can be split into two parts by cutting one bond. In a reducible diagram, the integration can be done independently on both sides of the bond which is cut in the reduction (Fig. 3.2.). Using the reduction of the reducible diagram to the simplest possible diagram, one can express a cluster integral as a combination of irreducible cluster integrals. Here an irreducible cluster integral means the integral re-defined by

$$\beta_k = \frac{1}{k!V}\int\cdots\int \underset{\substack{\text{summation over those diagrams}\\ \text{which cannot be reduced}}}{\sum_{k+1\geq l>j\geq 1}} \prod f_{ij}\cdot d\boldsymbol{r}_1\cdots d\boldsymbol{r}_{k+1}.$$

β_1 and β_2 earlier in the problem are simple examples of β_k. One can verify the relation

$$b_l = \frac{1}{l^2}\sum_{\substack{n\\ \Sigma kn_k = l-1}}\prod_k \frac{(l\beta_k)^{n_k}}{n_k!}$$

and derive relation (3.34) [cf. J. E. Mayer and M. G. Mayer, *Statistical Mechanics* (John Wiley and Sons, New York, 1940)].

CHAPTER 4

APPLICATIONS OF FERMI- AND BOSE-STATISTICS

This chapter is intended to provide a deeper understanding of Fermi- and Bose-statistics and to develop further applications to real physical problems. Fermi- and Bose-statistics exhibit characteristic features which are different from classical Boltzmann statistics when the condition

$$\frac{N}{V} \gtrsim \frac{(2\pi mkT)^{\frac{3}{2}}}{h^3},$$

is fulfilled. In other words, quantum mechanical effects appear at low temperatures and high densities. Particularly important applications are those involving electrons in solids.

Fundamental Topics

§ 4.1. FUNDAMENTAL FORMULAE OF FERMI-STATISTICS

In a fermion system † the average occupation number of a one-particle state τ is given by

$$\bar{n}_\tau = \frac{1}{e^{(\varepsilon_\tau - \mu)/kT} + 1} \tag{4.1}$$

(see § 1.15), where ε_τ is the energy of the quantum state τ and μ is the chemical potential. For the Fermi distribution, μ is sometimes called the *Fermi level* or the *Fermi potential* (sometimes the *Fermi energy*, which is not very suitable but which is becoming more common). If the total energy of the system is simply the sum of the one-particle energies, then the average total energy E is

$$E = \sum_\tau \varepsilon_\tau \bar{n}_\tau = \sum_\tau \frac{\varepsilon_\tau}{e^{(\varepsilon_\tau - \mu)/kT} + 1}. \tag{4.2}$$

The total number of particles in the system is

$$N = \sum_\tau \bar{n}_\tau = \sum_\tau \frac{1}{e^{(\varepsilon_\tau - \mu)/kT} + 1} \tag{4.3}$$

† Particles which obey Fermi-statistic are called fermions.

and the free energy is given by

$$F = N\mu - kT \sum_{\tau} \log(1 + e^{-(\varepsilon_\tau - \mu)/kT}). \tag{4.4}$$

Equations (4.1)–(4.3) can be interpreted in the following different ways:

1) When T and μ are given (when the system is in contact with a heat reservoir and a particle reservoir), (4.1) gives the average occupation number of each one-particle state, (4.2) the average total energy, and (4.3) the average total number of particles in the system.

2) When T and N are given, μ is determined from (4.3) as a function of T and N. Then (4.1) and (4.2) give the average occupation number and the average total energy, respectively.

3) When E and N are given, T and μ can be found from equations (4.2) and (4.3) as functions of E and N. Then (4.1) will give the average occupation number.

One-particle state density: Let V be the volume of the space in which the particles are confined. When V is made larger and larger, the one-particle levels become more and more densely distributed. For a volume sufficiently large the number of states which have energies between ε and $\varepsilon + \Delta\varepsilon$ may be written as

$$D(\varepsilon)\Delta\varepsilon$$

which defines the one-particle state density $D(\varepsilon)$.

In terms of the state density, equations (4.2)–(4.4) can be written as

$$E = \int \varepsilon f(\varepsilon) D(\varepsilon)\, d\varepsilon, \tag{4.2'}$$

$$N = \int f(\varepsilon) D(\varepsilon)\, d\varepsilon, \tag{4.3'}$$

$$F = N\mu - kT \int D(\varepsilon)\, d\varepsilon \log(1 + e^{-(\varepsilon - \mu)/kT}), \tag{4.4'}$$

where

$$f(\varepsilon) = \frac{1}{e^{\beta(\varepsilon - \mu)} + 1} \qquad (\beta = 1/kT). \tag{4.5}$$

§ 4.2. FERMI DISTRIBUTION FUNCTION

The function $f(\varepsilon)$ defined by (4.5) is called the Fermi distribution function. Figure 4.1 shows the form of this function when plotted against ε. In the limit of $T \to 0$, this looks like a sharp step, but at finite temperatures the step is broadened with widths of about kT on both sides of $\varepsilon = \mu$. Figure 4.2 shows examples of $D(\varepsilon) f(\varepsilon)$. $\varepsilon_0 (= 0)$ is the lowest energy of the one-particle states.

(a) $T = 0°$ K: The value μ_0 of μ at $0°$ K is determined by

$$N = \int_{\varepsilon_0}^{\mu_0} D(\varepsilon)\, d\varepsilon = \text{the shaded area of curve a.}$$

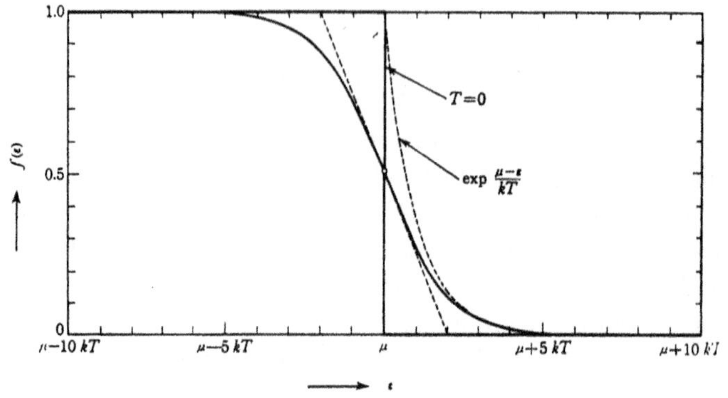

Fig. 4.1. Fermi distribution function.

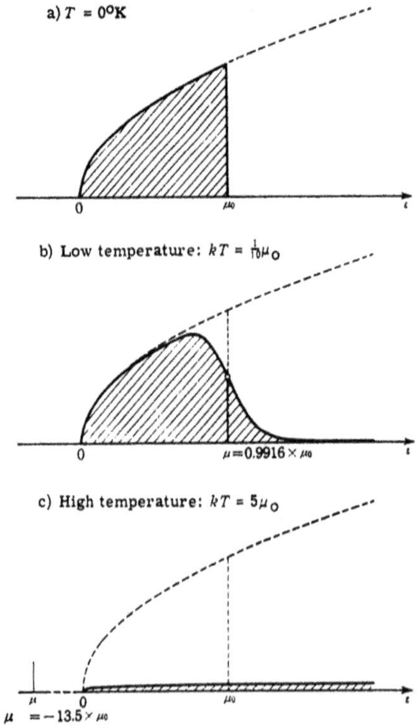

Fig. 4.2. ($\varepsilon_0 = 0$).

μ_0 is the Fermi level (Fermi potential) at 0° K, and

$$E_0 = \int_{\varepsilon_0}^{\mu_0} \varepsilon D(\varepsilon)\, d\varepsilon$$

is the lowest energy of the total system.

(b) $T \neq 0, T \ll T_0, T_0 = (\mu_0 - \varepsilon_0)/k$ (low temperatures): The value of μ is determined by the condition that

$$N = \int D(\varepsilon) f(\varepsilon)\, d\varepsilon = \text{the shaded area of curve b.}$$

Thus μ is a function of T and is called the Fermi level at T° K. At low temperatures, μ is greater than ε_0 as is indicated in Figure 4.2.

(c) $T \neq 0, T \gg T_0$ (high temperatures): At sufficiently high temperatures μ becomes smaller than ε_0 as indicated in Figure 4.2c. In this case, the Fermi distribution may be replaced by the Boltzmann distribution (classical limit).

Formulae for degenerate cases: Let $g(\varepsilon)$ be a function which is continuous and repeatedly differentiable at $\varepsilon = \mu$, and furthermore which varies rather slowly in the interval $|\varepsilon - \mu| \lesssim kT$. Then one has

$$\int_{\varepsilon_0}^{\infty} g(\varepsilon) f(\varepsilon)\, d\varepsilon = \int_{\varepsilon_0}^{\mu} g(\varepsilon)\, d\varepsilon + \frac{\pi^2 (kT)^2}{6} g'(\mu) + \frac{7\pi^4}{360} (kT)^4 g'''(\mu) + \cdots \tag{4.6a}$$

or

$$-\int_{\varepsilon_0}^{\infty} \varphi(\varepsilon) \frac{df}{d\varepsilon}\, d\varepsilon = \varphi(\mu) + \frac{\pi^2}{6} (kT)^2 \varphi''(\mu) + \frac{7\pi^4}{360} (kT)^4 \varphi^{(IV)}(\mu) + \cdots, \tag{4.6b}$$

with $\varphi' = g$, $\quad \varphi(\varepsilon_0) = 0$.

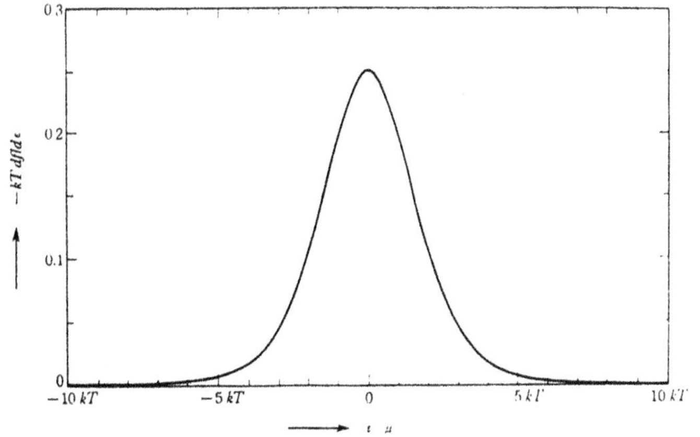

Fig. 4.3. Graph of $-kT\, df/d\varepsilon$.

Proof: Consider the integral

$$I = \int_{\varepsilon_0}^{\infty} f(\varepsilon)\frac{d\varphi}{d\varepsilon}d\varepsilon = f(\varepsilon)\varphi(\varepsilon)\Big|_{\varepsilon_0}^{\infty} - \int_{\varepsilon_0}^{\infty}\varphi(\varepsilon)\frac{df}{d\varepsilon}d\varepsilon = -\int_{\varepsilon_0}^{\infty}\varphi(\varepsilon)\frac{df}{d\varepsilon}d\varepsilon \tag{4.7}$$

where we assumed that $\varphi(\varepsilon_0) = 0$. Now the function

$$\frac{df}{d\varepsilon} = -\frac{\beta}{\{e^{\beta(\varepsilon-\mu)}+1\}\{e^{-\beta(\varepsilon-\mu)}+1\}}$$

takes on only very small values outside of the region $|\varepsilon - \mu| \lesssim kT$ (see Fig. 4.3). One may thus expand $\varphi(\varepsilon)$ as

$$\varphi(\varepsilon) = \varphi(\mu) + (\varepsilon-\mu)\varphi'(\mu) + \tfrac{1}{2}(\varepsilon-\mu)^2\varphi''(\mu) + \cdots,$$

insert it into (4.7), and integrate term by term.

In integration the lower limit ε_0 of the integrals may be reduced to $-\infty$ without any appreciable change of the results (strong degeneracy). The necessary formulae for integration are

$$\int_{-\infty}^{\infty}\frac{df}{d\varepsilon}d\varepsilon = -1,$$

$$\int_{-\infty}^{\infty}(\varepsilon-\mu)^n\frac{df}{d\varepsilon}d\varepsilon = 0 \qquad (n = \text{odd})$$

$$\int_{-\infty}^{\infty}(\varepsilon-\mu)^n\frac{df}{d\varepsilon}d\varepsilon = -\frac{1}{kT}\int_{-\infty}^{\infty}\frac{(\varepsilon-\mu)^n e^{\beta(\varepsilon-\mu)}}{\{e^{\beta(\varepsilon-\mu)}+1\}^2}d\varepsilon$$

$$= -(kT)^n\int_{-\infty}^{\infty}\frac{x^n e^x\,dx}{(e^x+1)^2}$$

$$= -2(kT)^n n!(1-2^{-n+1})\zeta(n) \qquad (n = \text{even})\dagger,$$

where

$$\zeta(n) = \sum_{l=1}^{\infty} l^{-n}, \qquad \zeta(2) = \frac{\pi^2}{6}, \qquad \zeta(4) = \frac{\pi^4}{90}$$

Footnote see pag. 233.

are the Riemann ζ functions. With these formulae put into (4.7), one gets

$$I = \varphi(\mu) + \sum_{r=1}^{\infty} 2\cdot(1 - 2^{1-2r})\zeta(2r)(kT)^{2r}\varphi^{(2r)}(\mu).$$

By putting $g = \varphi'(\varepsilon)$ one obtains (4.6a).

Fermi potential μ (T) at low temperatures: The formula (4.6) has many applications which will be given as problems. One useful result is

$$\mu = \mu_0\left\{1 - \frac{\pi^2}{6}\frac{\mathrm{d}\log D(\mu_0)}{\mathrm{d}\log\mu_0}\left(\frac{kT}{\mu_0}\right)^2 + \cdots\right\}. \tag{4.8}$$

§ 4.3. ELECTRONIC ENERGY BANDS IN CRYSTALS

Free electrons have a simple state density function:

$$D(\varepsilon) = 2\frac{4\pi V}{h^3}p^2\frac{\mathrm{d}p}{\mathrm{d}\varepsilon} \qquad (\varepsilon = p^2/2m, \text{ the factor 2 being the spin degeneracy})$$

$$= \frac{8\pi V}{h^3}(2m^3\varepsilon)^{\frac{1}{2}} \tag{4.9}$$

which is a parabola as shown in Figure 4.4.

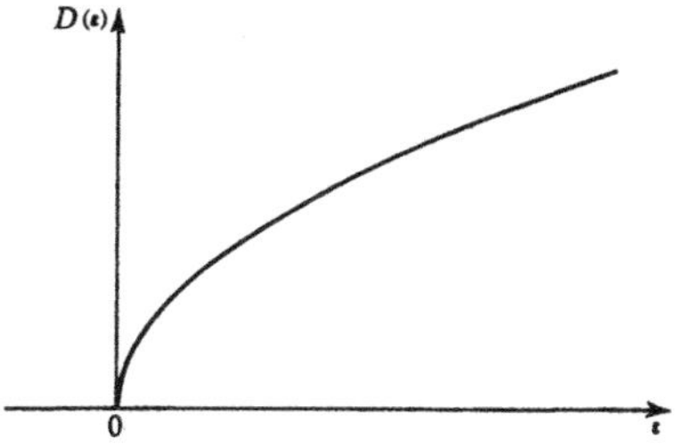

Fig. 4.4. Density of states curve for free electrons.

---

†
$$\int_0^\infty \frac{x^n e^x dx}{(e^x+1)^2} = -\int_0^\infty x^n \frac{d}{dx}\left(\frac{1}{e^x+1}\right) dx = n\int_0^\infty \frac{x^{n-1}}{e^x+1} dx$$

$$= n\sum_{k=0}^{\infty}(-1)^k\int_0^\infty x^{n-1}e^{-(k+1)x}dx = n!\sum_{k=0}^{\infty}\frac{(-1)^k}{(k+1)^n}$$

$$= n!\left\{\sum_{l=\text{odd}}\frac{1}{l^n} - \sum_{l=\text{even}}\frac{1}{l^n}\right\} = n!\left\{\sum_{l=1}^{\infty}\frac{1}{l^n} - 2\sum_{l=\text{even}}\frac{1}{l^n}\right\}$$

$$= n!\left(1 - \frac{2}{2^n}\right)\sum_{l=1}^{\infty}\frac{1}{l^n} = n!\,(1 - 2^{1-n})\,\zeta(n).$$

Electrons in crystalline solids have more complicated state densities, being influenced by the periodicity of the crystal. As a consequence the state density curve will generally split into many groups as shown schematically by Figure 4.5. This kind of structure is called the *band structure* of crystalline

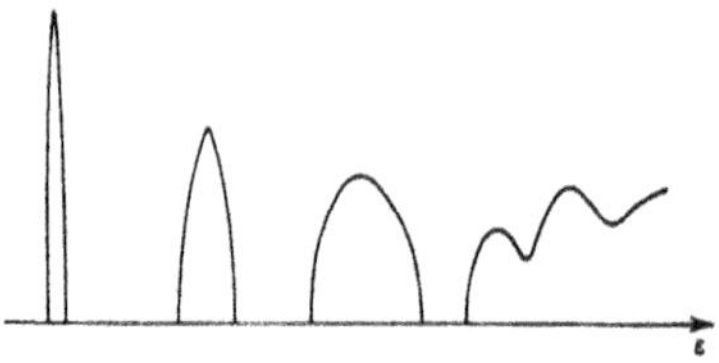

Fig. 4.5.

electrons. The detailed structure of bands is characteristic of each particular solid.

Detailed explanation of the reason why such band structures are produced is beyond the scope of the present book, so that only a brief note may be added here. As is well known, electrons in an isolated atom have discrete energy levels. When the atoms are brought together to form a crystal, each electron will no longer belong to a particular atom but it is shared, so to speak, by all the atoms in the crystal. Therefore, electrons in a crystal have the dual nature of atomic and of free electron character, and so the discrete atomic energy levels are broadened to have a certain width of energy. This qualitatively explains the origin of bands. Generally, the low energy bands have a clear correspondence to atomic levels, but the higher bands easily overlap so that the correspondence becomes not so simple.

§ 4.4. HOLES

$$\bar{n}'_\tau = 1 - \bar{n}_\tau = \frac{1}{e^{-(\varepsilon_\tau - \mu)/kT} + 1} \tag{4.10}$$

represents the probability that the one-particle state is not occupied. Taking the condition in which all the states are occupied as the standard, we may say that the state τ is occupied by a *hole* when it is not occupied by an electron. In this way, holes may be considered as particles if the particles are fermions.

For a system of electrons, if a one-electron state with momentum $\boldsymbol{p}$ and energy $\varepsilon(\boldsymbol{p})$ is not occupied, there exists a hole with

$$\left.\begin{array}{ll} \text{charge:} & e' = +|e|, \\ \text{momentum:} & \boldsymbol{p}' = -\boldsymbol{p}, \\ \text{energy:} & \varepsilon'(\boldsymbol{p}') = \text{const.} - \varepsilon(\boldsymbol{p}), \\ \text{Fermi potential:} & \mu' = \text{const.} - \mu. \end{array}\right\} \tag{4.11}$$

The statistical distribution of the holes is then described by the Fermi distribution function

$$\bar{n}'_{p'} = \frac{1}{\exp\{(\varepsilon'_{p'} - \mu)/kT\} + 1}. \tag{4.12}$$

Figures 4.6 and 4.7 show two situations where a certain band is fully occupied

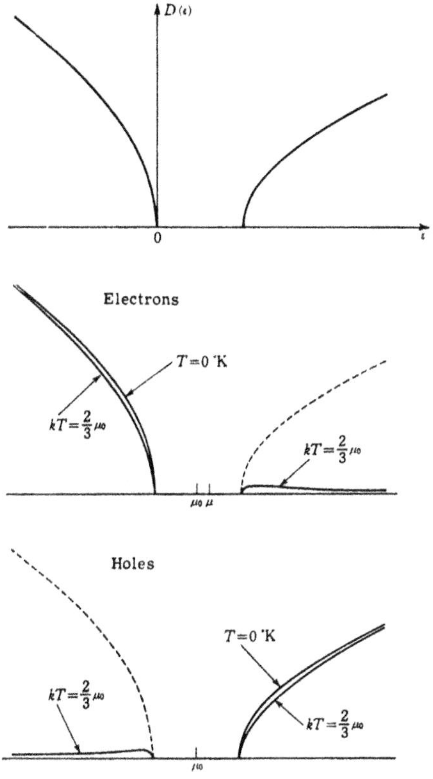

Fig. 4.6.

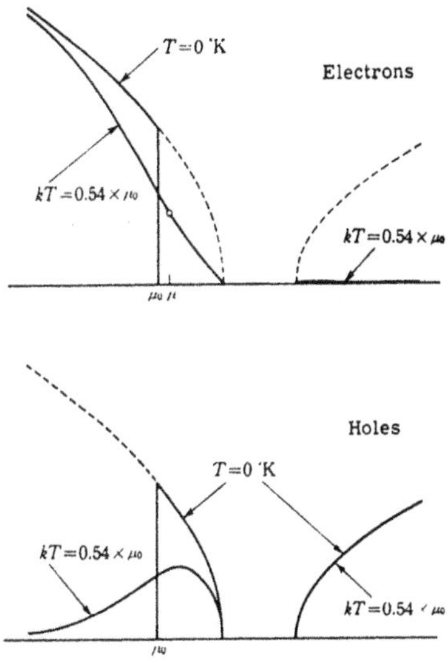

Fig. 4.7.

or nearly filled at 0° K. At higher temperatures electrons are distributed both in the lower and in the upper bands. These distributions are represented in the figure either in terms of electrons or holes. It is convenient to use the concept of electron distribution for the upper band and that of holes for the lower band.

§ 4.5. SEMICONDUCTORS

One of the important applications of Fermi-statistics is the problem of semiconductors. Examples are given as problems, but some fundamental concepts are introduced here.

Intrinsic semiconductors are those materials which have completely filled bands (*filled bands,* sometimes called *valence bands*) and completely vacant

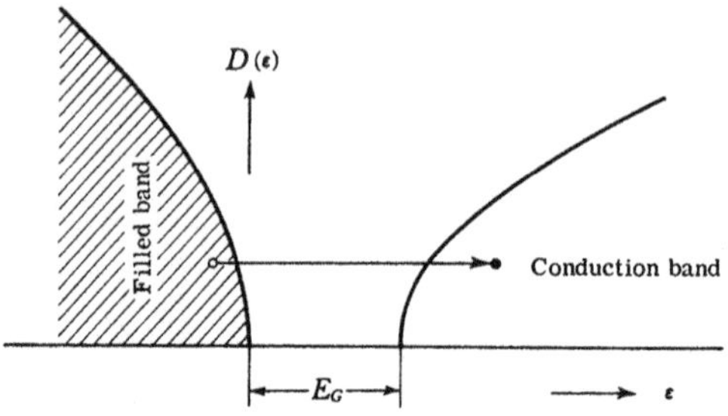

Fig. 4.8. Intrinsic semiconductor.

bands (sometimes called *conduction bands*) at 0° K. At finite temperatures electrons are excited from the lower bands to the upper bands. Both the electrons and the holes thus created in the upper and lower bands give rise to electrical conduction. Therefore a crystal which is an insulator at 0° K becomes a conductor at higher temperatures. The temperature at which conduction appears depends on the *energy gap* E_G between the filled band and the conduction band.

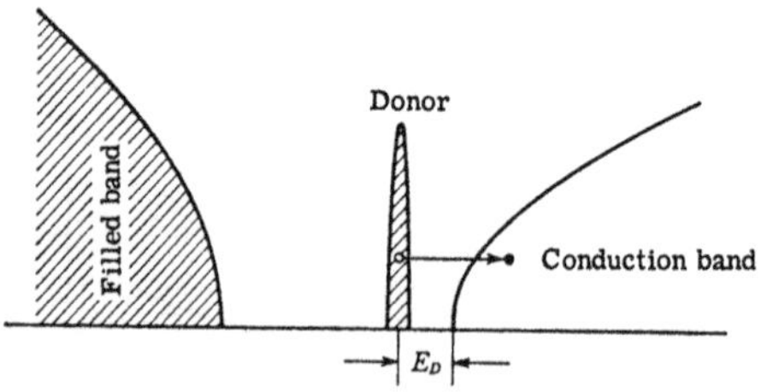

Fig. 4.9.

Impurity semiconductors: Some crystals with a relatively large energy gap may exhibit good conductivity at relatively low temperatures only when they contain impurities. The reason is that the impurities produce new levels in the energy gap between the filled band and the conduction band. There are two types of such impurity semiconductors:

N-type semiconductors have impurity levels which are filled by electrons at 0° K. At higher temperatures electrons are excited from these levels to the conduction band and electric conduction appears due to these conduction electrons. These impurity levels are called *donors*. Example: Ge containing P or As (see Fig. 4.9).

P-type semiconductors have impurity levels which are not occupied by electrons at 0° K. At finite temperatures, electrons are excited from the filled band to these levels, thus producing holes in the filled band. These holes give rise to conduction. These impurity levels are called *acceptors*. Examples: Ge containing B or Al (see Fig. 4.10).

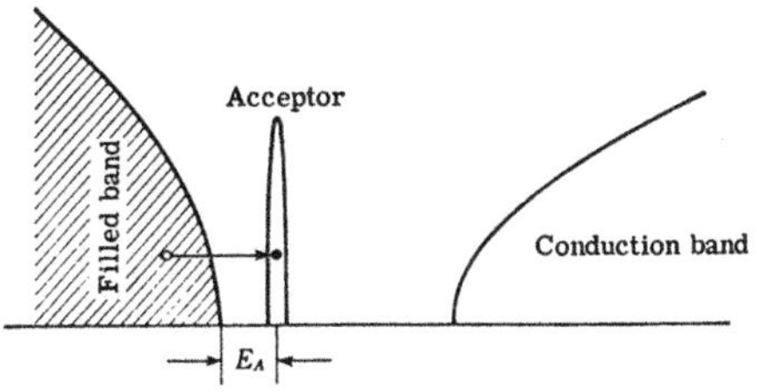

Fig. 4.10.

Effective mass of electrons and holes: The energy of an electron in the conduction band or that of a hole in the filled band is generally a rather complicated function of its momentum. Usually, however, one deals only with electrons which are at the bottom of the conduction band or holes at the top of the filled band, for which simple approximations of the type

$$\begin{aligned} \varepsilon(\boldsymbol{p}) &= E_G + \frac{1}{2m_e^*}\boldsymbol{p}^2, \qquad &\text{(electrons)} \\ \varepsilon'(\boldsymbol{p}') &= \frac{1}{2m_h^*}\boldsymbol{p}'^2, \qquad &\text{(holes)} \end{aligned} \tag{4.13}$$

may be used. This implies that electrons in the conduction band behave like free electrons with a mass m_e^* and holes in the filled bands behave like free positrons with a mass m_h^* and charge $|e|$. The m_e^* and m_h^* are called the *effective masses*. More generally, the effective mass is a tensor quantity and is a function of the momentum. When (4.13) is assumed, the state density $D(\varepsilon)$ becomes a parabola as function of the energy just like that for free electrons (Fig. 4.11).

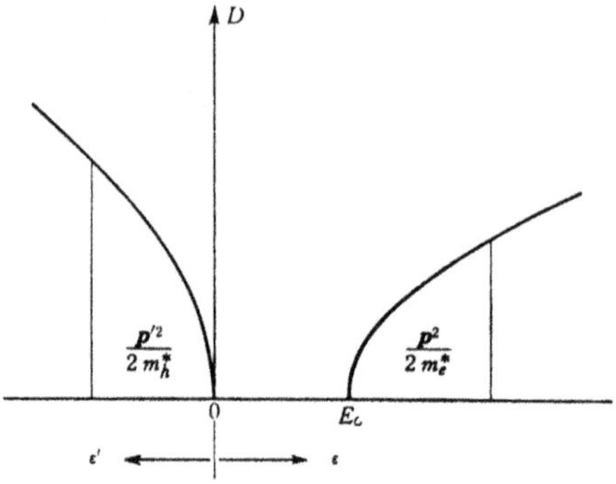

Fig. 4.11.

§ 4.6. BOSE-STATISTICS. LIQUID HELIUM

For bosons one has the formulae

$$\bar{n}_\tau = \frac{1}{e^{(\varepsilon_\tau - \mu)/kT} - 1}, \tag{4.14}$$

$$E = \sum \varepsilon_\tau \bar{n}_\tau = \sum_\tau \frac{\varepsilon_\tau}{e^{(\varepsilon_\tau - \mu)/kT} - 1}, \tag{4.15}$$

$$N = \sum \bar{n}_\tau = \sum_\tau \frac{1}{e^{(\varepsilon_\tau - \mu)/kT} - 1}, \tag{4.16}$$

$$F = N\mu + kT \sum_\tau \log(1 - e^{-(\varepsilon_\tau - \mu)/kT}) \tag{4.17}$$

(see § 1.15). In most cases the particle number N is given so that equation (4.16) determines μ as a function of T and N/V. The function

$$f(\varepsilon) = \frac{1}{e^{(\varepsilon - \mu)/kT} - 1} \tag{4.18}$$

is called the *Bose distribution function.* As shown in Fig. 4.12, $f(\varepsilon)$ becomes infinite as $\varepsilon \to \mu$. Therefore, if the lowest value of the one-particle energy is chosen as zero, one must have

$$\mu \leqq 0. \tag{4.19}$$

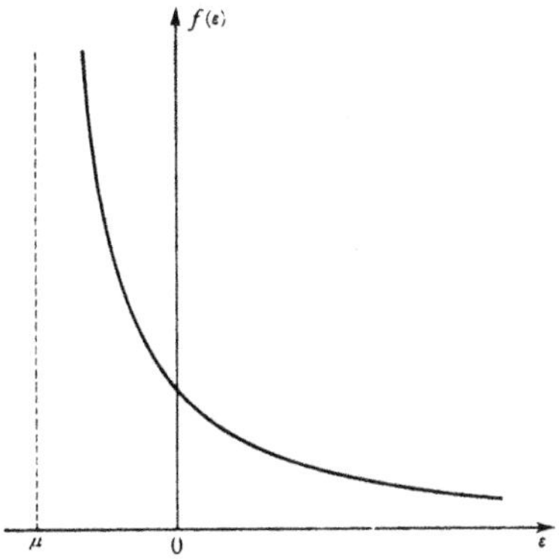

Fig. 4.12. Bose distribution function.

When the condition $|\mu| \gg kT$ is satisfied, the system is nondegenerate so that the Bose-statistics may be replaced by Boltzmann statistics. When the condition

$$|\mu| \sim kT \tag{4.20}$$

is fulfilled, there is a remarkable degeneracy. This condition may be expressed as

$$\frac{N}{V} \gtrsim \frac{(2\pi mkT)^{\frac{3}{2}}}{h^3}$$

for free particles. Only helium atoms can satisfy this condition among really existent bosons since their mass is light and the density of liquid helium is, on the one hand, sufficiently high for the left-hand side to be large and, on the other hand, sufficiently low for it to be considered to be a gas.

Liquid helium II. Helium gas liquefies at 4.22° K. When this is cooled down it makes a second order phase change at 2.19° K, where it reveals an anomalous specific heat. The phase above this transition temperature is called liquid helium I and that below it liquid helium II. Liquid helium II is a very peculiar fluid and is in fact a quantum liquid. That is to say, some quantum effects show up on a macroscopic scale. A detailed explanation of this is out of place here, but it will be only remarked, in passing, that this transition (λ-transition) is often interpreted essentially as a result of a *Bose-Einstein condensation* which is the strongest degeneracy effect of a boson system.

Examples

1. Calculate the Fermi potential μ and the internal energy E of an ideal Fermi gas composed of particles of spin $\frac{1}{2}$ up to terms of the order T^4 when the degeneracy is sufficiently high.

SOLUTION

The state density $D(\varepsilon)$ for a free particle in a box of volume V is given in problem 2.6 as

$$D(\varepsilon) = \frac{2V}{h^3}\frac{\mathrm{d}}{\mathrm{d}\varepsilon}\int_0^\varepsilon 4\pi p^2\,\mathrm{d}p \qquad (p^2 = 2m\varepsilon)$$

$$= 2\cdot 2\pi V\left(\frac{2m}{h^2}\right)^{\frac{3}{2}}\varepsilon^{\frac{1}{2}}, \tag{1}$$

where the factor 2 has appeared due to the spin degeneracy. At 0° K the energy states are completely occupied up to the level $\varepsilon = \mu_0$ which is determined by the relation

$$\int_0^{\mu_0} D(\varepsilon)\,\mathrm{d}\varepsilon = 4\pi V\left(\frac{2m}{h^2}\right)^{\frac{3}{2}}\int_0^{\mu_0}\sqrt{\varepsilon}\,\mathrm{d}\varepsilon = \frac{8\pi}{3}V\left(\frac{2m}{h^2}\mu_0\right)^{\frac{3}{2}} = N, \tag{2}$$

or

$$\mu_0 = \frac{h^2}{2m}\left(\frac{3N}{8\pi V}\right)^{\frac{2}{3}},$$

where N is the total number of particles. Hence one has

$$D(\varepsilon) = \tfrac{3}{2}N\varepsilon^{\frac{1}{2}}/\mu_0^{\frac{3}{2}}. \tag{3}$$

At finite temperatures one should use the formula (4.3):

$$\int_0^\infty f(\varepsilon)D(\varepsilon)\,\mathrm{d}\varepsilon = \tfrac{3}{2}N\mu_0^{-\frac{3}{2}}\int_0^\infty \varepsilon^{\frac{1}{2}}f(\varepsilon)\,\mathrm{d}\varepsilon = N. \tag{4}$$

Using the formula (4.6a) with $\varepsilon_0 = 0$, $g(\varepsilon) = \varepsilon^{\frac{1}{2}}$ one obtains

$$\left(\frac{\mu}{\mu_0}\right)^{\frac{3}{2}}\left\{1 + \frac{\pi^2}{8}\left(\frac{kT}{\mu}\right)^2 + \frac{7\pi^4}{640}\left(\frac{kT}{\mu}\right)^4 + \cdots\right\} = 1, \tag{5}$$

or

$$\mu = \mu_0 \left\{ 1 + \frac{\pi^2}{8}\left(\frac{kT}{\mu}\right)^2 + \frac{7\pi^4}{640}\left(\frac{kT}{\mu}\right)^4 + \cdots \right\}^{-\frac{2}{3}}$$

$$= \mu_0 \left\{ 1 - \frac{\pi^2}{12}\left(\frac{kT}{\mu}\right)^2 + \frac{\pi^4}{720}\left(\frac{kT}{\mu}\right)^4 + \cdots \right\}. \quad (6)$$

In the first approximation one obtains

$$\mu = \mu_0 \left\{ 1 - \frac{\pi^2}{12}\left(\frac{kT}{\mu_0}\right)^2 + \cdots \right\},$$

and in the second approximation one has

$$\mu = \mu_0 \left\{ 1 - \frac{\pi^2}{12}\left(\frac{kT}{\mu_0}\right)^2 - \frac{\pi^4}{80}\left(\frac{kT}{\mu_0}\right)^4 + \cdots \right\}. \quad (7)$$

For the internal energy E one has

$$E = \int_0^\infty \varepsilon D(\varepsilon) f(\varepsilon)\, \mathrm{d}\varepsilon = \frac{3}{2}\frac{N}{\mu_0^{\frac{3}{2}}} \int_0^\infty \varepsilon^{\frac{3}{2}} f(\varepsilon)\, \mathrm{d}\varepsilon \quad (8)$$

by (4.2′). Using (4.6a) with $g(\varepsilon) = \varepsilon^{\frac{3}{2}}$, one obtains

$$E = \frac{3}{5} N \left(\frac{\mu}{\mu_0}\right)^{\frac{3}{2}} \mu \left\{ 1 + \frac{5\pi^2}{8}\left(\frac{kT}{\mu}\right)^2 - \frac{7\pi^4}{384}\left(\frac{kT}{\mu}\right)^4 + \cdots \right\}, \quad (9)$$

so that

$$E = \frac{3}{5} N\mu_0 \left\{ 1 + \frac{5\pi^2}{12}\left(\frac{kT}{\mu_0}\right)^2 - \frac{\pi^4}{16}\left(\frac{kT}{\mu_0}\right)^4 + \cdots \right\} \quad (10)$$

by equation (7).

NOTE: In Figure 4.17 the graphs of $\mu = \mu_0$ and $\mu = \mu_0 (1 - \frac{1}{12}\pi^2 (kT/\mu_0)^2)$ are given.

2. Show that the specific heat C_V of an ideal Fermi gas is given by

$$C_V = \tfrac{1}{3}\pi^2 k^2 T D(\mu_0)$$

at sufficiently low temperatures. Here $D(\varepsilon)$ is the state density for one particle.

SOLUTION

From equation (4.3′) one obtains, at 0° K

$$\int_0^{\mu_0} D(\varepsilon)\,\mathrm{d}\varepsilon = N\,, \tag{1}$$

and, at T° K

$$\int_0^{\infty} D(\varepsilon)f(\varepsilon)\,\mathrm{d}\varepsilon = N\,. \tag{2}$$

Using (4.6a), one obtains

$$N = \int_0^{\mu} D(\varepsilon)\,\mathrm{d}\varepsilon + \tfrac{1}{6}\pi^2 (kT)^2 D'(\mu) + \cdots. \tag{3}$$

Subtracting (1) from (3) we have

$$\int_{\mu_0}^{\mu} D(\varepsilon)\,\mathrm{d}\varepsilon + \frac{\pi^2}{6}(kT)^2 D'(\mu) + \cdots = 0\,. \tag{4}$$

At sufficiently low temperatures, one can expect $\mu - \mu_0 \ll \mu_0, \mu$ and hence one can approximate equation (4) by

$$(\mu - \mu_0)D(\mu_0) + \tfrac{1}{6}\pi^2 (kT)^2 D'(\mu_0) + \cdots = 0\,.$$

Hence one has

$$\mu \fallingdotseq \mu_0 - \tfrac{1}{6}\pi^2 (kT)^2 \left(\frac{\mathrm{d}}{\mathrm{d}\varepsilon}\log D(\varepsilon)\right)_{\varepsilon=\mu_0} \tag{5}$$

On the other hand, one obtains

$$E = \int_0^{\infty} \varepsilon f(\varepsilon) D(\varepsilon)\,\mathrm{d}\varepsilon = \int_0^{\mu} \varepsilon D(\varepsilon)\,\mathrm{d}\varepsilon + \tfrac{1}{6}\pi^2 (kT)^2 \left(\frac{\mathrm{d}}{\mathrm{d}\varepsilon}\varepsilon D(\varepsilon)\right)_{\varepsilon=\mu} + \cdots \tag{6}$$

by the use of (4.6a). Here one can also use the approximation

$$\int_0^{\mu} \varepsilon D(\varepsilon)\,\mathrm{d}\varepsilon \fallingdotseq \int_0^{\mu_0} \varepsilon D(\varepsilon)\,\mathrm{d}\varepsilon + (\mu - \mu_0)\mu_0 D(\mu_0) \fallingdotseq \int_0^{\mu_0} \varepsilon D(\varepsilon)\,\mathrm{d}\varepsilon - \tfrac{1}{6}\pi^2 (kT)^2 \mu_0 D'(\mu_0)\,.$$

So, finally, one has

$$E \fallingdotseq \int_0^{\mu_0} \varepsilon D(\varepsilon)\,\mathrm{d}\varepsilon + \tfrac{1}{6}\pi^2 (kT)^2 D(\mu_0), \tag{7}$$

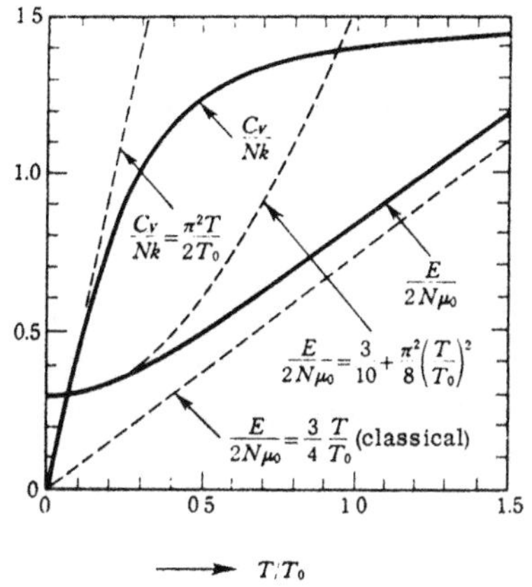

Fig. 4.13.

and hence

$$C_V = (\partial E/\partial T)_V \fallingdotseq \tfrac{1}{3}\pi^2 k^2 T D(\mu_0).$$

If one adopts expression (4.9) for $D(\varepsilon)$, the formulae $D(\mu_0) = 3N/2\mu_0$ and

$$C_V \fallingdotseq \tfrac{1}{2}\pi^2 Nk \frac{T}{T_0} \qquad (T_0 = \mu_0/k) \tag{8}$$

follow immediately.

3. Consider an intrinsic semiconductor which has an energy gap of width E_G, conduction electrons of density n and holes of density p. Show that the following relation holds:

$$n = p = 2\left\{\frac{2\pi(m_e m_h)^{\frac{1}{2}} kT}{h^2}\right\}^{\frac{3}{2}} e^{-E_G/2kT}$$

and also that the Fermi potential of the electron system is given by

$$\mu = \tfrac{1}{2}E_G + \tfrac{3}{4}kT \log\frac{m_h}{m_e}$$

if the conduction electron and the hole are assumed to behave as free particles of effective mass m_e and m_h respectively. Here the origin of energy is taken at the top of the filled band and the assumption $E_G \gg kT$ is made. Estimate the value of n $(= p)$ for the case: $E_G = 0.7$ eV, $T = 300°$ K, $m_h = m_e = m$.

SOLUTION

The total number N of electrons is given by

$$N = \sum_i \frac{1}{e^{\beta(\varepsilon_i - \mu)} + 1} + \sum_j \frac{1}{e^{\beta(\varepsilon_j - \mu)} + 1}, \tag{1}$$

where ε_i is an energy level in the conduction band and ε_j is an energy level in the filled band. In the case of an intrinsic semiconductor the total number of electronic states in the full band is equal to N, that is $\sum_j 1 = N$. Hence one has, from equation (1)

$$\sum_i \frac{1}{e^{\beta(\varepsilon_i - \mu)} + 1} = \sum_j \left\{1 - \frac{1}{e^{\beta(\varepsilon_j - \mu)} + 1}\right\} = \sum_j \frac{1}{e^{\beta(-\varepsilon_j + \mu)} + 1}. \tag{2}$$

This equation shows the equality of the number of conduction electrons (the left hand side) and the number of holes in the filled band (the right hand side), that is,

$$n = p. \tag{3}$$

ε_i and ε_j can be approximately written as

$$\varepsilon_i = E_G + \frac{p^2}{2m_e}, \qquad \varepsilon_j = -\frac{p^2}{2m_h} \tag{4}$$

where p is the momentum of the particle, and n and p are given by

$$n = \frac{2}{h^3} \int \frac{dp}{e^{\beta(E_G - \mu + p^2/2m_e)} + 1}, \tag{5}$$

$$p = \frac{2}{h^3} \int \frac{dp}{e^{\beta(\mu + p^2/2m_h)} + 1}. \tag{6}$$

At 0° K, n and p are both equal to zero. At temperatures which are not very high, the relations $E_G - \mu \gg kT$, and $\mu \gg kT$ still hold and the magnitude of n and p can be considered very small. Hence one can assume the properties of non-degeneracy for both conduction electrons and holes. Then one obtains

$$n \sim \frac{2}{h^3} \int e^{-\beta(E_G - \mu + p^2/2m_e)} \, dp = 2 \cdot \left(\frac{2\pi m_e kT}{h^2}\right)^{\frac{3}{2}} e^{-(E_G - \mu)/kT}, \tag{7}$$

$$p \sim \frac{2}{h^3} \int e^{-\beta(\mu + p^2/2m_h)} \, dp = 2 \cdot \left(\frac{2\pi m_h kT}{h^2}\right)^{\frac{3}{2}} e^{-\mu/kT}. \tag{8}$$

From equations (7), (8) and (3) one can determine $e^{\mu/kT}$ as

$$e^{\mu/kT} = \left(\frac{m_h}{m_e}\right)^{\frac{3}{4}} e^{E_G/2kT}. \tag{9}$$

Hence one has from (7) and (8)

$$n = p = 2\cdot\left[\frac{2\pi(m_e m_h)^{\frac{1}{2}} kT}{h^2}\right]^{\frac{3}{2}} e^{-E_G/2kT}. \tag{10}$$

From (9) one has

$$\mu = \tfrac{1}{2}E_G + \tfrac{3}{4}kT\log\frac{m_h}{m_e}. \tag{11}$$

The Fermi potential which is determined by (11) lies very near the middle of the energy gap provided that the value of $\log(m_h/m_e)$ is of the order of unity and that the temperature is well below the value of E_G/k. For $E_G = 0.7$ eV E_G/k is 0.81×10^4 °K. Hence at ordinary temperatures the relations $E_G/k \gg T$, $E_G - \mu \gg kT$, and $\mu \gg kT$ are satisfied. Putting $T = 300°$ K and

$$m_e = m_h = m,$$

one obtains $n = p = 4.83 \times 10^{15} \times (300)^{\frac{3}{2}} e^{-0.41 \times 10^4/300} \sim 1.6 \times 10^{13}\,\text{cm}^{-3}$.

DIVERTISSEMENT 10

Electron theory of metals. The electron theory of metals was initiated by Drude and Lorentz [H. A. Lorentz, *Theory of Electrons* (Teubner, Leipzig, 1909)]. This classical theory assumed two kinds of electrons in a metal: free electrons and bound electrons. This assumption was given a sound basis many years later by the band theory in the modern quantum theory of solids. The free electron model succeeded in explaining the good electrical and heat conductivities of metals, but at the same time it introduced a serious headache for the classical electron theory, because each electron should then contribute $\frac{3}{2}k$ to the heat capacity according to the equipartition law of classical statistical mechanics. This contradicts the well-known Dulong-Petit's law. This difficulty was similar to that of the Rayleigh-Jeans law of black body radiation. Unlike the latter, the heat capacity difficulty could not be resolved by the quantum theory of Planck, but could only be overcome by the birth of quantum mechanics and the introduction of the idea of Fermi-statistics.

We owe the modern electron theory to A. Sommerfeld [see for instance, Sommerfeld and Bethe, *Elektronentheorie der Metalle, Handbuch der Physik*, 24/2 (1933)] who reformulated the whole of Drude-Lorentz's electron theory by means of quantum mechanics. Besides the solution of the heat capacity difficulty, another brilliant success of such modern electron theory was achieved by W. Pauli who gave the first theory of the spin paramagnetism of free electron gases (Zeits. f. Physik **41** (1926) 81).

4. Consider an n-type semiconductor whose donor levels lie E_D below the bottom of the conduction band. Let N_D, n_D and n be the number of donors, the number of electrons in the donor levels, and the number of conduction electrons, per unit volume respectively. Derive the relation

$$\frac{n(N_D - n_D)}{n_D} = \tfrac{1}{2} N_c \mathrm{e}^{-E_D/kT}$$

and interpret its physical meaning. Here it is assumed that the donor level cannot be occupied by two electrons at the same time, that the system of conduction electrons is not degenerate, and that $N_c = 2(2\pi m^* kT/h^2)^{\frac{3}{2}}$, in which m^* is the effective mass of the conduction electrons.

SOLUTION

One considers a sample of unit volume. The free energy F of the electrons which are on the donor levels is given by

$$F = -n_D E_D - kT \log\left\{\frac{N_D!}{n_D!(N_D - n_D)!} \cdot 2^{n_D}\right\} \tag{1}$$

$$= -n_D E_D - kT\left\{n_D \log 2 - n_D \log \frac{n_D}{N_D} - (N_D - n_D) \log \frac{N_D - n_D}{N_D}\right\} \tag{2}$$

where the number in the logarithm in the second term of (1) means the total number of configurations of n_D electrons (+ or − spin) distributed over N_D donor levels. In (2) this factor is simplified by the use of Stirling's formula. The chemical potential μ of the electrons in the donor levels is given by

$$\mu = \frac{\partial F}{\partial n_D} = -E_D - kT\left\{\log 2 - \log \frac{n_D}{N_D - n_D}\right\}, \tag{3}$$

so that one has

$$\frac{n_D}{N_D - n_D} = 2 \cdot \mathrm{e}^{(E_D + \mu)/kT} \tag{4}$$

or

$$n_D = \frac{N_D}{\frac{1}{2}\mathrm{e}^{-(E_D + \mu)/kT} + 1}. \tag{5}$$

On the other hand, the number of conduction electrons n is given by

$$n = \frac{2}{h^3} \int \frac{\mathrm{d}\boldsymbol{p}}{\mathrm{e}^{(\varepsilon - \mu)/kT} + 1} \qquad (\varepsilon = p^2/2m^*) \tag{6}$$

where the factor 2 is due to the spin degeneracy. On account of the assumption of weak degeneracy, equation (6) can be written as

$$n = \frac{2}{h^3}\int e^{-(\varepsilon-\mu)/kT}\,dp = 2\cdot\left(\frac{2\pi m^* kT}{h^2}\right)^{\frac{3}{2}}\cdot e^{\mu/kT} = N_c\,e^{\mu/kT}. \tag{7}$$

Eliminating μ from (4) and (7), one has

$$\frac{n(N_D - n_D)}{n_D} = \tfrac{1}{2} N_c\, e^{-E_D/kT}. \tag{8}$$

The principle of the above reasoning is nothing but the determination of the dissociative equilibrium shown by the following equation,

$$D \rightleftarrows D^+ + e \tag{9}$$

where D means the donor with its electron, D^+ the ionized donor and e the conduction electron. If one uses the notations $[D] = n_D, [D^+] = N_D - n_D$ and $[e] = n$, equation (8) has the form of the law of mass action,

$$\frac{[D^+][e]}{[D]} = K(T). \tag{10}$$

Alternative solution

The grand partition function Ξ_D for a system of electrons which are distributed over N_D donor levels can be set up as follows. The possible microscopic states for a donor level are the empty state, the state occupied by an electron of + spin and that of − spin. The energies of these occupied states are both equal to $-E_D$. All donor levels are assumed independent with each other. One has

$$\Xi_D = \{1 + 2\lambda\, e^{E_D/kT}\}^{N_D}, \tag{11}$$

where $\lambda = e^{\mu/kT}$ is the absolute activity of the electron. The average number of electrons distributed over N_D donor levels is given by $\lambda\, \partial \log \Xi_D/\partial\lambda$ and this can be put equal to n_D,

$$n_D = \lambda \frac{\partial \log \Xi_D}{\partial \lambda} = \frac{2\lambda\, e^{E_D/kT}}{1 + 2\lambda\, e^{E_D/kT}} N_D = \frac{N_D}{\frac{1}{2}e^{-(E_D+\mu)/kT} + 1}.$$

This is the same as equation (5). After this, one can proceed in the same way as in the former solution. The above procedure is analogous to the one by which the Fermi distribution is derived from the $T-\mu$ canonical distribution.

Note: The origin of the difference of (5) from the usual Fermi distribution is the assumption that only one electron can enter into the donor level at the same time. If one assumes that two electrons of different spins can occupy independently one donor level, one gets instead of equation (5),

$$n_D = \frac{2N_D}{e^{-(E_D+\mu)/kT}+1} \qquad \text{(2 is the spin degeneracy)} \tag{12}$$

and then

$$\frac{n(2N_D - n_D)}{n_D} = N_c e^{-E_D/kT}. \tag{13}$$

5. Consider an ideal Bose gas composed of N particles in a volume V, and let N and N' be the number of particles in the lowest one-particle state (momentum $p = 0$) and of the number of particles in the higher states ($p \neq 0$) respectively. Show that when the temperature falls below a temperature T_c, N_0 suddenly becomes comparable with the total number N and that in this region the chemical potential μ is equal to zero (Bose-Einstein condensation).

SOLUTION

Let ε_i be the one-particle energy levels. One finds

$$N = \sum_i \frac{1}{e^{\beta(\varepsilon_i-\mu)}-1} \qquad (\beta = 1/kT) \tag{1}$$

or equivalently,

$$N = N_0 + N', \qquad N_0 = \frac{g_0}{e^{-\beta\mu}-1}, \qquad N' = \sum_{i\neq 0} \frac{1}{e^{\beta(\varepsilon_i-\mu)}-1} \tag{2}$$

where g_0 is the degeneracy of the lowest energy state $\varepsilon_0 = 0$ and can be put equal to unity. Concerning N', one can treat the equation in the usual way (cf. problem 1.6), using the state density for the free particle $D(\varepsilon) = 2\pi V(2m/h^2)^{\frac{3}{2}}\varepsilon^{\frac{1}{2}}$

$$\begin{aligned} N' &= 2\pi V\left(\frac{2m}{h^2}\right)^{\frac{3}{2}} \int_0^\infty \frac{\varepsilon^{\frac{1}{2}}\,d\varepsilon}{e^{\beta(\varepsilon-\mu)}-1} \\ &= V\left(\frac{2\pi mkT}{h^2}\right)^{\frac{3}{2}} \frac{2}{\sqrt{\pi}} \int_0^\infty \frac{x^{\frac{1}{2}}\,dx}{e^{x-\beta\mu}-1} \\ &= V\left(\frac{2\pi mkT}{h^2}\right)^{\frac{3}{2}} F_{\frac{3}{2}}(\alpha), \qquad (\alpha = -\beta\mu). \end{aligned} \tag{3}$$

This expression when integrated does not contain the state $\varepsilon_0 = 0$ because the density of states ($\sim \sqrt{\varepsilon}$) becomes zero at that point. This is the reason why N_0 is inserted separately in equation (2). The function $F_{\frac{3}{2}}(\alpha)$ has the form

$$F_{\frac{3}{2}}(\alpha) \equiv \frac{2}{\sqrt{\pi}} \int_0^\infty \frac{x^{\frac{1}{2}}\,dx}{e^{x+\alpha} - 1} = \sum_{n=1}^\infty e^{-n\alpha}/n^{\frac{3}{2}} \tag{4}$$

and can be defined only in the region $\alpha = -\beta\mu > 0$, as can be seen from the expansion shown above. $F_{\frac{3}{2}}(\alpha)$ is a monotonically decreasing function:

$$F_{\frac{3}{2}}(\alpha) \leqq F_{\frac{3}{2}}(0) = \sum_{n=1}^\infty n^{-\frac{3}{2}} \fallingdotseq 2.612 (= \zeta(\tfrac{3}{2})). \tag{5}$$

Hence one has from equation (3),

$$N' < V\left(\frac{2\pi mkT}{h^2}\right)^{\frac{3}{2}} \times 2.612 = N'_{max}(T). \tag{6}$$

This means that there is an upper limit for the number of particles in the higher states and that this limit decreases with decreasing temperature. When $N'_{max}(T)$ becomes smaller than the total number N, the remaining $N - N'_{max}(T)$ particles must enter into the lowest state $\varepsilon_0 = 0$. The critical temperature T_c is defined by the relation

$$2.612\left(\frac{2\pi mkT_c}{h^2}\right)^{\frac{3}{2}} = \frac{N}{V}. \tag{7}$$

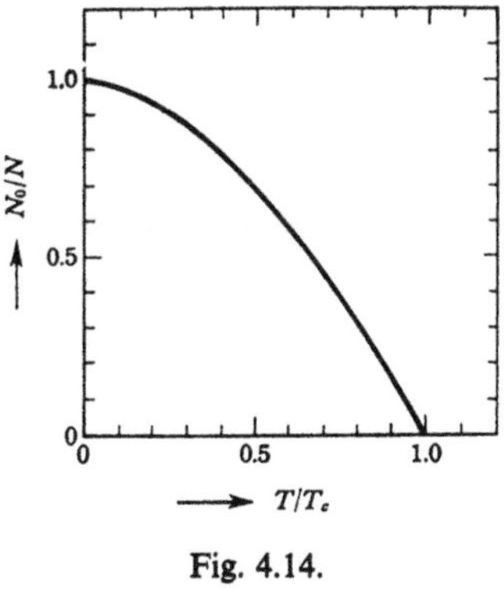

Fig. 4.14.

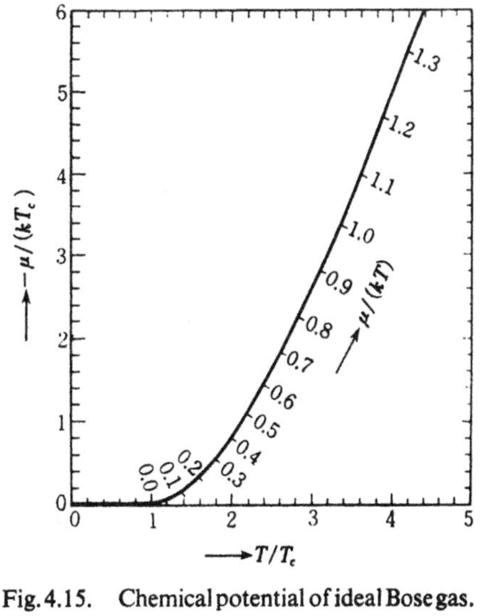

Fig. 4.15. Chemical potential of ideal Bose gas.

Below this critical temperature μ remains equal to zero, and N' and N_0 are given by

$$\left.\begin{aligned} N' &= N(T/T_c)^{\frac{3}{2}}, \\ N_0 &= N\{1-(T/T_c)^{\frac{3}{2}}\}. \end{aligned}\right\} \tag{8}$$

NOTE: At temperatures $T > T_c$, the equation

$$N = N' = V\left(\frac{2\pi mkT}{h^2}\right)^{\frac{3}{2}} F_{\frac{3}{2}}(-\beta\mu) \tag{9}$$

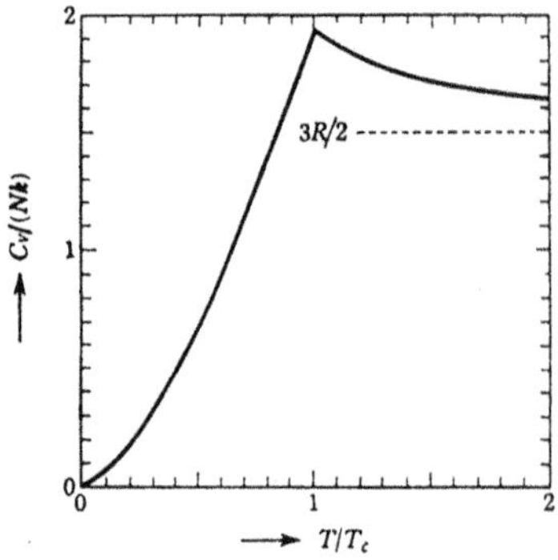

Fig. 4.16. Specific heat of ideal Bose gas.

has a solution with $\mu < 0$. In this case $N_0 = O(1)$ and can be neglected compared to N' $(= O(N))$. When T decreases to T_c, μ approaches zero, and at temperatures $T < T_c$, we have $\mu = 0$ and $N_0 = O(N)$ [cf. F. London, *Superfluids II* (John Wiley & Sons, New York, 1954)].

6. Show that the chemical potential of photon gas is equal to zero.

SOLUTION

The number of photons in a container is not a constant but only its average value is determined as a function of the container volume V and the temperature T. This is because the photons (light) are emitted and absorbed by the matter inside or by the surface of the walls of the container. Hence one should omit the condition that the total number N = constant in the procedure for the derivation of the Bose distribution explained in problem (1.31). Accordingly, the chemical potential which is to be introduced as a Lagrange multiplier drops out from the Bose distribution. This is equivalent to putting $\mu = 0$ in equation (4.14).

NOTE: The equilibrium value of the total number of photons can be determined by the condition of minimizing the free energy $F(T, V, N)$ if one uses a method which allows an approximate calculation of the canonical partition function by its maximum term

$$(\partial F/\partial N)_{T,V} = 0.$$

This is merely the relation $\mu = 0$.

ALTERNATIVE SOLUTION

Electromagnetic waves confined in a container can be regarded as superpositions of the normal modes of oscillation. Let ν_i be the frequency of the i-th normal mode and n_i be the quantum number for that mode which can be treated as a quantized harmonic oscillator. Then

$$E(n_0, n_1, \cdots) = \sum_i n_i h\nu_i \tag{1}$$

is the energy of the electromagnetic wave which is in the quantum state specified by $(n_0, n_1, \ldots)$. Here the zero point energy of the oscillator is omitted by the adjustment of the energy zero. By (1), one can consider n_i as the number of photons with energy $h\nu_i$. Hence the canonical partition function for this photon gas is given by

$$Z(T, V) = \sum_{n_0=0}^{\infty} \sum_{n_1=0}^{\infty} \cdots \exp\left\{-\frac{E(n_0, n_1, \cdots)}{kT}\right\} = \prod_i \{1 - e^{-h\nu_i/kT}\}^{-1} \tag{2}$$

and the average value of n_j is given by

$$\bar{n}_j = \sum_{n_j=0}^{\infty} n_j e^{-n_j h\nu_j/kT} \Big/ \sum_{n_j=0}^{\infty} e^{-n_j h\nu_j/kT} = \frac{1}{e^{h\nu_j/kT} - 1}. \tag{3}$$

One can interpret $Z(T, V)$ as the grand partition function and $\bar{n}_j$ as the distribution function for an ideal gas with $\mu = 0$ respectively.

Problems

A

1. Let the density of states of the electrons in some sample be assumed to be a constant D for $\varepsilon > 0$ ($D = 0$ for $\varepsilon < 0$) and the total number of electrons be equal to N.

(i) Calculate the Fermi potential μ_0 at 0° K.

(ii) Derive the condition that the system is non-degenerate.

(iii) Show that the specific heat is proportional to T when the system is highly degenerate.

2. At finite temperatures the Fermi distribution $f(\varepsilon)$ can be represented in a rough approximation by a broken line as shown in the figure. Give an elementary explanation of the origin of the linear specific heat at low temperatures by the use of this approximation.

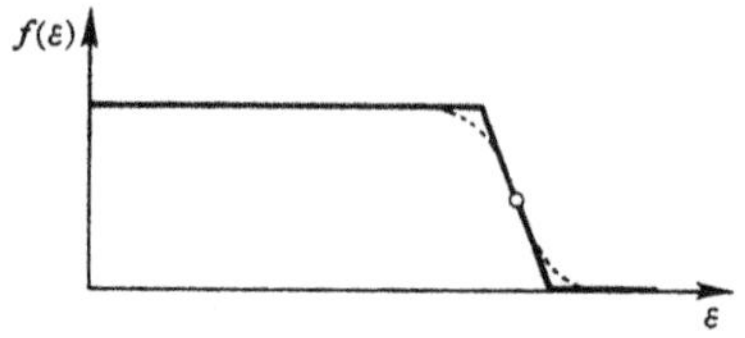

3. Let some physical quantity I of a system of electrons be expressed by an integral containing the Fermi distribution $f(\varepsilon)$:

$$I = \int_0^\infty g(\varepsilon)f(\varepsilon)\,d\varepsilon \equiv \int_0^\infty \psi(\varepsilon)D(\varepsilon)f(\varepsilon)\,d\varepsilon \qquad (g = \psi D)$$

where D is the density of states. Derive the following formulae when the degeneracy is sufficiently strong:

$$\left(\frac{\partial I}{\partial T}\right)_\mu = \tfrac{1}{3}\pi^2 k^2 T g'(\mu_0) + O(T^3),$$

$$\left(\frac{\partial I}{\partial \mu}\right)_T = g(\mu_0) + O(T^2),$$

$$\left(\frac{\partial I}{\partial T}\right)_N = \tfrac{1}{3}\pi^2 k^2 T \psi'(\mu_0)D(\mu_0) + O(T^3).$$

(*Hint:* Follow the procedure beginning with (4.7).)

4. An electron in a magnetic field $\boldsymbol{H}$ has an energy $\pm\,\mu_B H$ according to whether the spin magnetic moment is parallel or anti-parallel to the field. Calculate the spin paramagnetic susceptibility of a system of free electrons at an 0° K where the degeneracy is complete.

5. Show that the spin paramagnetic susceptibility of a system of electrons at arbitrary temperature is given by

$$\chi = 2\mu_B^2 \int^\infty D'(\varepsilon)f(\varepsilon)\,d\varepsilon$$

where D is the density of states for one electron per unit volume without the spin degeneracy. Derive general formulas for the cases where the degeneracy is very strong and very weak, respectively, and apply them to a system of free electrons.

6. It is considered that in metals there are some electrons which can migrate freely. The atomic heat of a Na crystal (atomic volume $= 24\,\text{cm}^3/\text{mol}$) is expected to be equal to $4.5R$ if one assumes the presence of one free electron per atom and the law of equipartition of energy as in classical statistics. Show the details of this reasoning, then explain why the atomic heat of the metals usually obeys the law of Dulong-Petit and why the contribution of the free electrons seems to be almost zero, in contrast to the conclusion given above.

7. Consider the spin paramagnetic susceptibility of a free electron gas. Explain the physical meaning of the qualitative difference between its value in the case of strong degeneracy and the case of weak degeneracy. Explain also the meaning of the ratio of these values. Put the main stress on a clear explanation of the physical meaning of the relation.

8. Estimate the electronic specific heat and the spin paramagnetic susceptibility (per unit mass) of Li and Na. Assume that the valence electrons of both metals can be regarded as free electrons and use $0.534\,\text{g/cm}^3$ and $0.97\,\text{g/cm}^3$ as the densities of Li and Na, respectively.

9. Show that the equation of state of an ideal Fermi gas can be written as

$$pV = \tfrac{2}{3}U$$

and derive the formula for the compressibility when the degeneracy is strong. Estimate the compressibility of Na crystals. Assume that Na crystal has one free electron per atom, and use the following values: atomic weight $= 23$, density $= 0.97\,\text{g/cm}^3$.

10. Let the internal potential of free electrons in a metal be $-w$, and the Fermi

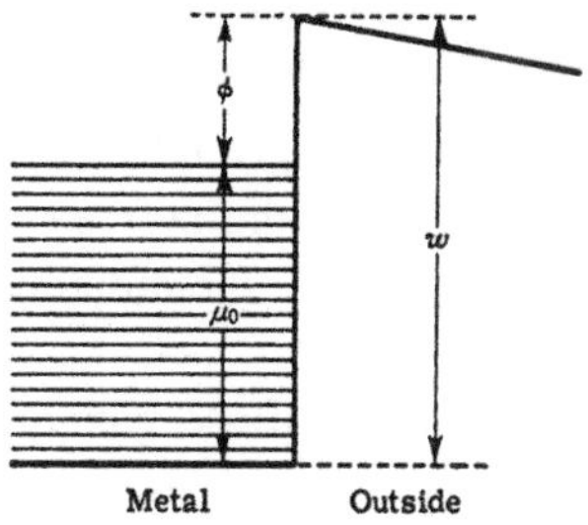

potential μ_0 at 0° K be $-\phi$ measured from the free vacuum. At finite temperatures electrons having higher energies at the upper tail of the Fermi distribution can escape to the exterior. When an appropriate potential difference is applied between an anode and this metal (taken as a cathode), it is possible to collect all of the electrons escaping from the metal. Show that the resulting thermionic current I per unit area of the metal surface can be given by the Richardson formula

$$I = AT^2 e^{-\phi/kT}$$

(ϕ is usually of the order of 1 eV and is larger than kT by a factor of 10^2).

11. Consider a sample of a metal which contains N atoms and assume that an energy band of this metal which can accomodate $2N$ electrons is being occupied by $2N - N'$ electrons. Show that the contribution of these electrons to the thermal properties of the metal is equivalent to that of an electron gas for which the energy levels are $-\varepsilon_i$ and the chemical potential is $-\mu$, where ε_i and μ are the energy levels and the chemical potential of the actual system of electrons.

12. In example 4, calculate the density of conduction electrons and their Fermi potential at sufficiently low and high temperatures respectively. (*Hint:* Use the condition of electrical neutrality $n = N_D - n_{D'}$.) Assuming the electrical conductivity σ of this semiconductor to be given by

$$\sigma = \frac{e^2 n}{m^*} \tau$$

with a constant relaxation time τ, draw the curve $\log \sigma$ versus $1/kT$. What information can one get from the tangent of such a curve?

13. Consider a semiconductor which has N acceptors (per cm^3) whose levels are E_A above the top of the filled band. Find the temperature dependence of the density of holes which are created in the filled band. Here it is assumed that the density of acceptors is so small that the holes are not degenerate and that each acceptor level can accommodate only one electron.

14. The energy spectrum of the photon is given by $E(q) = \hbar cq$, $q = |\boldsymbol{q}|$, where $\boldsymbol{q}$ is the wave vector. Calculate the Helmholtz free energy, the entropy and the internal energy of this photon gas, and the radiation pressure.

15. Show that a two-dimensional ideal Bose gas does not exhibit a Bose-Einstein condensation.

B

16. Calculate the specific heat C_V and the spin paramagnetic susceptibility χ of a free electron gas at high temperatures near the degeneracy temperature and show how they approach the classical limit.

17. In a uniform magnetic field $\boldsymbol{H}$ taken along the z-direction, the projection of the orbital motion of an electron on the xy-plane makes a circular motion with angular frequency $\omega_0 = eH/mc$ on account of the Lorentz force $e\,|\boldsymbol{V} \times \boldsymbol{H}|/c$. This circular motion can be regarded as a quantized harmonic oscillation. Hence we can see that the energy level of this electron is given by the following formula

$$E(l, p_z) = \frac{e\hbar}{mc} H\,(l + \tfrac{1}{2}) + \frac{1}{2m} p_z^2 \qquad (l = 0, 1, 2, \cdots)$$

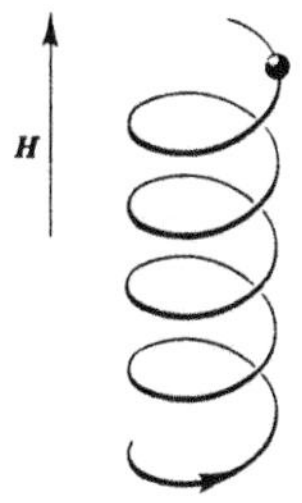

Motion of an electron in a magnetic field.

where $p_z^2/2m$ is the kinetic energy of the translational motion in the z-direction. Assume the temperature to be so high that this electron can be treated by Boltzman statistics, and show that its magnetic moment is given by

$$M = -\,N\mu_{\rm B} L\left(\frac{\mu_{\rm B} H}{kT}\right), \qquad L(x) = \coth x - \frac{1}{x},$$

where $\mu_{\rm B} = e\hbar/2mc$ is the Bohr magneton. The spin of the electrons and the Coulomb interactions between electrons are neglected.

(*Hint:* Note that the energy level specified by l has a high degeneracy which originates from the coalescence of the levels in a zero field. The energy levels in zero magnetic field which lie between the limits,

$$2\mu_{\rm B} H l < (p_x^2 + p_y^2)/2m < 2\mu_{\rm B} H(l + 1)$$

coalesce into the level $2\mu_{\rm B} H\,(l + \tfrac{1}{2})$ in the magnetic field H.)

18. Show that the magnetic susceptibility χ of a non-degenerate electron gas is given by

$$\chi = \frac{n(\mu_B^2 - \frac{1}{3}\mu_B^{*2})}{kT}.$$

Here n is the electron density, μ_B is the Bohr magneton and $\mu_B^* = e\hbar/2m^*c$ is the effective Bohr magneton for the orbital motion, m^* being the effective mass of the electron. The Coulomb interactions between electrons can be ignored.

19. Show that the photo-current I which is induced by a light source having frequency ν and unit intensity is given by

$$I \propto AT^2\phi\left(\frac{h[\nu - \nu_0]}{kT}\right),$$

$$\phi(\delta) = \int_0^\infty \log(1 + e^{\delta - y})\, dy,$$

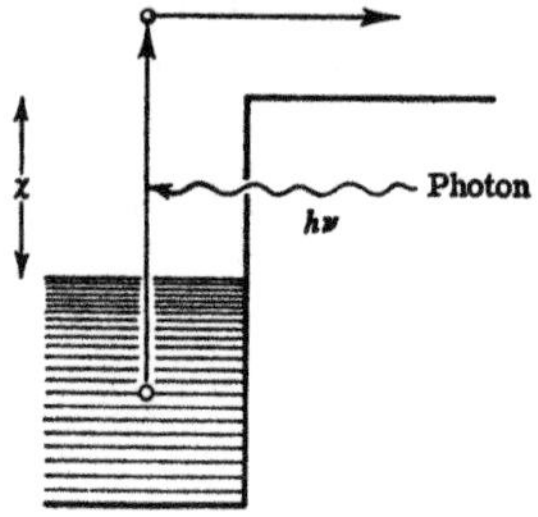

where ν_0 is the threshold frequency of the photo-electric effect defined by $h\nu_0 = \chi$ (χ = work function) and A is a constant. Further, derive the following expansion formula for ϕ (δ):

$$\phi = \begin{cases} e^{\delta} - \dfrac{e^{2\delta}}{2^2} + \dfrac{e^{3\delta}}{3^2} - \cdots, & (\delta \leqq 0) \\[2ex] \dfrac{\pi^2}{6} + \dfrac{\delta^2}{2} - \left\{ e^{-\delta} - \dfrac{e^{-2\delta}}{2^2} + \dfrac{e^{-3\delta}}{3^2} - \cdots \right\} & (\delta \geqq 0). \end{cases}$$

NOTE: The external photo-electric effect in a metal occurs when a conduction electron in the metal absorbs a photon and acquires enough additional energy to escape from the surface. This is shown schematically in the figure. Let the components of the momentum of the electron which are parallel to the metallic surface be assumed to remain unchanged by the absorption of a photon.

20. Consider a sample of semiconducting Ge which contains $N_D = 10^{15}$ cm^{-3} donors and $N_A = 10^{14}$ cm^{-3} acceptors. The donor levels lie $E_D = 0.04$ eV below the bottom of the conduction band. The graph shows the Fermi level μ

of this semiconductor as a function of temperature T, μ being measured from the middle of the gap between the filled band and the conduction band.

Explain why this relation between μ and T holds. For simplicity assume that the effective mass of the electron in the conduction band, m^*, is given by $m^* = 0.4\,m$.

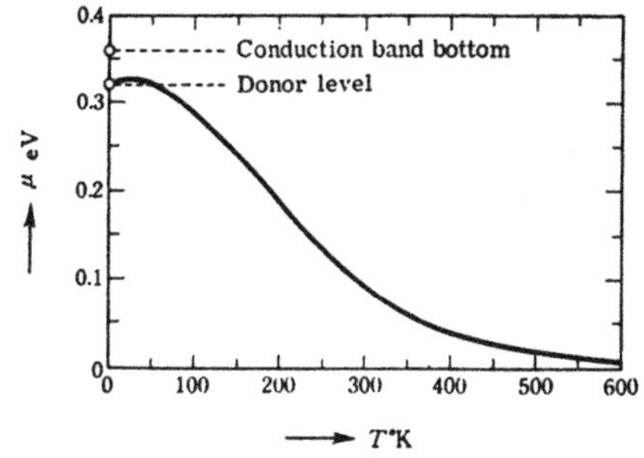

21. If one considers electron spin, an electron in a donor level may have two states. The Coulomb interaction between electrons, however, prevents a donor level from accommodating two electrons at the same time. Find the spin magnetic susceptibility of the system of electrons in the donor levels under the above assumptions.

22. Show that the internal energy of an ideal Bose gas is given by

$$E = \tfrac{3}{2}kTV\left(\frac{2\pi mkT}{h^2}\right)^{\frac{3}{2}} \sum_{l=1}^{\infty} \frac{e^{l\mu/kT}}{l^{\frac{5}{2}}}$$

when the degeneracy is weak. Here μ is the chemical potential. Find similar expansion formulae for the equation of state, the entropy and the Helmholtz free energy.

C

23. In the treatment of high energy fermions of spin $\frac{1}{2}$, one must consider relativistic effects. The energy ε is given by

$$\varepsilon = c\sqrt{p^2 + (mc)^2}$$

where $\boldsymbol{p}$ is the momentum, m is the rest mass and c is the velocity of light in a vacuum. Show that the average values of the total number N, the total energy and the pressure are given by the following formulae:

$$N = 8\pi \frac{m^3c^3}{h^3} V \int_0^\infty \frac{\sinh^2\theta \cosh\theta \, d\theta}{e^{-\beta\mu+\beta mc^2 \cosh\theta}+1},$$

$$E = 8\pi \frac{m^4c^5}{h^3} V \int_0^\infty \frac{\sinh^2\theta \cosh^2\theta \, d\theta}{e^{-\beta\mu+\beta mc^2 \cosh\theta}+1},$$

$$p = \frac{8\pi}{3} \frac{m^4c^5}{h^3} \int_0^\infty \frac{\sinh^4\theta \, d\theta}{e^{-\beta\mu+\beta mc^2 \cosh\theta}+1},$$

where the new variable θ is introduced by the relation $|\boldsymbol{p}| = mc \sinh\theta$, $\beta = 1/kT$ and μ is the chemical potential which includes the rest energy mc^2. Estimate the values of the above quantities at 0° K.

24. Assume that the relativistic ideal Fermi gas introduced in the previous problem is completely degenerate. Derive the non-relativistic relation

$$E \fallingdotseq E_0 + K = N\left(mc^2 + \frac{3}{5}\frac{|\boldsymbol{p}_0|^2}{2m}\right), \qquad pV \fallingdotseq \tfrac{2}{3}K$$

for the case

$$\frac{h}{mc}\left(\frac{3n}{8\pi}\right)^{\frac{1}{3}} \ll 1,$$

and also the extreme-relativistic relation

$$E \fallingdotseq K = N\tfrac{3}{8}|\boldsymbol{p}_0|c, \qquad pV \fallingdotseq \tfrac{1}{3}K$$

for the case

$$\frac{h}{mc}\left(\frac{3n}{8\pi}\right)^{\frac{1}{3}} \gg 1.$$

Here $n = N/V$ is the number density of particles and $\boldsymbol{p}_0$ is the momentum (Fermi momentum) at which the energy becomes equal to the chemical potential at 0° K.

25. Find the chemical potential and the heat capacity of the extreme relativistic ideal Fermi gas (spin $\frac{1}{2}$) which is highly degenerate.

26. Consider an ideal Bose gas composed of particles which have internal degrees of freedom. It is assumed for simplicity that only the first excited level ε_1 of those internal levels has to be taken into account besides the ground state level $\varepsilon_0 = 0$. Determine the Bose-Einstein condensation temperature of this gas as a function of the energy ε_1.

27. Find the first few terms in the virial expansion of an ideal quantum gas when the degeneracy is weak.

28. Derive the density matrix of an ideal Fermi gas in the q-representation. Neglect the spin.

29. Find the magnetic susceptibility due to the orbital motion of the electrons in problem 17, under the assumptions of strong degeneracy $kT \ll \mu$ and of weak field $\mu_B H \ll kT$. [*Hint:* Use Euler's summation formula

$$\sum_{l=a}^{b-1} f(l+\tfrac{1}{2}) \doteqdot \int_a^b f(x)\,\mathrm{d}x - \tfrac{1}{24}\{f'(b) - f'(a)\},$$
$$(f(x+\tfrac{1}{2}) - f(x-\tfrac{1}{2}) - f'(x) \ll f(x)).]$$

30. In the magnetic susceptibility of the electron system treated in problem 29, oscillatory terms appear as a function of H when the magnetic field becomes fairly strong ($kT < \mu_B H \ll \mu$). Derive the expression for these oscillatory terms (de Haas-van Alphen effect). (*Hint:* As shown in (2.6), the density of states $D(\varepsilon)$ can be determined by the inverse Laplace transformation

$$D(\varepsilon) = \frac{1}{2\pi \mathrm{i}} \int_{c-\mathrm{i}\infty}^{c+\mathrm{i}\infty} \mathrm{e}^{\beta\varepsilon} Z(\beta)\,\mathrm{d}\beta,$$

where $Z(\beta)$ is the partition function of one electron in Boltzmann statistics, $\beta = 1/kT$ and $c\ (> 0)$ is the convergence abscissa of $Z(\beta)$. It is assumed that ε is positive. Use the result of problem 17.)

Solutions

1. (i) At 0° K, the energy levels are occupied by electrons up to $\varepsilon = \mu_0$. Hence one has

$$D\mu_0 = N, \quad \text{thus} \quad \mu_0 = N/D. \tag{1}$$

(ii) The chemical potential μ is determined by the relation

$$N = D\int_0^\infty \frac{d\varepsilon}{e^{\beta(\varepsilon-\mu)} + 1}. \tag{2}$$

The condition which guarantees that there is no degeneracy is

$$e^{-\beta\mu} \gg 1. \tag{3}$$

When this is satisfied, one has from equation (2)

$$\frac{N}{D} = \int_0^\infty e^{-\beta(\varepsilon-\mu)}\,d\varepsilon = \frac{e^{\beta\mu}}{\beta}.$$

Now condition (3) becomes equivalent to

$$N/DkT \ll 1. \tag{4}$$

This means that the total number N of electrons is very small compared to the number of electrons that can be accommodated in the energy range of width kT, i.e.

$$N \ll DkT = \text{the number of states in the energy interval } kT \tag{5}$$

is the required condition.

(iii) For $\beta\mu \gg 1$, one has from equation (2)

$$N = D\left[\int_0^\mu d\varepsilon - \int_0^\mu \left\{1 - \frac{1}{e^{\beta(\varepsilon-\mu)} + 1}\right\} d\varepsilon + \int_\mu^\infty \frac{d\varepsilon}{e^{\beta(\varepsilon-\mu)} + 1}\right]$$

$$= D\left\{\mu - \int_0^\mu \frac{d\varepsilon}{e^{-\beta(\varepsilon-\mu)} + 1} + \int_\mu^\infty \frac{d\varepsilon}{e^{\beta(\varepsilon-\mu)} + 1}\right\}$$

$$\doteqdot D\left(\mu - \int_0^\infty \frac{dy}{e^{\beta y} + 1} + \int_0^\infty \frac{dy}{e^{\beta y} + 1}\right) = D\mu \tag{6}$$

and similarly the internal energy E is given by

$$E = \int_0^\infty \frac{\varepsilon D \,\mathrm{d}\varepsilon}{\mathrm{e}^{\beta(\varepsilon-\mu)}+1}$$

$$= D\left\{\int_0^\mu \varepsilon \,\mathrm{d}\varepsilon - \int_0^\mu \frac{\varepsilon \,\mathrm{d}\varepsilon}{\mathrm{e}^{-\beta(\varepsilon-\mu)}+1} - \int_\mu^\infty \frac{\varepsilon \,\mathrm{d}\varepsilon}{\mathrm{e}^{\beta(\varepsilon-\mu)}+1}\right\}$$

$$\fallingdotseq D\left\{\tfrac{1}{2}\mu^2 + 2(kT)^2 \int_0^\infty \frac{x \,\mathrm{d}x}{\mathrm{e}^x+1}\right\}$$

$$= \tfrac{1}{2}D\mu^2 + \tfrac{1}{6}\pi^2 D(kT)^2 . \tag{7}$$

The first term of (7) does not depend on the temperature as can be seen from equation (6). Hence one gets

$$\frac{\mathrm{d}E}{\mathrm{d}T} \fallingdotseq \tfrac{1}{3}\pi^2 Dk^2 T \tag{8}$$

at low temperatures.

2. $D(\varepsilon)f(\varepsilon)$ is shown by the broken line in the figure. If one adopts the approximation shown in this problem, $D(\varepsilon)f(\varepsilon)$ takes the form represented by the heavy line in the figure. The difference ΔE between the internal energy at $T°$ K

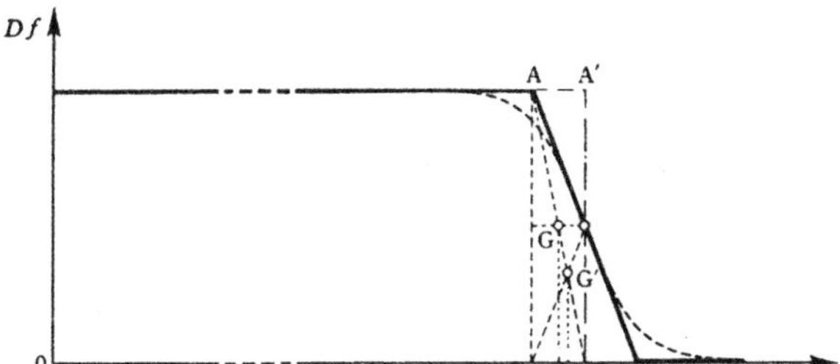

and the one at 0° K corresponds to the difference between the coordinates of the centers of gravity G′ of the triangle ABC and G of the rectangle AA′B′B on the ε-axis. Putting BC $= 2\alpha kT$, where α means a constant of the order of unity, one finds the coordinates of G and G′ as

$$\text{G:} \quad \text{OB} + \tfrac{1}{2}\text{BB}' = \text{OB} + \tfrac{1}{2}\alpha kT,$$

$$\text{G}': \quad \text{OB} + \tfrac{1}{3}\text{BC} = \text{OB} + \tfrac{2}{3}\alpha kT.$$

Hence one obtains

$$\Delta E = \alpha kTD \times \tfrac{1}{6}\alpha kT = \tfrac{1}{6}\alpha^2 D(kT)^2\,, \tag{1}$$

where αkTD means the number of electrons contained in the region AA′B′B. If one chooses AC to be the tangent to the Fermi distribution curve, one has $2\alpha kT = 1/|f'(\mu)| = 4kT$. Hence one obtains from equation (1)

$$C = \frac{\mathrm{d}\Delta E}{\mathrm{d}T} = \tfrac{4}{3}Dk^2T \tag{2}$$

(the exact value of the coefficient in (2) is $\frac{1}{3}\pi^2$ as given in problem 1).

3. From (4.5), one has

$$\left(\frac{\partial f}{\partial T}\right)_\mu = \frac{\varepsilon-\mu}{kT^2}\frac{e^{\beta(\varepsilon-\mu)}}{[e^{\beta(\varepsilon-\mu)}+1]^2} = -\frac{\varepsilon-\mu}{T}f'(\varepsilon)\,, \qquad \left(\frac{\partial f}{\partial \mu}\right)_T = -f'(\varepsilon)\,.$$

Proceeding in a similar way as in (4.7), one obtains

$$\left(\frac{\partial I}{\partial T}\right)_\mu = -\frac{1}{T}\int_0^\infty (\varepsilon-\mu)g(\varepsilon)f'(\varepsilon)\,\mathrm{d}\varepsilon = \tfrac{1}{6}\pi^2k^2T\left[\frac{\mathrm{d}^2}{\mathrm{d}\varepsilon^2}(\varepsilon-\mu)g\right]_{\varepsilon=\mu} + \cdots$$

$$= \tfrac{1}{3}\pi^2k^2Tg'(\mu) + \mathrm{O}(T^3)\,, \tag{1}$$

and

$$\left(\frac{\partial I}{\partial \mu}\right)_T = -\int_0^\infty g(\varepsilon)f'(\varepsilon)\,\mathrm{d}\varepsilon = g(\mu) + \mathrm{O}(T^2)\,. \tag{2}$$

For $g = D$, one has $I = N$. From the relation

$$\mathrm{d}N = \left(\frac{\partial N}{\partial \mu}\right)_T\mathrm{d}\mu + \left(\frac{\partial N}{\partial T}\right)_\mu\mathrm{d}T = 0$$

one finds

$$\left(\frac{\partial \mu}{\partial T}\right)_N = -\tfrac{1}{3}\pi^2k^2T\frac{D'(\mu)}{D(\mu)} + \mathrm{O}(T^3)\,. \tag{3}$$

Considering relation (3), one finds that in the functions g' and g in equations (1) and (2), μ can be replaced by its value μ_0 at 0° K. One also gets

$$\left(\frac{\partial I}{\partial T}\right)_N = \left(\frac{\partial I}{\partial \mu}\right)_T\left(\frac{\partial \mu}{\partial T}\right)_N + \left(\frac{\partial I}{\partial T}\right)_\mu$$

$$= -g(\mu)\tfrac{1}{3}\pi^2k^2T\frac{D'(\mu)}{D(\mu)} + \tfrac{1}{3}\pi^2k^2Tg'(\mu) + \mathrm{O}(T^3)$$

$$= \tfrac{1}{3}\pi^2k^2T\psi'(\mu_0)D(\mu_0) + \mathrm{O}(T^3)\,. \tag{4}$$

4. The energy of the electron is given by

$$\varepsilon = \frac{p^2}{2m} \pm \mu_B H \tag{1}$$

where the $\pm$ signs correspond to the two directions of the spin magnetic moment. At 0° K the electrons under consideration occupy energy levels up to the Fermi potential μ_0. Hence the kinetic energy $p^2/2m$ of the electron with + spin ranges from 0 to $\mu_0 - \mu_B H$ and that of − spin from 0 to $\mu_0 + \mu_B H$. Their numbers are given by

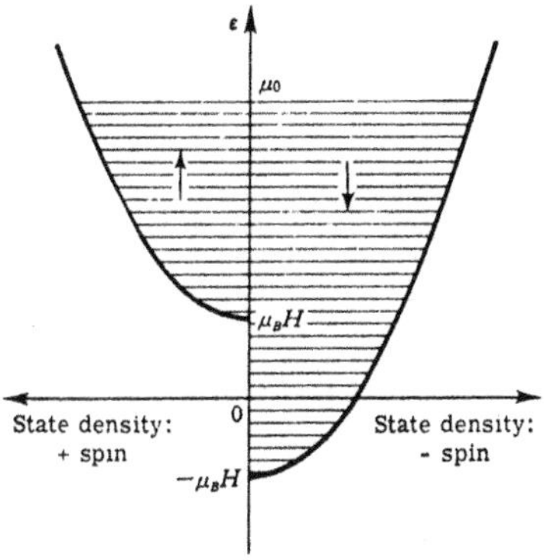

$$N_+ = \frac{4\pi V}{3h^3} \cdot p_+^3 \quad \left(\frac{1}{2m} p_+^2 = \mu_0 - \mu_B H\right), \tag{2}$$

$$N_- = \frac{4\pi V}{3h^3} \cdot p_-^3 \quad \left(\frac{1}{2m} p_-^2 = \mu_0 + \mu_B H\right). \tag{3}$$

The total magnetic moment is given by

$$M = -\mu_B(N_+ - N_-) = -\frac{4\pi V}{3h^3} \mu_B(p_+^3 - p_-^3)$$

$$= \mu_B \frac{4\pi V}{3h^3} [\{2m(\mu_0 + \mu_B H)\}^{\frac{3}{2}} - \{2m(\mu_0 - \mu_B H)\}^{\frac{3}{2}}].$$

Assuming $\mu_0 \gg \mu_B H$, one has

$$M = 3H\mu_B^2 \frac{4\pi V}{3h^3} (2m\mu_0)^{\frac{3}{2}}/\mu_0 + \cdots = \frac{3}{2} \frac{\mu_B^2 N}{\mu_0} H + \cdots, \tag{4}$$

where $N = 2 \cdot 4\pi V (2m\mu_0)^{\frac{3}{2}}/3h^2$ has been used. Hence one obtains finally

$$\chi = \tfrac{3}{2} n\mu_B^2/\mu_0 = \tfrac{3}{2} n\mu_B^2/kT_0.$$

5. As in the previous problem, the magnetic moment per unit volume is given by

$$M = \mu_B \left\{ \int \frac{D(\varepsilon)\,d\varepsilon}{e^{\beta(\varepsilon - \mu_B H - \mu)} + 1} - \int \frac{D(\varepsilon)\,d\varepsilon}{e^{\beta(\varepsilon + \mu_B H - \mu)} + 1} \right\}$$

$$= \mu_B \int \{D(\varepsilon + \mu_B H) - D(\varepsilon - \mu_B H)\} f(\varepsilon)\,d\varepsilon. \tag{1}$$

Using the Taylor expansion of D for small $\mu_B H$, one obtains to the first order in H

$$\chi = \frac{M}{H} = 2\mu_B^2 \int_0^\infty D'(\varepsilon) f(\varepsilon)\, d\varepsilon . \tag{2}$$

(i) At 0° K, $f(\varepsilon) = 0$ for $\varepsilon > \mu_0$. Then one has

$$\chi_0 = 2\mu_B^2 \int_0^{\mu_0} D'(\varepsilon)\, d\varepsilon = 2\mu_B^2 D(\mu_0) . \tag{3}$$

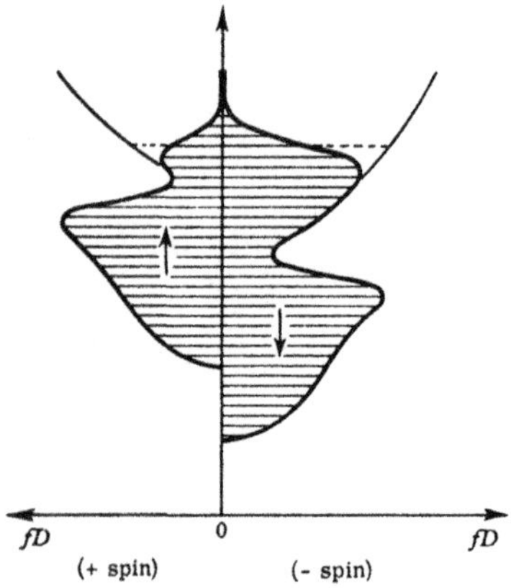

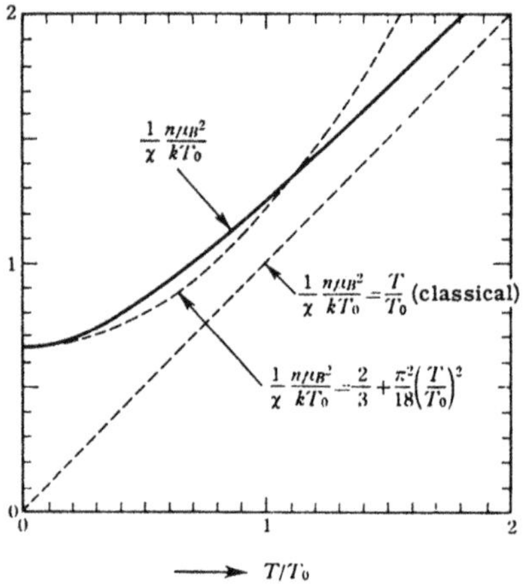

(ii) In the case of a high degree of degeneracy, one can use the results in problem 3. Putting $\psi = 2\mu_B^2 D'/D$, one obtains

$$\left(\frac{\partial \chi}{\partial T}\right)_N = \frac{2\pi^2}{3} \mu_B^2 k^2 T D(\mu_0) \cdot \left(\frac{D'}{D}\right)' = \frac{2\pi^2 k^2 T}{3} \mu_B^2 D(\mu_0) \frac{d^2 \log D(\mu_0)}{d\mu_0^2} \tag{4}$$

and this, together with equation (3), leads to

$$\chi = 2\mu_B^2 D(\mu_0) \left\{ 1 + \tfrac{1}{6}\pi^2 k^2 T^2 \frac{d^2 \log D(\mu_0)}{d\mu_0^2} + \cdots \right\} . \tag{5}$$

(iii) In the non-degenerate case, one can put $f(\varepsilon) = e^{-\beta(\varepsilon - \mu)}$. Then one has

$$\int_0^\infty D'(\varepsilon) e^{-\beta(\varepsilon-\mu)}\, d\varepsilon = \left[D e^{-\beta(\varepsilon-\mu)} \right]_0^\infty + \beta \int_0^\infty D(\varepsilon) e^{-\beta(\varepsilon-\mu)}\, d\varepsilon = \tfrac{1}{2}\beta n ,$$

thus

$$\chi = n\mu_B^2 / kT , \tag{6}$$

where n is the number of electrons per unit volume. In the case of free electrons, one has $D(\varepsilon) = C\varepsilon^{\frac{1}{2}}$ and then

$$\int_0^{\mu_0} D(\varepsilon)\,\mathrm{d}\varepsilon = \tfrac{2}{3} C\mu_0^{\frac{3}{2}} = \tfrac{1}{2} n,$$

$$\frac{\mathrm{d}^2 \log D(\mu_0)}{\mathrm{d}\mu_0^2} = \frac{1}{2}\frac{\mathrm{d}^2 \log \mu_0}{\mathrm{d}\mu_0^2} = -\frac{1}{2\mu_0^2}.$$

Hence equation (5) becomes

$$\chi = \frac{3}{2}\frac{n\mu_B^2}{\mu_0}\left\{1 - \frac{\pi^2}{12}\frac{k^2T^2}{\mu_0^2} + \cdots\right\}$$

$$= \frac{3}{2}\frac{n\mu_B^2}{kT_0}\left\{1 - \frac{\pi^2}{12}\left(\frac{T}{T_0}\right)^2 + \cdots\right\}, \qquad (7)$$

where T_0 is the degeneracy temperature defined by $T_0 = \mu_0/k$.

6. The system of free electrons considered as a classical ideal gas has a molar specific heat at constant volume (C_V) equal to $\frac{3}{2}R$ as a consequence of the law of equipartition of energy applied to its translational degrees of freedom. On the other hand, lattice vibrations of the metalic crystal contribute $3R$ to C_V, because they have $3N_0 - 6 \simeq 3N_0$ (N_0 = Avogadro's number) degrees of freedom per mole and can be treated as a system of $3N_0$ harmonic oscillators which obey classical statistics. Adding up these two contributions, one can expect $4.5R$ as the atomic heat of the metal.

If the system of electrons is considered as a degenerate Fermi gas, the electrons occupy the conduction band of the metal up to a level of the order of the Fermi potential $\mu_0 = kT_0$ ($\gg kT$). Thermal energy which is of the order of kT cannot excite the electrons in the lower levels to empty upper levels, because of the Pauli principle. Only the electrons near the Fermi potential can absorb the thermal energy kT and jump up freely to empty levels, because the degenerate Fermi distribution function falls off sharply from 1 to 0 within an energy range of width kT around the Fermi potential. Consequently the number of electrons which can excite thermally is of the order $N \times T/T_0$, and their contribution to the atomic heat can be expected to be of the order of $\frac{3}{2} R \times T/T_0$ and will be negligible when $T \ll T_0$. Assuming the density of states (4.9), one can estimate μ_0 as

$$\mu_0 = \frac{h^2}{2m}\left(\frac{3N}{8\pi V}\right)^{\frac{2}{3}} = \frac{h^2}{8m}\left(\frac{3N_0}{\pi V_A}\right)^{\frac{2}{3}} = 4.166 \times 10^{-11} V_A^{-\frac{2}{3}}\,\mathrm{erg} \qquad (1)$$

where V_A refers to the atomic volume. For Na, one puts $V_A = 24 \text{ cm}^2$ and gets $\mu_0 = 5.0 \times 10^{-12}$ erg $(= 3.0 \text{ eV})$, $T_0 = \mu_0/k = 3.6 \times 10^{4\circ}$ K. Accordingly the relation $T \ll T_0$ is well satisfied at ordinary temperatures.

NOTE: The effective mass m^* of the conduction electron of Na is 0.98 m.

7. In the non-degenerate case, the spin paramagnetic susceptibility is given by (6) of the solution to problem 5,

$$\chi_\infty = n\mu_B^2/kT, \tag{1}$$

and in the strongly degenerate case, it is given by (2) of that solution:

$$\chi_0 = \tfrac{3}{2}n\mu_B^2/kT_0 . \tag{2}$$

Now their ratio

$$\frac{\chi_0}{\chi_\infty} = \frac{3}{2}\frac{T}{T_0} \tag{3}$$

can be regarded as representing the ratio of the number of electrons which can receive thermal energy to the total number of electrons which compose the highly degenerate Fermi gas. In other words, if one assumes that $n' = n \times \frac{3}{2} T/T_0$ electrons can be treated as classical (no degeneracy) and have a magnetic susceptibility given by (1), one obtains the results given by (2). The origin of this behavior is the *Pauli principle*. An electron of energy ε cannot absorb a quantum of thermal excitation if the state of the energy $\varepsilon + kT$ is already filled. Electrons which occupy the energy region of width kT around the Fermi potential can be freely excited, however.

NOTE: Equation (1) is a special case of the Langevin-Debye formula

$$\chi_s = n\mu_{\text{eff}}^2/3kT, \qquad s = \tfrac{1}{2}, \qquad g = 2 .$$

(Cf. (5) in the solution to problem 2.9.)

8. The formulae for the ideal Fermi gas can be used here.

$$C_V = N\pi^2k^2T/2\mu_0 \qquad \text{(Example 2)}$$

$$\chi_s = 3\,N\mu_B^2/2\mu_0 \qquad \text{(Problem 4)}$$

$$\mu_0 = \frac{h^2}{2m}\{3N/8\pi V\}^{\frac{2}{3}} \quad \text{(Example 1)},$$

where N is the number of electrons per unit mass. In terms of the density ρ, the molecular weight M, the number of valence electrons z and Avogadro's number N_0, one has $N = zN_0/M$ and $N/V = \rho N$, so that

$$c_V = \frac{4}{3^{\frac{2}{3}}}\pi^{\frac{8}{3}}\frac{k^2 m N_0^{\frac{1}{3}}}{h^2}\rho^{-\frac{2}{3}}\frac{z^{\frac{1}{3}}}{M^{\frac{1}{3}}}T \doteqdot 0.3248 \times 10^{-4}\frac{z^{\frac{1}{3}}}{\rho^{\frac{2}{3}}M^{\frac{1}{3}}}T \qquad \text{cal/g·deg},$$

$$\chi_s = 4 \cdot 3^{\frac{1}{3}}\pi^{\frac{2}{3}}\frac{\mu_B^2 m N_0^{\frac{1}{3}}}{h^2}\rho^{-\frac{2}{3}}\frac{z^{\frac{1}{3}}}{M^{\frac{1}{3}}} \doteqdot 1.864^6 \times 10^{-6}\frac{z^{\frac{1}{3}}}{\rho^{\frac{2}{3}}M^{\frac{1}{3}}} \qquad \text{c.g.s.e.m.u/g.}$$

If one puts $z = 1$, $\rho = 0.534\ \text{g cm}^{-3}$, $M = 7$ for Li, and $z = 1$, $\rho = 0.97\ \text{g cm}^{-3}$, $M = 23$ for Na, one obtains

Li : $c_V = 0.2579 \times 10^{-4}\, T$ cal/g·deg, $\quad \chi_s = 1.4808 \times 10^{-6}$ c.g.s.e.m.u/g,
Na : $c_V = 0.1166 \times 10^{-4}\, T$ cal/g·deg, $\quad \chi_s = 0.6691 \times 10^{-6}$ c.g.s.e.m.u/g.

9. Using the grand partition function Ξ, one has

$$pV = kT\log\Xi = kT\int_0^\infty D(\varepsilon)\,d\varepsilon\log\{1 + e^{\beta(\mu-\varepsilon)}\}, \tag{1}$$

where $D(\varepsilon)$ is given by

$$D(\varepsilon) = C\varepsilon^{\frac{1}{2}}, \qquad \tfrac{2}{3}C\mu_0^{\frac{3}{2}} = N, \qquad \mu_0 = \frac{h^2}{2m}\left(\frac{3N}{8\pi V}\right)^{\frac{2}{3}} \tag{2}$$

for the free electron. Upon integrating by parts, one obtains

$$pV = (\tfrac{3}{2}NkT/\mu_0^{\frac{3}{2}})\int_0^\infty \varepsilon^{\frac{1}{2}}\,d\varepsilon\log\{1 + e^{\beta(\mu-\varepsilon)}\}$$

$$= (\tfrac{3}{2}NkT/\mu_0^{\frac{3}{2}})\left\{\left[\tfrac{2}{3}\varepsilon^{\frac{3}{2}}\log\{1 + e^{\beta(\mu-\varepsilon)}\}\right]_0^\infty + \tfrac{2}{3}\beta\int_0^\infty \varepsilon^{\frac{3}{2}}\,d\varepsilon/\{e^{\beta(\varepsilon-\mu)} + 1\}\right\}$$

$$= (N/\mu_0^{\frac{3}{2}})\int_0^\infty \varepsilon^{\frac{3}{2}}\,d\varepsilon f(\varepsilon) = \tfrac{2}{3}\int_0^\infty \varepsilon D(\varepsilon)f(\varepsilon)\,d\varepsilon = \tfrac{2}{3}E. \tag{3}$$

In the strongly degenerate case, E is given by (10) of example 1. Consequently one has

$$p = \frac{2}{5}\frac{\mu_0}{v} + \frac{\pi^2}{6}\frac{k^2T^3}{\mu_0 v} + \cdots, \tag{4}$$

and the compressibility κ is given by

$$\kappa = -\frac{1}{V}\left(\frac{\partial V}{\partial p}\right)_T = -\frac{1}{v(\partial p/\partial v)_T}$$
$$= \left\{\frac{2}{3}\frac{\mu_0}{v} + \frac{\pi^2 k^2 T^2}{18\ \mu_0 v} + \cdots\right\}^{-1} = \frac{3}{2}\frac{v}{\mu_0}\left\{1 - \frac{\pi^2}{18}\left(\frac{kT}{\mu_0}\right)^2 + \cdots\right\}, \tag{5}$$

considering the volume dependence of μ_0 ($\propto V^{-\frac{2}{3}}$). At 0° K, one has

$$\kappa_0 = \frac{3}{2}\frac{v}{\mu_0} = \frac{3}{2}\frac{V_A}{N_0}\bigg/\frac{h^2}{2m}\left(\frac{3N_0}{8\pi V_A}\right)^{\frac{2}{3}} = 0{,}597 \cdot V_A^{\frac{5}{3}} \times 10^{-13}\,\mathrm{cm}^2/\mathrm{dyne}$$

and, for $V_A = 23/0.97\ \mathrm{cm}^3$, this gives the value $\kappa_0 = 11.7 \times 10^{-12}\ \mathrm{cm}^2/\mathrm{dyne}$.

10. To escape from the metal, an electron must strike the surface from inside with a kinetic energy larger than w in the direction normal to the surface. Assuming that any electron which satisfies the above condition will certainly escape from the surface, one finds for the number of electrons leaving the surface per unit time and per unit aera:

$$\mathfrak{N} = \int\limits_{p_z > \sqrt{2mw}} \mathrm{d}p_z \cdot \frac{p_z}{m} \int\limits_{-\infty}^{\infty} \mathrm{d}p_x \int\limits_{-\infty}^{\infty} \mathrm{d}p_y \frac{2}{h^3} \frac{1}{\mathrm{e}^{(\varepsilon-\mu)/kT} + 1}, \qquad \varepsilon = \frac{p_x^2 + p_y^2 + p_z^2}{2m}, \tag{1}$$

where $p_z/m\,(= v_z)$ is the velocity in the z-direction, and $(2/h^3)\,(\mathrm{e}^{\beta(\varepsilon-\mu)} + 1)^{-1}$ is the number density of electrons which have an energy ε. Putting $p_x^2 + p_y^2 = p'^2$ one has

$$\mathfrak{N} = \int\limits_{p_z > \sqrt{2mw}} \mathrm{d}p_z \cdot \frac{p_z}{m} \int\limits_0^{\infty} 2\pi p'\,\mathrm{d}p' \frac{2}{h^3} \frac{1}{\exp\left\{\left(\dfrac{p_z^2 + p'^2}{2m} - \mu\right)\Big/ kT\right\} + 1}$$
$$= \frac{4\pi mkT}{h^3} \int\limits_{p_z > \sqrt{2mw}} \frac{p_z}{m}\mathrm{d}p_z \cdot \log\left[1 + \exp\left\{\left(\mu - \frac{p_z^2}{2m}\right)\Big/ kT\right\}\right]$$
$$= \frac{4\pi mkT}{h^3} \int\limits_w^{\infty} \mathrm{d}\varepsilon_z \log\left[1 + \mathrm{e}^{(\mu-\varepsilon_z)/kT}\right]. \tag{2}$$

Here one can assume $\mu(T) = \mu_0$ and $w - \mu_0 = \phi \gg kT$, and hence one can expand the integrand in terms of $e^{\beta(\mu-\varepsilon_z)} \ll 1$. Retaining only the first term, one has

$$\mathfrak{N} = \frac{4\pi mkT}{h^3}\int_w^\infty d\varepsilon\, e^{-(\varepsilon-\mu_0)/kT}$$

$$= \frac{4\pi m(kT)^2}{h^3} e^{(\mu_0-w)/kT} = \frac{4\pi m(kT)^2}{h^3} e^{-\phi/kT}. \tag{3}$$

The current $J = e\mathfrak{N}$ is given by

$$J = \frac{4\pi mek^2}{h^3} T^2 e^{-\phi/kT}. \tag{4}$$

NOTE: Strictly speaking, this phenomenon is not a process exhibiting thermal equilibrium. We have assumed, however, that the number of electrons which strike the surface of the metal per unit time from inside is equal to the value at thermal equilibrium, notwithstanding the fact that electrons are flowing out steadily through the surface (saturated thermionic current). This number can also be given by the number $\mathfrak{N}'$ of electrons which strike the surface of the metal per unit time from outside the metal and that outside. The density of the electron gas outside the metal can be considered to be very small. Hence we can assume Boltzmann statistics so that the pressure p is given by

$$\frac{p}{kT} = 2\left(\frac{2\pi mkT}{h^2}\right)^{\frac{3}{2}} \cdot e^{-\phi/kT} \tag{5}$$

where 2 is the spin degeneracy. Here the energy zero is taken at the zero of the kinetic energy, and hence the chemical potential has become equal to $-\phi$. The relation between $\mathfrak{N}'$ and p is given by (2) in the solution to problem 6.4:

$$\mathfrak{N}' = \frac{p}{(2\pi mkT)^{\frac{1}{2}}}. \tag{6}$$

Combining these two equations, one obtains the same expression for $\mathfrak{N}'$ as in equation (3).

11. The thermal properties of the free electron gas composed of $2N - N'$ electrons which occupy the energy band under consideration are determined by the Helmholtz free energy given by

$$F = (2N - N')\mu - kT\sum_i \log\{1 + e^{\beta(\mu-\varepsilon_i)}\}, \tag{1}$$

$$2N - N' = kT\frac{\partial}{\partial\mu}\left[\sum_i \log\{1 + e^{\beta(\mu-\varepsilon_i)}\}\right]. \tag{2}$$

One can transform these two equations quite formally as follows:

$$F = (2N - N')\mu - kT\sum_i \log\{1 + e^{\beta(\varepsilon_i - \mu)}\} - \sum_i (\mu - \varepsilon_i)$$
$$= \sum_i \varepsilon_i - N'\mu - kT\sum_i \log[1 + e^{\beta\{-\mu-(-\varepsilon_i)\}}], \tag{3}$$

and

$$2N - N' = \sum_i \frac{1}{1 + e^{\beta(\varepsilon_i - \mu)}} = \sum_i \left\{1 - \frac{1}{1 + e^{\beta(\mu - \varepsilon_i)}}\right\} = 2N - \sum_i \frac{1}{1 + e^{\beta(\mu - \varepsilon_i)}}, \tag{4}$$

or

$$N' = \sum_i \frac{1}{1 + e^{\beta\{-\varepsilon_i - (-\mu)\}}}. \tag{5}$$

The term $\sum_i \varepsilon_i$ in equation (3) is a constant which has nothing to do with the thermal properties and which can be removed by the adjustment of the energy zero. As one can see easily, (3) and (5) are nothing but the equations which determine the Helmholtz free energy of the system of electrons which have an energy $-\varepsilon_i$ while the chemical potential is $-\mu$.

NOTE: This can be regarded as an ideal Fermi gas composed of positive holes (see (4.4)).

12. From equation (8) of example 4, one has

$$\frac{n^2}{N_D - n} = \tfrac{1}{2} N_c e^{-E_D/kT} \tag{1}$$

under the condition of electrical neutrality $n = N_D - n_D$. From this one can determine n as a function of the temperature T and the donor density N_D, as follows:

$$\frac{n}{N_D} = \frac{N_c}{4N_D} e^{-E_D/kT}\left\{-1 + \sqrt{1 + \frac{8N_D}{N_c} e^{E_D/kT}}\right\}. \tag{2}$$

Substituting this into (7) of example 4, one has

$$\mu = kT \log\frac{n}{N_c} = -E_D + kT\log\left[\tfrac{1}{4}\left\{-1 + \sqrt{1 + \frac{8N_D}{N_c} e^{E_D/kT}}\right\}\right] \tag{3}$$
$$(-\mu \gg kT).$$

At sufficiently low temperatures the condition

$$e^{E_D/kT} \gg N_c/(8N_D) \tag{4}$$

is always fulfilled. Consequently one obtains

$$n = (\tfrac{1}{2} N_D N_c)^{\frac{1}{2}} e^{-E_D/2kT} = N_D \left(\frac{N_c}{2N_D}\right)^{\frac{1}{2}} e^{-E_D/2kT} \ll N_D, \tag{5}$$

$$\mu = -\tfrac{1}{2} E_D + \tfrac{1}{2} kT \log \frac{N_D}{2N_c}, \tag{6}$$

where the factors 1 in the square root and -1 in the bracket in (2) and (3) have been neglected. At low temperatures, the second term in (6) is small and μ is nearly equal to $-\frac{1}{2} E_D$. That is, the Fermi potential is at the middle of the gap between the bottom of the conduction band and the donor level, corresponding to a very slight ionization of the donors. At sufficiently high temperatures, N becomes large and $e^{E_D/kT}$ approaches unity, hence

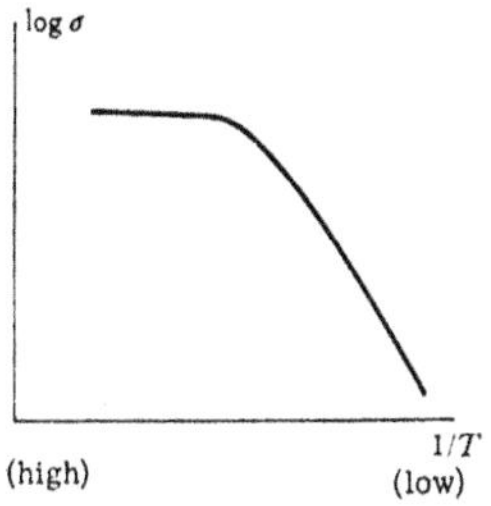

$$(8\, N_D/N_c) e^{E_D/kT} \ll 1 \tag{7}$$

is satisfied. In this case, one has

$$n = N_D \left\{ 1 - \frac{2N_D}{N_c} e^{E_D/kT} + \cdots \right\}, \tag{8}$$

$$\mu = kT \left\{ \log \frac{N_D}{N_c} - \frac{2N_D}{N_c} e^{E_D/kT} + \cdots \right\}, \tag{9}$$

that is, the Fermi potential becomes lower than the donor level and the donors are ionized almost completely. Using (5) or (8), one finds the following for the conductivity:

$$\text{low temperatures} \qquad \sigma = \frac{e^2\tau}{m^*}\left(\frac{2\pi m^* kT}{h^2}\right)^{\frac{3}{4}} N_D^{\frac{1}{2}} e^{-E_D/2kT}, \tag{10}$$

$$\text{high temperatures} \qquad \sigma = \frac{e^2\tau}{m^*} N_D. \tag{11}$$

The behavior of $\log \sigma$ against $1/T$ is shown in the figure. The temperature dependence of σ given by (10) is mainly determined by the exponential factor, because $T^{\frac{3}{4}}$ varies slowly. Consequently one has

$$\partial \log \sigma \Big/ \partial \left(\frac{1}{T}\right) \sim -\tfrac{1}{2} E_D \tag{12}$$

in the low temperature range. From this the value of E_D can be determined.

NOTE: The relationship between the donor level and the Fermi potential can be found in the following way. Putting $\mu/kT = x$, one rewrites (5) and (7) of example 4 as

$$n/N_c = e^x \tag{13}$$

and

$$\frac{N_D - n_D}{N_c} = \frac{N_D/N_c}{1 + 2\mathrm{e}^{E_D/kT}\,\mathrm{e}^x}. \tag{14}$$

One may determine $\mu\,(= xkT)$ graphically from the intersection of the curves of $\log n/N_c$ and $\log (N_D - n_D)/N_c$ as a function of $x\ (< 0)$. $\log n/N_c$ is linear in x. Strictly speaking, this is not true for $x \sim 0$ or $x > 0$, because in those cases the system is nearly or highly degenerate, and hence relation (13) does not hold. In (14), the unity in the denominator can be neglected for the case $x + E_D/kT \gg -\log 2$: on the other hand, the denominator is nearly

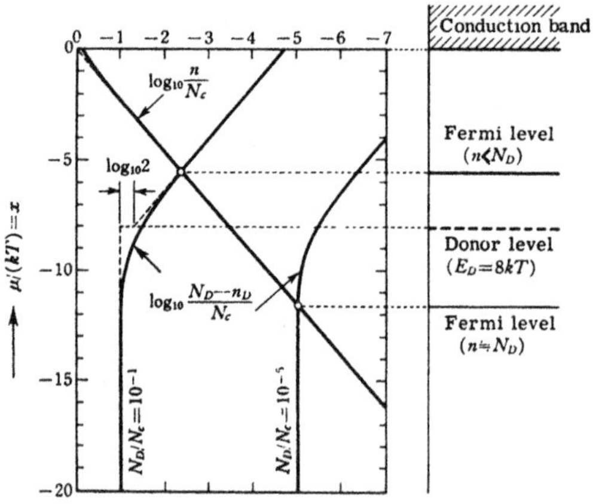

equal to 1 for $x + E_D/kT \ll -\log 2$. Consequently one finds the curve ($\log (N_D - n_D)/N_c$) to have the behavior shown in the figure. This figure gives two curves which correspond to two different values of N_D/N_c.

13. One can proceed in the same way as in example 4. The average number of electrons which occupy the acceptor levels is given by

$$n_A = \frac{N_A}{\frac{1}{2}\mathrm{e}^{(E_A-\mu)/kT} + 1} \tag{1}$$

(see (5) of example 4), taking the origin of the energy at the top of the filled band. On the other hand, the number of electrons in the filled band is given by

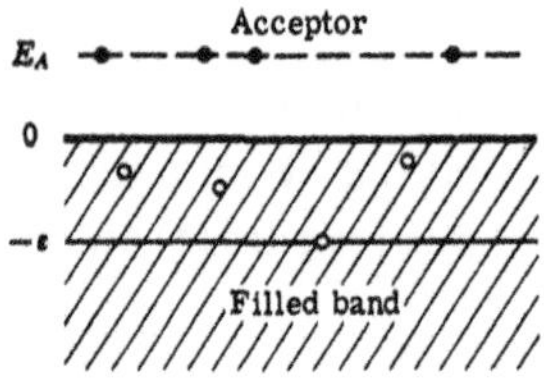

$$n = \int_0^\infty \frac{2D(\varepsilon)\,\mathrm{d}\varepsilon}{\mathrm{e}^{-(\varepsilon+\mu)/kT} + 1}, \tag{2}$$

where $D(\varepsilon)$ is the state density at $-\varepsilon$, not including spin degeneracy. The number of positive holes in the filled band is given by

$$p = \int_0^\infty 2\left\{1 - \frac{1}{e^{-(\varepsilon+\mu)/kT} + 1}\right\} D(\varepsilon)\,d\varepsilon = \int_0^\infty \frac{2D(\varepsilon)\,d\varepsilon}{e^{(\varepsilon+\mu)/kT} + 1} = \int_0^\infty \frac{2D(\varepsilon)\,d\varepsilon}{e^{(\varepsilon-\mu')/kT} + 1}, \quad (3)$$

as can be seen easily. In the last expression of (3) the quantity μ', defined as $\mu' = -\mu$, has been used. This is the Fermi distribution for holes of energy ε and the Fermi potential μ'. The number of holes in the acceptor levels is given by

$$p_A = N_A - n_A = \frac{N_A}{1 + 2e^{-(E_A-\mu)/kT}} = \frac{N_A}{2e^{-(E_A+\mu')/kT} + 1}. \quad (4)$$

Assuming no degeneracy for holes, one gets

$$p = 2\left(\frac{2\pi m_h kT}{h^2}\right)^{\frac{3}{2}} e^{\mu'/kT} \equiv N_c e^{\mu'/kT} \quad (5)$$

from (3), and

$$\frac{p(N_A - p_A)}{p_A} = 2N_c e^{-E_A/kT} \quad (6)$$

from (4). Equation (4) represents the law of mass action for the reaction $A \to A^- + h$ (h = hole).

Introducing the condition of electrical neutrality $p = N_A - p_A$, one obtains

$$p = (2N_c N_A)^{\frac{1}{2}} e^{-E_A/(2kT)} = 2N_A^{\frac{1}{2}} \left(\frac{2\pi m_h kT}{h^2}\right)^{\frac{3}{4}} e^{-E_A/(2kT)},$$

at low temperatures and

$$p \sim N_A$$

at high temperatures.

NOTE: Notice the effect of spin degeneracy on the hole distribution. In the denominator of equation (4), the factor 2 appears instead of $\frac{1}{2}$ as in equation (5) of solution to example 4. A donor level has a spin weight 2 for accommodation of an electron. On the contrary, an acceptor level has no spin degeneracy for the accommodation of the hole. If one removes the restriction of at most one electron per donor or acceptor level and assumes the accommodation of two electrons (+, − spin) in one acceptor level to be possible, one obtains

$$n_A = \frac{2N_A}{e^{(E_A-\mu)/kT} + 1}, \qquad p_A = \frac{2N_A}{e^{-(E_A+\mu')/kT} + 1}$$

instead of (1) and (2). These are completely symmetrical with the donor case.

14. The Helmholtz free energy $F(=N\mu - pV)$ of the photon gas is equal to $-pV$ because the chemical potential μ of the photon gas is equal to zero, as shown in example 6. Here p is the radiation pressure and V is the volume of the radiation field. On the other hand, using the grand partition function Ξ given in (2) of example 6, one has

$$pV = kT \log \Xi = -kT \sum_i \log\{1 - e^{-h\nu_i/kT}\}. \tag{1}$$

Here the summation over i is to be interpreted as that over the wave vectors $\boldsymbol{q}$, and this can be replaced by an integral. Taking into account the weight 2 which comes from the two different directions of polarization of the photon, one obtains

$$pV = -kT \int_0^\infty \log\{1 - e^{-\hbar cq/kT}\} \cdot \frac{2\cdot 4\pi q^2\,dq}{(2\pi)^3} V$$

$$= -\frac{(kT)^4 V}{\pi^2(\hbar c)^3}\int_0^\infty \log(1 - e^{-x})\cdot x^2\,dx = \frac{(kT)^4 V}{3\pi^2(\hbar c)^3}\int_0^\infty \frac{x^3\,dx}{e^x - 1}, \tag{2}$$

$$(x = \hbar cq/kT)$$

after integration by parts. The above integral turns out to have the value $\frac{1}{15}\pi^4$.

Hence one has, finally

$$pV = -F = V\frac{\pi^2(kT)^4}{45(\hbar c)^3} = \frac{4\sigma}{3c} VT^4, \tag{3}$$

where $\sigma = \pi^2 k^4/(60\hbar^3 c^2) = 5.672 \times 10^{-5}$ g sec^{-3} deg^{-4} is the Stefan-Boltzmann constant.

The entropy S is given by

$$S = \frac{16\sigma}{3c} VT^3, \tag{4}$$

and the internal energy $U(= F + TS)$ is given by

$$U = \frac{4\sigma}{c} VT^4 = 3pV. \tag{5}$$

15. The energy levels of free particles (i.e., ideal gas molecules) in a two-dimensional region of area $L_x \times L_y$ are given by

$$\varepsilon(k_x, k_y) = \hbar^2(k_x^2 + k_y^2)/2m,$$

$$k_x = 2\pi n_x/L_x, \qquad k_y = 2\pi n_y/L_y, \qquad n_x, n_y = 0, \pm 1, \cdots.$$

One can use a formula which is established generally in the case of Bose statistics, i.e.

$$N = \sum_i \frac{1}{e^{(\varepsilon_i - \mu)/kT} - 1}, \qquad (1)$$

where N is the total number of particles, μ is the chemical potential and the summation is to be extended over all possible energy levels. Assuming L_x and L_y to be sufficiently large, one can replace this summation by integration in $k_x k_y$ - space:

$$N = \frac{L_x L_y}{(2\pi)^2} \int \frac{dk_x \, dk_y}{\exp\left\{\left(\frac{\hbar^2 k^2}{2m} - \mu\right)\Big/ kT\right\} - 1}, \qquad (k^2 = k_x^2 + k_y^2). \qquad (2)$$

Hence one obtains

$$N = \frac{L_x L_y}{(2\pi)^2} \cdot 2\pi \int_0^\infty \frac{k\, dk}{\exp\left\{\left(\frac{\hbar^2 k^2}{2m} - \mu\right)\Big/ kT\right\} - 1} = 2\pi L_x L_y \frac{m}{h^2} \int_0^\infty \frac{d\varepsilon}{e^{(\varepsilon - \mu)/kT} - 1}, \qquad (3)$$

or

$$N = L_x L_y \frac{2\pi m k T}{h^2} \sum_{l=1}^{\infty} \frac{1}{l} e^{l\mu/kT}. \qquad (4)$$

From this equation, we know that one can always get a value of μ which is not of the order $1/N$. Hence we can conclude that there are no levels which are occupied by a number of molecules of order N, so that Bose-Einstein condensation does not occur.

Divertissement 11

Brownian Motion. Investigating the pollen of plants, the botanist, Robert Brown, discovered in the year 1828 for the first time that the pollen become dispersed in water in a great number of small particles which were perceived to have an uninterrupted and irregular swarming motion. In 1905, Einstein wrote: "In this paper it will be shown that according to the molecular-kinetic theory of heat, bodies of microscopically-visible size suspended in a liquid will perform movements of such magnitude that they can be easily observed in a microscope, on account of the molecular motions of heat. It is possible that the movements to be discussed here are identical with the so-called "Brownian molecular motion"; however, the information available to me regarding the latter is so lacking in precision, that I can form no judgment in the matter.

"If the movement discussed here can actually be observed (together with the laws relating to it that one would expect to find), then classical thermodynamics can no longer be looked upon as applicable with precision to bodies even of dimensions distinguishable in a microscope; an exact determination of actual atomic dimensions is then possible. On the other hand, had the prediction of this movement proved to be incorrect, a weighty argument would be provided against the molecular-kinetic conception of heat" A. Einstein, *Investigations on the Theory of the Brownian Motion*, edited by R. Fürth (Dover Pub.).

211111022111232300000011001011122334534221213202210222123222 3222222138422......

The above set of numbers is due to A. Westgren who made intermittent observation of the number of colloidal particles in a given element of volume in a colloidal solution under an ultramicroscope. Such a series of numbers can be perfectly analysed by Smoluchowski's theory of Brownian motion which appeared in 1906 following Einstein's pioneer work. Even in equilibrium, a physical system never stops its molecular thermal motion. This incessant motion of molecules, on the one hand, proves statistical mechanics to be a most powerful and necessary method of theoretical physics, and, on the other hand, it makes classical thermodynamics no longer useful for such Brownian motions. The stochastic theory of Brownian motion still remains one of the very fascinating subjects of theoretical physics, for which a useful selection of classical papers is available: *Selected Papers on Noise and Stochastic Processes*, edited by N. Wax (Dover).

16. Using the Fermi distribution function $f(\varepsilon) = 1/[\exp \beta(\varepsilon - \mu) + 1]$ and the state density given in (3) of the solution to example 1, one finds for the internal energy E and the spin paramagnetic moment M of the free electron gas:

$$E = N \tfrac{3}{2} \mu_0^{-\frac{3}{2}} \int_0^\infty \varepsilon^{\frac{3}{2}} f(\varepsilon)\, \mathrm{d}\varepsilon, \tag{1}$$

$$M = N\mu_\mathrm{B} \tfrac{3}{4} \mu_0^{-\frac{3}{2}} \int_0^\infty \varepsilon^{\frac{1}{2}} \{f(\varepsilon - \mu_\mathrm{B}H) - f(\varepsilon + \mu_\mathrm{B}H)\}\, \mathrm{d}\varepsilon \tag{2}$$

(cf. eq. (1) of solution to problem 5).

$$N = N \tfrac{3}{2} \frac{1}{\mu_0^{\frac{3}{2}}} \int_0^\infty \varepsilon^{\frac{1}{2}} f(\varepsilon)\, \mathrm{d}\varepsilon, \tag{3}$$

which gives the relation between N and μ, can be transformed, by the substitution $\beta\varepsilon = x$, into

$$\frac{2}{3}\left(\frac{\mu_0}{kT}\right)^{\frac{3}{2}} = \int_0^\infty \frac{x^{\frac{1}{2}}\, \mathrm{d}x}{\mathrm{e}^{x-\beta\mu} + 1} \quad \text{or} \quad \frac{2}{3}\left(\frac{T_0}{T}\right)^{\frac{3}{2}} = F_{\frac{1}{2}}\left(\frac{\mu}{kT}\right), \tag{4}$$

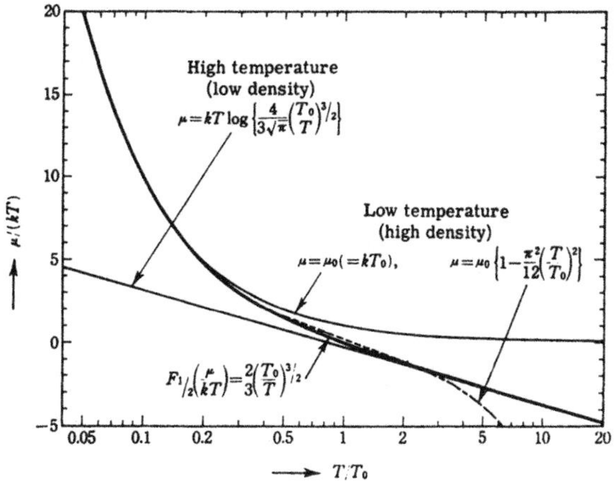

Fig. 4.17.

where $T_0 = \mu_0/k$ and $F_{\frac{1}{2}}(\beta\mu)$ is the function defined by the integral in (4). One finds that $F_{\frac{1}{2}}(\mu/kT) \to 0$ for $\mu \to -\infty$, $F_{\frac{1}{2}}(\mu/kT) \to \infty$ for $\mu \to \infty$, and also that $F_{\frac{1}{2}}$ is a monotonically increasing function of μ. If one defines a temperature T_1 by the relation

$$\frac{2}{3}\left(\frac{T_0}{T_1}\right)^{\frac{3}{2}} = F_{\frac{1}{2}}(0) = \int_0^\infty \frac{x^{\frac{1}{2}}\,\mathrm{d}x}{\mathrm{e}^x + 1} = \Gamma(\tfrac{3}{2})\cdot(1 - 2^{-\frac{1}{2}})\zeta(\tfrac{3}{2}), \qquad \zeta(\tfrac{3}{2}) = 2.612,$$

it is shown that $\mu < 0$ for $T > T_1$. Accordingly, at sufficiently high temperatures one can evaluate the integrals in (1), (2), and (3) using the expansion

$$f(\varepsilon) = \mathrm{e}^{\beta(\mu-\varepsilon)}/\{1 + \mathrm{e}^{\beta(\mu-\varepsilon)}\} = \sum_{l=1}^{\infty}(-1)^{l-1}\,\mathrm{e}^{l\beta(\mu-\varepsilon)}$$

obtaining the result:

$$E = N\tfrac{3}{2}\mu_0^{-\frac{3}{2}}(kT)^{\frac{5}{2}}\frac{3\sqrt{\pi}}{4}\sum_{l=1}^{\infty}(-1)^{l-1}\frac{\mathrm{e}^{l\beta\mu}}{l^{\frac{5}{2}}}, \tag{1'}$$

$$M = N\mu_{\mathrm{B}}\tfrac{3}{4}\mu_0^{-\frac{3}{2}}(kT)^{\frac{3}{2}}\frac{\sqrt{\pi}}{2}\sum_{l=1}^{\infty}(-1)^{l-1}\frac{\mathrm{e}^{l\beta\mu}}{l^{\frac{3}{2}}}\cdot 2\sinh(l\beta\mu_{\mathrm{B}}H), \tag{2'}$$

$$N = N\tfrac{3}{2}\mu_0^{-\frac{3}{2}}(kT)^{\frac{3}{2}}\frac{\sqrt{\pi}}{1}\sum_{l=1}^{\infty}(-1)^{l-1}\frac{\mathrm{e}^{l\beta\mu}}{l^{\frac{3}{2}}}. \tag{3'}$$

Considering the relation

$$\mu_0 = \frac{h^2}{2m}\left(\frac{3N}{8\pi V}\right)^{\frac{2}{3}}, \qquad \text{or} \quad N\tfrac{3}{2}\mu_0^{-\frac{3}{2}}(kT)^{\frac{3}{2}}\frac{\sqrt{\pi}}{2} = 2\left(\frac{2\pi mkT}{h^2}\right)^{\frac{3}{2}} V, \tag{4}$$

one knows that the first terms of (1′), (2′) and (3′) give the classical limits (non-degenerate values). Consequently, one can investigate the approach to the classical limit taking into account the second term ($l = 2$). From (3′) one has

$$2V\left(\frac{2\pi mkT}{h^2}\right)^{\frac{3}{2}}\left\{e^{\beta\mu} - \frac{1}{2^{\frac{3}{2}}}e^{2\beta\mu} + \cdots\right\} = N, \tag{5}$$

and putting

$$2\left(\frac{2\pi mkT}{h^2}\right)^{\frac{3}{2}} \equiv N_c, \qquad \eta \equiv \frac{N}{N_c V} = \frac{N}{2V}\frac{h^3}{(2\pi mkT)^{\frac{3}{2}}} \tag{6}$$

one obtains

$$e^{\beta\mu} - 2^{-\frac{3}{2}}e^{2\beta\mu} + \cdots = \eta .$$

At high temperatures ($\eta \ll 1$), this equation can be solved by successive approximations:

$$e^{\beta\mu} = \eta + 2^{-\frac{3}{2}}e^{2\beta\mu} + \cdots = \eta + 2^{-\frac{3}{2}}\eta^2 + \cdots . \tag{7}$$

Substituting this into (1′) and (2′), one obtains

$$E = \tfrac{3}{2}NkT\eta^{-1}\{e^{\beta\mu} - 2^{-\frac{5}{2}}e^{2\beta\mu} + \cdots\} = \tfrac{3}{2}NkT\{1 + 2^{-\frac{5}{2}}\eta + \cdots\}, \tag{8}$$

$$\begin{aligned} M &= N\mu_B\eta^{-1}\{e^{\beta\mu}\sinh(\beta\mu_B H) - 2^{-\frac{3}{2}}e^{2\beta\mu}\sinh(2\beta\mu_B H) + \cdots\} \\ &= N\mu_B\left[\sinh(\beta\mu_B H) + 2^{-\frac{3}{2}}\eta\{\sinh(\beta\mu_B H) - \sinh(2\beta\mu_B H)\} + \cdots\right]. \end{aligned} \tag{9}$$

Hence the specific heat and the magnetic susceptibility are given by

$$C_V = \tfrac{3}{2}Nk\{1 - 2^{-\frac{7}{2}}\eta + \cdots\}, \tag{10}$$

$$\chi = \frac{N\mu_B^2}{VkT}\{1 - 2^{-\frac{3}{2}}\eta + \cdots\}. \tag{11}$$

These correction terms are proportional to $(T_0/T)^{\frac{3}{2}}$ and tend to zero as the temperature increases.

17. Following the hint, the degeneracy of the l-th level can be obtained as follows. For simplicity, the system of electrons under consideration is assumed to be confined to a cube of volume L^3. In zero magnetic field ($H = 0$), the number of energy levels in the region $dp_x\,dp_y$ around p_x and p_y is given by

$$\frac{L^2}{h^2}\,dp_x\,dp_y\,. \tag{1}$$

In the presence of a magnetic field H in z direct ion, these levels coalesce into the harmonic oscillator-like levels as seen in the figure. The energy levels of the

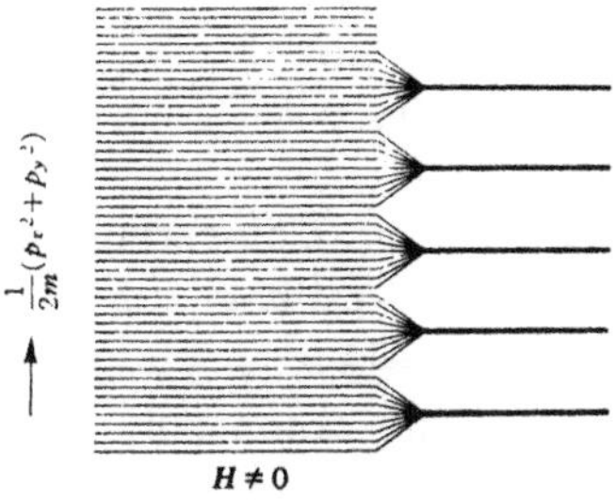

The scheme shows that the magnetic field H makes the energy levels of the motion perpendicular to H converge into ones of harmonic vibration.

kinetic motion perpendicular to the magnetic field coalesce into harmonic oscillator levels. Hence the degeneracy of each of these harmonic oscillator-like levels is obtained as

$$\frac{L^2}{h^2}\iint\limits_{2\mu_B Hl < \frac{p_x^2+p_y^2}{2m} < 2\mu_B H(l+1)} dp_x\,dp_y =$$

$$= \frac{L^2}{h^2}\int\limits_{2\mu_B Hl < \frac{p^2}{2m} < 2\mu_B H(l+1)} 2\pi p\,dp = \frac{L^2}{h^2}\Big[\pi p^2\Big]_{p^2/2m = 2\mu_B Hl}^{p^2/2m = 2\mu_B H(l+1)}$$

$$= \frac{L^2}{h^2}\pi\cdot 4\mu_B Hm = \frac{L^2 eH}{hc}\,. \tag{2}$$

Consequently the partition function Z_1 per electron is given by

$$Z_1 = \frac{L}{h}\int_{-\infty}^{\infty} dp_z \sum_{l=0}^{\infty} \frac{L^2 eH}{hc}\exp\left[-\beta\left\{2\mu_B H(l+\tfrac{1}{2}) + \frac{p_z^2}{2m}\right\}\right] \qquad (\beta = 1/kT)$$

$$= \frac{(2\pi mkT)^{\frac{1}{2}}}{h}\cdot L \times \frac{L^2 eH}{hc}\,\frac{e^{-\beta\mu_B H}}{1 - e^{-2\beta\mu_B H}}$$

$$= V\left(\frac{2\pi mkT}{h^2}\right)^{\frac{3}{2}} \frac{\mu_B H/kT}{\sinh(\mu_B H/kT)}\,. \tag{3}$$

Thus the partition function Z for N electrons is given by

$$Z = Z_1^N/N! . \tag{4}$$

Hence one has

$$M = kT\frac{\partial \log Z}{\partial H} = NkT\frac{\partial}{\partial H}\log\frac{\mu_B H/kT}{\sinh(\mu_B H/kT)} = -N\mu_B L(\mu_B H/kT). \tag{5}$$

NOTE: The equation of motion for an electron in a magnetic field H in the z-direction is given by

$$m\dot{v}_x = -\frac{eH}{c}v_y, \qquad m\dot{v}_y = \frac{eH}{c}v_x, \qquad m\dot{v}_z = 0.$$

Putting $eH/mc = \omega_0$, one has $\dot{v}_x = -\omega_0 v_y$, $\dot{v}_y = \omega_0 v_x$. Comparing these equations with the canonical equation of motion $\dot{p} = -\omega_0^2 q$, $\dot{q} = p$ of the harmonic oscillator whose Hamiltonian is given by $\mathrm{H} = \frac{1}{2}p^2 + \frac{1}{2}\omega_0^2 q^2$, one sees that v_x and v_y are dynamical variables which are canonically conjugate. Consequently the motion of the electron perpendicular to the magnetic field is quantized as a harmonic oscillator and the energy levels are given by $\hbar\omega_0(l+\frac{1}{2})$, $(l = 0, 1, 2, \ldots)$. The degeneracy $L^2 eH/hc$ for each of these levels can be interpreted as the uncertainty of the center of the circular motion of the electron.

18. This problem is a combination of problem 4 and the previous problem. The energy levels of an electron are given by

$$\varepsilon = \frac{1}{2m^*}p_z^2 + 2\mu_B^* H(l+\tfrac{1}{2}) \pm \mu_B H .$$

The partition function in the case of Boltzmann statistics is obtained from the solutions to problems 17 and 4 as

$$Z = \frac{1}{N!}Z_1^N,$$

$$Z_1(\beta) = V\left(\frac{2\pi m^*}{\beta h^2}\right)^{\frac{3}{2}} \frac{\beta\mu_B^* H}{\sinh(\beta\mu_B^* H)}\cdot 2\cosh(\beta\mu_B H)$$

so that

$$Z(\beta, H) = Z(\beta, 0)\cdot\left[\frac{\beta\mu_B^* H}{\sinh(\beta\mu_B^* H)}\cdot\cosh(\beta\mu_B H)\right]^N, \tag{1}$$

thus

$$M = kT\frac{\partial \log Z}{\partial H} = N\mu_B\tanh(\beta\mu_B H) - N\mu_B^* L(\beta\mu_B^* H). \tag{2}$$

Hence one has for $H \to 0$

$$\chi = \frac{M}{H} = \frac{N(\mu_B^2 - \frac{1}{3}\mu_B^{*2})}{kT}. \tag{3}$$

19. The system of electrons in a metal is assumed to be an ideal Fermi gas of particles having mass m^*. From the note given in the problem, one knows that an electron must have a velocity p_z/m^* in order to be able to come out of the metal ($z > 0$) after absorbing a photon of energy $h\nu$, which satisfies the condition

$$h\nu + \frac{p_z^2}{2m^*} > \chi + \mu, \qquad \text{or} \quad p_z > \sqrt{2m^*(\chi + \mu - h\nu)} = p_0 . \tag{1}$$

Let α be the probability that an electron which strikes the surface from inside with a momentum p_z and which satisfies condition (1) will absorb a photon. The flux S of electrons which satisfy (1), i.e. the number of electrons which strike a unit area of the surface per unit time with momentum p_z satisfying (1), is given by

$$S = \frac{2}{h^3}\int_{p_0}^{\infty} \frac{p_z}{m^*}\mathrm{d}p_z \int\int_{-\infty}^{\infty} \mathrm{d}p_x \mathrm{d}p_y \left\{\exp\frac{(p_x^2 + p_y^2 + p_z^2)/(2m^*) - \mu}{kT} + 1\right\}^{-1}, \tag{2}$$

which is similar to equation (1) of the solution to problem 10. Introducing planar polar coordinates and putting $(p_x^2 + p_y^2)/(2m^*kT) = u$, one has

$$S = \frac{4\pi m^* kT}{h^3}\int_{p_0}^{\infty} \mathrm{d}p_z \frac{p_z}{m^*}\int_0^{\infty} \mathrm{d}u\left[\exp\left\{u + \frac{p_z^2/(2m^*) - \mu}{kT}\right\} + 1\right]^{-1}$$

$$= \frac{4\pi m^* kT}{h^3}\int_{p_0}^{\infty} \mathrm{d}p_z \frac{p_z}{m^*}\left(-\log\left[1 + \exp\left\{-u - \frac{p_z^2/(2m^*) - \mu}{kT}\right\}\right]\right)\Bigg|_{u=0}^{u=\infty}$$

$$= \frac{4\pi m^* kT}{h^3}\int_{p_0}^{\infty} \mathrm{d}p_z \frac{p_z}{m^*}\log\left\{1 + \exp\frac{\mu - p_z^2/(2m^*)}{kT}\right\}. \tag{3}$$

Changing the integration variable to $y = \{p_z^2/(2m^*) - \mu - \chi + h\nu\}/kT$ in order to reduce the lower limit of integration to zero, one gets

$$S = \frac{4\pi m^*(kT)^2}{h^3}\int_0^{\infty} \mathrm{d}y \log\left[1 + \exp\left\{\frac{h(\nu - \nu_0)}{kT} - y\right\}\right] = AT^2\,\phi(\delta). \tag{4}$$

The current I is obtained by multiplying with the electronic charge $-e$ and the above mentioned probability α, that is,

$$I = -\alpha e S .$$

Expansion of the function $\phi(\delta)$ in (4) can be done as follows ($\delta = h(\nu - \nu_0)/kT$). For $\delta \leqq 0$, one has $e^{\delta - y} < 1$ and the integration can be performed term by term after expansion of the logarithm:

$$\phi(\delta) = \int_0^\infty \mathrm{d}y \log(1 + e^{\delta - y}) = \sum_{n=1}^\infty (-1)^{n-1} \frac{e^{n\delta}}{n} \int_0^\infty e^{-ny}\, \mathrm{d}y$$

$$= \sum_{n=1}^\infty \frac{(-1)^{n-1}}{n^2} e^{n\delta} . \tag{5}$$

For $\delta \geqq 0$, dividing the region of integration, one obtains

$$\phi(\delta) = \left(\int_0^\delta + \int_\delta^\infty \right) \mathrm{d}y \log(1 + e^{\delta - y}) = \int_0^\delta \mathrm{d}x \log(1 + e^x) + \int_0^\infty \mathrm{d}x \log(1 + e^{-x})$$

$$= \int_0^\delta \mathrm{d}x \{x + \log(1 + e^{-x})\} + \left[x \log(1 + e^{-x})\right]_0^\infty + \int_0^\infty \frac{x\, \mathrm{d}x}{e^x + 1} . \tag{6}$$

The last integral which is equal to ϕ (0) can be evaluated as $\frac{1}{12}\pi^2$ ($= \frac{1}{2}\zeta(2)$ as was shown in the second footnote in § 4.2). The first integral of (6) can be shown to be

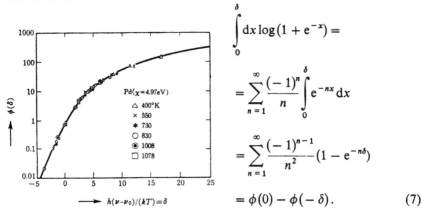

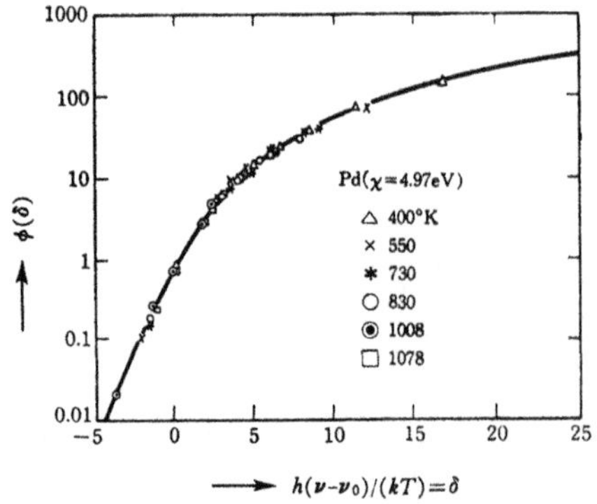

$$\int_0^\delta \mathrm{d}x \log(1 + e^{-x}) =$$

$$= \sum_{n=1}^\infty \frac{(-1)^n}{n} \int_0^\delta e^{-nx}\, \mathrm{d}x$$

$$= \sum_{n=1}^\infty \frac{(-1)^{n-1}}{n^2} (1 - e^{-n\delta})$$

$$= \phi(0) - \phi(-\delta) . \tag{7}$$

Hence one obtains, finally

$$\phi(\delta) = \tfrac{1}{2}\delta^2 + 2\phi(0) - \phi(-\delta) = \tfrac{1}{6}\pi^2 + \tfrac{1}{2}\delta^2 - \phi(-\delta), \tag{8}$$

where for $\phi(-\delta)$ one can use equation (5). The figure shows the behavior of $\phi(\delta)$.

NOTE: The external photoelectric effect of a metal near threshold frequency ν_0 can be explained well by this theory. The curve log (I/AT^2) versus $h(\nu - \nu_0)/kT$ is called the "Fowler Plot". Adjusting the observed values to this plot, one can get ν_0 and hence the work function χ.

20. (i) At 0° K, electrons take the configuration of minimum energy. The electrons in the donors D fall into the acceptor levels until the acceptors are all filled. One has the reaction equation

$$\mathrm{D} + \mathrm{A} \rightarrow \mathrm{D}^+ + \mathrm{A}^- + (E_\mathrm{G} - E_\mathrm{D} - E_\mathrm{A}),$$

where E_G, E_D and E_A are the energy gap between the bands, the energy difference between the bottom of the conduction band and the donor level, and the difference between the top of the filled band and the acceptor level, respectively. Now the electrons falling from the donors fill up the acceptors

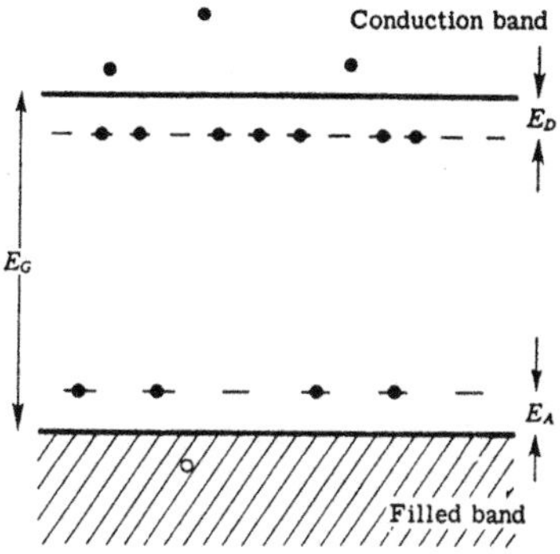

and the fraction of the empty donors will be about $\frac{1}{10}$ because N_D is about 10 times larger than N_A. In this configuration the Fermi level must lie at the donor level,

$$\mu(0°\mathrm{K}) = \tfrac{1}{2}E_\mathrm{G} - E_\mathrm{D}. \tag{1}$$

(ii) At sufficiently high temperatures, the electrons in the filled band can be excited to the conduction band. When the density of the holes in the filled band and the density of the electrons in the conduction band become very much larger than N_D and N_A, the effects of donors and acceptors can be neglected, and the sample shows characteristics similar to those of an intrinsic semiconductor. In this case the Fermi level comes in the middle of the energy gap E_G and one has

$$\mu(\infty) = 0, \tag{2}$$

as seen in example 3.

(iii) At temperatures between these two extreme cases ($T = 0$ and $T = \infty$), μ has a value between those given by (1) and (2) and this can be determined as follows. The following notation is used: n = the density of the conduction electrons; n_D = the density of the electrons in the donor levels; P_A = the density of the holes in the acceptor levels; p = the density of the holes in the filled band. Then one has

$$n = \int_0^\infty \frac{D_c(\varepsilon)\,d\varepsilon}{\exp\{(\varepsilon + \frac{1}{2}E_G - \mu)/kT\} + 1} \doteqdot g_c \left(\frac{2\pi m_e kT}{h^2}\right)^{\frac{3}{2}} \exp\{(\mu - \frac{1}{2}E_G)/kT\}$$

$$\equiv N_c \exp\{(\mu - \tfrac{1}{2}E_G)/kT\}, \tag{3}$$

$$n_D = \frac{N_D}{\frac{1}{2}\exp\{(-E_D + \frac{1}{2}E_G - \mu)/kT\} + 1}, \tag{4}$$

$$p = \int_0^\infty \left[1 - \frac{1}{\exp\{(-\frac{1}{2}E_G - \varepsilon' - \mu)/kT\} + 1}\right] D_v(\varepsilon')\,d\varepsilon'$$

$$= \int_0^\infty \frac{D_v(\varepsilon')\,d\varepsilon'}{\exp\{(\frac{1}{2}E_G + \varepsilon' + \mu)/kT\} + 1} \tag{5}$$

$$\doteqdot g_v \left(\frac{2\pi m_h kT}{h^2}\right)^{\frac{3}{2}} \exp\{-(\mu + \tfrac{1}{2}E_G)/kT\},$$

$$p_A = N_A - \frac{N_A}{\frac{1}{2}\exp\{(-\frac{1}{2}E_G + E_A - \mu)/kT\} + 1} \tag{6}$$

$$= \frac{N_A}{1 + 2\exp\{(\frac{1}{2}E_G - E_A + \mu)/kT\}}.$$

Here it is assumed that the conduction band consists of g_c single bands of effective mass m_c. (This is not exact, but is sufficient for a qualitative discussion.) Similar assumptions are made for the filled band, but a detailed discussion is unnecessary.

The condition of electrical neutrality – the density of the positive charge being equal to the density of the negative charge – can be rewritten as

$$N_D - n_D + p = N_A - p_A + n \tag{7}$$

considering $[D^+] = N_D - n_D$, $[A^-] = N_A - p_A$. [Conduction electron density] $= n$ and [Positive hole density] $= p$. In general, one obtains a fourth order equation for $\exp(\mu/kT)$ by substituting equations (3)–(6) in (7). In this problem, however, one can make a simplied analysis. At moderate temperatures, the number of holes in acceptors or in the filled band can be neglected. From equation (7) one has

$$n + n_D = N_D - N_A. \tag{8}$$

Putting

$$\exp\{(-\tfrac{1}{2}E_G + \mu)/kT\} = x$$

one obtains

$$N_c x + \frac{2N_D x}{e^{-E_D/kT} + 2x} = N_D - N_A \tag{9}$$

where $N_c = g_c \cdot (0.4)^{\frac{3}{2}} T^{\frac{3}{2}} \times 2.37 \times 10^{15}\,\mathrm{cm}^{-3}$, as can be seen from the definition in (3). Considering that $E_D/kT = 460/T$, $N_D = 10^{15}\,\mathrm{cm}^{-3}$, and $N_A = 10^{14}$ cm^{-3}, one finds that the temperature T_1 at which the relation

$$N_c(T_1) \sim 2N_D e^{E_D/kT_1}, \qquad \text{i.e.} \qquad N_c(T_1) e^{-E_D/kT_1} \sim 2N_D \tag{10}$$

holds is about 80° K. (Note that $\frac{3}{2}\log T_1 \sim 460/T_1$ approximately.) At $T < T_1$, one has $N_c \ll 2N_D e^{E_D/kT}$ and hence $n \ll n_D$ from (4). Neglecting $N_c x$ $(= n)$ in (9), one has, approximately,

$$\mu = \tfrac{1}{2}E_G - E_D + kT\log\frac{N_D - N_A}{2N_A} \qquad (T \ll T_1). \tag{11}$$

Considering that $(N_D - N_A)/2N_A \approx \frac{9}{2}$, one sees that the second term of (11) is positive and μ increases with the temperature. At $T > T_1$, n cannot be said to be smaller than n_D. Assuming that

$$N_c \gg 2N_D e^{-E_D/kT} \tag{12}$$

one can find, from (9),

$$x \sim (N_D - N_A)/N_c \sim N_D/N_c, \tag{13}$$

thus

$$\mu = \tfrac{1}{2}E_G - kT\log\frac{N_c}{N_D - N_A} \qquad (n \sim N_D - N_A).$$

Upon substituting (13), one has, from (12), $2x \ll \exp(-E_D/kT)$. The second term of (9) turns out to be smaller than the first term. Considering that

$N_D - N_A \sim N_D$, one can see that $x \sim N_D/N_c$ is the approximate solution of (9) for the case specified by (12). Consequently it is seen that μ decreases with increasing temperature ($T > T_1$) and also that μ should have a maximum in the range $T < T_1$. In fact, one can expect that μ should approach the middle of the gap between the donor level and the bottom of the conduction band if $N_D - N_A = N_D$ is assumed, because this corresponds to the case where only the donors are present (in example 4). In the temperature region from (11) up to the range where the maximum of μ occurs, one has $n < n_D$. This region is called a "bound region" because the donors still have a considerable number of electrons. In the region where (12) and (13) are satisfied, the donor levels become almost exhausted and the number n of conduction electrons will be nearly equal to the constant value $N_D - N_A$. This region is called the "saturation region". When the temperature rises still higher, the excitation of electrons from the filled band becomes important. This region is called the "intrinsic region". One can estimate the temperature of this region by putting the density of conduction electrons for the case of the intrinsic semi-conductor (example 3) equal to the above-mentioned saturated value N_D. Substituting $\frac{1}{2}E_G \sim 0.36\,\text{eV}$ in equation (10) of the solution to example 3, one has

$$n \sim \left(\frac{m_e m_h}{m}\right)^{\frac{3}{4}} \times 4.83 \times 10^{15} \times T^{\frac{3}{2}} e^{-4140/T}\,\text{cm}^{-3}.$$

Putting $n \sim 10^{15}\,\text{cm}^{-3}$, one gets $T^{\frac{3}{2}} \exp(-4140/T) \sim 1$, i.e. $\frac{3}{2} \times 2.3 \log_{10} T \sim$ $\sim 4140/T$, and hence $T \sim 450°$ K. Summarizing, one can say that the behavior of μ is as follows: At 0° K, μ coincides with the donor level. It increases with temperature and then decreases after passing a maximum at a temperature of several tens of degrees Kelvin. Then it approaches the middle of the gap between the conduction band and the filled band when the temperature increases to several hundreds of degrees Kelvin. The figure given in this problem shows the behavior mentioned above.

21. The grand partition function for electrons distributed over the N_D donors is given by

$$\begin{aligned} \Xi &= [1 + \lambda\{e^{(E_D - \mu_B H)/kT} + e^{(E_D + \mu_B H)/kT}\}]^{N_D} \\ &= \left[1 + 2\lambda e^{E_D/kT} \cosh \frac{\mu_B H}{kT}\right]^{N_D}, \end{aligned} \tag{1}$$

where the three terms in the bracket correspond to the empty state, the state occupied by an electron with positive spin and one with negative spin, respectively. Also $\lambda = \exp(\mu/kT)$ is the absolute activity of the electron. The origin of the chemical potential (Fermi potential) μ, i.e., the origin of the energy, is taken at the bottom of the conduction band. Hence the energy of an electron in the donor level is $-E_D + \mu_B H$ or $-E_D - \mu_B H$, according to whether the electron spin is positive or negative. One has from (1)

$$M = kT\frac{\partial \log \Xi}{\partial H} = N_D\mu_B\frac{2\lambda\, e^{E_D/kT}\sinh(\mu_B H/kT)}{1 + 2\lambda\, e^{E_D/kT}\cosh(\mu_B H/kT)}. \tag{2}$$

For small H, one gets

$$\chi = \frac{M}{H} = \frac{N_D\mu_B^2}{kT}\frac{1}{\frac{1}{2}e^{-(E_D+\mu)/kT} + 1} = \frac{n_D\mu_B^2}{kT} \tag{3}$$

per unit volume. This means that the n_D electrons in the donor levels orient their spins in the magnetic field independently. This can be easily understood (Langevin-Debye formula).

Alternative solution

One can proceed as in example 4.

$$F = n_+(-E_D + \mu_B H) + n_-(-E_D - \mu_B H) - kT\log\frac{N_D!}{n_+!\,n_-!\,(N_D - n_+ - n_-)!} \tag{4}$$

where n_+ and n_- are the numbers of donors with positive and negative spins, respectively. Using the Stirling formula, one can obtain

$$F = -(n_+ + n_-)E_D + (n_+ - n_-)\mu_B H + $$
$$+ kT\left\{n_+\log\frac{n_+}{N_D} + n_-\log\frac{n_-}{N_D} + (N_D - n_+ - n_-)\log\frac{N_D - n_+ - n_-}{N_D}\right\}. \tag{5}$$

$$\frac{\partial F}{\partial n_+} = \frac{\partial F}{\partial n_-} = \mu = \text{Fermi potential of electron} \tag{6}$$

and

$$M = -\left(\frac{\partial F}{\partial H}\right)_{T, n_+, n_-} = -(n_+ - n_-)\mu_B. \tag{7}$$

From (6), one gets

$$n_+ = \frac{\lambda\, e^{(E_D - \mu_B H)/kT}}{1 + 2\lambda\, e^{E_D/kT}\cosh(\mu_B H/kT)}, \qquad n_- = \frac{\lambda\, e^{(E_D + \mu_B H)/kT}}{1 + 2\lambda\, e^{E_D/kT}\cosh(\mu_B H/kT)} \tag{8}$$

and (2) from (7).

22. The internal energy is given by (4.15). Replacing the summation by an integration over $\varepsilon = p^2/2m$, one has

$$E = 2\pi V\left(\frac{2m}{h^2}\right)^{\frac{3}{2}} \int_0^\infty \frac{\varepsilon^{\frac{3}{2}}\,\mathrm{d}\varepsilon}{\mathrm{e}^{\beta(\varepsilon-\mu)}-1}, \qquad \beta = \frac{1}{kT}.$$

When the degeneracy is weak ($\mu < 0$), one can use the expansion in terms of $\mathrm{e}^{-\beta(\varepsilon-\mu)}$ ($\ll 1$):

$$\begin{aligned} E &= 2\pi V\left(\frac{2m}{h^2}\right)^{\frac{3}{2}} \sum_{l=1}^\infty \int_0^\infty \varepsilon^{\frac{3}{2}}\,\mathrm{e}^{l\beta(\mu-\varepsilon)}\,\mathrm{d}\varepsilon \\ &= 2\pi V\left(\frac{2m}{h^2}\right)^{\frac{3}{2}} (kT)^{\frac{5}{2}} \sum_{l=1}^\infty \frac{1}{l^{\frac{5}{2}}}\,\mathrm{e}^{l\beta\mu} \int_0^\infty x^{\frac{3}{2}}\,\mathrm{e}^{-x}\,\mathrm{d}x \\ &= \tfrac{3}{2} kTV\left(\frac{2\pi mkT}{h^2}\right)^{\frac{3}{2}} \sum_{l=1}^\infty \frac{\mathrm{e}^{l\mu/kT}}{l^{\frac{5}{2}}}. \end{aligned}$$

As in problem 17 one can verify $pV = \frac{2}{3}E$. One can obtain the expansions for F and S using the relations $F = N\mu - pV = N\mu - \frac{2}{3}E$, and $S = (E - F)/T = (\frac{5}{3}E - N\mu)/T$.

23. The number of states which have a momentum with magnitude between $p = |\boldsymbol{p}|$ and $p + \mathrm{d}p$ is given by

$$2\frac{V}{h^3}4\pi p^2\,\mathrm{d}p = \frac{8\pi V}{h^3}(mc)^3 \sinh^2\theta \cosh\theta\,\mathrm{d}\theta$$

where the spin weight 2 is included. Putting $\varepsilon = c(p^2 + m^2c^2)^{\frac{1}{2}} = mc^2\cosh\theta$, one has from (4.3), (4.2) and (4.4)

$$N = \sum \frac{1}{\mathrm{e}^{\beta(\varepsilon-\mu)}+1} = 8\pi V\left(\frac{mc}{h}\right)^3 \int_0^\infty \frac{\sinh^2\theta\cosh\theta\,\mathrm{d}\theta}{\exp\{\beta(mc^2\cosh\theta-\mu)\}+1}, \tag{1}$$

$$E = \sum \frac{\varepsilon}{\mathrm{e}^{\beta(\varepsilon-\mu)}+1} = 8\pi V\frac{m^4c^5}{h^3} \int_0^\infty \frac{\sinh^2\theta\cosh^2\theta\,\mathrm{d}\theta}{\exp\{\beta(mc^2\cosh\theta-\mu)\}+1}, \tag{2}$$

$$pV = N\mu - F = kT\sum \log(1 + \mathrm{e}^{-\beta(\varepsilon-\mu)}),$$

thus $$p = 8\pi\left(\frac{mc}{h}\right)^3 \int_0^\infty \frac{1}{\beta}\sinh^2\theta\cosh\theta\,\mathrm{d}\theta \log\left[1 + \exp\{-\beta(mc^2\cosh\theta - \mu)\}\right]$$

$$= \frac{8\pi}{3}\frac{m^4c^5}{h^3}\int_0^\infty \frac{\sinh^4\theta\,\mathrm{d}\theta}{\exp\{\beta(mc^2\cosh\theta - \mu)\} + 1}. \tag{3}$$

At $T = 0°$ K ($\beta \to \infty$), the Fermi distribution function becomes a step function. Hence, putting the chemical potential μ_0 as

$$\mu_0 = mc^2\cosh\theta_0, \tag{4}$$

one has

$$\frac{N}{V} = 8\pi\frac{m^3c^3}{h^3}\int_0^{\theta_0}\sinh^2\theta\cosh\theta\,\mathrm{d}\theta = \frac{8\pi}{3}\frac{m^3c^3}{h^3}\sinh^3\theta_0, \tag{5}$$

$$\frac{E}{V} = 8\pi\frac{m^4c^5}{h^3}\int_0^{\theta_0}\sinh^2\theta\cosh^2\theta\,\mathrm{d}\theta = \frac{\pi}{4}\frac{m^4c^5}{h^3}\{\sinh(4\theta_0) - 4\theta_0\}, \tag{6}$$

$$p = \frac{8\pi}{3}\frac{m^4c^5}{h^3}\int_0^{\theta_0}\sinh^4\theta\,\mathrm{d}\theta = \frac{\pi}{3}\frac{m^4c^5}{h^3}\left\{\frac{1}{4}\sinh(4\theta_0) - 2\sinh(2\theta_0) + 3\theta_0\right\}. \tag{7}$$

Defining the maximum momentum $\boldsymbol{p}_0$ (Fermi momentum), $|\boldsymbol{p}_0| = p_0 = mc\sinh\theta_0$, one obtains from (5)

$$\frac{N}{V} = \frac{8\pi}{3}\frac{p_0^3}{h^3}. \tag{8}$$

As may be seen clearly, μ_0 (or θ_0) is determined as a function of N/V from (5) or (8).

24. One can use the formulae (5), (6) and (7) of the previous problem which apply to the completely degenerate case. Putting

$$\frac{h}{mc}\left(\frac{3n}{8\pi}\right)^{\frac{1}{3}} = \sinh\theta_0$$

(from (5)), one can discuss the cases where $\sinh\theta_0 \ll 1$ and $\sinh\theta_0 \gg 1$ which correspond to $p_0 \ll mc$ and $p_0 \gg mc$, respectively. For $p_0 \ll mc$, one can use the non-relativistic approximation for the energy, $\varepsilon = mc^2 + p^2/2m$.

Hence the required results will be evident from examples 1 and 9. One can also derive these results using

$$\sinh(4\theta_0) - 4\theta_0 = \frac{1}{3!}(4\theta_0)^3 + \frac{1}{5!}(4\theta_0)^5 + \cdots \fallingdotseq \frac{32}{3}\left(\frac{p_0}{mc}\right)^3\left\{1 + \frac{3}{10}\left(\frac{p_0}{mc}\right)^2\right\},$$

$$\tfrac{1}{4}\sinh(4\theta_0) - 2\sinh(2\theta_0) + 3\theta_0 = \frac{1}{4\cdot 5!}(4\theta_0)^5 - \frac{2}{5!}(2\theta_0)^5 + \cdots \fallingdotseq \frac{8}{5}\left(\frac{p_0}{mc}\right)^5.$$

For $p_0 \gg mc$, one uses the extreme-relativistic formula $\varepsilon = cp$. Putting $p_0 \sim \frac{1}{2}\, mc \exp\theta_0$ ($\theta_0 \gg 1$), one can use the following approximations for equations (6) and (7) in the previous problem:

$$\sinh(4\theta_0) - 4\theta_0 \simeq \tfrac{1}{2}\exp(4\theta_0) = 8\left(\frac{p_0}{mc}\right)^4,$$

$$\tfrac{1}{4}\sinh(4\theta_0) - 2\sinh(2\theta_0) + 3\theta_0 \simeq \tfrac{1}{8}\exp(4\theta_0) = 2\left(\frac{p_0}{mc}\right)^4.$$

NOTE: The relation between the pressure and the kinetic energy K can be discussed as follows. For a gas confined in a container of volume V, the magnitude of the momentum $\boldsymbol{p}$ of the particle in a possible quantum state is proportional to $V^{-\frac{1}{3}}$. Hence for the non-relativistic case, $\varepsilon = mc^2 + p^2/2m$, one can see that the kinetic energy K is proportional to $V^{-\frac{2}{3}}$ apart from the rest mass energy mc^2; and then, from $p = -(\partial E/\partial V)_S$, the pressure $= 2K/3V$. For the extreme-relativistic case $\varepsilon = cp$, ε is proportional to $V^{-\frac{1}{3}}$, and the pressure is equal to $E/3V = K/3V$.

25. Putting $\varepsilon = cp$, as mentioned in the solution to problem 24, one has from (4.2) and (4.3)

$$\frac{N}{V} = \frac{8\pi}{h^3}\int_0^\infty \frac{p^2\,dp}{e^{\beta(\varepsilon-\mu)} + 1} = \frac{8\pi}{h^3c^3}\int_0^\infty \frac{\varepsilon^2\,d\varepsilon}{e^{\beta(\varepsilon-\mu)} + 1}, \tag{1}$$

$$\frac{E}{V} = \frac{8\pi}{h^3}\int_0^\infty \frac{\varepsilon p^2\,dp}{e^{\beta(\varepsilon-\mu)} + 1} = \frac{8\pi}{h^3c^3}\int_0^\infty \frac{\varepsilon^3\,d\varepsilon}{e^{\beta(\varepsilon-\mu)} + 1}. \tag{2}$$

For the strongly-degenerate case, formula (4.6a) can be used. Putting $g(\varepsilon) = \varepsilon^s$, one has

$$\int_0^\infty \varepsilon^s f(\varepsilon)\,d\varepsilon = \frac{\mu^{s+1}}{s+1} + \frac{(\pi kT)^2}{6}s\mu^{s-1} + \frac{7(\pi kT)^4}{360}s(s-1)(s-2)\mu^{s-3} + \cdots. \tag{3}$$

For $s = 2$ and 3, one gets

$$\frac{N}{V} = \frac{8\pi\mu^3}{3h^3c^3}\left\{1 + \left(\frac{\pi kT}{\mu}\right)^2\right\}, \tag{4}$$

$$\frac{E}{V} = \frac{2\pi\mu^4}{h^3c^3}\left\{1 + 2\left(\frac{\pi kT}{\mu}\right)^2 + \frac{7}{15}\left(\frac{\pi kT}{\mu}\right)^4\right\}. \tag{5}$$

From (4),

$$\mu = \mu_0\left\{1 - \frac{1}{3}\left(\frac{\pi kT}{\mu_0}\right)^2 + \cdots\right\}, \qquad \mu_0 = hc\left(\frac{3n}{8\pi}\right)^{\frac{1}{3}} \qquad \left(n = \frac{N}{V}\right), \tag{6}$$

is obtained. Substituting into (5), one obtains

$$\frac{E}{V} = \frac{2\pi\mu_0^4}{h^3c^3}\left\{1 + \frac{2}{3}\left(\frac{\pi kT}{\mu_0}\right)^2 + \cdots\right\}, \quad \text{thus} \quad \frac{E}{N} = \tfrac{3}{4}\mu_0\left\{1 + \frac{2}{3}\left(\frac{\pi kT}{\mu_0}\right)^2 + \cdots\right\}, \tag{7}$$

and hence for the heat capacity at constant volume

$$C_V = \left(\frac{\partial E}{\partial T}\right)_{V,N} = Nk\frac{\pi^2k^2T}{\mu_0} + \cdots = N\frac{2\pi^{\frac{7}{3}}k^2T}{3^{\frac{1}{3}}hcn^{\frac{1}{3}}} + \cdots. \tag{8}$$

The first term of C_V shows the same temperature dependence as in the non-relativistic case, but the coefficient becomes

$$\frac{(3n)^{\frac{1}{3}}}{2\pi}\frac{h}{mc}$$

times that of the non-relativistic case and thus will be very much greater. The calculation of the heat capacity at constant pressure will be left to the reader.

NOTE: The problem of the relativistic fermions treated above can be applied to the theory of the structure of a star which is composed of high energy particles.

26. One can proceed as in example 5. The energy of a particle will be the sum of the kinetic energy $\varepsilon_p = p^2/2m$ and the energy of the internal motion ε_j $(= 0, \varepsilon_1)$. One has, corresponding to equation (1) in example 5,

$$N = \sum_j\sum_p \frac{1}{e^{\beta(\varepsilon_p + \varepsilon_j - \mu)} - 1} = \sum_p \frac{1}{e^{\beta(\varepsilon_p - \mu)} - 1} + \sum_p \frac{1}{e^{\beta(\varepsilon_p + \varepsilon_1 - \mu)} - 1}. \tag{1}$$

At temperatures above the Bose-Einstein condensation temperature, the sum over ε_p can be replaced by the integral to give

$$N' = V\left(\frac{2\pi mkT}{h^2}\right)^{\frac{3}{2}} \{F_{\frac{3}{2}}(-\beta\mu) + F_{\frac{3}{2}}(-\beta(\mu - \varepsilon_1))\} \tag{2}$$

where the function $F_{\frac{3}{2}}(x)$ is defined in equation (3) of example 5. At $\mu = 0$, the particles which are in the lowest state of internal motion begin to condense to the state of zero kinetic energy ($p = 0$). The particles in the excited internal state do not take part in the condensation. The condensation temperature T_c is determined by the condition $N' = N$. At $T < T_c$, N' becomes smaller than N and $N - N'$ particles condense to the zero state.

$$\frac{N}{V} = \left(\frac{2\pi mkT_c}{h^2}\right)^{\frac{3}{2}} \{F_{\frac{3}{2}}(0) + F_{\frac{3}{2}}(\varepsilon_1/kT_c)\}. \tag{3}$$

From (4) of the solution to example 5, one has

$$F_{\frac{3}{2}}(\beta\varepsilon_1) = \sum_{n=1}^{\infty} e^{-n\beta\varepsilon_1}/n^{\frac{3}{2}}.$$

For $\varepsilon_1 \to \infty$ ($F_{\frac{3}{2}}(\beta\varepsilon_1) \ll 1$), one obtains T_c^0 from (3) reduced to

$$2.612\left(\frac{2\pi mkT_c^0}{h^2}\right)^{\frac{3}{2}} = \frac{N}{V}. \tag{4}$$

Using T_c^0, one has

$$\left(\frac{T_c^0}{T_c}\right)^{\frac{3}{2}} = 1 + \frac{1}{2.612}(e^{-\varepsilon_1/kT_c} + \cdots) \tag{5}$$

and assuming $e^{-\varepsilon_1/kT_c} \ll 1$

$$T_c = T_c^0\left\{1 - \frac{2}{3}\cdot\frac{1}{2.612}e^{-\varepsilon_1/kT_c} + \cdots\right\} \fallingdotseq T_c^0\{1 - 0.255\,e^{-\varepsilon_1/kT_c^0} + \cdots\}.$$

T_c is lower than T_c^0 which is naturally to be expected.

27. For the weakly-degenerate case, for a Bose gas μ (< 0) is not very near to zero and for a Fermi gas $\mu < 0$. One can calculate $pV = kT \log \Xi$, $N = kT\,\partial \log \Xi/\partial\mu$, $\Xi = \prod_i (1 \mp e^{\beta(\mu - \varepsilon_i)})^{\mp 1}$ (here and in the following the upper

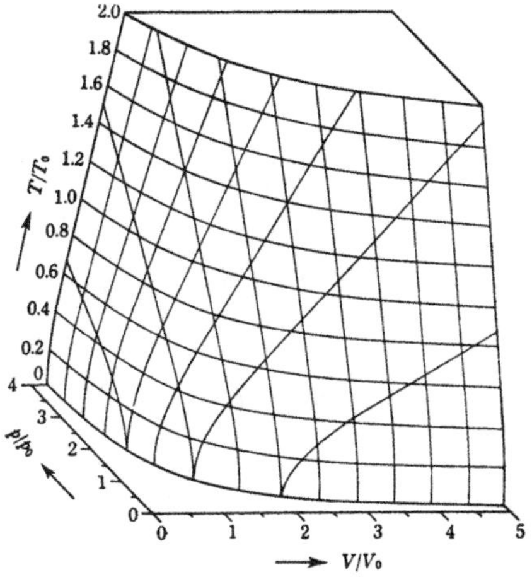

(a) Ideal Fermi-Dirac gas.

$$p_0V_0 = \frac{2}{5}NkT_0, \; kT_0 = \frac{\hbar^2}{2m}\left(3\pi^2\frac{N}{V_0}\right)^{\frac{2}{3}}.$$

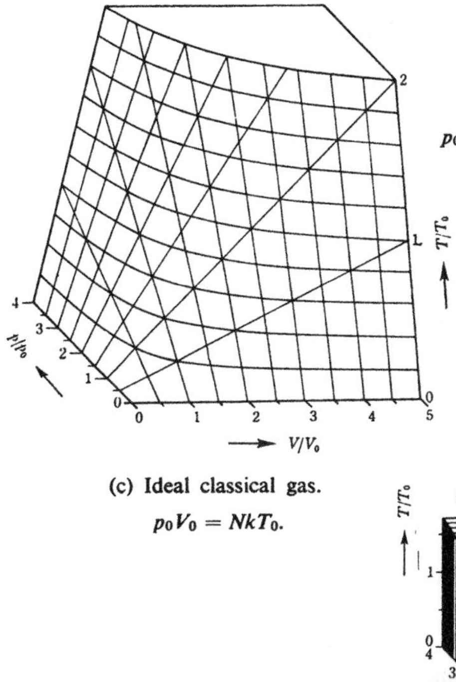

(c) Ideal classical gas.

$$p_0V_0 = NkT_0.$$

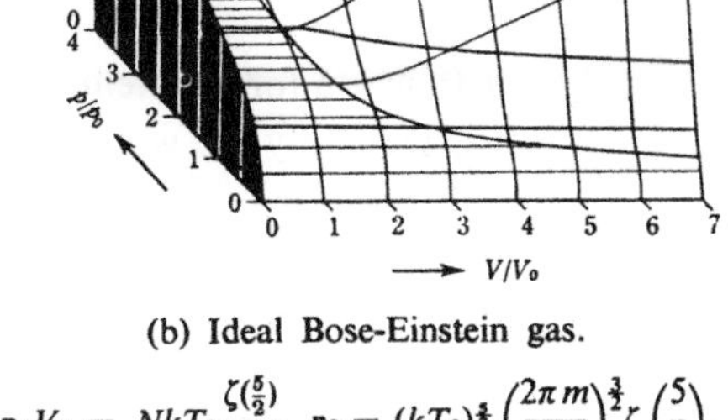

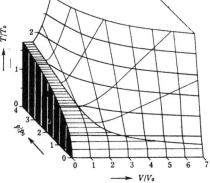

(b) Ideal Bose-Einstein gas.

$$p_0V_0 = NkT_0\frac{\zeta(\frac{5}{2})}{\zeta(\frac{3}{2})}, \; p_0 = (kT_0)^{\frac{5}{2}}\left(\frac{2\pi m}{h^2}\right)^{\frac{3}{2}}\zeta\left(\frac{5}{2}\right).$$

Fig. 4.18.

sign is for the Bose gas and the lower one is for the Fermi gas) by expanding in terms of $e^{\beta(\mu-\varepsilon)}$ (< 1):

$$pV = \mp kT \sum_i \log\{1 \mp e^{\beta(\mu-\varepsilon_i)}\} = kT \sum_{n=1}^{\infty} (\pm 1)^{n-1} \frac{\lambda^n}{n} C_n, \tag{1}$$

$$N = \sum_{n=1}^{\infty} (\pm 1)^{n-1} \lambda^n C_n, \tag{2}$$

$$C_n = \sum_i e^{-n\beta\varepsilon_i} = 2\pi g V \left(\frac{2m}{h^2}\right)^{\frac{3}{2}} \int_0^{\infty} \varepsilon^{\frac{1}{2}} e^{-n\beta\varepsilon} d\varepsilon = \frac{gV}{\lambda_T^3} \frac{1}{n^{\frac{3}{2}}}, \tag{3}$$

where g is the weight for the internal degrees of freedom and $\lambda_T = h/(2\pi mkT)^{\frac{1}{2}}$. From (3) and (2), one obtains

$$N = \frac{gV}{\lambda_T^3} \sum_{n=1}^{\infty} (\pm 1)^{n-1} \frac{\lambda^n}{n^{\frac{3}{2}}}. \tag{2'}$$

Assuming the power series in $x = N\lambda_T^3/gV$ for

$$\lambda = a_1 x + a_2 x^2 + a_3 x^3 + \cdots \tag{4}$$

one can determine a_1, a_2 ... successively from (2′): $a_1 = 1$, $a_2 = \mp \frac{1}{2^{\frac{3}{2}}}$, $a_3 = \frac{1}{4} - \frac{1}{3^{\frac{3}{2}}}$, ... Substituting this value of λ into (1), one obtains

$$\begin{aligned} pV &= VkT\left[1 \mp \frac{1}{2^{\frac{5}{2}}} \frac{\lambda_T^3 N}{gV} + \left(\frac{1}{8} - \frac{2}{3^{\frac{5}{2}}}\right)\left(\frac{\lambda_T^3 N}{gV}\right)^2 \mp \cdots\right] \\ &= NkT\left[1 \mp 2^{-\frac{5}{2}} \frac{n}{n_c} + \left(\frac{1}{8} - \frac{2}{3^{\frac{5}{2}}}\right)\left(\frac{n}{n_c}\right)^2 \mp \cdots\right], \end{aligned} \tag{5}$$

where we put $n_c = g(2\pi mkT/h^2)^{\frac{3}{2}}$.

NOTE: Figure 4.18 shows the surfaces for the equation of state of a quantum gas.

28. The total wave function for a system of N fermions which occupy the one-particle states specified by the wave vectors $\boldsymbol{k}_1, \boldsymbol{k}_2, \ldots, \boldsymbol{k}_N$ is given by

$$\Psi_{\boldsymbol{k}_1,\ldots,\boldsymbol{k}_N}(\boldsymbol{r}_1, \cdots, \boldsymbol{r}_N) = \frac{1}{\sqrt{N!V^N}} \sum_P (-1)^P \exp\left(i \sum_{j=1}^{N} \boldsymbol{k}_j \cdot \boldsymbol{r}_{Pj}\right), \tag{1}$$

where P means the permutation which operates on the numbering of the particles (1, 2, ..., N). The number j goes to Pj. The symbol $\sum_p$ means the summation over all permutations and $(-1)^P$ is $+1$ for even permutations and -1 for odd permutations. The matrix element of the density matrix in the canonical ensemble is given by

$$\langle \boldsymbol{r}_1, \cdots, \boldsymbol{r}_N | \rho | \boldsymbol{r}'_1, \cdots, \boldsymbol{r}'_N \rangle =$$

$$= \frac{1}{Z} \sum_{\boldsymbol{k}_1, \ldots, \boldsymbol{k}_N}' \exp\left(-\frac{\hbar^2}{2mkT} \sum_{j=1}^{N} \boldsymbol{k}_j^2 \right) \Psi_{\boldsymbol{k}_1, \ldots, \boldsymbol{k}_N}(\boldsymbol{r}_1, \cdots, \boldsymbol{r}_N) \Psi^*_{\boldsymbol{k}_1, \ldots, \boldsymbol{k}_N}(\boldsymbol{r}'_1, \cdots, \boldsymbol{r}'_N), \tag{2}$$

where Z is the partition function and the summation means the sum over all different quantum states. The state $(\boldsymbol{k}_1, \boldsymbol{k}_2, \ldots, \boldsymbol{k}_N)$ and the one which is generated by the permutation of the numberings of $\boldsymbol{k}_j$ should be considered as the same state. This restriction on the summation can be removed by dividing by $N!$. Substitution of (1) in (2) gives

$$\langle \boldsymbol{r}_1, \cdots, \boldsymbol{r}_N | \rho | \boldsymbol{r}'_1, \cdots, \boldsymbol{r}'_N \rangle =$$

$$= \frac{1}{Z(N!)^2 V^N} \sum_P \sum_{P'} (-1)^P (-1)^{P'} \sum_{\boldsymbol{k}_1, \ldots, \boldsymbol{k}_N} \exp\left(-\frac{\hbar^2}{2mkT} \sum_{j=1} \boldsymbol{k}_j^2 \right) \times$$

$$\times \exp\left\{ \mathrm{i} \sum_{j=1}^{N} \boldsymbol{k}_j \cdot (\boldsymbol{r}_{Pj} - \boldsymbol{r}'_{P'j}) \right\}. \tag{3}$$

The summation over $\boldsymbol{k}$ can be replaced by an integration for large V:

$$\frac{1}{V} \sum_{\boldsymbol{k}} \exp\left\{ -\frac{\hbar^2 \boldsymbol{k}^2}{2mkT} + \mathrm{i}\boldsymbol{k} \cdot (\boldsymbol{r}_{Pj} - \boldsymbol{r}'_{P'j}) \right\} =$$

$$= \frac{1}{(2\pi)^3} \iiint_{-\infty}^{\infty} \exp\left\{ -\frac{\hbar^2 \boldsymbol{k}^2}{2mkT} + \mathrm{i}\boldsymbol{k} \cdot (\boldsymbol{r}_{Pj} - \boldsymbol{r}'_{P'j}) \right\} \mathrm{d}k_x \mathrm{d}k_y \mathrm{d}k_z$$

$$= \frac{1}{\lambda_T^3} \exp\left(-\frac{\pi}{\lambda_T^2} |\boldsymbol{r}_{Pj} - \boldsymbol{r}'_{P'j}|^2 \right), \tag{4}$$

where $\lambda_T = 2\pi\hbar/\sqrt{2\pi mkT}$. Decomposing the permutation P' into P and Q, one can replace the summation over P' by a summation over Q.

$$\langle \boldsymbol{r}_1, \cdots, \boldsymbol{r}_N | \rho | \boldsymbol{r}'_1, \cdots, \boldsymbol{r}'_N \rangle =$$

$$= \frac{1}{Z(N!)^2 \lambda_T^{3N}} \sum_P \sum_Q (-1)^Q \exp\left(-\frac{\pi}{\lambda_T^2} \sum_j |\boldsymbol{r}_{Pj} - \boldsymbol{r}'_{PQj}|^2 \right), \tag{5}$$

where $(-1)^{P'} = (-1)^P (-1)^Q$ is used. The exponential factor becomes independent of P because it involves the summation over all j, and hence Pj can be replaced by j,

$$\langle \boldsymbol{r}_1, \cdots, \boldsymbol{r}_N | \rho | \boldsymbol{r}'_1, \cdots, \boldsymbol{r}'_N \rangle = \frac{1}{ZN!\lambda_T^{3N}} \sum_Q (-1)^Q \exp\left(-\frac{\pi}{\lambda_T^2} \sum_{j=1}^N |\boldsymbol{r}_j - \boldsymbol{r}'_{Qj}|^2 \right). \tag{6}$$

Now the partition function Z (= Tr ρ) is given by

$$Z = \frac{1}{N!} \sum_Q (-1)^Q \frac{1}{\lambda_T^{3N}} \int \cdots \int_V \exp\left(-\frac{\pi}{\lambda_T^2} \sum_{j=1}^N |\boldsymbol{r}_j - \boldsymbol{r}_{Qj}|^2 \right) \mathrm{d}\boldsymbol{r}_1 \, \mathrm{d}\boldsymbol{r}_2 \cdots \mathrm{d}\boldsymbol{r}_N. \tag{7}$$

A set of numbers ($Q1$, $Q2$, ..., QN) is divided into subgroups corresponding to the decomposition of the permutation Q into cyclic permutations. The numbers which belong to a certain subset are connected by the exponent in (7). The numbers in the different subsets are not connected. For instance, by

$$\mathrm{P}\begin{pmatrix} 1, 2, 3, 4, 5 \\ 3, 1, 2, 5, 4 \end{pmatrix},$$

one has the subset (1, 2, 3) connected by $|\boldsymbol{r}_1 - \boldsymbol{r}_3|^2 + |\boldsymbol{r}_3 - \boldsymbol{r}_2|^2 + |\boldsymbol{r}_2 - \boldsymbol{r}_1|^2$ and the subset (4,5) connected by $|\boldsymbol{r}_4 - \boldsymbol{r}_5|^2 + |\boldsymbol{r}_5 - \boldsymbol{r}_4|^2$. The subset which has l numbers contributes the factor

$$J_l = \frac{1}{\lambda_T^{3l}} \int \cdots \int_V \exp\left\{ -\frac{\pi}{\lambda_T^2} (|\boldsymbol{r}_1 - \boldsymbol{r}_2|^2 + |\boldsymbol{r}_2 - \boldsymbol{r}_3|^2 + \cdots \right.$$

$$\left. + |\boldsymbol{r}_{l-1} - \boldsymbol{r}_l|^2 + |\boldsymbol{r}_l - \boldsymbol{r}_1|^2) \right\} \mathrm{d}\boldsymbol{r}_1 \, \mathrm{d}\boldsymbol{r}_2 \cdots \mathrm{d}\boldsymbol{r}_l \tag{8}$$

to the integration in (7). If one considers the last $\boldsymbol{r}_1$ as $\boldsymbol{r}_1{}'$ the integral in (8) is a convolution of the Gauss transformation. For sufficiently large V ($\gg \lambda_T^3$) one can extend the limits of integration to infinity and use the formula

$$\int_{-\infty}^{\infty} \frac{\mathrm{e}^{-(x-y)^2/a}}{\sqrt{a\pi}} \frac{\mathrm{e}^{-(y-z)^2/b}}{\sqrt{b\pi}} \mathrm{d}y = \frac{\mathrm{e}^{-(x-z)^2/(a+b)}}{\sqrt{(a+b)\pi}} \qquad (a, b > 0). \tag{9}$$

After that one puts $\boldsymbol{r}_1 = \boldsymbol{r}_1'$ to obtain

$$J_l = \frac{V}{\lambda_T^3 l^{\frac{3}{2}}} \qquad (V \gg \lambda_T^3). \tag{10}$$

Let N_l be the numbers of the subsets of l numbers which appear in the decomposition according to Q, then

$$N = \sum_{l=1}^{N} l N_l . \tag{11}$$

A set of numbers $(N_1, N_2, \ldots)$ corresponds to a given Q. Then one has from (7)

$$Z = \frac{1}{N!}\sum_{Q}(-1)^Q \prod_l J_l^{N_l} = \frac{1}{N!}\sum_{Q}(-1)^Q \prod_l \left(\frac{V}{\lambda_T^3 l^{\frac{5}{2}}}\right)^{N_l} . \tag{12}$$

For a given $(N_1, N_2, \ldots)$ one cannot determine Q uniquely. The total number of ways of dividing N numbered objects into N_1 subsets of objects 1, N_2 of objects 2, ... N_l of objects l is given by $N!/\prod_l (l!)^{N_l} N_l!$. Taking into account the order of objects in each subset, one sees that

$$N!/\prod_l \{l^{N_l} \cdot N_l!\}$$

permutations contribute the same factor $\prod_l \{V/\lambda_T^3 l^{\frac{5}{2}}\}^{N_l}$ to (12). Every permutation Q can be decomposed into a succession of exchanges of 2 numbers. A cyclic permutation of l numbers is decomposed into $l-1$ exchanges, so one has $(-1)^Q = (-1)^{\Sigma_l (l-1) N_l}$. Therefore (12) may be written as

$$Z' = \sum_{(N_1, N_2, \ldots)}' \prod_{l=1}^{N} \frac{1}{N_l!}\left\{\frac{(-1)^{l-1} V}{l^{\frac{5}{2}} \lambda_T^3}\right\}^{N_l} , \tag{13}$$

where the summation should be taken over all possible sets $(N_1, N_2, \ldots)$.

NOTE 1: In order to check the validity of formula (12), one can calculate the grand partition function for this case:

$$\sum_{N=0}^{\infty} e^{\mu N/kT} \sum_{(N_1, N_2, \ldots)}' \prod_{l=1}^{N} \frac{1}{N_l!}\left\{\frac{(-1)^{l-1} V}{l^{\frac{5}{2}} \lambda_T^3}\right\}^{N_l} =$$

$$= \prod_{l=1}^{\infty} \sum_{N_l=0}^{\infty} \frac{1}{N_l!}\left\{\frac{(-1)^{l-1} V}{l^{\frac{5}{2}} \lambda_T^3} e^{\mu l/kT}\right\}^{N_l}$$

$$= \exp\left\{\frac{V}{\lambda_T^3}\sum_{l=1}^{\infty}\frac{(-1)^{l-1}}{l^{\frac{5}{2}}} e^{\mu l/kT}\right\} = \exp\left[\sum_k \sum_{l=1}^{\infty} \frac{(-1)^{l-1}}{l}\{e^{(\mu-\varepsilon_k)/kT}\}^l\right]$$

$$= \exp\sum_k \log\{1 + e^{(\mu-\varepsilon_k)/kT}\} = \prod_k \{1 + e^{(\mu-\varepsilon_k)/kT}\}, \tag{14}$$

where formula (10) has been used for $\sum e^{-\beta l \varepsilon_k}$.

NOTE 2: For an ideal Bose gas, $(-1)^Q$ in (6) and hence $(-1)^{l-1}$ in (13) disappear from the formulae.

29. The energy levels and their degeneracy are given in the solution to problem 17. From (4.4) the Helmholtz free energy (including the potential energy in the magnetic field) of the system of electrons is given by

$$F = N\mu - kT\frac{4\pi V m\mu_B H}{h^3}\sum_{l=0}^{\infty}\int_{-\infty}^{\infty}\log\left\{1+\exp\frac{\mu - 2\mu_B H(l+\frac{1}{2}) - p_z^2/2m}{kT}\right\}\mathrm{d}p_z. \tag{1}$$

For $\mu_B H \ll kT$, the change of the integrand corresponding to the change of l by 1 is very small, and hence the use of Euler's formula is permissible. Then

$$F \risingdotseq N\mu - \frac{4\pi m V}{h^3}\mu_B H kT\int_0^{\infty}\mathrm{d}x\int_{-\infty}^{\infty}\mathrm{d}p_z\log\left\{1+\exp\frac{\mu - 2\mu_B Hx - p_z^2/2m}{kT}\right\} +$$

$$+\frac{\pi m V}{3h^3}(\mu_B H)^2\int_{-\infty}^{\infty}\frac{\mathrm{d}p_z}{1+\exp\left[\{p_z^2/2m-\mu\}/kT\right]}. \tag{2}$$

In the first integral, transform the integration variables to p_z, $\varepsilon = 2\mu_B Hx + p_z^2/2m$ and integrate over p_z. In the second integral, put $p_z^2/2m = \varepsilon$. One has, then,

$$F \risingdotseq N\mu - \frac{2\pi(2m)^{\frac{3}{2}}}{h^3}kTV\int_0^{\infty}\log\{e^{(\mu-\varepsilon)/kT}+1\}\sqrt{\varepsilon}\,\mathrm{d}\varepsilon +$$

$$+\frac{\pi(2m)^{\frac{1}{2}}}{6h^3}(\mu_B H)^2 V\int_0^{\infty}\frac{\varepsilon^{-\frac{1}{2}}\,\mathrm{d}\varepsilon}{e^{(\varepsilon-\mu)/kT}+1}. \tag{3}$$

The first and second terms give the expression for the free energy under zero field conditions in which the chemical potential is replaced by that existing in the presence of the magnetic field. The third term is the lowest order correction term due to the quantization into the Landau level. Considering $(\partial F/\partial\mu)_{T,V,N} = 0$, the susceptibility for given N can be obtained from

$$\chi = -\frac{1}{VH}\left(\frac{\partial F}{\partial H}\right)_{T,V,N,\mu} \tag{4}$$

Hence one sees that only the third makes a contribution to χ. For $\mu \gg kT$, retaining only the lowest order in (kT/μ), one obtains

$$\chi \doteqdot -\frac{2\pi(2m)^{\frac{3}{2}}\mu^{\frac{1}{2}}}{3h^3}\mu_{\mathrm{B}}^2 . \tag{5}$$

Replacing μ by μ_0 and inserting the spin factor 2, one can see that the susceptibility given by (5) is equal to $-\frac{1}{3}$ times that given in problem 4. The negative sign means diamagnetism.

NOTE: For a detailed calculation by means of Euler's summation formula, see R. B. Dingle: Proc. Roy. Soc. **A211** (1952) 500–516. Formula (5) (inserting the spin factor 2) was first given by Landau.

30. Integrating twice by parts, one has from (4.4′)

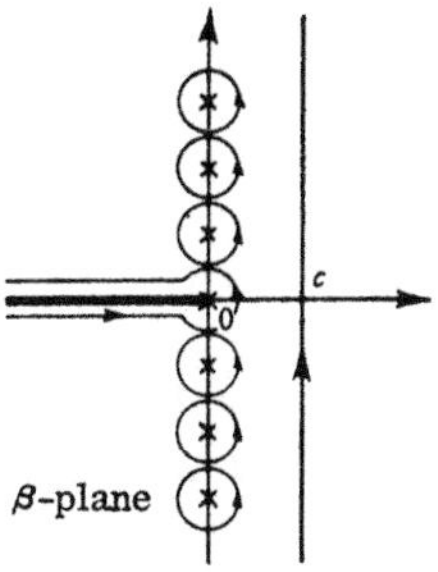

$$F - N\mu = \int_0^\infty \Omega_1(\varepsilon)\frac{\partial f}{\partial \varepsilon}\mathrm{d}\varepsilon ,$$

$$\Omega_1(\varepsilon) = \frac{1}{2\pi \mathrm{i}}\int_{c-\mathrm{i}\infty}^{c+\mathrm{i}\infty} \frac{\mathrm{e}^{\beta\varepsilon}}{\beta^2} Z(\beta)\,\mathrm{d}\beta . \tag{1}$$

This allows the integration over the energy to be done very easily. From equation (3) in the solution to problem 17, one gets

$$\Omega_1(\varepsilon) = V\frac{(2\pi m)^{\frac{3}{2}}\mu_{\mathrm{B}}H}{h^3}\frac{1}{2\pi\mathrm{i}}\int_{c-\mathrm{i}\infty}^{c+\mathrm{i}\infty}\frac{\mathrm{e}^{\beta\varepsilon}\,\mathrm{d}\beta}{\beta^{\frac{5}{2}}\sinh(\beta\mu_{\mathrm{B}}H)} . \tag{2}$$

The integrand has a branch point at $\beta = 0$ and first order poles at $\beta = \mathrm{i}l\pi/\mu_{\mathrm{B}}H$ ($l = \pm 1, \pm 2, \ldots$). The cut is chosen on the negative real axis. The path of integration can be deformed to the circles around the poles and the contour around the cut shown in the figure. The contour integral around the cut gives a smooth function of ε and this will be omitted from our considerations. The integral around the poles can be evaluated from the residues and this gives an oscillatory function of ε. Putting $\beta\mu_{\mathrm{B}}H = \mathrm{i}l\pi + z$, one has

$$\Omega_1^{\mathrm{osc}}(\varepsilon) = V\frac{(2\pi m)^{\frac{3}{2}}(\mu_{\mathrm{B}}H)^{\frac{3}{2}}}{h^3}\sum_{l=\pm1,\pm2,\ldots}\frac{(-1)^l}{2\pi\mathrm{i}}\oint\frac{\mathrm{e}^{\mathrm{i}l\pi\varepsilon/\mu_{\mathrm{B}}H}\,\mathrm{e}^{\varepsilon z/\mu_{\mathrm{B}}H}}{(\mathrm{i}l\pi + z)^{\frac{5}{2}}\sinh z}\mathrm{d}z . \tag{3}$$

Since the z-integration is performed around the origin, the approximation $\sinh z \fallingdotseq z$ and $z = 0$ can be made in the other factors, and one obtains

$$\Omega_1^{\mathrm{osc}}(\varepsilon) = V\frac{(2\pi m)^{\frac{3}{2}}(\mu_{\mathrm{B}}H)^{\frac{5}{2}}}{h^3} \sum_{l=\pm 1,\, \pm 2,\, \dots} \frac{(-1)^l}{(\mathrm{i}l\pi)^{\frac{5}{2}}} \mathrm{e}^{\mathrm{i}l\pi\varepsilon/\mu_{\mathrm{B}}H}$$

$$= -V\frac{(2\pi m)^{\frac{3}{2}}}{h^3}\sum_{l=1}^{\infty}\frac{(\mu_{\mathrm{B}}H)^{\frac{5}{2}}}{(\pi l)^{\frac{5}{2}}}(-1)^l\, 2\cos\left(\frac{l\pi\varepsilon}{\mu_{\mathrm{B}}H} - \frac{\pi}{4}\right). \tag{4}$$

Substituting this into the first equation of (1), one carries out the integration over ε. Note that $\partial f/\partial\varepsilon$ has a peak of width kT around $\varepsilon = \mu$. Each term of (4) undergoes oscillations with a period of the order $\mu_{\mathrm{B}}H/l$. Hence if $\mu_{\mathrm{B}}H \ll kT$, cancellation occurs and the value of the integral is small. For $kT < \mu_{\mathrm{B}}H$, the integral gives an appreciable value and the de Haas-van Alphen effect appears. At $kT \sim \mu_{\mathrm{B}}H$, the integration should be done carefully:

$$I = 2\int_0^{\infty} \cos\left(\frac{l\pi\varepsilon}{\mu_{\mathrm{B}}H} - \frac{\pi}{4}\right)\frac{\partial f}{\mathrm{d}\varepsilon}\mathrm{d}\varepsilon$$

$$= -\Re\left[\mathrm{e}^{(\mathrm{i}l\pi\mu/\mu_{\mathrm{B}}H)-\mathrm{i}\pi/4}\frac{2}{kT}\int_0^{\infty}\frac{\mathrm{e}^{\mathrm{i}l\pi(\varepsilon-\mu)/\mu_{\mathrm{B}}H}}{\{\mathrm{e}^{(\varepsilon-\mu)/2kT} + \mathrm{e}^{-(\varepsilon-\mu)/2kT}\}^2}\right]$$

$$= -\Re\left\{\mathrm{e}^{(\mathrm{i}l\pi\mu/\mu_{\mathrm{B}}H)-\mathrm{i}\pi/4}\int_{-\mu/2kT}^{\infty}\frac{\mathrm{e}^{2\pi\mathrm{i}lkTz/\mu_{\mathrm{B}}H}}{\cosh^2 z}\mathrm{d}z\right\}. \tag{5}$$

The lower limit of the integral can be replaced by $-\infty$ because one always has $\mu \gg kT$. Then the path of integration can be closed by the upper semicircle and the integral can be evaluated by the residues of the second order poles at $z = \mathrm{i}\,(n + \frac{1}{2})\pi$, $n = 0, 1, 2, \dots$. Replacing z by $y = z - \mathrm{i}\,(n + \frac{1}{2})\pi$, and putting $\cosh z = \mathrm{i}\,(-1)^n \sinh y$, one obtains

$$\int_{-\infty}^{\infty}\frac{\mathrm{e}^{2\pi\mathrm{i}lkTz/\mu_{\mathrm{B}}H}}{\cosh^2 z}\mathrm{d}z = -\sum_{n=0}^{\infty}\mathrm{e}^{-\pi^2 lkT(2n+1)/\mu_{\mathrm{B}}H}\oint\frac{\mathrm{e}^{2\pi\mathrm{i}lkTy/\mu_{\mathrm{B}}H}}{\sinh^2 y}\mathrm{d}y$$

$$= \frac{2\pi^2 lkT/\mu_{\mathrm{B}}H}{\sinh\{\pi^2 lkT/\mu_{\mathrm{B}}H\}}, \tag{6}$$

thus

$$I = -\frac{2\pi^2 lkT/\mu_{\mathrm{B}}H}{\sinh\{\pi^2 lkT/\mu_{\mathrm{B}}H\}}\cos\left(\frac{l\pi\mu}{\mu_{\mathrm{B}}H} - \frac{\pi}{4}\right). \tag{7}$$

Substituting this into (4) and (1), one gets

$$F_{\text{osc}} - N\mu = V\frac{(2\pi m)^{\frac{3}{2}}}{h^3}\frac{2kT}{\sqrt{\pi}}(\mu_B H)^{\frac{3}{2}}\sum_{l=1}^{\infty}\frac{(-1)^l}{l^{\frac{3}{2}}}\frac{\cos\{(l\pi\mu/\mu_B H) - \frac{1}{4}\pi\}}{\sinh\{\pi^2 lkT/\mu_B H\}}. \tag{8}$$

Hence the oscillatory term in the susceptibility is given by

$$\chi_{\text{osc}} = -\frac{1}{VH}\left(\frac{\partial F_{\text{osc}}}{\partial H}\right)_{T,V,N,\mu} \doteqdot \mu_B^2\frac{(2\pi m\mu)^{\frac{3}{2}}}{h^3}\frac{2\sqrt{\pi}kT}{(\mu_B H)^2}\left(\frac{\mu_B H}{\mu}\right)^{\frac{1}{2}}\times$$

$$\times\sum_{l=1}^{\infty}\frac{(-1)^l}{l^{\frac{1}{2}}}\frac{\sin\{(l\pi\mu/\mu_B H) - \frac{1}{4}\pi\}}{\sinh\{\pi^2 lkT/\mu_B H\}}, \tag{9}$$

where the terms of higher order in $\mu_B H/kT$ and μ/kT are neglected.

NOTE: Using the density of the electron, $n \doteqdot 4\pi\ (2m\mu)^{\frac{3}{2}}\ (3h^3)$ (not including the spin weight), one has from (9)

$$\chi_{\text{osc}} \doteqdot -\frac{n\mu_B^2}{kT}\frac{3\pi}{2}\left(\frac{kT}{\mu_B H}\right)^2\left(\frac{\mu_B H}{\mu}\right)^{\frac{1}{2}}\frac{\sin\{(\pi\mu/\mu_B H) - \frac{1}{4}\pi\}}{\sinh\{\pi^2 kT/\mu_B H\}}, \tag{10}$$

taking only the term $l = 1$. This shows that $\chi_{\text{osc}}/(H^{\frac{3}{2}}\sinh[\pi^2 kT/\mu_B H])$ is a periodic function of $1/H$ with period $2\mu_B/\mu$. The condition $kT < \mu_B H$ for the appearance of this term is also the condition for $\sinh\{\pi^2 kT/\mu_B H\}$ to remain small. The intensity of the magnetic field which satisfies $\mu_B H = \pi^2 kT$ is $H = 1.469 \times 10^5 \times T$ oersted (T in °K). This shows that observation will be difficult except at low temperatures. Due to the presence of the factor $(\mu_B H/\mu)^{\frac{1}{2}}$ in (10), the oscillatory term is very small compared to the constant term calculated in problem 29, and observation requires very high accuracy.

CHAPTER 5

STRONGLY INTERACTING SYSTEMS

So far we have been dealing mostly with weakly interacting systems, that is, such systems in which the interactions of the particles are so weak that their motion can be regarded as nearly free (example: dilute gases) or such systems in which some simple decomposition of the motion is possible as normal modes (example: lattice vibrations in crystals). These examples are typical and have many actual application. On the other hand, there are a good many (and perhaps the most interesting) cases where the interactions between the particles are so strong that such simplications are no longer possible. Ferromagnetism and phase change in general are good examples. A rigorous treatment of such systems is extremely difficult, but various methods of approximation can be devised which in fact reveal the essential points of the physical phenomena due to such strong interactions. In this chapter, some of the most typical problems will be treated, and these are to serve as a guide to more advanced studies.

Fundamental Topics†

§ 5.1. MOLECULAR FIELD APPROXIMATION

Self-consistent molecular field: The following method is a most useful and simple approximation in treating a system consisting of a number of interacting particles. Let us now observe a representative particle in the system. It experiences forces exerted by other particles surrounding it. This interaction force field is approximated by an average field or a *molecular field* (although it depends on the states of the surrounding particles and so realizes various values probabilistically). Then the representative particle is treated by statistical mechanics. From this one may in turn determine the average field which this particle exerts on its neighbors. If the particles are identical, the average field so calculated must coincide with that assumed before. This requirement of *self-consistency* serves as a condition to determine the molecular field, so that the statistical properties of the system are fixed and the thermodynamic properties can be calculated.

† cf. C. Kittel, *Introduction to Solid State Physics* (John Wiley & Sons, New York, 1956), R. H. Fowler and E. A. Guggenheim, *Statistical Thermodynamics* (Cambridge Univ. Press, 1939).

This approach is very basic and is used not only in statistical many-body problems but also in quantum-mechanical many-body problems. In quantum mechanics, the Hartree or Fock approximation is an example.

Ising model: A good example illustrating this concept is provided by a simple model of ferromagnetism. Each atom in a crystal is supposed to have a magnetic moment μ_0 which may point upwards or downwards as shown in Figure 5.1. These magnetic moments are called *Ising spins*. The states of each Ising spin are represented by a variable σ_j ($j = 1, 2, \dots N$; N being the total number of atoms) which takes the values 1 or -1. Neighboring spins in the crystal have an interaction J (exchange interaction) whose sign depends on whether the spins are parallel or anti-parallel, as follows:

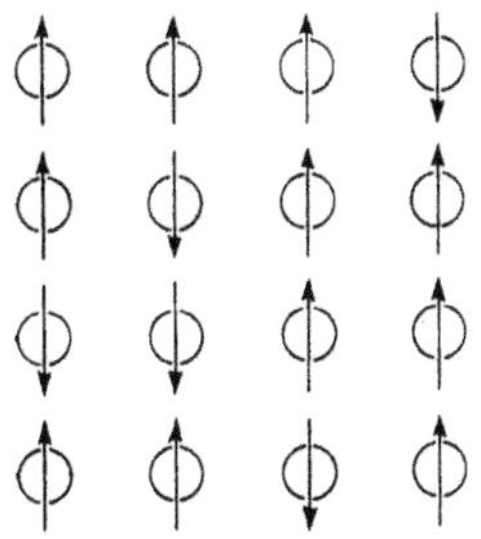

Fig. 5.1.

$$J_{++} = J_{--} = -J, \qquad J_{+-} = J.$$

Therefore the interaction energy of the spins can be expressed by

$$\mathscr{H}_{\text{int}} = -\sum_{<ij>} J\sigma_i\sigma_j, \tag{5.1}$$

where the summation is taken over interacting neighboring pairs. If $J > 0$, then neighboring spins tend to be parallel, so that there is a possibility that ferromagnetism will appear. If $J < 0$, then neighboring spins tend to be antiparallel, so that antiferromagnetism may arise (see § 5.3).

The Weiss field: When an external magnetic field is applied to this system, each atom feels a field which is composed of this external field and the exchange interaction with its neighbors. The latter, in fact, is a fluctuating field, but is now replaced by an average field which is equivalent to a magnetic field H' called the *molecular field* or the *Weiss field*. (Weiss was the first to introduce this idea.) On this assumption one takes

$$H_{\text{eff}} = H + H' \tag{5.2}$$

as the effective field acting on the spin under consideration. When the spin system has no magnetization one has to put $H' = 0$. Therefore one may generally assume that

$$H' = qM, \tag{5.3}$$

that is, that the molecular field H' is proportional to the intensity of the magnetization M. The constant q is called the *molecular field constant*. Under the effective field H_{eff} given by (5.2), the average magnetic moment in the direction of H_{eff} is given by (see example 10, Chapter 1),

$$\bar{\mu} = \mu_0 \tanh \frac{\mu_0 H_{\text{eff}}}{kT}. \tag{5.4}$$

Hence, if the number of atoms per unit volume is denoted by n, the magnetization M is

$$M \equiv n\bar{\mu} = n\mu_0 \tanh \left\{ \frac{\mu_0}{kT}(H + qM) \right\}, \tag{5.5}$$

by (5.2) and (5.3). This determines H' in a self-consistent manner.

Let z be the number of neighboring spins around a given spin, $\bar{z}_+$ the average number of upward spins, and $\bar{z}_-$ the average number of downward spins among these z spins. Then one has

$$\bar{z}_+ - \bar{z}_- = zM/M_\infty, \qquad \text{where} \qquad M_\infty = n\mu_0 \tag{5.6}$$

since $\bar{z}_+/z$ and $\bar{z}_-/z$ are the fractions of upward and downward spins in the system. Further, one may put

$$\mu_0 H' = J(\bar{z}_+ - \bar{z}_-) = zJM/M_\infty, \quad \text{thus} \quad q = zJ/\mu_0 M_\infty \tag{5.7}$$

because $\pm\, \mu_0 H'$ represents the average interaction of a spin with its neighbors. Therefore (5.5) can be written as

$$\frac{M}{M_\infty} = \tanh\left(\frac{\mu_0 H}{kT} + \frac{zJM}{kTM_\infty} \right). \tag{5.8}$$

In particular, the spontaneous magnetization $M_s(T)$ in the absence of an external field ($H = 0$) is determined by the equation

$$X = \tanh\left(\frac{zJ}{kT} X \right), \qquad \frac{M_s}{M_\infty} \equiv X. \tag{5.9}$$

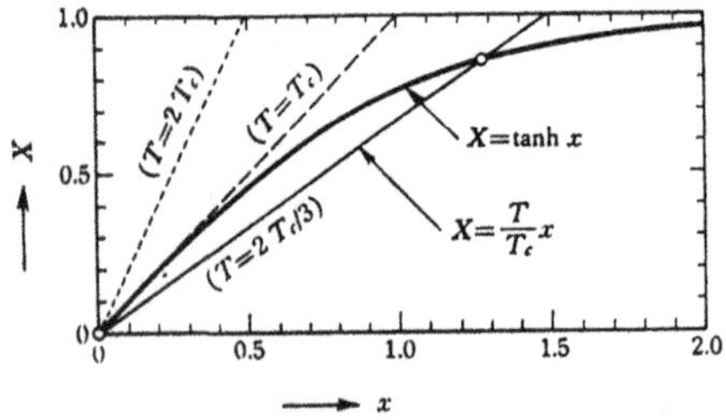

Fig. 5.2.

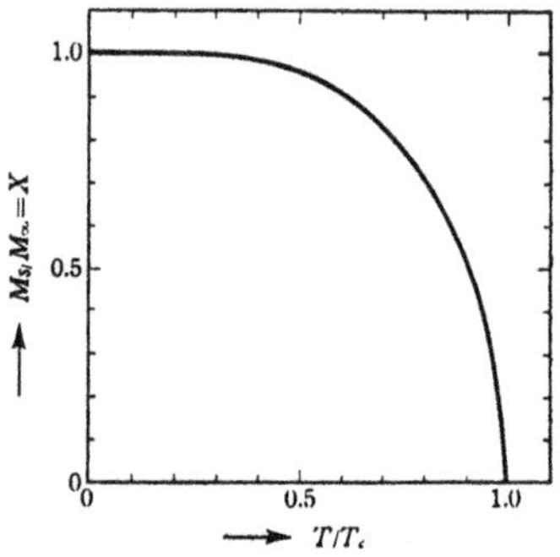

Fig. 5.3.

This means that the intersection of the two graphs

$$X = \tanh x, \qquad X = \frac{kT}{zJ}x \equiv \frac{T}{T_C}x, \qquad (T_C = zJ/k) \tag{5.10}$$

as shown in Figure 5.2 gives $x(T)$ and so $X(T) = M_s(T)/M_\infty$. M_s is also called the saturation magnetization. M_∞ is that at 0° K.

§ 5.2. BRAGG-WILLIAMS APPROXIMATION

The molecular field approximation was introduced in the previous section on the basis of physical intuition, but it can also be derived by statistical-mechanical reasoning as follows: Let N be the total number of atoms, and N_+ and N_- the numbers of atoms with upward spins or downward spins. If these positive or negative spins make an ideal mixture, then the number of the configurations of distributing these spins in the lattice is obviously given by

$$W = \binom{N}{N_+} = \frac{N!}{N_+!N_-!} \qquad (N = N_+ + N_-).$$

Therefore the entropy becomes

$$S = k \log W = -k\left\{N_+ \log\frac{N_+}{N} + N_- \log\frac{N_-}{N}\right\}$$

$$= -kN\{\tfrac{1}{2}(1+X)\log\tfrac{1}{2}(1+X) + \tfrac{1}{2}(1-X)\log\tfrac{1}{2}(1-X)\} \tag{5.11}$$

by virtue of the Stirling formula. Here X is defined by

$$\frac{N_+}{N} = \tfrac{1}{2}(1+X), \qquad \frac{N_-}{N} = \tfrac{1}{2}(1-X). \tag{5.12}$$

Now there are $\frac{1}{2}zN$ pairs of neighboring spins in the crystal, among which N_{++} are $++$ pairs, N_{--} are $--$ pairs, and N_{+-} are $+-$ pairs. Then the interaction energy (5.1) can be written as

$$E = -J(N_{++} + N_{--} - N_{+-}). \tag{5.13}$$

When N_+ and N_- are given, the numbers N_{++}, N_{--} are not determined uniquely, but may take on various values. If they are averaged simply (see note at the end of this page, they may be put as

$$\left.\begin{aligned} \bar{N}_{++} &= \tfrac{1}{2} zN_+p_+ = \tfrac{1}{2} zN_+ \frac{N_+}{N} = \tfrac{1}{8} z N(1+X)^2 , \\ \bar{N}_{+-} &= zN_+p_- \quad = zN_+ \frac{N_-}{N} \quad = \tfrac{1}{4} z N(1-X^2), \\ \bar{N}_{--} &= \tfrac{1}{2} zN_-p_- = \tfrac{1}{2} zN_- \frac{N_-}{N} = \tfrac{1}{8} z N(1-X)^2 . \end{aligned}\right\} \tag{5.14}$$

Here $p_+ = N_+/N$ and $p_- = N_-/N$ are the probabilities that a particular site on the lattice is occupied by a + spin or by a − spin. The factor $\frac{1}{2}$ appearing in the expressions of N_{++} and N_{--} takes account of the fact that each + + or − − pair is counted twice in zN_+p_+. In the approximation that N_{++} etc. in (5.13) are replaced by those average numbers given by (5.14), the total energy becomes

$$E = -\tfrac{1}{2} z JNX^2 . \tag{5.15}$$

Considering (5.11), one then has

$$\begin{aligned} F &= E - TS \\ &= -\tfrac{1}{2} z NJX^2 + NkT\left\{\tfrac{1}{2}(1+X)\log\tfrac{1}{2}(1+X) + \tfrac{1}{2}(1-X)\log\tfrac{1}{2}(1-X)\right\}. \end{aligned} \tag{5.16}$$

The equilibrium value (the most probable value) of X is determined by the condition, $\partial F/\partial X = 0$, or

$$zNJX = \tfrac{1}{2} NkT \log\frac{1+X}{1-X}, \qquad \text{or} \qquad \frac{zJ}{kT} X = \tfrac{1}{2} \log\frac{1+X}{1-X}, \tag{5.17}$$

or

$$\tanh\left(\frac{zJ}{kT} X\right) = X . \tag{5.18}$$

The last equation is the same as (5.9). Therefore one can say that the molecular field approximation is equivalent to the approximation which assumes the entropy to be the mixing entropy of an ideal mixture and the energy to be the average energy simply averaged over all possible configurations. These assumptions for entropy and energy were first used for the problems of alloys. This method is called Bragg and Williams' approximation (Bragg and Williams, Proc. Roy. Soc. A **145** (1934) 699). These assumptions are also used for the theory of regular solutions (see problem 6).

NOTE: The average involved in (5.14) is simply the average over all possible configurations of + and − spins in the lattice, giving equal weight to each of the configurations. The last assumption is, in fact, not true. If J is positive, $++$ and $--$ pairs have lower energies than $+-$ pairs, so that neighbors around a $+$ spin are more likely to be $+$ spins rather than $-$ spins. In (5.14), this tendency of aggregation is overlooked.

§ 5.3. CO-OPERATIVE PHENOMENA

Ferromagnetism, discussed in the preceding section, is a typical example of so-called co-operative phenomena, which means, generally, the appearance of a certain ordering in a system due to the interactions of particles in the system. Spontaneous magnetization is a kind of order which is maintained by the interaction of atoms. It is maximum at $T = 0°$ K, decreases with increasing temperature, and vanishes at a critical temperature (Curie temperature) T_C (see Fig. 5.3). Above the critical temperature the system becomes paramagnetic, because the ordering is disturbed by increasing thermal agitation so that the order loses the ability to maintain itself. In thermodynamic language, the entropy term in the free energy becomes dominant at higher temperatures, which favor configurations with greater randomness and with larger entropy. At low temperatures, the energy term overcomes the entropy term so that ordered states are favored.

Order-disorder transition: Ordering phenomena are rather common when the interaction is strong between the units in a system. For example:

Order-disorder change of alloys: In a ZnCu alloy (AB type alloy) for example, Zn atoms and Cu atoms are randomly mixed at high temperatures, but they become ordered at low temperatures. In the ordered states the atoms arrange themselves regularly on the lattice as shown in Figure 5.4. This type of ordered lattice is called a regular lattice (*ordered lattice, or superlattice*). This structure is proved experimentally by X-ray or neutron diffraction. The *order parameter* X is defined in the following way. Take the perfectly ordered lattice as the standard, and divide the lattice sites into an a-sublattice and a b-sublattice, which accommodate A atoms and B atoms, respectively, at absolute zero. At higher temperatures (or also at zero temperature if the composition is not ideal) the ordering is not perfect, so that there are, for instance, B atoms on the sublattice a. The distribution of A and B atoms among the sublattices is represented by the numbers,

$$\left.\begin{aligned}\begin{bmatrix}\mathrm{A}\\ \mathrm{a}\end{bmatrix} &= \frac{N}{4}(1+X), & \begin{bmatrix}\mathrm{B}\\ \mathrm{a}\end{bmatrix} &= \frac{N}{4}(1-X),\\ \begin{bmatrix}\mathrm{A}\\ \mathrm{b}\end{bmatrix} &= \frac{N}{4}(1-X), & \begin{bmatrix}\mathrm{B}\\ \mathrm{b}\end{bmatrix} &= \frac{N}{4}(1+X)\end{aligned}\right\} \tag{5.19}$$

where

$$\begin{bmatrix} \mathrm{A} \\ \mathrm{a} \end{bmatrix},$$

for example, is the number of A atoms on the sublattice a. N is the total number of lattice sites, and $\frac{1}{2}N$ is that of a sublattice. When X takes the value 1 or -1, the ordering is perfect (interchange of a and b sublattices means a change of the sign of X), whereas $X = 0$ means perfect disorder. If the interaction is such that AB pairs have lower energies than AA or BB pairs, then perfect ordering should be realized at absolute zero. The degree of ordering decreases at higher temperatures and eventually vanishes at a certain critical temperature. Figure 5.4 shows a few examples of ordered lattices.

Antiferromagnetism: If the exchange interaction J is negative, then anti-

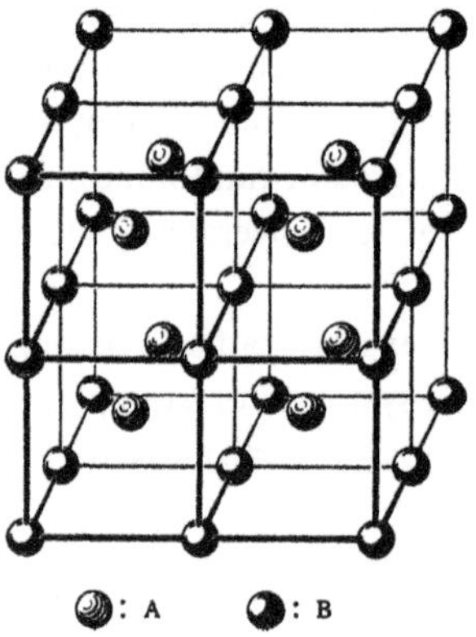

Fig. 5.4 (a). AB–type.

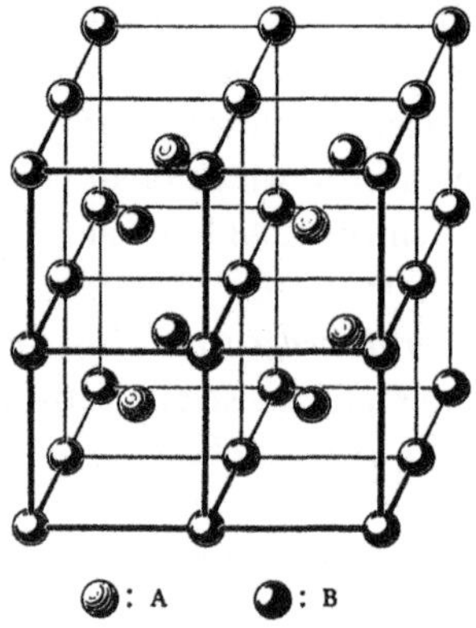

Fig. 5.4 (b). AB_3–type.

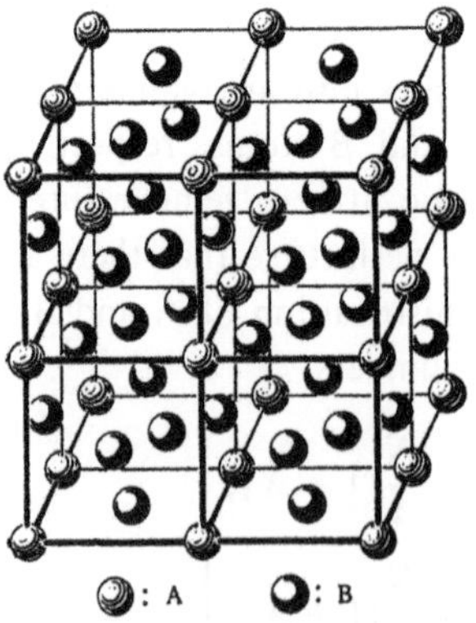

Fig. 5.4 (c). AB_3-type.

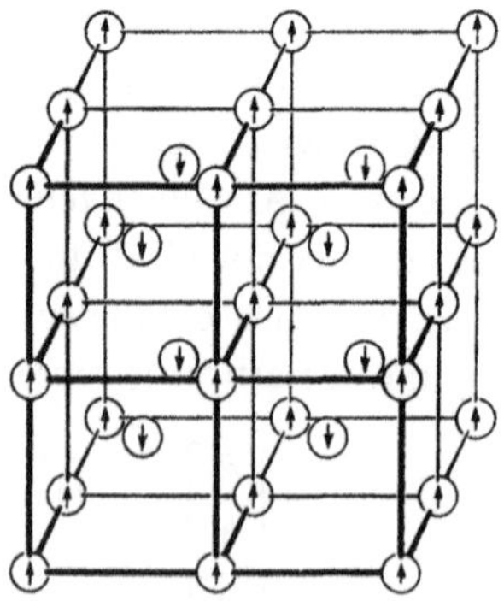

Fig. 5.5. Antiferromagnetic spin arrangement.

parallel spins are more stable than parallel spins. Therefore an alternating spin arrangement will be realized at low temperatures as shown, for example, in Figure 5.5. In this case, the lattice sites are divided into two sublattices. In each sublattice the spins are directed in the same direction, but the two groups have exactly opposite directions. As a result, there appears no net magnetization†. The crystal is paramagnetic, but it has peculiar features which distinguish it from ordinary paramagnets. At a certain critical temperature (Néel temperature) the ordering is lost and the crystal becomes an ordinary paramagnet. The ordering which occurs below the Néel temperature can be verified by neutron diffraction experiments.

More generally, in many cases the transition from one phase to another may be interpreted as some kind of order-disorder transition. It is not, however, always easy to find out which variable properly represents the state of the order. When such a variable is found, one has to use a certain approximation in order to calculate the free energy as a function of such a variable. The appearance of superconductivity, and the lambda transition of liquid helium are well-known examples.

§ 5.4. AVERAGE POTENTIAL IN CHARGED PARTICLE SYSTEMS

A rigorous treatment is also very difficult for a system in which the particles are interacting through Coulomb forces. The characteristic feature of the Coulomb interaction is its long range, being proportional to $1/r$. This makes necessary a difference in treatment from the previous examples of magnetic spins or atoms in alloys. It is, however, still possible to introduce the concept of an average field which is determined by the distribution of charged particles and which in turn governs the distribution itself. This is an approximation which corresponds exactly to the molecular field.

Consider, for example, a classical system of charged particles. The charge density at a point $\boldsymbol{r}$ in the space is given by

$$\rho(\boldsymbol{r}) = \overline{\sum_i e_i \delta(\boldsymbol{r} - \boldsymbol{r}_i)} = \sum_s e_s n_{s0} \exp\{-e_s \varphi(\boldsymbol{r})/kT\} \tag{5.20}$$

where s indicates the species of the particle, e_s being the charge, n_{s0} is the number density of the particle s when $\varphi = 0$, $n_{s0} \exp\{-e_s\varphi(\boldsymbol{r})/kT\}$ is the particle number density at $\boldsymbol{r}$, and T is the temperature. Now, the potential $\varphi(\boldsymbol{r})$ is determined by the Poisson equation

$$\Delta\varphi(\boldsymbol{r}) = -4\pi\rho(\boldsymbol{r}). \tag{5.21}$$

---

† In actual antiferromagnets, there are many types of spin arrangement, not only the simplest example as explained here.

Solving (5.20) and (5.21) together under suitable boundary conditions, one is able to find φ and the distribution of the particles. The thermodynamic properties of the system will be then determined.

§ 5.5. DEBYE-HÜCKEL THEORY

Debye and Hückel applied the idea mentioned in the previous section to ionic solutions and succeeded in deriving thermodynamic properties of ionic solutions. Let $i = 1, 2, \ldots$ be the species of ions in the solution. The average number density of the i-species is denoted by n_i and the charge by

$$e_i = |e| z_i, \ldots,$$
$$(\sum_i e_i n_i = 0: \text{condition for electrical neutrality}), \tag{5.22}$$

where z_i is the valency of the ion. The free energy of the total system actually contains the interactions between the solvent molecules and that between the ions and the solvent molecules, but we consider in the following only the part F_{el} which comes from the ionic interactions.

Now let us number the ions by α, β, ... and suppose that the charge on the α-th ion is increased by δe_α. In this procedure the following work has to be done from the outside: (1) the work δw_α to bring up the charge δe_α against the repulsion due to the charge e_α on the α-th ion, and (2) the work done against the interaction of δe_α with other ions surrounding α. The second amount of work is exactly given by $\psi_\alpha \delta e_\alpha$, which is related to the interaction free energy F_{el} by

$$\frac{\partial F_{el}}{\partial e_\alpha} = \psi_\alpha . \tag{5.23}$$

In order to find ψ_α, consider the distribution of ions around the α-th ion. The average potential $\psi(r)$ around the α-th ion is determined by

$$\Delta\psi(r) = -\frac{4\pi}{D}\rho(r) = -\frac{4\pi}{DV}\sum_\beta e_\beta \exp\{-e_\beta\psi(r)/kT\} \tag{5.24}$$
$$= -\frac{4\pi}{D}\sum_i e_i n_i \exp\{-e_i\psi(r)/kT\},$$

following the process of approximation explained in the previous section. In (5.24), D is the dielectric constant of the solvent, V is the volume of the system and the summation is made over all ions except the α-th ion. Assuming, for simplicity, that $e\psi(r) \ll kT$, keeping only the first order term of ψ at the right hand side (the zero-th order term vanishes because of the neutrality condition), one has

$$\Delta\psi(r) = \kappa^2\psi(r), \qquad \kappa^2 = \frac{4\pi}{DkT}\sum_i e_i^2 n_i. \tag{5.25}$$

The solution of this equation satisfying the condition $\psi(\infty) = 0$ is $\psi(r) = Ae^{-\kappa r}/r$. Now the ions have a finite size so that neighboring ions cannot be closer than a certain distance a. Since there are no other ions nor solvent molecules within the distance a, $r < a$, the potential there must be e_α/r. The constant A of the above solution is fixed by the continuity condition of the electric displacement at $r = a$, i.e.,

$$-D\frac{\partial\psi}{\partial r}\bigg|_{r=a} = \frac{DA}{r^2}e^{-\kappa r}(1 + \kappa r)\bigg|_{r=a} = \frac{e_\alpha}{a^2}.$$

Thus the solution of (5.25) is given by

$$\psi(r) = \frac{e_\alpha}{Dr}\frac{e^{-\kappa(r-a)}}{1 + \kappa a}. \tag{5.26}$$

In particular, the potential at $r = a$ becomes

$$\psi(a) = \frac{e_\alpha}{Da}\frac{1}{1 + \kappa a}. \tag{5.27}$$

The energy $\psi(a)\delta e_\alpha$ is the sum of the work δw_α and $\psi_\alpha\delta e_\alpha$ mentioned above. The work δw_α is exactly that amount of work to be done when the charge e_α is brought from infinity up to an ion (conductor) with a radius a and a charge e_α embedded in a dielectric medium having a dielectric constant D; this is evidently equal to $(e_\alpha/Da)\delta e_\alpha$. Therefore ψ_α in (5.23) is obtained as

$$\psi_\alpha = \psi(a) - \frac{e_\alpha}{Da} = -\frac{e_\alpha}{D}\frac{\kappa}{1 + \kappa a}. \tag{5.28}$$

Using (5.23), the interaction free energy F_{el} is obtained by the integration

$$F_{\text{el}} = \sum_\alpha \int_0^1 \psi_\alpha(\lambda)e_\alpha \, d\lambda, \tag{5.29}$$

where $\psi_\alpha(\lambda)$ means the value of ψ_α (5.28) when all e_α, e_β, ... are replaced by λe_α, λe_β, ... Carrying out the integration, one gets

$$F_{\text{el}} = -\sum_\alpha \frac{n_\alpha e_\alpha^2}{D}\kappa\int_0^1 \frac{\lambda^2 \, d\lambda}{1 + \lambda\kappa a} = -\sum \frac{n_\alpha e_\alpha^2}{3D}\kappa g(\kappa a), \tag{5.30}$$

$$g(x) \equiv \frac{3}{x^3}\{\log(1 + x) - x + \tfrac{1}{2}x^2\} = 1 - \tfrac{3}{4}x + \tfrac{3}{5}x^2 - \cdots.$$

If we assume that $\kappa a \ll 1$, (5.30) can be simplified to

$$F_{el} = -\sum_{\alpha} n_\alpha e_\alpha^2 \kappa/3D = -\frac{1}{3}\left(\frac{\sum n_\alpha e_\alpha^2}{D}\right)^{\frac{3}{2}}\left(\frac{4\pi}{kT}\right)^{\frac{1}{2}}. \tag{5.31}$$

This is the free energy per unit volume. When there are N ions in the solution in a volume V and if all of them are ionized with $+z$ or $-z$ valencies, then (5.31) may be written as

$$\frac{F_{el}}{NkT} = -\tfrac{2}{3}\pi^{\frac{1}{2}}\left\{\frac{N^{\frac{1}{3}}z^2e^2}{DkT}\right\}^{\frac{3}{2}}. \tag{5.32}$$

§ 5.6. DISTRIBUTION FUNCTIONS IN A PARTICLE SYSTEM

Consider a system of N identical particles confined in a volume V. The configurational distribution function of these N particles may be denoted by $F_N(\boldsymbol{r}_1, \dots, \boldsymbol{r}_N) \equiv F_N\{N\}$. This distribution function must be symmetrical in the coordinates $\boldsymbol{r}_1, \boldsymbol{r}_2, \dots$ and is normalized,

$$\int \cdots \int F_N\{N\}\, \mathrm{d}\{N\} = 1\,, \tag{5.33}$$

where $\{N\}$ represents the set of the coordinates $\boldsymbol{r}_1, \dots, \boldsymbol{r}_N$. By integrating $F\{N\}$ over $N - n$ particles we are left with the *n-body (reduced) distribution function*

$$\begin{aligned} F_n(\boldsymbol{r}_1, \cdots, \boldsymbol{r}_n) \equiv F_n\{n\} &= \frac{N!}{(N-n)!}\int \cdots \int F_N\{N\}\, \mathrm{d}\{N-n\} \\ &\left(\equiv \frac{N!}{(N-n)!}\int \cdots \int F_N(\boldsymbol{r}_1, \cdots, \boldsymbol{r}_N)\, \mathrm{d}\boldsymbol{r}_{n+1} \cdots \mathrm{d}\boldsymbol{r}_N\right). \end{aligned} \tag{5.34}$$

The *one-body (one-particle) distribution function* in fluids (liquids or gases) is simply a constant,

$$F_1(\boldsymbol{r}) = N \int \cdots \int F_N\{N\}\, \mathrm{d}\{N-1\} \equiv \rho(\boldsymbol{r}) = \text{const.} \equiv \rho_1 \tag{5.35}$$

which represents the number density. The *two-body distribution function* may be written as

$$F_2(\boldsymbol{r}_1, \boldsymbol{r}_2) = N(N-1)\int \cdots \int F_N\{N\}\, \mathrm{d}\{N-2\} \equiv \rho_1^2 g(\boldsymbol{r}) \qquad (\boldsymbol{r} = \boldsymbol{r}_2 - \boldsymbol{r}_1) \tag{5.36}$$

which defines *the correlation function* $g(\boldsymbol{r})$, i.e., $g(\boldsymbol{r})\, \mathrm{d}\boldsymbol{r}$ is the probability that a particle is found in the volume element $\mathrm{d}\boldsymbol{r}$ surrounding $\boldsymbol{r}$ when there is a particle at the origin $\boldsymbol{r} = 0$.

Let $U\{N\}$ be the potential function of the molecular interactions. For a classical system of particles, the distribution function $F\{N\}$ is explicity expressed by

$$F_N\{N\} = \frac{\lambda^N}{N!} \exp\left[(F - U\{N\})/kT\right] \tag{5.37}$$

$$\left[\lambda = \left(\frac{2\pi mkT}{h^2}\right)^{\frac{3}{2}}, \qquad F = \text{Helmholtz free energy}\right].$$

Then the n-particle distribution function is

$$F_n\{n\} = \frac{\lambda^N}{(N-n)!} \int \cdots \int \exp\left[(F - U\{N\})/kT\right] \mathrm{d}\{N-n\}. \tag{5.38}$$

Most physical quantities to be observed for a particle system are sum functions which have the form

$$A = \sum \{n\}_N a\{n\}. \tag{5.39}$$

Here $a\{n\}$ is a symmetric function with respect to n particle coordinates. The summation in (5.39) is over all possible choices of n out of N particles. Averages of sum functions of this type can be expressed in terms of the reduced distribution functions, namely

$$\langle A \rangle = \frac{1}{n!} \int \cdots \int a\{n\} F_n\{n\} \,\mathrm{d}\{n\}. \tag{5.40}$$

Examples

1. Consider a ferromagnetic crystal, each atom of which has a magnetic moment $g\mu_B \boldsymbol{S}$. Assume that there is an *exchange interaction* $-2J\boldsymbol{S}_j \cdot \boldsymbol{S}_l$ between each atom and its nearest neighbors with a positive exchange integral J, where the suffices represent the sites of the spins. This interaction causes the spins to align parallel at low temperatures. This is a simple model of ferromagnetic crystals, and is usually called the *Heisenberg model*. Now, derive the paramagnetic susceptibility χ as a function of T at high temperatures and obtain an expression for the Curie temperature T_C in terms of J, $S\,(=|\boldsymbol{S}|)$ and the number of nearest neighbors, z. Use the molecular field approximation.

SOLUTION

Consider an atom of spin $\boldsymbol{S}_0$ and its nearest neighbors with spins $\boldsymbol{S}_1, \ldots, \boldsymbol{S}_z$. The part of the energy that depends on the central spin $\boldsymbol{S}_0$ is

$$U = -2J\boldsymbol{S}_0 \cdot \sum_{m=1}^{z} \boldsymbol{S}_m - g\mu_B \boldsymbol{H} \cdot \boldsymbol{S}_0 \tag{1}$$

where $\boldsymbol{H}$ is the external field. In the molecular field approximation, one replaces the surrounding spins $\boldsymbol{S}_m$ by their average value $\bar{\boldsymbol{S}}$ and writes (1) as

$$U \simeq -2zJ\bar{\boldsymbol{S}}\cdot\boldsymbol{S}_0 - g\mu_B\boldsymbol{H}\cdot\boldsymbol{S}_0 = -g\mu_B(\boldsymbol{H}+q\boldsymbol{M})\cdot\boldsymbol{S}_0 . \tag{2}$$

Here $\boldsymbol{M}$ is the intensity of magnetization:

$$\boldsymbol{M} = ng\mu_B\bar{\boldsymbol{S}} \tag{3}$$

with n representing the number of spins per unit volume of the crystal, and

$$q = 2zJ/ng^2\mu_B{}^2 \tag{4}$$

is called the molecular field constant. Taking $\boldsymbol{H} + q\boldsymbol{M} = \boldsymbol{H}'$ as the appropriate molecular field, as suggested by the form of equation (2), one can easily calculate the average value $\bar{\boldsymbol{S}}_0$ of $\boldsymbol{S}_0$ (see Chapter 2, problem 9). Since the spins are all equivalent, $\bar{\boldsymbol{S}}_0$ must be equal to $\bar{\boldsymbol{S}}$. This condition of self-consistency, together with (3), gives an equation for determining $\boldsymbol{M}$ as a function of $\boldsymbol{H}$. At sufficiently high temperatures $g\mu_B H' \ll kT$, and one only has to calculate up to the first order in H'. If one takes the z-axis in the direction of $\boldsymbol{H}$, one will have a non-vanishing value only for the z-component of M:

$$\begin{aligned} M &= ng\mu_B \sum_{m=-S}^{S} m \exp\{\beta g\mu_B m(H+qM)\} \Big/ \sum_{m=-S}^{S} \exp\{\beta g\mu_B m(H+qM)\} \\ &= n\beta(g\mu_B)^2(H+qM)\sum_{m=-S}^{S} m^2/(2S+1) + \cdots \\ &= \frac{n(g\mu_B)^2}{kT}\tfrac{1}{3}S(S+1)(H+qM) + \cdots . \end{aligned} \tag{5}$$

Therefore, one finds for M, to first order in H,

$$M\left\{1 - \frac{2}{3}\frac{zJS(S+1)}{kT}\right\} = \frac{n(g\mu_B)^2}{3kT}S(S+1)H ,$$

or

$$M = \frac{n(g\mu_B)^2S(S+1)}{3\,k(T-T_C)}H \equiv \chi H , \tag{6}$$

where

$$T_C = \tfrac{2}{3}zJS(S+1)/k . \tag{7}$$

From (6) one has for the paramagnetic susceptibility

$$\chi = n(g\mu_B)^2S(S+1)/\{3k(T-T_C)\}$$

which is valid at $T > T_C$. T_C, as given by (7), is the temperature at which $\chi \to \infty$, that is, the Curie temperature.

NOTE 1: Expression (6) is referred to as *Curie-Weiss' law*. Plotting $1/\chi$ against T will give a straight line, as in Figure 5.6. For most real substances, however, the curves deviate from a straight line in the neighborhood of T_C, as illustrated by the dotted curve in the figure, which is for nickel. If one draws the tangent to the observed $1/\chi$ curve at a temperature very much higher than T_C, it will, therefore, intersect the T-axis at T_C' which is different from T_C. This temperature T_C' is sometimes called the paramagnetic Curie temperature.

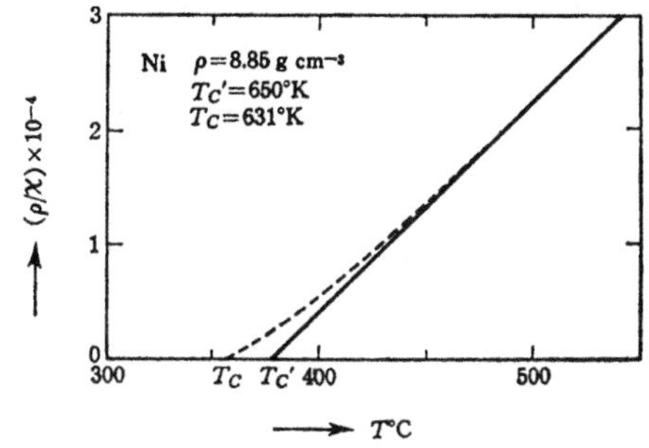

Fig. 5.6.

The ferromagnetic Curie temperature, in contrast, is the temperature at which the value of $1/\chi$ actually goes to zero. Any difference between T_C and T_C' indicates the inadequacy of the molecular field approximation.

NOTE 2: The coefficient

$$n(g\mu_B)^2 S(S+1)/3k = C \tag{8}$$

which appears in relation (6) is called *Curie's constant*. Equation (7) can then be written as $T_C = qC$. One could have written (5) in a more general fashion as

$$M = ng\mu_B \bar{S}_z = ng\mu_B S B_S\left(\frac{g\mu_B S(H+qM)}{kT}\right) \tag{9}$$

(see Chapter 2, problem 9). Equation (5) is the first term in the expansion of the right hand side of (9).

2. A binary alloy AB forms a superlattice below a critical temperature T_C. Derive an expression for the free energy as a function of X, the degree of long range order. Examine in detail its behavior near the point T_C. Show that the equation for determining X below T_C is of the form (5.17). (Use the Bragg-

Williams approximation. Take account of nearest neighbor interactions only. Assume that the lattice can be decomposed into two sublattices a and b which penetrate one another in such a way that any site on a sublattice is surrounded by neighbors belonging to the other sublattice.)

SOLUTION

Let the interaction energy of the nearest neighbor pairs AA, AB and BB be v_{AA}, v_{AB} and v_{BB}, respectively. Further, let the number of sites belonging to the sublattices a and b be denoted by [a] and by [b], respectively, and let the number of A atoms which belong to the sublattice a be denoted by $\begin{bmatrix}\mathrm{A}\\\mathrm{a}\end{bmatrix}$ and the number of the nearest neighbor pairs of the type AA, that is a pair of A atoms located on one of the a-sites and its neighboring b-site, be denoted by $\begin{bmatrix}\mathrm{A}&\mathrm{A}\\\mathrm{a}&\mathrm{b}\end{bmatrix}$, and so forth. In the Bragg-Williams approximation one substitutes appropriate average values in the same way as in (5.14) (see, however, the note at the end of § 5.2):

$$\begin{aligned}
\begin{bmatrix}\mathrm{A}&\mathrm{A}\\\mathrm{a}&\mathrm{b}\end{bmatrix} &= \begin{bmatrix}\mathrm{A}\\\mathrm{a}\end{bmatrix} z \begin{bmatrix}\mathrm{A}\\\mathrm{b}\end{bmatrix}\Big/[\mathrm{b}] = \frac{N}{4}(1+X)z\frac{N}{4}(1-X)\Big/\frac{N}{2} = \frac{zN}{8}(1-X^2),\\
\begin{bmatrix}\mathrm{A}&\mathrm{B}\\\mathrm{a}&\mathrm{b}\end{bmatrix} &= \begin{bmatrix}\mathrm{A}\\\mathrm{a}\end{bmatrix} z\tfrac{1}{2}(1+X) = \frac{zN}{8}(1+X)^2,\\
\begin{bmatrix}\mathrm{B}&\mathrm{A}\\\mathrm{a}&\mathrm{b}\end{bmatrix} &= \begin{bmatrix}\mathrm{B}\\\mathrm{a}\end{bmatrix} z\tfrac{1}{2}(1-X) = \frac{zN}{8}(1-X)^2,\\
\begin{bmatrix}\mathrm{B}&\mathrm{B}\\\mathrm{a}&\mathrm{b}\end{bmatrix} &= \begin{bmatrix}\mathrm{B}\\\mathrm{a}\end{bmatrix} z\tfrac{1}{2}(1+X) = \frac{zN}{8}(1-X^2).
\end{aligned} \tag{1}$$

Hence the interaction energy is given approximately by

$$\begin{aligned}
E &= v_{AA}\begin{bmatrix}\mathrm{A}&\mathrm{A}\\\mathrm{a}&\mathrm{b}\end{bmatrix} + v_{AB}\left\{\begin{bmatrix}\mathrm{A}&\mathrm{B}\\\mathrm{a}&\mathrm{b}\end{bmatrix} + \begin{bmatrix}\mathrm{B}&\mathrm{A}\\\mathrm{a}&\mathrm{b}\end{bmatrix}\right\} + v_{BB}\begin{bmatrix}\mathrm{B}&\mathrm{B}\\\mathrm{a}&\mathrm{b}\end{bmatrix}\\
&= \frac{zN}{8}\left[v_{AA}(1-X^2) + v_{AB}\{(1+X)^2+(1-X)^2\} + v_{BB}(1-X^2)\right]\\
&= \frac{zN}{8}(v_{AA}+2v_{AB}+v_{BB}) + \frac{z}{4}NX^2\left(v_{AB} - \frac{v_{AA}+v_{BB}}{2}\right)\\
&= E_0 - \frac{z}{4}NX^2 v, \qquad v \equiv \frac{v_{AA}+v_{BB}}{2} - v_{AB} > 0.
\end{aligned} \tag{2}$$

The entropy of the system is obtained in the same way as in (5.11):

$$S = k\log\frac{[\mathrm{a}]!}{\begin{bmatrix}\mathrm{A}\\ \mathrm{a}\end{bmatrix}!\begin{bmatrix}\mathrm{B}\\ \mathrm{a}\end{bmatrix}!}\frac{[\mathrm{b}]!}{\begin{bmatrix}\mathrm{A}\\ \mathrm{b}\end{bmatrix}!\begin{bmatrix}\mathrm{B}\\ \mathrm{b}\end{bmatrix}!} = k\log\left[\frac{\frac{N}{2}!}{\frac{N}{4}(1+X)!\frac{N}{4}(1-X)!}\right]^2$$

$$= -Nk\{\tfrac{1}{2}(1+X)\log\tfrac{1}{2}(1+X)+\tfrac{1}{2}(1-X)\log\tfrac{1}{2}(1-X)\} \qquad (3)$$

so that the free energy becomes

$$F(X,T) = E_0 - \frac{z}{4}NX^2v + NkT\{\tfrac{1}{2}(1+X)\log\tfrac{1}{2}(1+X) + \\ + \tfrac{1}{2}(1-X)\log\tfrac{1}{2}(1-X)\}. \qquad (4)$$

The entropy as a function of X has a maximum value of $Nk \log 2$ at $X = 0$ and vanishes at $X = \pm 1$. In Figure 5.7, F is plotted against X for given values

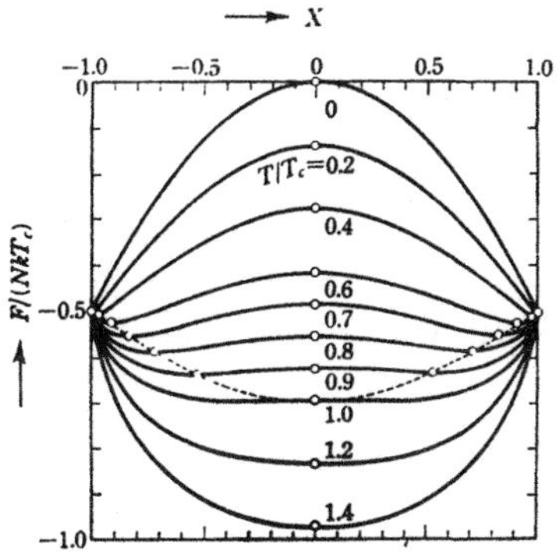

Fig. 5.7. ($E_0 = 0$).

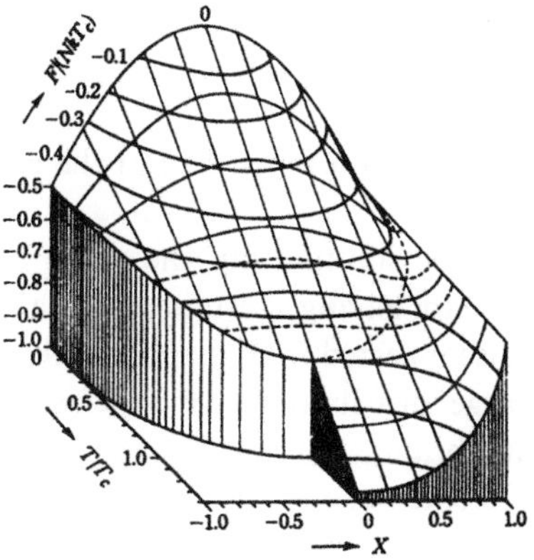

Fig. 5.8. ($E_0 = 0$).

of T/T_C, and Figure 5.8 shows its relief mapping. In order to discuss the behavior of F in the neighborhood of $X = 0$, we expand its expression, (4), in powers of X, up to terms of order X^4:

$$F = E_0 - NkT\log 2 - \tfrac{1}{4}zNvX^2 + NkT\{\tfrac{1}{2}X^2 + \tfrac{1}{12}X^4 + \cdots\} \\ = E_0 - NkT\log 2 + \tfrac{1}{2}N(kT - \tfrac{1}{2}zv)X^2 + \tfrac{1}{12}NkTX^4 + \cdots. \qquad (5)$$

One sees from (5) that the coefficient of X^2 is positive or negative according as

$$T \gtrless T_C \equiv zv/2k. \qquad (6)$$

This means that for $T > T_C$, F has a minimum at $X = 0$, while for $T < T_C$, F has a maximum at $X = 0$ and minima at $X \gtrless 0$. These values of X at which F becomes a minimum are determined by

$$\frac{\partial F}{\partial X} = 0,$$

or

$$\frac{zv}{2kT} X = \tfrac{1}{2} \log \frac{1+X}{1-X}. \tag{7}$$

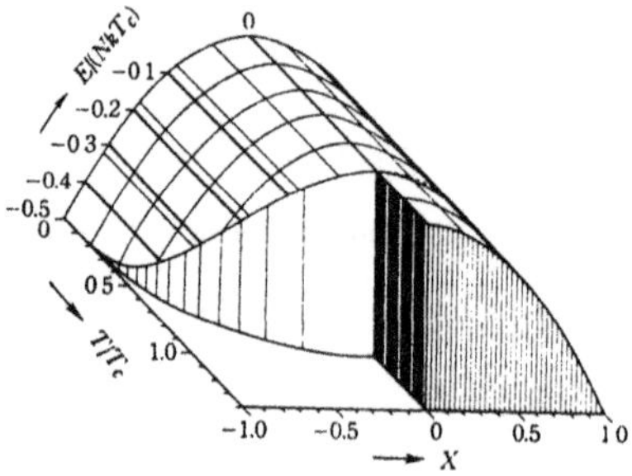

Fig. 5.9. ($E_0 = 0$).

Figure 5.9 is a relief map of the energy.

3. An ionic solution fills the space between the electrodes of a parallel plate capacitor having a certain potential difference V_0. The capacitor once charged, is disconnected from its potential source. Obtain an expression for the space charge distribution which exists after the system has reached thermal equilibrium. Assume, for the sake of simplicity, that the applied potential difference is so small that $eV_0 \ll kT$.

SOLUTION

The density of the positive ions should be equal to that of the negative ions on the central plane bisecting the space between the capacitor plates. (Let this plane be $x = 0$.) Writing the density as $n_+ = n_- = n$, putting the potential there equal to zero and using the dielectric constant D of the solution, we can write (5.20) and (5.21), which determine the distribution of the space charge, as

$$\frac{d^2\varphi}{dx^2} = -\frac{4\pi\rho}{D}, \tag{1}$$

and

$$\rho(x) = e(n_+ - n_-) = en\left\{\exp\left(-\frac{e\varphi}{kT}\right) - \exp\left(\frac{e\varphi}{kT}\right)\right\} = -2en \sinh\frac{e\varphi}{kT}. \tag{2}$$

Substituting (2) in (1) gives

$$\frac{d^2\varphi}{dx^2} = \frac{8\pi en}{D} \sinh \frac{e\varphi}{kT} \tag{3}$$

which can be approximated by

$$\frac{d^2\varphi}{dx^2} = \frac{8\pi e^2 n}{DkT}\varphi \tag{4}$$

under the assumption that $e\varphi \ll kT$ so that $\sinh(e\varphi/kT) \approx e\varphi/kT$. Put

$$\frac{DkT}{4\pi e^2 n} = l^2 \tag{5}$$

so that (4) takes the form

$$\varphi'' = 2\varphi/l^2 . \tag{6}$$

The general solution of (6) is given by

$$\varphi = A \sinh(\sqrt{2}x/l) + B \cosh(\sqrt{2}x/l) .$$

The boundary condition that $\varphi = 0$ at $x = 0$ leads to $B = 0$, while the other condition that $\varphi = \pm\frac{1}{2} V_0$ at $x = \pm a$ determines the value of A:

$$\varphi = \frac{V_0}{2} \sinh \frac{\sqrt{2}x}{l} \Big/ \sinh \frac{\sqrt{2}a}{l} . \tag{7}$$

One thus obtains the distribution of space charge, according to (2),

$$\rho = -2e^2 n\varphi/kT$$

$$= -\frac{e^2 n V_0}{kT} \sinh \frac{\sqrt{2}x}{l} \Big/ \sinh \frac{\sqrt{2}a}{l} \tag{8}$$

(see Fig. 5.10).

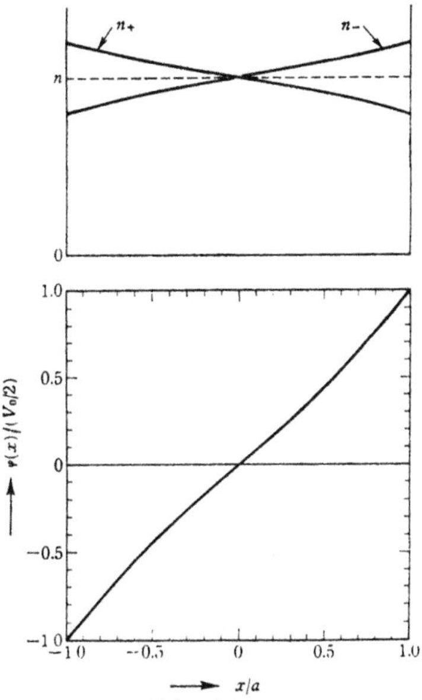

Fig. 5.10.

NOTE: The length l, as defined by (5), characterizes the decay of the potential and is called the Debye length. It is the inverse of κ which is defined by (5.25). As will be seen from (5.26), the potential due to a point charge has a range of about this length.

4. Show that the n-particle distribution function which is defined by (5.38) satisfies the following equation:

$$\frac{\partial}{\partial \boldsymbol{r}_i} F_n\{n\} + \frac{1}{kT}\frac{\partial U\{n\}}{\partial \boldsymbol{r}_i} F_n\{n\} + \frac{1}{kT}\int \frac{\partial \phi(\boldsymbol{r}_i, \boldsymbol{r}_{n+1})}{\partial \boldsymbol{r}_i} F_{n+1}\{n+1\}\,\mathrm{d}\boldsymbol{r}_{n+1} = 0$$

where

$$U\{n\} = \sum_{1 \leqq i < j \leqq n} \phi(\boldsymbol{r}_i, \boldsymbol{r}_j) \qquad (n = 2, \cdots, N).$$

SOLUTION

Differentiating

$$D_N\{N\} \equiv \exp[\beta(F - U\{N\})] \tag{1}$$

with respect to $\boldsymbol{r}_i$, one obtains

$$\frac{\partial D_N\{N\}}{\partial \boldsymbol{r}_i} + \beta \frac{\partial U\{N\}}{\partial \boldsymbol{r}_i} D_N\{N\} = 0. \tag{2}$$

In accordance with the defining equation (5.38), i.e.,

$$F_n\{n\} = \frac{\lambda^N}{(N-n)!}\int \cdots \int D_N\{N\}\,\mathrm{d}\{N-n\}$$

we multiply (2) by the factor $\lambda^N/(N-n)!$ and integrate it with respect to $\boldsymbol{r}_{n+1}, \ldots, \boldsymbol{r}_N$. Since we can write

$$\frac{\partial U\{N\}}{\partial \boldsymbol{r}_i} = \frac{\partial}{\partial \boldsymbol{r}_i} U\{n\} + \frac{\partial}{\partial \boldsymbol{r}_i}\sum_{j=n+1}^{N} \phi(\boldsymbol{r}_i, \boldsymbol{r}_j),$$

we obtain the following

$$\frac{\partial F_n\{n\}}{\partial \boldsymbol{r}_i} + \beta \frac{\partial U\{n\}}{\partial \boldsymbol{r}_i} \cdot F_n\{n\} +$$
$$+ \beta \sum_{j=n+1}^{N} \int \cdots \int \frac{\partial \phi(\boldsymbol{r}_i, \boldsymbol{r}_j)}{\partial \boldsymbol{r}_i} \frac{\lambda^N}{(N-n)!} D_N\{N\}\,\mathrm{d}\boldsymbol{r}_{j+1} \cdots \mathrm{d}\boldsymbol{r}_N = 0.$$

The integration involved in the last term with respect to all the coordinates but one, namely $\boldsymbol{r}_k$ ($k = n+1, \ldots, N$), gives the function $F_{n+1}(\boldsymbol{r}_1, \ldots, \boldsymbol{r}_n, \boldsymbol{r}_k)/(N-n)$. The factorial $(N-n)!$, in the last term is replaced by $(N-n-1)!$ because the summation in front of the integration sign cancels out the factor $(N-n)$. Therefore the function in the integrand is represented by $F_{n+1}\{n+1\}$ by definition; that is

$$\frac{\partial F_n\{n\}}{\partial \boldsymbol{r}_i} + \beta \frac{\partial U\{n\}}{\partial \boldsymbol{r}_i} F_n\{n\} + \beta \int \frac{\partial \phi(\boldsymbol{r}_i, \boldsymbol{r}_{n+1})}{\partial \boldsymbol{r}_i} F_{n+1}\{n+1\}\,\mathrm{d}\boldsymbol{r}_{n+1} = 0.$$

DIVERTISSEMENT 12

Statistical mechanics of co-operative phenomena. The molecular field approximation or Bragg-Williams' approximation, the method of Kirkwood and Bethe's approximation were mentioned in this chapter. These theoretical methods are very interesting as well as important in the statistical-mechanical treatment of strongly-interacting systems. However, it is forbiddingly difficult to construct an exact theory, or even to proceed further in these approximations. Progress of a sort has been made only for mathematical models such as the system of Ising spins, and hence it has been so far hardly applicable to real physical problems. In physical applications, we almost always have to be satisfied with the crudest approximation methods such as the molecular field approximation. Actually, such crude theories have proved to be extremely useful to obtain at least a qualitative understanding of a number of experimental facts, and therefore the molecular field approximation method is one of the most powerful ways of approach. There are, however, some cases where this approximation turns out to be insufficient. For example, the anomalous heat capacity of ferromagnets or alloys has a residual part above the transition temperature, which cannot be explained so long as one adheres to this approximation. This phenomenon comes from the fact that there still remains a certain degree of local order even above the critical temperature T_C; e.g., in the neighborhood of a plus atom the probability of finding plus atoms is larger than of finding minus atoms. In Bethe's approximation, which was the point of problem 20, one takes account of this effect by use of the degree of short-range order in addition to that of long-range order. In fact, we get $\bar{N}_{++}$ and $\bar{N}_{--}$ larger than $\bar{N}_{+-}$.

The figure below shows several specific heat curves for a two-dimen-

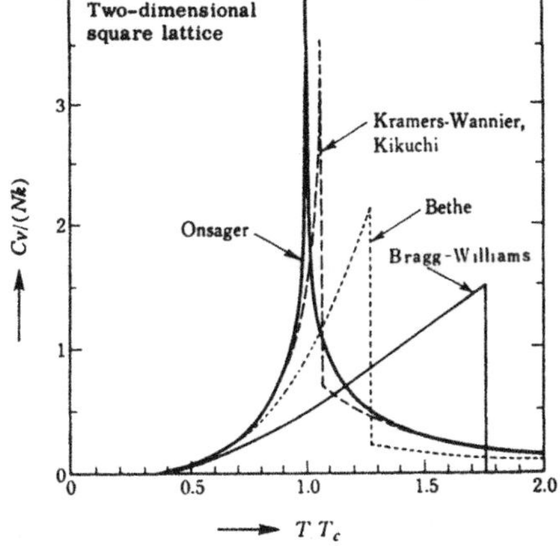

sional (square) Ising spin lattice which were calculated using various methods of approximation.

This system can also be treated exactly, and the result is included in the figure. The exact calculation is due to L. Onsager (Phys. Rev. 65 (1944) 117; the reader is also referred to a recent review of the Ising spin problem by Domb; C. Domb, Advances in Physics 9 (1960) 149, 245) and shows that the specific heat goes to infinity like $\log|T - T_C|^{-1}$ in the neighborhood of T_C. For general three-dimensional crystals, no one has ever succeeded in making an exact calculation. In this sense, the statistical mechanics of phase transitions has not been established in a rigorous way for real substances. Onsager's solution has been the only example, though a somewhat unrealistic one which proves a discontinuous phase change. Very recently, another instructive example has been treated by Kac, Uhlenbeck and Hemmer; their model is a one-dimensional gas in which molecules are interacting with each other by a long-range attraction and a short-range repulsion. A rigorous calculation is made possible by taking the limit of very weak and very long range attractive forces. They showed that in this limit the equation of state is found to coincide with the well-known van der Waals equation. This model is again unrealistic but it is extremely interesting.

Problems

A

1. Calculate the specific heat of the Ising model of ferromagnetism by use of the molecular field approximation. Examine, in particular, its behavior near the Curie point and also at sufficiently low temperatures where $T \ll T_C$. What will be the entropy change $\Delta S = S(T) - S(0)$ $(T > T_C)$?

2. Derive the equation for the spontaneous magnetization of the Heisenberg model of ferromagnetism (cf. example 1) using the molecular field approximation. Examine, in particular, the behavior of the magnetization in the neighborhood of T_C and in the low temperature limit: $T \to 0$.

3. Consider an antiferromagnetic crystal in which each atom has a spin of magnitude S. Unlike a Heisenberg ferromagnet, neighboring spins have an exchange interaction $2\,|J|\,\boldsymbol{S}_j \cdot \boldsymbol{S}_i$ which tends to align the spins antiparallel. The crystal structure is assumed to be such that the whole lattice can be divided into two interpenetrating sublattices (like an ideal AB type alloy). Therefore The spins belonging to each sublattice tend to align parallel, whereas they align antiparallel between different sublattices (Van Vleck's model of an antiferromagnet). Now calculate the paramagnetic susceptibility χ at higher temperatures than T_C, using the molecular field approximation. Assume that the molecular field prevailing in the a sublattice is $-q_2\boldsymbol{M}_a - q_1\boldsymbol{M}_b$, and that in the b sublattice it is $-q_2\boldsymbol{M}_b - q_1\boldsymbol{M}_a$.

4. Calculate the anomalous heat capacity of a binary alloy of AB type by use of the Bragg-Williams approximation.

5. Derive an equation for determining the degree of long range order for an alloy of AB type which has a ratio of atomic concentrations different from 1 : 1. Apply the Bragg-Williams approximation. Find the relation between the transition temperature and the ratio of atomic concentrations. (*Hint:* Let the total numbers of A and B atoms be denoted by N_A and N_B, respectively, and assume that $N_A < N_B$. Hence $x_A = N_A/(N_A + N_B) < \frac{1}{2}$. Define the degree of long range order, X, by

$$\begin{bmatrix}\mathrm{A}\\ \mathrm{a}\end{bmatrix} = \frac{N}{2}x_A(1 + X), \qquad \begin{bmatrix}\mathrm{A}\\ \mathrm{b}\end{bmatrix} = \frac{N}{2}x_A(1 - X),$$

$$\begin{bmatrix}\mathrm{B}\\ \mathrm{a}\end{bmatrix} = \frac{N}{2}(x_B - x_A X), \qquad \begin{bmatrix}\mathrm{B}\\ \mathrm{b}\end{bmatrix} = \frac{N}{2}(x_B + x_A X),$$

where $x_B = 1 - x_A$ and $N = N_A + N_B$).

6. Find the free energy and the chemical potentials of each component of a two-component solution, assuming a lattice model for a liquid and using the Bragg-Williams approximation. Only nearest neighbor interactions may be assumed to be present. Note, in particular, the difference between miscible and inmiscible cases.

7. Suppose each atom of a crystal can take any one of r internal states, which are designated by σ, and that the interaction between an atom which is in state σ' and its nearest neighbor which is in σ'' is given by $u(\sigma', \sigma'')$. (For simplicity, assume $u(\sigma', \sigma'') = u(\sigma'', \sigma')$.) One can treat this system in the following approximate way, which is an analogue of the Bragg-Williams method: let the probability that an atom is in state σ be $f(\sigma)$, and write the interaction energy as

$$U = \tfrac{1}{2} N z \sum_{\sigma'} \sum_{\sigma''} u(\sigma', \sigma'') f(\sigma') f(\sigma'')$$

where the number of nearest neighbors is z. Then the entropy, in accordance with (1.101), can be written as

$$S = -Nk \sum_{\sigma} f(\sigma) \log f(\sigma)$$

and $f(\sigma)$ is to be determined so as to minimize the free energy $F = U - TS$. What, then, is the equation for determining $f(\sigma)$?

8. Apply the method stated in problem 7 to the Ising model which was explained in § 5.1.

9. Electrons escape from a metal and are attracted by the positive charge induced on the surface, thus forming a distribution of space charge near the surface of the metal (see the figure). Assuming that the density of the electrons outside the metal is sufficiently small, show that it obeys the relation:

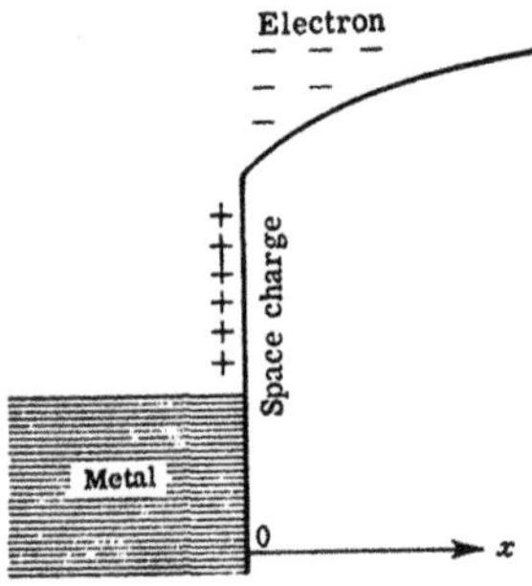

$$n(x) \propto (x_0 + x)^{-2},$$

where $n(x)$ is the density of electrons at the point x from the surface and x_0 is a constant.

10. If one regards the free energy of a solution of electrolyte F as a function of the ionic charges e_α, e_β, ..., it should satisfy the relation:

$$\frac{\partial}{\partial e_\beta}\frac{\partial F}{\partial e_\alpha} = \frac{\partial}{\partial e_\alpha}\frac{\partial F}{\partial e_\beta}.$$

Show that the $\psi_\alpha = \partial F/\partial e_\alpha$, which was defined by (5.23) and given by (5.28) in the Debye-Hückel theory, satisfy the above condition.

11. Define the n-particle distribution function of a pure liquid by (5.34), then show that $F_1(\boldsymbol{r})$ is the number density of the particles and that the average $\langle U \rangle$ of the sum of two-particle interactions

$$U = \sum_{i<j}\sum u(\boldsymbol{r}_i - \boldsymbol{r}_j)$$

is given by

$$\langle U \rangle = \frac{N^2}{2V}\int u(\boldsymbol{r})g(\boldsymbol{r})\,d\boldsymbol{r}$$

where $g(\boldsymbol{r}_1 - \boldsymbol{r}_2)$ is defined by

$$F_2(\boldsymbol{r}_1, \boldsymbol{r}_2) = n^2 g(\boldsymbol{r}_1 - \boldsymbol{r}_2)$$

and represents the spatial correlation of two particles.

12. Show that the pressure of a gas consisting of particles which interact according to a two-body central force interaction

$$\sum_{i<j}\sum u(r_{ij})$$

is given by

$$p = nkT - \frac{2\pi}{3} n^2 \int_0^\infty \frac{\mathrm{d}u(r)}{\mathrm{d}r} g(r) r^3 \,\mathrm{d}r$$

so long as classical statistical mechanics is valid. Here $g(r)$ is the two-particle correlation function as defined in the preceding problem. [*Hint:* Perform the transformation $r_i \to \alpha r_i$ on all the co-ordinates in the partition function Q_N, as defined by (3.19), and represent the derivative of Q_N with respect to V by that with respect to α. In this manner derive a formula for p (virial theorem) and transform it appropriately.]

B

13. One defines the Néel temperature T_N of an antiferromagnetic material as the critical point below which both of the sublattices have spontaneous magnetizations M_a and M_b. Find the Néel point for the Van Vleck model which was described in problem 3. Use the molecular field approximation.

14. Find the magnetic susceptibility for the Van Vleck model, which was treated in problems 3 and 13, at temperatures $T < T_N$. Again use the molecular field approximation. (The magnetizations of sublattices below the Néel temperature are equal in magnitude but opposite in direction: $\boldsymbol{M}_a = -\boldsymbol{M}_b = \boldsymbol{M}'$. Notice that there is a difference between the case when an external field $\boldsymbol{H}$ is applied perpendicularly to $\boldsymbol{M}'$ (which defines the perpendicular susceptibility $\chi_\perp$) and the case when $\boldsymbol{H}$ is parallel to $\boldsymbol{M}'$ (which defines the parallel susceptibility $\chi_\parallel$).

15. The energy of an Ising spin system in a magnetic field H is, according to (5.1), given by

$$\mathscr{H} = -g\mu_B H \sum_{j=1}^{N} \sigma_j - J \sum \sigma_i \sigma_j .$$

If one has succeeded in expressing the partition function of this system

$$Z_I = \sum_{\sigma 1 = \pm 1} \cdots \sum_{\sigma_N = \pm 1} \exp(-\mathscr{H}/kT)$$

as a function of H and T, then one can derive the partition function of the binary alloy, which was expounded in § 5.3, as a function of X and T. Explain this relation. (Call A atoms on a, or B on b, r-atoms and consider them to correspond to positive spins, and call the other atoms w-atoms and consider them to correspond to negative spins. Then Z_I is the grand partition function when one allows an indefinite number of positive or negative spins.)

16. 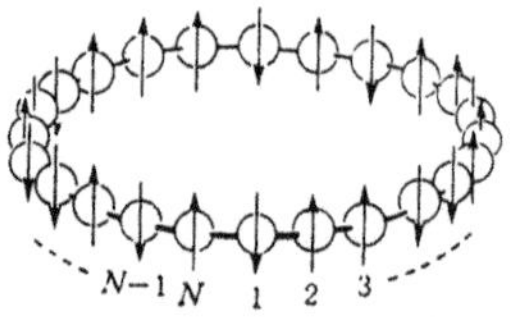

Suppose that N Ising spins are arranged along a ring. Assume that the energy of this system is given by

$$\mathscr{H} = -J \sum_{j=1}^{N} \sigma_j \sigma_{j+1} \qquad (\sigma_{N+1} \equiv \sigma_1).$$

Show that the expression, where σ_i takes the value $+1$ or -1.

$$F = -NkT \log\{2\cosh(J/kT)\}$$

is an exact expression of its free energy and find the relation between the specific heat and the temperature. (Notice that $\exp(\alpha\sigma_i\sigma_j) = \cosh\alpha + \sigma_i\sigma_j \sinh\alpha$.)

17. Consider a solution of solvent molecules A and polymer solute molecules B. Suppose that a molecule B is a chain consisting of pieces of r structural units linked successively and free to turn. Suppose further that a structural unit of B is of the same size as of the A molecule, as illustrated by the figure. Show that the number of ways of putting the polymers at N sites is given approximately by

$$W = \frac{z^{N_2}(z-1)^{N_2(r-2)}}{N_2!\sigma^{N_2}} N^{-(r-1)N_2} r^{rN_2} \left\{\frac{(N/r)!}{(N_1/r)!}\right\}^r$$

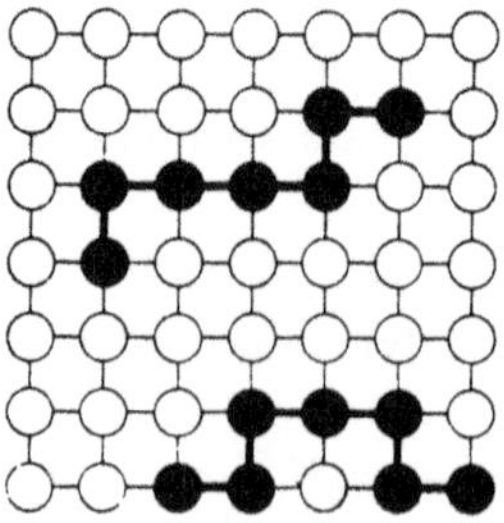

where N_1 is the number of the solvent molecules and N_2 is that of the solute molecules, so that $N = N_1 + rN_2$ is the total number of lattice sites (Flory's approximation). Here z denotes the number of nearest neighbor sites and σ is a symmetry number which has the value 2 when the polymer is symmetrical with respect to its center and 1 otherwise. From W above, derive the expression for the mixing entropy which can be written as

$$\Delta S = -k(N_1 \log \phi_1 + N_2 \log \phi_2)$$

where

$$\phi_1 \equiv \frac{N_1}{N_1 + rN_2}, \qquad \phi_2 \equiv \frac{rN_2}{N_1 + rN_2}$$

are the volume concentrations of the solvent and the solute.

18. Suppose that positive charge is uniformly distributed with density $n_0 e$ and on this background electrons are distributed with an average number density of n_0. When an extra charge ze is introduced at a certain point (which may be taken as the origin of the coordinate system) in this medium then the electrons will be redistributed. Derive the expressions of the electronic distribution and the potential for the following two cases: (i) when the temperature is so high that the electron gas is non-degenerate and (ii) when it is 0° K and the gas is completely degenerate. (Since the equation for determining the electronic distribution will turn out to be non-linear, treat it by means of an appropriate linear approximation.)

C

19. According to the correspondence mentioned in problem 15, the partition function for a binary alloy can be written as

$$Z(X) = \sum_{(X)} \exp\{\beta \sum_{<ij>} \sigma_i \sigma_j\} \equiv W(X) \cdot \langle \exp(\beta A) \rangle$$

with $\beta = v/2kT$. Here $\sum_{(X)}$ means the summation over the values ± 1 of every Ising spin σ_j under the condition that the degree of long-range order be X and $\langle A \rangle$ is an average defined by

$$\langle A \rangle = \frac{\sum_{(X)} A(\sigma_1, \cdots, \sigma_N)}{\sum_{(X)} 1} \equiv \frac{\sum_{(X)} A(\sigma_1, \cdots, \sigma_N)}{W(X)}.$$

Hence $W(X) = \sum_{(X)} 1$. Calculate $\log Z(X)$ up to terms of β^2 by expanding the expression $\langle \exp(\beta A) \rangle$ in powers of β and calculating $\langle A^n \rangle$ in elementary fashion (term by term), and thus find an approximate value of the transition temperature T_C of order-disorder transformation. (*Kirkwood's method of approximation.*)

20. A method of improving the molecular field approximation may be the following one for the case of Ising spin system: one retains the interaction of a single spin σ_0 and its surrounding z spins $\sigma_1, \ldots, \sigma_z$, i.e., $J\sigma_0 \sum \sigma_j$ and takes into account the rest only by way of a molecular field H' which is assumed to act on the z spins $\sigma_1, \ldots, \sigma_z$. That is, one assumes that the energy of the $z+1$ spins is given by

$$\mathscr{H}_{z+1} = -\mu_0 H\sigma_0 - \mu_0(H+H')\sum_{j=1}^{z}\sigma_j - J\sum_{j=1}^{z}\sigma_0\sigma_j .$$

The molecular field H' is to be determined by the condition that the average $\bar{\sigma}_0$ of σ_0 is equal to the average $\bar{\sigma}_j$ of σ_j ($j = 1, 2, \ldots, z$) (the condition of self-consistency). Find the Curie point T_C of the Ising spin system in this way (*Bethe's approximation*).

21. Find the average number of neighboring pairs in the system of Ising spins, using Bethe's method of approximation described in the preceding problem. Let them be denoted by $\bar{N}_{++}$, $\bar{N}_{--}$ and $\bar{N}_{+-}$ for the cases $++$, $--$ and $+-$ pairs, respectively, and show that they satisfy the following relation:

$$\frac{\bar{N}_{+-}{}^2}{\bar{N}_{++}\bar{N}_{--}} = 4\,e^{-4J/kT} .$$

22. The s-particle distribution function $F_s\{s\} = F_s(\boldsymbol{r}_1, \ldots, \boldsymbol{r}_s)$ for a gas, which was defined in (5.34), satisfies

$$\lim_{\text{all}\,|\boldsymbol{r}_{ij}|\to\infty} F_s(\boldsymbol{r}_1, \cdots, \boldsymbol{r}_s) = v^{-s}$$

where $v^{-1} = N/V$ indicates the number density of the particles. One can put

$$F_s(\boldsymbol{r}_1, \cdots, \boldsymbol{r}_s) = v^{-s} g_s(\boldsymbol{r}_1, \cdots, \boldsymbol{r}_s)$$

and expands the function g in such a way that

$$g_s(\boldsymbol{r}_1, \cdots, \boldsymbol{r}_s) = g_s^{(0)} + v^{-1} g_s^{(1)} + v^{-2} g_s^{(2)} + \cdots$$

and can calculate, in principle, all the terms successively according to the equation derived in example 4. Prove this statement and apply it to the formula for pressure which was given in problem 12 and find the espressions for the second and the third virial coefficients.

Solutions

1. By (5.15) $E = -\frac{1}{2}zNJX^2$. Hence the heat capacity is given by

$$C = -zNJX\frac{dX}{dT}. \tag{1}$$

Now X is determined by (5.18), i.e.,

$$\tanh\left(\frac{T_C}{T}X\right) = X. \tag{2}$$

Differentiating (2) with respect to T:

$$\frac{1}{\cosh^2(T_CX/T)}\left(-\frac{T_C}{T^2}X + \frac{T_C}{T}\frac{dX}{dT}\right) = \frac{dX}{dT},$$

and putting $T_CX/T = x$ gives

$$\frac{dX}{dT} = -\frac{x}{T}\Bigg/\left(\cosh^2 x - \frac{T_C}{T}\right). \tag{3}$$

Therefore, by using the relation $zJ/k = T_C$, equation (1) can be expressed in termsof x

$$C = Nkx^2\Bigg/\left\{\cosh^2 x - \frac{x}{\tanh x}\right\} \tag{4}$$

and (2) becomes

$$T/T_C = \tanh x/x. \tag{5}$$

Equations (4) and (5) together are parametric expressions giving C as a function of T/T_C. The point $x = 0$ corresponds to $T = T_C$, and $x \to \infty$ corresponds to $T \to 0$.

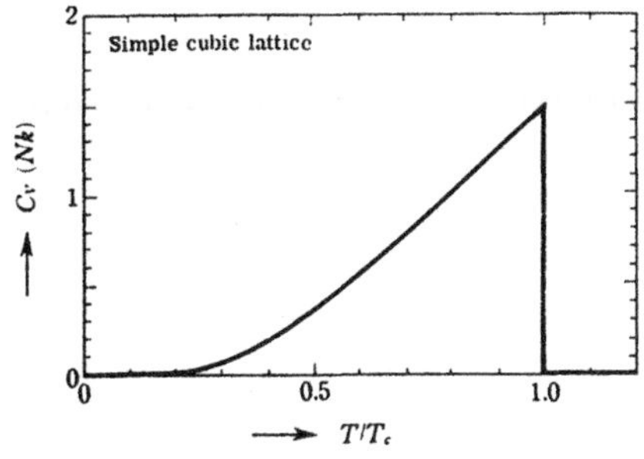

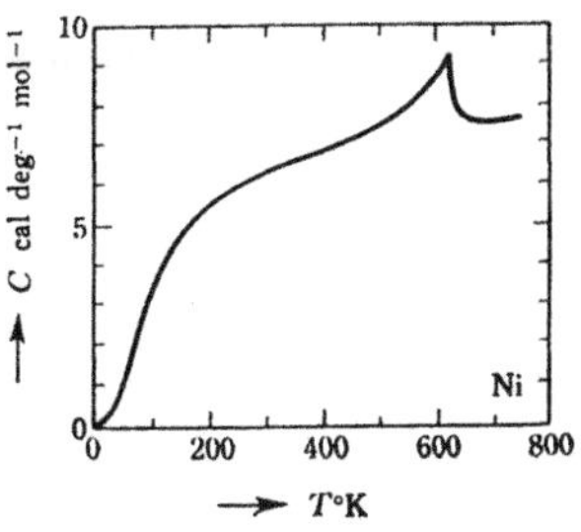

In the neighborhood of T_C in particular, one can apply the expansion formula for tanh directly to (2) and obtain

$$\frac{1}{3}\left(\frac{T_C}{T}\right)^3 X^2 + \cdots = \frac{T_C}{T} - 1, \qquad \text{or} \quad X^2 = 3\left(\frac{T_C}{T} - 1\right) + \cdots = 3\frac{T_C - T}{T_C} + \cdots \tag{6}$$

Substituting this for X in (1) gives

$$C(T \to T_C - 0) = \frac{3}{2}\frac{zNJ}{T_C} = \tfrac{3}{2}Nk. \tag{7}$$

At T greater than T_C, X is always zero so that C identically vanishes. Therefore, the specific heat changes discontinuously in the manner shown in the figure above. At lower temperature, $T \ll T_C$ $(x \to \infty)$, (5) can be written as $x \sim T_C/T$: while in (4), $x/\tanh x$ can be neglected as compared with $\cosh^2 x$, so that

$$C \cong 4Nkx^2 e^{-2x} = 4Nk\left(\frac{T_C}{T}\right)^2 \exp(-2T_C/T). \tag{8}$$

This shows that C rapidly approaches zero as $T \to 0$. The entropy variation can be obtained by integrating C/T with respect to T. Noting that $x = 1$ at $T = 0$, the integration gives

$$S(T) - S(0) = \int_0^T \frac{C}{T}\mathrm{d}T = -zUJ\int_{T=0}^{T} \frac{X}{T}\mathrm{d}X$$

by the use of equation (1). Expressing T as a function X by (2), one proceeds as follows:

$$S(T) - S(0) = \int_0^T \frac{C}{T}\mathrm{d}T = -zNJ\int_{T=0}^{T} \frac{X}{T}\mathrm{d}X = z\frac{NJ}{T_C}\int_X^1 \tfrac{1}{2}\log\left(\frac{1+X}{1-X}\right)\mathrm{d}X$$

$$= \tfrac{1}{2}Nk\left[(1+X)\log(1+X) + (1-X)\log(1-X)\right]_X^1$$
$$= \tfrac{1}{2}Nk\{2\log 2 - (1+X)\log(1+X) - (1-X)\log(1-X)\}$$
$$= -Nk\{\tfrac{1}{2}(1+X)\log\tfrac{1}{2}(1+X) + \tfrac{1}{2}(1-X)\log\tfrac{1}{2}(1-X)\}. \tag{9}$$

Here, X appears as a function T according to (2). In particular,

$$S(T > T_C) - S(0) = Nk\log 2, \tag{10}$$

a result which could have been anticipated.

NOTE 1: From $E = -\frac{1}{2}zNJX^2$ and from S of (9), one can obtain the free energy F as a function T and X, and determine the equilibrium X according to the condition $(\partial F/\partial X)_T = 0$ (cf. example 2). This is, in fact, the Bragg-Williams method (§ 5.2).

NOTE 2: The anomalous heat capacity considered here together with contributions from the other degrees of freedom make up the observed heat capacity in real substances. The figure above shows the observed values for nickel.

2. If one puts

$$\boldsymbol{H}' = \boldsymbol{H} + q\boldsymbol{M} \tag{1}$$

for the molecular field, $\boldsymbol{M}$ will turn out to be parallel to $\boldsymbol{H}'$ and its magnitude is given by (9) of example 1. Since the spontaneous magnetization is by definition the value which M takes when $H \cong 0$, it is determined by the equation

$$M = ng\mu_B S B_S\left(q\frac{g\mu_B SM}{kT}\right). \tag{2}$$

With the aid of the expansion formula (equation (4) of problem 9, Chapter 2):

$$B_S(x) = \frac{S+1}{3S}x - \frac{1}{45}\frac{(S+1)\{(S+1)^2+S^2\}}{2S^3}x^3 + \cdots. \tag{3}$$

Equation (2) can be written in the form

$$M = \frac{a}{T}M - \frac{b}{T^3}M^3 + \cdots. \tag{4}$$

Therefore, the non-zero solution is given by

$$M^2 \cong \frac{T^3}{b}\left(\frac{a}{T} - 1\right) \simeq \frac{T_C^2}{b}(T_C - T) \tag{5}$$

or

$$M \sim A\sqrt{T_C - T}, \qquad A = T_C/\sqrt{b},$$

where we have put $a = T_C$. From (5) we see that T_C is the ferromagnetic Curie point and (5) is valid only for $T < T_C$. On the other hand, for $x \gg 1$, $B_S(x) = 1 - (1/S)\,e^{-x/S}$. Therefore, (2) can be written approximately as

$$M = M_0\left\{1 - \frac{1}{S}\exp\left(-\frac{g\mu_B}{kT}qM_0\right)\right\} \tag{6}$$

for $T \to 0$. Here $M_0 = ng\mu_B S$ means the saturation magnetization at absolute zero.

NOTE: The figure shows the calculated value of M/M_0 plotted against T/T_C for the cases: $S = \frac{1}{2}$, $S = 1$ and $S \to \infty$. The general trend of the observed values for Ni and Fe follows the curve for $S = \frac{1}{2}$. However, at the lowest temperatures, the experimental data are not in agreement with (6), but are known to follow better the equation

$$M = M_0 \left\{ 1 - \left(\frac{T}{T_0} \right)^{\frac{3}{2}} \right\}$$

which is derivable from the spin wave theory of ferromagnetism (c.f. note appended to the solution to problem 19, Chapter 2).

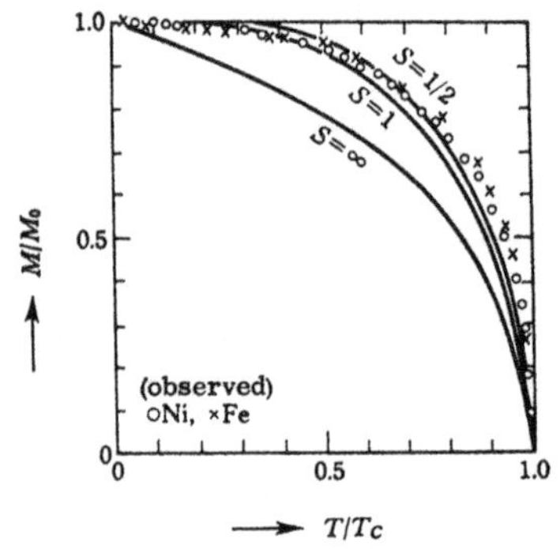

3. Follow the same method of solution as used in example 1 and problem 2. Let the magnetizations of the sublattices a and b be denoted by $\boldsymbol{M}_a$ and $\boldsymbol{M}_b$, respectively, so that

$$\boldsymbol{M}_a = \tfrac{1}{2} ng\mu_B \boldsymbol{S}_a, \qquad \boldsymbol{M}_b = \tfrac{1}{2} ng\mu_B \boldsymbol{S}_b, \tag{1}$$

where n represents the number of atoms per unit volume. One then puts

$$\boldsymbol{H}'_a = \boldsymbol{H} - q_2 \boldsymbol{M}_a - q_1 \boldsymbol{M}_b, \qquad \boldsymbol{H}'_b = \boldsymbol{H} - q_2 \boldsymbol{M}_b - q_1 \boldsymbol{M}_a \tag{2}$$

for the molecular fields prevailing at a and b, respectively, and obtains

$$\boldsymbol{M}_a = \frac{n}{2} g\mu_B \boldsymbol{S}_a = \frac{n}{2} \frac{(g\mu_B)^2}{3kT} S(S+1) \boldsymbol{H}'_a = \frac{C}{2T} (\boldsymbol{H} - q_2 \boldsymbol{M}_a - q_1 \boldsymbol{M}_b), \tag{3}$$

$$\boldsymbol{M}_b = \frac{n}{2} g\mu_B \boldsymbol{S}_b = \frac{n}{2} \frac{(g\mu_B)^2}{3kT} S(S+1) \boldsymbol{H}'_b = \frac{C}{2T} (\boldsymbol{H} - q_1 \boldsymbol{M}_a - q_2 \boldsymbol{M}_b) \tag{4}$$

for the magnetizations of the two sublattices, similar to (5) in the solution to example 1. Here $C = n(g\mu_B)^2 S(S+1)/3k$. The total magnetization is, from (3) and (4),

$$M = \frac{C}{T}H - \frac{C}{2T}(q_1 + q_2)(M_a + M_b) = \frac{C}{T}H - \frac{C}{2T}(q_1 + q_2)M,$$

or

$$M = \frac{C}{T+\theta}H, \qquad \text{where} \qquad \theta = \tfrac{1}{2}(q_1 + q_2)C. \tag{5}$$

Here, q_1, and q_2 are given explicitly by

$$q_1 = 2z_1 |J_1| / \tfrac{1}{2} n g^2 \mu_B^2, \qquad q_2 = 2z_2 |J_2| / \tfrac{1}{2} n g^2 \mu_B^2,$$

where z_1 and z_2 are, respectively, the number of nearest neighbors on the other and on the same sublattice, while J_1 and J_2 are the corresponding interaction energies.

NOTE: Compare (5) with the Curie-Weiss law, equation (6) of the solution to example 1. The spins in an anti-ferromagnetic substance tend to align anti-parallel, so that the susceptibility comes out smaller than that of Curie's law, C/T.

4. Follow the method of solution used in example 2 and problem 1. By equations (2) and (7) of the solution to example 2,

$$C = \frac{dE}{dT} = -\tfrac{1}{2} z N v X \frac{dX}{dT} \tag{1}$$

$$\frac{zv}{2kT} X = \tfrac{1}{2} \log \frac{1+X}{1-X}. \tag{2}$$

Putting $zv/2k = T_C$, one rewrites (2) as

$$\frac{T_C}{T} X = \tfrac{1}{2} \log \frac{1+X}{1-X}, \qquad \text{or} \qquad \tanh\left(\frac{T_C}{T} X\right) = X. \tag{3}$$

Substituting $\frac{1}{2} z$ for J in (1) and (2) of the solution to problem 1, one obtains precisely (1) and (3). Therefore the heat capacity behaves just as in the figure for solution 1. The transition at T_C is of the second kind, since there is a discontinuity of heat capacity there.

5. From the definition of X (as suggested in the hint), $X = 0$ represents the completely disordered state and $X = 1$ the completely ordered state. Follow the method of solution used in example 2. By analogy with equation (2) of example 2, one obtains for the energy

$$
\begin{aligned}
E &= v_{AA}\begin{bmatrix} A & A \\ a & b \end{bmatrix} + v_{AB}\left\{\begin{bmatrix} A & B \\ a & b \end{bmatrix} + \begin{bmatrix} B & A \\ a & b \end{bmatrix}\right\} + v_{BB}\begin{bmatrix} B & B \\ a & b \end{bmatrix} \\
&= v_{AA}\tfrac{1}{2}zN\,x_A^2(1-X^2) + v_{AB}\tfrac{1}{2}zN\,\{x_A(1+X)(x_B+x_AX) + \\
&\qquad + x_A(1-X)(x_B-x_AX)\} + v_{BB}\tfrac{1}{2}zN\,(x_B^2 - x_A^2X^2) \\
&= \tfrac{1}{2}zN\,\{v_{AA}x_A^2 + 2v_{AB}x_Ax_B + v_{BB}x_B^2\} - zNx_A^2vX^2\,. \qquad (1)
\end{aligned}
$$

The entropy, on the other hand, is given by

$$
\begin{aligned}
S &= k\log\frac{[a]!}{\begin{bmatrix} A \\ a \end{bmatrix}!\begin{bmatrix} B \\ a \end{bmatrix}!}\,\frac{[b]!}{\begin{bmatrix} A \\ b \end{bmatrix}!\begin{bmatrix} B \\ b \end{bmatrix}!} \\
&= k\log\frac{(\tfrac{1}{2}N)!}{\tfrac{1}{2}N\,x_A(1+X)!\,\tfrac{1}{2}N(x_B-x_AX)!}\,\frac{(\tfrac{1}{2}N)!}{\tfrac{1}{2}N\,x_A(1-X)!\,\tfrac{1}{2}N(x_B+x_AX)!} \\
&\equiv -\tfrac{1}{2}Nk\,\{x_A(1+X)\log x_A(1+X) + (x_B-x_AX)\log(x_B-x_AX) + \\
&\qquad + x_A(1-X)\log x_A(1-X) + (x_B+x_AX)\log(x_B+x_AX)\}\,. \qquad (2)
\end{aligned}
$$

The defining equation of X, which is derived from $\partial(E-TS)/\partial X = 0$, is

$$\frac{zx_Av}{kT}X = \tfrac{1}{4}\log\left(\frac{1+X}{1-X}\cdot\frac{x_B+x_AX}{x_B-x_AX}\right), \qquad (3)$$

which can also be written as

$$\frac{X}{x_B+x_AX^2} = \tanh\frac{2zx_Av}{kT}X \equiv \tanh\left(\frac{4x_AT_C^0}{T}X\right). \qquad (4)$$

In the region $X = 0$ the foregoing can be written as

$$X\left\{\frac{4x_AT_C^0}{T} - \frac{1}{x_B}\right\} - X^3\left\{-\frac{x_A}{x_B^2} + \frac{1}{3}\left(\frac{4x_AT_C^0}{T}\right)^3\right\} + \cdots = 0, \qquad (5)$$

where $T_C^0 = zv/2k$ is the transition point for the case $x_A = x_B = \frac{1}{2}$. The transition point is that temperature where the coefficient of the first term in

(5) changes its sign. Hence

$$T_C = 4x_A x_B T_C^0 = 4x_A(1 - x_A)T_C^0 .$$

It attains its maximum at $x_A = \frac{1}{2}$, as shown in the figure on the right.

NOTE: This figure is a phase diagram of order-disorder transitions. Since the coefficient of the second term in (5) is positive at $T < T_C$:

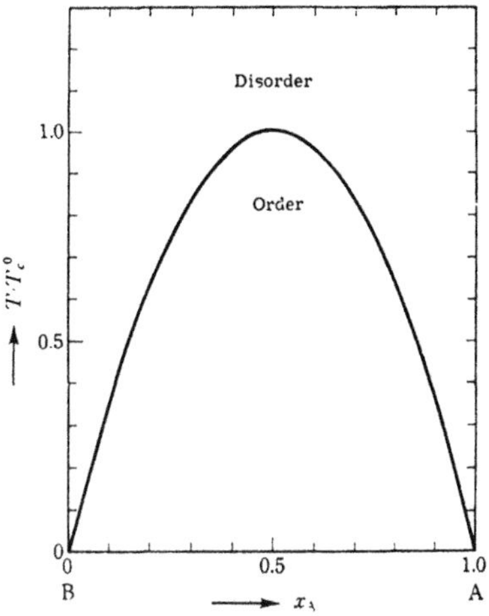

$$\frac{1}{3}\left(\frac{4x_A T_C^0}{T}\right)^3 - \frac{x_A}{x_B^2} > \frac{1}{3}\left(\frac{4x_A T_C^0}{T_C}\right)^3 - \frac{x_A}{x_B^2} =$$

$$= \frac{1}{3}\frac{1}{x_B^3} - \frac{x_A}{x_B^2}$$

$$= (1 - 3x_A x_B)/(3x_B^3) > \tfrac{1}{4}/3x_B^3 > 0$$

the equation has non-zero solutions for X at temperatures below T_C.

6. Let the number of the nearest neighbor sites around a site be z, and the number of component molecules be N_A and N_B respectively. Then the total energy is, according to the assumption,

$$E = v_{AA}N_{AA} + v_{AB}N_{AB} + v_{BB}N_{BB} , \tag{1}$$

where N_{AA}, etc. represent the number of AA nearest neighboring pairs, etc., and v_{AA} etc. represent the energies of AA neighboring pairs etc., respectively. In the Bragg-Williams approximation one assumes, similar to (5.14), that

$$N_{AA} = \tfrac{1}{2}zN_A \times \frac{N_A}{N_A + N_B}, \qquad N_{BB} = \tfrac{1}{2}zN_B \times \frac{N_B}{N_A + N_B},$$

$$N_{AB} = zN_A \times \frac{N_B}{N_A + N_B},$$

so that (1) is given by

$$E = \frac{z}{2}\frac{1}{N_A + N_B}\{v_{AA}N_A^2 + 2v_{AB}N_A N_B + v_{BB}N_B^2\} . \tag{2}$$

When the components A and B exist separately, the energy is given by

$$E_0 = \tfrac{1}{2}z\, v_{AA}N_A + \tfrac{1}{2}z\, v_{BB}N_B . \tag{3}$$

Inserting this into (2) gives

$$E = E_0 - zv\frac{N_A N_B}{N_A + N_B}, \qquad v = \tfrac{1}{2}(v_{AA} + v_{BB}) - v_{AB}. \tag{4}$$

The entropy in the present approximation is given by

$$S = k\log\frac{(N_A + N_B)!}{N_A!\,N_B!} = -k\left\{N_A\log\frac{N_A}{N_A + N_B} + N_B\log\frac{N_B}{N_A + N_B}\right\} \tag{5}$$

(see (5.11)), so that the free energy is

$$F = E_0 - Nzvx_A x_B + NkT\{x_A\log x_A + x_B\log x_B\}, \tag{6}$$

where $N = N_A + N_B$.

From this one can derive the chemical potentials μ_A and μ_B as

$$\left.\begin{aligned} \mu_A &= \frac{\partial F}{\partial N_A} = \tfrac{1}{2}z\,v_{AA} - zvx_B^2 + kT\log x_A, \\ \mu_B &= \frac{\partial F}{\partial N_B} = \tfrac{1}{2}z\,v_{BB} - zvx_A^2 + kT\log x_B. \end{aligned}\right\} \tag{7}$$

When $v > 0$, which means that molecules of different species can approach more easily, the energy decreases by mixing. Since, moreover, the increase of entropy by mixing stabilizes the state of mixture, A and B will mix at all temperatures.

When $v < 0$, on the other hand, which means that molecules of the same species have more affinity, the energy increases by mixing. What stabilizes the state of mixture, then, is the increase of entropy by mixing, which effect becomes smaller when the temperature is lower. Therefore, a solution of a certain concentration is stable only above a certain critical temperature, while below that point there occurs a separation of the solution into two phases.

Introducing x by

$$x_A = \tfrac{1}{2}(1 + x), \qquad x_B = \tfrac{1}{2}(1 - x) \tag{8}$$

one can transform (6) into

$$\begin{aligned} F &= E_0 - \tfrac{1}{4}N\,zv(1 - x^2) + \\ &+ NkT\{\tfrac{1}{2}(1 + x)\log\tfrac{1}{2}(1 + x) + \tfrac{1}{2}(1 - x)\log\tfrac{1}{2}(1 - x)\}. \end{aligned} \tag{9}$$

This expression has the same form as (4) in the solution to example 2. Plotting this F as a function x, one gets a curve which is similar to that in Figure 5.7, for the case $v < 0$. If T is less than T_C, which is defined by

$$T < T_C \equiv z\,|v|/2k \tag{10}$$

there are two minima at $x = \pm x'$. Let these two values of x give (x'_A, x'_B) and (x''_A, x''_B) through relation (8). Then these are the concentrations of the two phases which separate out of the solution. If, on the other hand, T is greater than T_C, the components mix for any concentration. The functional dependence of x' on T is given by the minimizing condition of (9), which as a matter of fact is the same as (7) in the solution to example 2.

NOTE 1: Any solution which is accompanied by an entropy change of the form (5) by mixing was called *regular* by Hildebrand.

NOTE 2: In the case of a crystalline mixture, there is the possibility of an ordered lattice occurring for the case $v > 0$, which was discussed in example 2.

NOTE 3: It may seem unrealistic to treat a liquid by a lattice model. Even though this objection is quite valid in general, the treatment is not so poor as long as only the effect of mixing is considered.

7. For the sake of simplicity, let us assume that $u(\sigma', \sigma'')$ is symmetric with respect to σ' and σ'', and further that all the atoms are equivalent. The variation of free energy when f is varied by the amount δf is given by

$$\begin{aligned}\delta F &= \delta U - T\delta S \\ &= \tfrac{1}{2}Nz\sum_{\sigma'}\sum_{\sigma''}u(\sigma',\sigma'')\{f(\sigma')\delta f(\sigma'') + f(\sigma'')\delta f(\sigma')\} + NkT\sum_{\sigma}\delta f(\sigma)\log f(\sigma) \\ &\qquad + NkT\sum_{\sigma}\delta f(\sigma) \\ &= Nz\sum_{\sigma'}\sum_{\sigma''}u(\sigma',\sigma'')f(\sigma'')\cdot\delta f(\sigma') + NkT\sum_{\sigma'}\log f(\sigma')\cdot\delta f(\sigma'),\end{aligned} \tag{1}$$

where the symmetry of u, $u(\sigma', \sigma'') = u(\sigma'', \sigma')$, and the normalization condition of f:

$$\sum_{\sigma}\delta f = 0 \tag{2}$$

have been taken into account. The equilibrium f is to be determined by

$$\delta F + A\sum_{\sigma'}\delta f(\sigma') = 0, \tag{3}$$

where A is a Lagrange multiplier. Inserting (1) into (3) gives

$$z\sum_{\sigma''}u(\sigma',\sigma'')f(\sigma'') + kT\log f(\sigma') + \alpha = 0, \qquad (\alpha = A/N)$$

or

$$f(\sigma') = C\exp\left[-\frac{z}{kT}\sum_{\sigma''}u(\sigma',\sigma'')f(\sigma'')\right] \tag{4}$$

where C is a normalization constant. The foregoing is just the equation we are seeking.

NOTE 1: Without making such an assumption concerning the symmetry of u, one proceeds in the following way: Looking upon one atom as the center, which shall be denoted by 0, and numbering its neighbors by l, one should write the interaction energy as

$$U = \tfrac{1}{2} N \sum_{l=1}^{z} \sum_{\sigma_0} \sum_{\sigma_l} u_{0l}(\sigma_0, \sigma_l) f(\sigma_0) f(\sigma_l)$$

since in general the interaction will depend on the direction of the line which connects the neighbor pair, as well as their internal states. Varying this one gets

$$\delta U = \tfrac{1}{2} N \sum_{l=1}^{z} \sum_{\sigma_0} \sum_{\sigma_l} u_{0l}(\sigma_0, \sigma_l) \{ f(\sigma_l)\, \delta f(\sigma_0) + f(\sigma_0)\, \delta f(\sigma_l) \} . \tag{5}$$

If the crystal consists only of equivalent sites, it has necessarily a symmetry of inversion. Let the site which is derived by the inversion from l be denoted by $\bar{l}$, then $u_{0l}(\sigma, \sigma') = u_{0\bar{l}}(\sigma' \sigma)$. Therefore (5) can be transformed to

$$\begin{aligned} \delta U &= \tfrac{1}{2} N \{ \sum_{l=1}^{z} \sum_{\sigma} \sum_{\sigma'} u_{0l}(\sigma, \sigma') f(\sigma')\, \delta f(\sigma) + \sum_{l=1}^{z} \sum_{\sigma} \sum_{\sigma'} u_{0l}(\sigma', \sigma) f(\sigma')\, \delta f(\sigma) \} \\ &= \tfrac{1}{2} N \{ \sum_{l=1}^{z} \sum_{\sigma} \sum_{\sigma'} \{ u_{0l}(\sigma, \sigma') f(\sigma') + u_{0\bar{l}}(\sigma, \sigma') f(\sigma') \}\, \delta f(\sigma) \} \\ &= N \sum_{l=1}^{z} u_{0l}(\sigma, \sigma') f(\sigma') \cdot \delta f(\sigma) . \end{aligned}$$

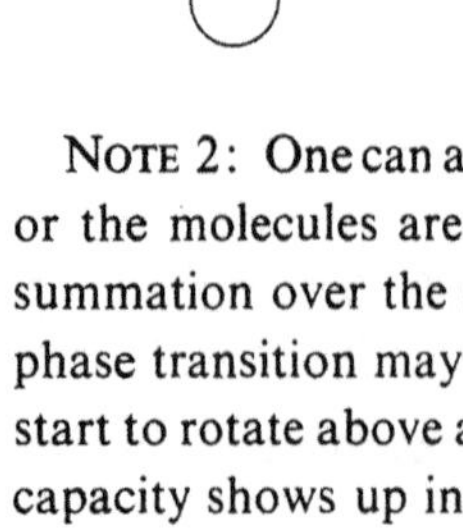

Notice that the summation over l is equivalent to that over $\bar{l}$. Therefore one obtains an equation for f,

$$f(\sigma) = C \exp\left[-\frac{1}{kT} \sum_{l=1}^{z} u_{0l}(\sigma, \sigma') f(\sigma') \right], \tag{6}$$

which is a more general form of equation (4).

NOTE 2: One can apply this method to the case where the states of the atoms or the molecules are continuous. In that case one has only to replace the summation over the states by an integration. As an example, the rotational phase transition may be mentioned. In certain molecular crystals, molecules start to rotate above a certain critical temperature, so that an anomalous heat capacity shows up in a similar way as in the alloys of AB type.

8. Ising's model is the case where the variable σ takes only the values ± 1 and the interaction is given explicitly by

$$u(\sigma, \sigma') = -J\sigma\sigma'. \tag{1}$$

Any function $f(\sigma)$ of σ takes only two values, $f(1)$ and $f(-1)$. Hence we can represent it by

$$f(\sigma) = \tfrac{1}{2}\{f(1) + f(-1)\} + \sigma\tfrac{1}{2}\{f(1) - f(-1)\} \equiv a + b\sigma.$$

One writes, therefore, for the probability $f(\sigma)$ in the solution to the foregoing problem,

$$f(\sigma) = \tfrac{1}{2}(1 + X\sigma) \tag{2}$$

so that (4) in the foregoing solution becomes

$$\begin{aligned}\tfrac{1}{2}(1 + X\sigma) &= C\exp\left[\frac{zJ}{2kT}\sum_{\sigma'}\sigma\sigma'(1 + X\sigma')\right]\\ &= C\exp\left\{\frac{zJ\sigma}{kT}X\right\},\end{aligned} \tag{3}$$

where use has been made of the identities: $\sum_{\sigma'}\sigma' = 0$, $\sum_{\sigma'}(\sigma')^2 = \sum 1 = 2$. From (3), it follows that

$$\sum_{\sigma}\tfrac{1}{2}(1 + X\sigma) = 1 = C(e^{zJX/kT} + e^{-zJX/kT}),$$

$$\sum_{\sigma}\tfrac{1}{2}\sigma(1 + X\sigma) = X = C(e^{zJX/kT} - e^{-zJX/kT}),$$

or

$$X = \tanh(zJX/kT),$$

a result which is identical with (5.9).

9. Put $x = 0$ at the surface of the metal and let $\varphi(x)$ be the electric potential. If the electron gas outside the metal is so rarefied that it can be treated classically, then the electronic density at x is given by

$$n(x) = n_0\exp\{e\varphi(x)/kT\} = n_0\exp\{-V/kT\}, \tag{1}$$

where the potential is assumed to be zero at $x = 0$. The potential energy for an electron is given by $V(x) = -e\varphi(x)$. This $V(x)$ increases as x increases from 0 toward infinity. Then $V(\infty) = \infty$, because $n(\infty) = 0$, and $V'(\infty) = 0$, because the electric field should vanish as $x \to \infty$. In terms of V, equation (5.12), i.e.,

$$\frac{d^2\varphi}{dx^2} = 4\pi en(x) = 4\pi en_0\exp\{e\varphi(x)/kT\}$$

can be written as

$$V'' = -4\pi e^2 n_0 e^{-V/kT}. \tag{2}$$

Multiplying the foregoing by V' and integrating it, using the boundary conditions given above, one obtains

$$\tfrac{1}{2}(V')^2 = 4\pi n_0 e^2 kT \mathrm{e}^{-V/kT},$$

from which it follows that

$$V' = (8\pi n_0 e^2 kT)^{\frac{1}{2}} \mathrm{e}^{-V/2kT}.$$

Integrating this result once again, one gets

$$\mathrm{e}^{V/2kT} = \left(\frac{2\pi n_0 e^2}{kT}\right)^{\frac{1}{2}} (x + x_0). \tag{3}$$

Since we have assumed that $V(0) = 0$, on substituting $x = 0$ in (3), one obtains the value of the integration constant x_0:

$$x_0 = \left(\frac{kT}{2\pi n_0 e^2}\right)^{\frac{1}{2}}. \tag{4}$$

Hence (3) can be rewritten as $\exp(V/2kT) = x/x_0 + 1$, from which

$$V = 2kT \log \frac{x + x_0}{x_0}. \tag{5}$$

On substituting this in (1), one finally obtains:

$$n(x) = n_0 \left(\frac{x_0}{x + x_0}\right)^2. \tag{6}$$

NOTE: The constant x_0 here is a counterpart of the Debye length which was introduced in the note to example 3.

10. From (5.28) we have

$$\frac{\partial \psi_\alpha}{\partial e_\beta} = -\frac{e_\alpha}{D} \frac{\mathrm{d}}{\mathrm{d}\kappa}\left(\frac{\kappa}{1 + \kappa a}\right) \cdot \frac{\partial \kappa}{\partial e_\beta}. \tag{1}$$

By differentiating the definition (5.25), i.e.,

$$\kappa^2 = \frac{4\pi}{DkTV} \sum_\gamma e_\gamma^2,$$

with respect to e_β, one obtains

$$\frac{\partial \kappa}{\partial e_\beta} = \frac{1}{2\kappa} \frac{\partial \kappa^2}{\partial e_\beta} = \frac{4\pi e_\beta}{\kappa DkTV}.$$

Substituting this in (1), one sees that (1) is symmetric with respect to e_α and e_β: that is,

$$\partial\psi_\alpha/\partial e_\beta = \partial\psi_\beta/\partial e_\alpha .$$

NOTE: The integration of (5.29) to obtain F_{el} is justified by the fulfillment of this symmetry relation. If this relation turns out inconsistent with a certain approximation method, then that method may not be satisfactory. [See for example: Fowler-Guggenheim; *Statistical Thermodynamics* (Cambridge Univ. Press, 1939) pp. 392, 409.]

11. The number density n at the position $\boldsymbol{r}$ is the average of the quantity

$$n(\boldsymbol{r}) = \sum_{i=1}^{N} \delta(\boldsymbol{r} - \boldsymbol{r}_i) .$$

Taking account of the fact that the distribution function $F_N\{N\}$ is symmetric with respect to the coordinates $\boldsymbol{r}_1, \ldots, \boldsymbol{r}_N$, one can calculate this average as follows:

$$\begin{aligned} n = \langle n(\boldsymbol{r})\rangle &= \int\cdots\int \sum_{i=1}^{N} \delta(\boldsymbol{r} - \boldsymbol{r}_i)\cdot F_N\{N\}\,\mathrm{d}\{N\} \\ &= N\int\cdots\int \delta(\boldsymbol{r}-\boldsymbol{r}_1) F_N\{N\}\,\mathrm{d}\{N\} \\ &= N\int\delta(\boldsymbol{r}-\boldsymbol{r}_1)\mathrm{d}\boldsymbol{r}_1\int\cdots\int F_N\{N\}\,\mathrm{d}\boldsymbol{r}_2\cdots\mathrm{d}\boldsymbol{r}_N = \int\delta(\boldsymbol{r}-\boldsymbol{r}_1)\mathrm{d}\boldsymbol{r}_1 F_1(\boldsymbol{r}_1) \\ &= F_1(\boldsymbol{r}) . \end{aligned} \tag{1}$$

In the case of a liquid, n is a constant independent of $\boldsymbol{r}$, and hence $F_1(\boldsymbol{r})$ does not depend on $\boldsymbol{r}$.

In quite a similar fashion one can calculate the average of the interaction

$$U = \sum_{i<j}\sum u(\boldsymbol{r}_i, \boldsymbol{r}_j) .$$

Since the summation here gives $\binom{N}{2}$ terms of the same form,

$$\begin{aligned} \langle U\rangle &= \int\cdots\int \sum_{i<j}\sum u(\boldsymbol{r}_i, \boldsymbol{r}_j) F_N\{N\}\,\mathrm{d}\{N\} \\ &= \binom{N}{2}\iint u(\boldsymbol{r}_1, \boldsymbol{r}_2)\,\mathrm{d}\boldsymbol{r}_1\,\mathrm{d}\boldsymbol{r}_2 \int\cdots\int F_N\{N\}\,\mathrm{d}\boldsymbol{r}_3\cdots\mathrm{d}\boldsymbol{r}_N \\ &= \tfrac{1}{2}\iint u(\boldsymbol{r}_1, \boldsymbol{r}_2)\,\mathrm{d}\boldsymbol{r}_1\,\mathrm{d}\boldsymbol{r}_2 F_2(\boldsymbol{r}_1, \boldsymbol{r}_2) . \end{aligned} \tag{2}$$

Replacing $u(\boldsymbol{r}_1, \boldsymbol{r}_2)$ by $\delta(\boldsymbol{r} - \boldsymbol{r}_1)\,\delta(\boldsymbol{r}' - \boldsymbol{r}_2)$ gives

$$\langle \nu\nu'\rangle \equiv \langle \sum_i\sum_j \delta(\boldsymbol{r}-\boldsymbol{r}_i)\,\mathrm{d}v \times \delta(\boldsymbol{r}'-\boldsymbol{r}_j)\,\mathrm{d}v'\rangle = F_2(\boldsymbol{r}, \boldsymbol{r}')\,\mathrm{d}v\,\mathrm{d}v' . \tag{3}$$

This is the average of the product of the numbers of the particles which exist at the same time in the volume elements $\mathrm{d}v$ and $\mathrm{d}v'$ around $\boldsymbol{r}$ and $\boldsymbol{r}'$ respectively. Hence the equation

$$\frac{F_2(\boldsymbol{r},\boldsymbol{r}')\,\mathrm{d}v'}{F_1(\boldsymbol{r})} = \frac{F_2(\boldsymbol{r},\boldsymbol{r}')\,\mathrm{d}v'}{n}$$

gives us the expectation value for the number of particles one will find in $\mathrm{d}v'$ when we know that there is a particle situated at $\boldsymbol{r}$. This value approaches $n\mathrm{d}v'$ in the case of a liquid as $|\boldsymbol{r}' - \boldsymbol{r}|$ becomes sufficiently large. Hence

$$\lim_{|\boldsymbol{r}-\boldsymbol{r}'|\to\infty} F_2(\boldsymbol{r},\boldsymbol{r}') = n^2 .$$

Therefore, one can define a function g which represents the spatial correlation of the particles by

$$F_2(\boldsymbol{r},\boldsymbol{r}') = n^2 g(\boldsymbol{r}' - \boldsymbol{r}) . \tag{4}$$

From the limiting property mentioned above,

$$\lim_{|\boldsymbol{r}|\to\infty} g(\boldsymbol{r}) = 1 .$$

In terms of g, (2) can be rewritten as

$$U = \tfrac{1}{2}\int \mathrm{d}\boldsymbol{r}_1\, n^2 \int u(\boldsymbol{r}_2 - \boldsymbol{r}_1) g(\boldsymbol{r}_2 - \boldsymbol{r}_1)\,\mathrm{d}\boldsymbol{r}_2$$
$$= \tfrac{1}{2} V n^2 \int u(\boldsymbol{r}) g(\boldsymbol{r})\,\mathrm{d}\boldsymbol{r} = \frac{N^2}{2V}\int u(\boldsymbol{r}) g(\boldsymbol{r})\,\mathrm{d}\boldsymbol{r} . \tag{5}$$

12. Making the transformation stated in the hint, one finds

$$Q_N = \frac{1}{N!}\int\cdots\int_V \exp\left\{-\frac{1}{kT}U(\boldsymbol{r}_1,\cdots,\boldsymbol{r}_N)\right\}\mathrm{d}\boldsymbol{r}_1\cdots\mathrm{d}\boldsymbol{r}_N$$
$$= \frac{\alpha^{3N}}{N!}\int\cdots\int_{V'} \exp\left\{-\frac{1}{kT}U(\alpha\boldsymbol{r}_1',\cdots,\alpha\boldsymbol{r}_N')\right\}\mathrm{d}\boldsymbol{r}_1'\cdots\mathrm{d}\boldsymbol{r}_N' . \tag{1}$$

Since $V = \alpha^3 V'$

$$\frac{\partial Q_N}{\partial V} = \left(\frac{\partial Q_N}{\partial\alpha}\Big/\frac{\mathrm{d}V}{\mathrm{d}\alpha}\right)_{\alpha=1} = \left[\frac{\partial Q_N}{\partial\alpha}\Big/\frac{3V}{\alpha}\right]_{\alpha=1} = \frac{1}{3V}\left[\frac{\partial Q_N}{\partial\alpha}\right]_{\alpha=1} . \tag{2}$$

Therefore

$$\frac{\partial Q_N}{\partial V} = \frac{N}{V} Q_N - \frac{1}{3N!\,V} \int \cdots \int \frac{1}{kT} \sum_{i=1}^{N} \frac{\partial U}{\partial \boldsymbol{r}_i} \cdot \boldsymbol{r}_i \exp\left\{-\frac{U}{kT}\right\} \mathrm{d}\boldsymbol{r}_1 \cdots \mathrm{d}\boldsymbol{r}_N$$

$$= \frac{N}{V} Q_N - \frac{1}{3kTN!\,V} \int \cdots \int \sum\sum_{i<j} r_{ij} \frac{\mathrm{d}u}{\mathrm{d}r_{ij}} \exp\left\{-\frac{U}{kT}\right\} \mathrm{d}\boldsymbol{r}_1 \cdots \mathrm{d}\boldsymbol{r}_N, \tag{3}$$

where use has been made in the second line of the fact that

$$U = \sum_{i<j} u(r_{ij}).$$

Substituting the normalized distribution function (5.37), i.e.,

$$F_N\{N\} = \frac{1}{Q_N N!} \exp\left\{-\frac{1}{kT} U\right\}$$

in the foregoing, and using definition (5.34), one obtains

$$p = kT \frac{\partial \log Q_N}{\partial V} = nkT - \frac{1}{3V} \int \cdots \int \sum\sum_{i<j} r_{ij} \frac{\mathrm{d}u}{\mathrm{d}r_{ij}} F_N\{N\}\, \mathrm{d}\{N\}$$

$$= nkT - \frac{1}{6V} \iint \mathrm{d}\boldsymbol{r}_1\, \mathrm{d}\boldsymbol{r}_2\, r_{12} \frac{\mathrm{d}u}{\mathrm{d}r_{12}} F_2(\boldsymbol{r}_1, \boldsymbol{r}_2)$$

$$= nkT - \frac{N^2}{6V^3} \iint \mathrm{d}\boldsymbol{r}_1\, \mathrm{d}\boldsymbol{r}_2\, r_{12} \frac{\mathrm{d}u}{\mathrm{d}r_{12}} g(\boldsymbol{r}_2 - \boldsymbol{r}_1)$$

$$= nkT - \frac{n^2}{6} \int r \frac{\mathrm{d}u}{\mathrm{d}r} g(r)\, \mathrm{d}\boldsymbol{r}$$

$$= nkT - \frac{2\pi}{3} n^2 \int_0^\infty \frac{\mathrm{d}u}{\mathrm{d}r} g(r) \cdot r^3\, \mathrm{d}r.$$

13. One proceeds in the same way as in the solutions to problems 2 and 3. Equation (2) in the solution to problem 2 determines the sublattice magnetizations. The magnetization vectors are parallel to their respective molecular fields:

$$\begin{aligned} \boldsymbol{M}_\mathrm{a} \,//\, \boldsymbol{H}'_\mathrm{a} &= \boldsymbol{H} - q_2 \boldsymbol{M}_\mathrm{a} - q_1 \boldsymbol{M}_\mathrm{b}, \\ \boldsymbol{M}_\mathrm{b} \,//\, \boldsymbol{H}'_\mathrm{b} &= \boldsymbol{H} - q_1 \boldsymbol{M}_\mathrm{b} - q_2 \boldsymbol{M}_\mathrm{a} \end{aligned} \tag{1}$$

and their magnitudes are given by

$$M_\mathrm{a} = \tfrac{1}{2} n\, g\mu_\mathrm{B} S B_S(g\mu_\mathrm{B} S H'_\mathrm{a}/kT), \qquad M_\mathrm{b} = \tfrac{1}{2} n\, g\mu_\mathrm{B} S B_S(g\mu_\mathrm{B} S H'_\mathrm{b}/kT). \tag{2}$$

If one puts $\boldsymbol{H} = 0$ in the foregoing, then $\boldsymbol{M}_a$ and $\boldsymbol{M}_b$ are antiparallel to each other and their magnitudes are given by

$$M_a = -\tfrac{1}{2} n\, g\mu_B S B_S\left(\frac{g\mu_B S}{kT}(q_2 M_a + q_1 M_b)\right), \quad M_b = -\tfrac{1}{2} n\, g\mu_B S B_S\left(\frac{g\mu_B S}{kT}(q_1 M_a + q_2 M_b)\right) \tag{3}$$

respectively. Expanding the right hand side of the foregoing series under the assumption of small M so that one can use the expansion formula (4) of the solution to problem 9, Chapter 2, one finds

$$\left.\begin{aligned} M_a &= -\frac{C}{2T}(q_2 M_a + q_1 M_b) + \gamma(q_2 M_a + q_1 M_b)^3 + \cdots, \\ M_b &= -\frac{C}{2T}(q_1 M_a + q_2 M_b) + \gamma(q_1 M_a + q_2 M_b)^3 + \cdots \end{aligned}\right\} \tag{4}$$

(cf. (4) of the solution to problem 2, and (3) and (4) of the solution to problem 3, Chapter 5), where γ is a certain positive constant. Substituting $M_a = -M_b = M'$ in (4) gives

$$M'\left\{1 - \frac{C}{2T}(q_1 - q_2)\right\} = -\gamma(q_1 - q_2)M'^3 + \dots. \tag{5}$$

If one has

$$q_1 > q_2, \tag{6}$$

then at temperatures

$$T < T_N \equiv \tfrac{1}{2}C(q_1 - q_2) \tag{7}$$

the coefficients of both sides of (5) are negative, so that there is a real solution $M' \neq 0$. Therefore, (6) is a condition for an anti-ferromagnetic spin ordering and the Néel temperature is given by T_N of (7).

NOTE 1: T_N of (7) is different from θ of (5), solution to problem 3, so long as $q_2 \neq 0$.

NOTE 2: If $q_1 < q_2$, contrary to (6), there is no real colution $M' \neq 0$ at $T < T_N$ as can be seen from (5). That is, there can be no anti-ferromagnetic ordering. Since the condition $0 < q_1 < q_2$ means that the interaction in the same sub-lattice a has a tendency to align its spins anti-parallel, it is evident that the ordering presupposed could not occur. (If the spins in a have a tendency to align parallel, then $q_2 < 0$: if the spins of a and b tend to align anti-parallel, then $q_1 > 0$. Therefore, $q_1 - q_2 > 0$.)

14. Consider first the case $\boldsymbol{M}' \perp \boldsymbol{H}$. Then the molecular fields prevailing at the sublattices a and b are given, respectively, by

$$\begin{aligned} \boldsymbol{H}'_a &= \boldsymbol{H} - q_2\boldsymbol{M}_a - q_1\boldsymbol{M}_b, \\ \boldsymbol{H}'_b &= \boldsymbol{H} - q_1\boldsymbol{M}_a - q_2\boldsymbol{M}_b. \end{aligned} \tag{1}$$

Since

$$\boldsymbol{H}'_a // \boldsymbol{M}_a, \qquad \boldsymbol{H}'_b // \boldsymbol{M}_b \tag{2}$$

the relation between these vectors will be as shown in the figure. Note that $-q_1\boldsymbol{M}_b + \boldsymbol{H}$ is parallel to $\boldsymbol{M}_a$, and $-q_1\boldsymbol{M}_a + \boldsymbol{H}$ is parallel to $\boldsymbol{M}_b$. If one writes θ for the angle shown in the figure and also puts $|\boldsymbol{M}_a| = M'$, then one has $2\theta = H/M'q_1$. The total magnetization $\boldsymbol{M} = \boldsymbol{M}_a + \boldsymbol{M}_b$ lies parallel to $\boldsymbol{H}$ and its magnitude is given by

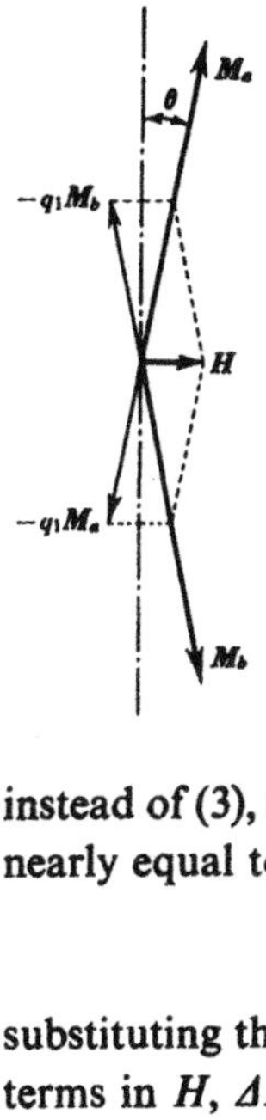

$$M = 2M'\theta = H/q_1 .$$

Hence the perpendicular susceptibility $\chi_\perp$ is

$$\chi_\perp = 1/q_1 . \tag{3}$$

Next consider the case $\boldsymbol{M}' // \boldsymbol{H}$. Here one has to deal with

$$\left.\begin{aligned} M_a &= \tfrac{1}{2} n\, g\mu_B S B_S\left(\frac{g\mu_B S}{kT}(H - q_2M_a - q_1M_b)\right), \\ M_b &= \tfrac{1}{2} n\, g\mu_B S B_S\left(\frac{g\mu_B S}{kT}(H - q_1M_a - q_2M_b)\right) \end{aligned}\right\} \tag{4}$$

instead of (3), solution to problem 13. H is so small that M_a and M_b are both nearly equal to M'. Putting, therefore,

$$M_a = M' + \Delta M_a, \qquad M_b = -M' + \Delta M_b, \tag{5}$$

substituting these in (4), expanding the function B_S and retaining first order terms in H, ΔM_a or ΔM_b, one finds the following two equations:

$$\left.\begin{aligned} \Delta M_a &= \frac{n}{2k}\frac{g^2\mu_B^2S^2}{T} B'_S\left(\frac{g\mu_B S}{kT}(q_1 - q_2)M'\right) \times \\ &\qquad \times (-q_2\Delta M_a - q_1\Delta M_b + H), \\ \Delta M_b &= \frac{n}{2k}\frac{g^2\mu_B^2S^2}{T} B'_S\left(\frac{g\mu_B S}{kT}(q_1 - q_2)M'\right) \times \\ &\qquad \times (-q_1\Delta M_a - q_2\Delta M_b + H). \end{aligned}\right\} \tag{6}$$

Since the increment in magnetization ΔM in the direction of M' is the algebraic sum of ΔM_a and ΔM_b, one has only to add the two equations of (6) to obtain

$$\Delta M \equiv \Delta M_a + \Delta M_b = \frac{C}{2T}\frac{3S}{S+1}B_S'\cdot\{-(q_1+q_2)\Delta M + 2H\},$$

thus

$$\chi = \frac{\Delta M}{H} = \frac{3CS}{S+1}B_S' \Bigg/ \left\{T + \tfrac{1}{2}(q_1+q_2)\frac{3CS}{S+1}B_S'\right\}, \tag{7}$$

where for the sake of simplicity we put

$$B_S' \equiv B_S'\left(\frac{g\mu_B S}{kT}(q_1-q_2)M'\right). \tag{8}$$

As $T \to 0$, $B_S' \to 0$, and hence $\chi_{//} \to 0$. On the other hand, as $T \to T_N$, $M' \to 0$ and hence $B_S'' \to (S+1)/3S$. This means that

$$\chi_{//} \to \frac{C}{T_N + \frac{1}{2}(q_1+q_2)C} = \frac{C}{\frac{1}{2}(q_1-q_2)C + \frac{1}{2}(q_1+q_2)C} = \frac{1}{q_1},$$

a result which coincides with (3).

NOTE: At $T > T_N$, χ is given by (5), solution to problem 3, that is

$$\chi = \frac{C}{\{T + \frac{1}{2}(q_1+q_2)C\}},$$

a result which coincides with $\chi_{//}$ and $\chi_\perp$ at $T = T_N$. Hence, if one plots the susceptibility against temperature one obtains a figure such as the one shown. The observed susceptibilities of real anti-ferromagnetic substances (e.g., MnF_2) follow a general trend in agreement with this result.

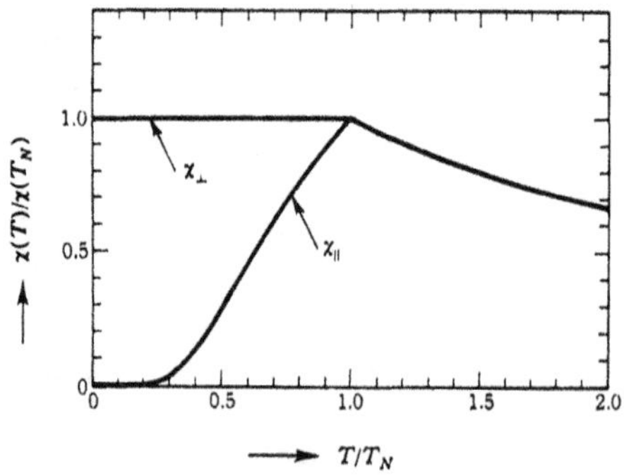

15. We may classify the atoms of a binary alloy into two species such that A atoms situated on the sublattice a, and B atoms on the sublattice b are called r atoms (right atoms), and the B's on a and the A's on b are called w atoms (wrong atoms). Then the interaction between neighboring r atoms or w atoms is $-J$ while that between an r atom and a w atom is J. Therefore, if r is made to correspond to an Ising spin of $\sigma = 1$ and w to a spin of $\sigma = -1$, the interaction energy of the alloy can be written in the form

$$\mathscr{H}_{\text{alloy}} = -J \sum\nolimits^{\langle ij \rangle} \sigma_i \sigma_j . \tag{1}$$

The definition of the degree of long-range order X can be written as

$$X = \left(\begin{bmatrix} \mathrm{A} \\ \mathrm{a} \end{bmatrix} + \begin{bmatrix} \mathrm{B} \\ \mathrm{b} \end{bmatrix} - \begin{bmatrix} \mathrm{B} \\ \mathrm{a} \end{bmatrix} - \begin{bmatrix} \mathrm{A} \\ \mathrm{b} \end{bmatrix} \right) \Big/ N = (N_{\mathrm{r}} - N_{\mathrm{w}})/N \equiv (N_+ - N_-)/N .$$

From (1) and (2) one can write the partition function of an alloy in the form

$$Z_{\mathrm{A}}(X, T) = \sum\nolimits_{N_- - N_+ = NX} \exp(-\mathscr{H}_{\text{alloy}}/kT), \tag{2}$$

where the summation is to be made over the values ± 1 of each of the $\sigma_i (i = 1, \ldots, N)$ but with the restriction that

$$N_+ - N_- = \sum_j \sigma_j \tag{3}$$

should be equal to the given value NX.

On the other hand, the partition function of the Ising spin system can be transformed in the following way:

$$\begin{aligned} Z_{\mathrm{I}} &= \sum \exp\left\{ -\frac{g\mu_{\mathrm{B}}H}{kT} \sum \sigma_j - \frac{J}{kT} \sum\nolimits^{\langle ij \rangle} \sigma_i \sigma_j \right\} \\ &= \sum_{N'=0}^{N} \exp(-\xi N') \sum\nolimits_{N_+ - N_- = N'} \exp\left\{ -\frac{J}{kT} \sum\nolimits^{\langle ij \rangle} \sigma_i \sigma_j \right\} \\ &= \sum_{N'=0}^{N} \lambda^{N'} Z_{\mathrm{A}}(X, T) \qquad (X = N'/N), \end{aligned}$$

where χ signifies

$$\lambda \equiv \mathrm{e}^{-\xi} \equiv \exp(-g\mu_{\mathrm{B}}H/kT).$$

Therefore, if one knows Z_{I} as a function of H and T, and hence as a function of λ and T, then one can derive $Z_A(X, T)$ as the coefficients of λ^{NX} in the power series expansion for the function Z_I. That is,

$$Z_A(X, T) = \frac{1}{2\pi i} \oint Z_I(\lambda, T) \frac{d\lambda}{\lambda^{NX+1}} \cdot$$

This Cauchy intergration is to be carried out along a path which circumscribes the origin.

16. Putting $J/kT = \beta$, one can write the partition function in the following way:

$$Z = \sum_{\sigma_1 = \pm 1} \cdots \sum_{\sigma_N = \pm 1} \exp\{\beta \sum \sigma_j \sigma_{j+1}\}$$

$$= \sum_{\sigma_1 = \pm 1} \cdots \sum_{\sigma_N = \pm 1} \prod (\cosh\beta + \sigma_j \sigma_{j+1} \sinh\beta), \tag{1}$$

where use has been made of the relation

$$\exp(a\sigma\sigma') = \begin{Bmatrix} e^a & (\sigma\sigma' = 1) \\ e^{-a} & (\sigma\sigma' = -1) \end{Bmatrix} = \cosh a + \sigma\sigma' \sinh a$$

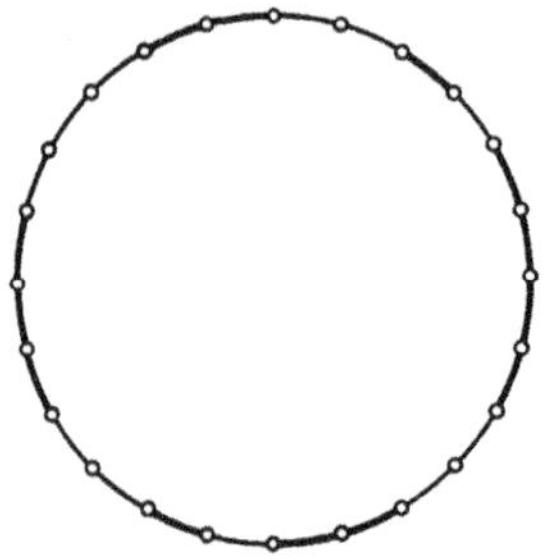

which is valid because the product $\sigma\sigma'$ takes on only one of the two values $\pm$ 1. If one expands the product in (1), one gets a sum of terms, each of which is a product of the form $(\cosh\beta)^{N-p}$ $(\sinh\beta)^p$ $(\sigma_i\ \sigma_{i+1})$ ... $(\sigma_j\ \sigma_{j+1})$. These terms may be represented graphically by heavy and light links making up the ring, the heavy one representing, say, the factor $\sigma\sigma' \sinh\beta$ and the light one representing the other factor $\cosh\beta$. If, in this representation, there is a site which is a joint between a heavy and a light link, then its spin occurs only once and the summation, to be taken over the two values ($\pm$ 1) of the spin will make the product vanish. As for a site which is a joint between two heavy links, its spin occurs squared and may be replaced by 1 because $\sigma^2 = 1$. It will then be clear that in order that a product makes a contribution, the chain in heavy links, if it exists should have no ends. Therefore, the only terms that do not vanish are the first term $(\cosh\beta)^N$ and the last term $(\sinh\beta)^N$. Hence

$$Z = 2^N \{(\cosh\beta)^N + (\sinh\beta)^N\}. \tag{2}$$

As long as $\beta \neq \infty$ ($T \neq 0$), $\cosh\beta > \sinh\beta$. Hence if $N \gg 1$, $(\cosh\beta)^N \gg (\sinh\beta)^N$, so that one can put

$$Z = 2^N (\cosh\beta)^N. \tag{3}$$

From this one derives for the free energy

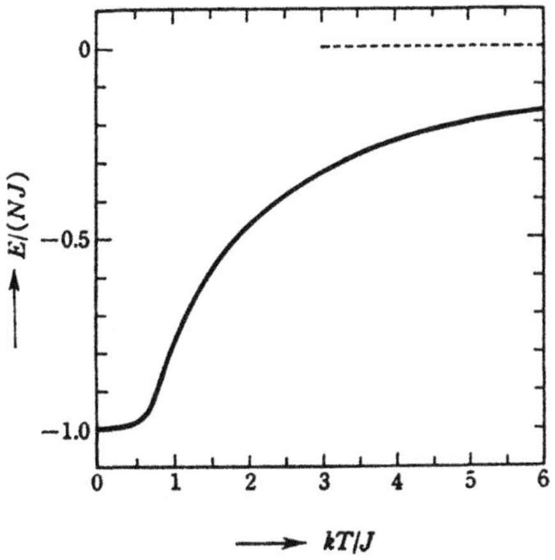

$$F = -NkT \log\left\{2\cosh\frac{J}{kT}\right\}, \quad (4)$$

so that the energy and the heat capacity are

$$E = \frac{\partial}{\partial T^{-1}}\frac{F}{T}$$

$$= -NJ\tanh\frac{J}{kT},$$

$$C = \frac{\mathrm{d}E}{\mathrm{d}T}$$

$$= \frac{NJ^2}{kT^2}\Big/\cosh^2\frac{J}{kT}.$$

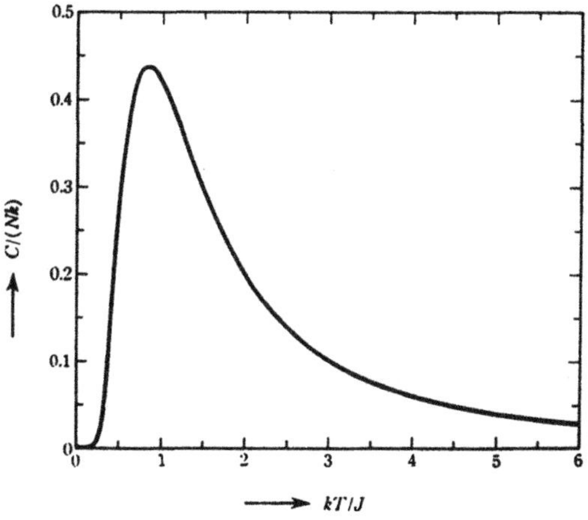

Plots of these quantities against T are shown in the figures.

17. Assume that the polymer molecules are numbered and are placed on an array of sites one by one according to their numbers. If the number of ways to place the l-th polymer molecule after one has placed the first $l-1$ molecules is called ν_l, then the number of ways to place all of the polymer molecules on the lattice is given by $\nu_1\,\nu_2\,\ldots\,\nu_{N_2}$. However, in reality, the polymer molecules are all of equivalent structure, so that one cannot actually distinguish one configuration from another which can be derived through a permutation of the polymers. Hence, to get the correct number one must divide the foregoing by the total number of permutations:

$$W = \frac{1}{N_2!}\prod_{l=1}^{N_2} v_l. \tag{1}$$

In order next to get an expression for v_l, one assumes the structural units of each of the polymers to be numbered along the chain as 1, 2, ..., r and places them one by one according to their numbers. One can place the first unit on any vacant site. The number of ways to place it is, therefore, $N - r(l-1)$. The second unit has to be placed on one of the vacant z sites which are the nearest neighbors to the first unit. If one makes the approximation that one can replace the probability of finding a neighboring site occupied by the polymers one has already placed, by the probability one would get if the whole structural units of the $l-1$ polymers were distributed randomly over the lattice, one obtains $z[1 - r(l-1)/N]$ for the number of the ways. The third unit has to be placed on one of the vacant sites which are the nearest neighbors to the second unit. One of these has already been occupied by the first unit. Taking this into account and proceeding in the same way as before, one obtains $(z-1)[1 - r(l-1)/N]$ ways of placing the third unit. The fourth unit may happen to be placed on a site which is the nearest neighbor to the first. However, let us neglect this possibility. Then the situation is quite the same as for the third. Therefore one finally obtains

$$v_l = \frac{N}{\sigma} z(z-1)^{r-2}\left[1 - \frac{r(l-1)}{N}\right]^r. \tag{2}$$

In this expression we have introduced a factor σ^{-1}. This is because it does not matter from which end of a polymer molecule one starts the above procedure. Substituting (2) in (1) gives

$$\begin{aligned} W &= \frac{z^{N_2}(z-1)^{N_2(r-2)}}{N_2!\,\sigma^{N_2}} N^{N_2} \prod_{l=1}^{N_2}\left[1 - \frac{r(\ -1)}{N}\right]^r \\ &= \frac{z^{N_2}(z-1)^{N_2(r-2)}}{N_2!\,\sigma^{N_2}} N^{-(r-1)N_2}\left\{r^{N_2}\frac{(N/r)!}{(N_1/r)!}\right\}^r. \end{aligned}$$

This is the expression which is given in the problem. We have used here the relation $N = N_1 + rN_2$.

When one has placed all the polymers in the way described above, one only has to fill in all the remaining vacancies with solute molecules. Therefore the configuration of the whole solution is determined solely by the configuration of the polymers and hence the entropy of the solution is given by $S = k \log W$. Inserting here the above expression of W and employing Stirling's formula one obtains

$$S(N_1, N_2) = N_2 k \log \frac{z(z-1)^{r-2}}{\sigma \mathrm{e}^{r-1}} - k\left\{N_1 \log \frac{N_1}{N} + N_2 \log \frac{N_2}{N}\right\}. \qquad (3)$$

The entropy before mixing is $S(0, N_2) + S(N_1, 0)$, which, inserted with (3), becomes

$$\begin{aligned} S(0, N_2) + S(N_1, 0) &= S(0, N_2) \\ &= N_2 k \log \frac{z(z-1)^{r-2}}{\sigma \mathrm{e}^{r-1}} + k N_2 \log r . \end{aligned}$$

Therefore the mixing entropy, which is the increase in entropy by mixing, is given by

$$\begin{aligned} \Delta S &= -k\left(N_1 \log \frac{N_1}{N} + N_2 \log \frac{rN_2}{N}\right) \\ &= -k(N_1 \log \phi_1 + N_2 \log \phi_2). \end{aligned} \qquad (4)$$

NOTE: Thermodynamic properties of polymer solutions may be treated on the basis of the mixing entropy derived here. The details are given in solution to problem 39, Chapter 4, vol. 1.

18. It is evident that both the distribution of electrons and the electrostatic potential is spherically symmetric around the point charge. Note that $\varphi(r)$ obeys Poisson's equation

$$\nabla^2 \varphi = 4\pi e(n - n_0) \qquad (1)$$

everywhere (except at the origin) and satisfies the boundary conditions:

$$\varphi(r) \sim ze/r \qquad \text{as} \qquad r \to 0 \qquad (2)$$

and

$$\varphi(r) \to 0 \qquad \text{as} \qquad r \to \infty . \qquad (2')$$

(i) According to Boltzmann's distribution law

$$n = n_0 \exp(e\varphi/kT), \qquad (3)$$

where we have taken n_0 as a constant factor because $n(r)$ should tend to n_0 as $r \to \infty$. Substituting this in (1) gives

$$\nabla^2 \varphi = 4\pi e n_0 \left[\exp(e\varphi/kT) - 1\right].$$

Let us look for the solution in a region sufficiently far from the point charge. Since $e\varphi/kT \ll 1$ there, we can expand the exponential function and retain terms up to those of the second order. Thus we find

$$\nabla^2\varphi = \kappa^2\varphi, \qquad \kappa^2 = \frac{4\pi n_0 e^2}{kT}. \tag{4}$$

The spherically symmetric solution which satisfies (2′) is

$$\varphi(r) = A\frac{\mathrm{e}^{-\kappa r}}{r}. \tag{5}$$

From (2) it follows that $A = ze$ and hence finally

$$\varphi(r) = \frac{ze}{r}\exp(-\kappa r). \tag{5′}$$

The distribution of electrons can be derived in the following way:
From (1) and (4) we have $4\pi e(n - n_0) = \kappa^2\varphi$. Substituting (5′) gives

$$n = n_0 + z\kappa^2\,\mathrm{e}^{-\kappa r}/4\pi r.$$

(ii) Sufficiently far away from the origin there are many electrons in a region across which the potential energy varies by a very much smaller amount than the average kinetic energy. Under these circumstances one can estimate the number of electronic states by putting it equal to the volume of the phase space. Since at 0° K the electrons occupy the lowest possible states,

$$n(r) = 2\cdot\frac{4\pi}{3}P(r)^3/h^3, \tag{6}$$

where the factor 2 has been introduced in order to take into account the two possible states of the electronic spin. $P(r)$ here means the maximum momentum the electrons have in the neighborhood of the point r. This quantity is determined through the condition (of equilibrium) that the maximum energy of the electrons (which is the Fermi potential at 0° K) is constant throughout:

$$\frac{1}{2m}P(r)^2 - e\varphi = \frac{1}{2m}P_\infty^2. \tag{7}$$

By use of the condition $n(\infty) = n_0$ and of (6), it follows that

$$n_0 = \frac{8\pi}{3h^3}P_\infty^3. \tag{8}$$

Eliminating $P(r)$ from (6) and (7) gives

$$n(r) = \frac{8\pi}{3h^3}\left[P_\infty^2 + 2me\varphi(r)\right]^{\frac{3}{2}}. \tag{9}$$

Expanding the right hand side of (9) in powers of $2me\,\varphi/P_\infty^2$, which may be assumed to be far smaller than 1 except in regions very near the origin, retaining terms up to the first order in φ and taking (8) into account, one obtains

$$n(r) = n_0 + \frac{me}{\pi\hbar^2}\left(\frac{3n_0}{\pi}\right)^{\frac{1}{3}}\varphi. \tag{9'}$$

Substituting this in (1) gives

$$\nabla^2\varphi = q^2\varphi, \qquad q^2 = 4\frac{me^2}{\hbar^2}\left(\frac{3n_0}{\pi}\right)^{\frac{1}{3}}. \tag{10}$$

The solution is, therefore, expression (5) or (5′) with κ replaced by q. One can obtain the electronic distribution by just substituting this φ in (9′).

NOTE: From expression (5′) one can see that the point charge ze is screened by the electrons. The characteristic inverse length appearing there is called the *screening constant*. The factor κ of (4) is *Debye's screening constant* and q of (10) is *Thomas-Fermi's screening constant*. The fact is, however, that though the method described in (ii) is that of Thomas-Fermi, it was Mott who first derived (10). So one might better call q Mott's screening constant. He developed the analysis described in (ii) in order to understand the behavior of the electric field arounds an impurity atom which is substituted in a metal. If the host metal is a mono-valent one, such as Cu, Ag or Au and the substitute atom is a di-valent one, such as Zn, or a tri-valent one, such as Al, one has to use the values $z = 1$ or 2 respectively. The numerical value of $1/q$ in these examples is about 0.6 Å. (See Mott and Jones, *Theory of Metals and Alloys*, 1936, p. 86.) It has been noticed by many authors that this Thomas-Fermi screening theory is not sufficient to account for the actual behavior of many alloys. It seems that the wave nature of metallic electrons can by no means be neglected.

19. First, $W(X)$ has already been calculated in example 2 as

$$W(X) = \textstyle\sum_{(X)} 1 = \left[(\tfrac{1}{2}N)!/\{\tfrac{1}{4}N(1+X)!\,\tfrac{1}{4}N(1-X)!\}\right]^2. \tag{1}$$

Next, the expansion of $\langle \exp(\beta A) \rangle$ is easily seen to be

$$\langle \exp(\beta A)\rangle = 1 + \beta \textstyle\sum_{(ij)} \langle\sigma_i\sigma_j\rangle + \tfrac{1}{2}\beta^2 N \textstyle\sum_{(ij)(kl)} \langle\sigma_i\sigma_j\sigma_k\sigma_l\rangle + \cdots. \tag{2}$$

Now in order to evaluate the average, it is helpful to assume that i, k, ... pertain to the sublattice a and j, l, ... pertain to the sublattice b, which by no means makes any restriction. Then, σ_i and σ_j can vary independently, so that

$$\langle \sigma_i \sigma_j \rangle = \langle \sigma_i \rangle \langle \sigma_j \rangle = X^2 .$$

There are three cases for the third term: (i) $i = k, j = l$, (ii) $i \neq k, j = l$ or $i = k, j \neq l$, and (iii) $i \neq k, j \neq l$. The number of terms for each of these cases is: (i) $Q = \frac{1}{2} Nz$, (ii) $Q(z - 1)$, and for (iii), Q^2 minus (i) and (ii), namely, $Q^2 - Q - 2(z-1)Q = Q(Q - 2z + 1)$. The average values in the three

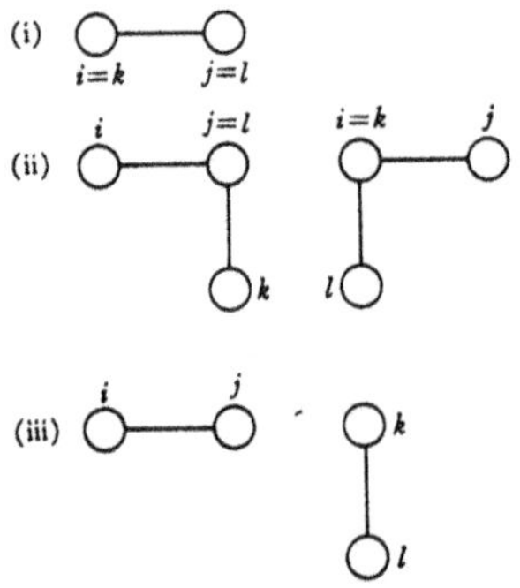

cases above are: (i) just 1, because $\sigma_i^2 = 1$, (ii) for a pair of spins i, k which are on the same sublattice

$$\langle \sigma_i \sigma_k \rangle = \frac{\frac{1}{2}N_+(\frac{1}{2}N_+ - 1)}{\frac{1}{2}N(\frac{1}{2}N - 1)} + \frac{\frac{1}{2}N_-(\frac{1}{2}N_- - 1)}{\frac{1}{2}N(\frac{1}{2}N - 1)} - 2\frac{\frac{1}{2}N_+\frac{1}{2}N_-}{\frac{1}{2}N(\frac{1}{2}N - 1)} \tag{3}$$

$$= X^2 - \frac{1 - X^2}{\frac{1}{2}N - 1} \simeq X^2 - \frac{2}{N}(1 - X^2) \equiv X^2 - \delta$$

and the same for $\langle \sigma_1 \sigma_l \rangle$, (iii) $\langle \sigma_i \sigma_j \sigma_k \sigma_l \rangle = \langle \sigma_i \sigma_k \rangle \langle \sigma_j \sigma_l \rangle$.

(3) is also applicable here. Thus the third term, evaluated exactly up to $O(N)$ turns out to be

$$\sum_{(ij)(kl)} \langle \sigma_i \sigma_j \sigma_k \sigma_l \rangle =$$
$$= Q + 2Q(z-1)(X^2 - \delta) + Q(Q - 2z + 1)(X^2 - \delta)^2$$
$$= Q + 2Q(z-1)X^2 + Q^2 X^4 - 2Q^2 X^2 \delta - Q(2z-1)X^4 + O(1)$$
$$= Q^2 X^4 + Q\{1 + 2(z-1)X^2 - 2zX^2(1 - X^2) - (2z-1)X^4\} + O(1)$$
$$= Q^2 X^4 + Q(1 - X^2)^2 .$$

Therefore the final result is

$$Z(X) = W(X)\{1 + \beta\tfrac{1}{2}NzX^2 + \tfrac{1}{2}\beta^2(\tfrac{1}{2}NzX^2)^2 + \tfrac{1}{2}\beta^2\tfrac{1}{2}Nz(1-X^2)^2 + \cdots\},$$

from which it follows that

$$F \equiv -kT\log Z = -N\left\{\frac{z}{4}vX^2 + \frac{zv^2}{16kT}(1-X^2)^2 + \cdots\right\}$$
$$+ NkT\{\tfrac{1}{2}(1+X)\log\tfrac{1}{2}(1+X) + \tfrac{1}{2}(1-X)\log\tfrac{1}{2}(1-X)\}.$$

One determines the value of the degree of long-range order by means of the equation:

$$\frac{\partial F}{\partial X} = 0, \quad \text{thus} \quad \tfrac{1}{2}\log\frac{1+X}{1-X} = X\left\{\frac{zv}{2kT} - \frac{zv^2}{4(kT)^2}(1-X^2) + \cdots\right\}.$$

Its limiting form as $X \to 0$ is

$$1 = \frac{zv}{2kT_{\mathrm{C}}} - \frac{zv^2}{4(kT_{\mathrm{C}})^2} + \cdots,$$

which determines the transition temperature, T_{C}. Its solution, when treated as an algebraic equation of second order in $(zv/2kT_{\mathrm{C}})$, gives

$$T_{\mathrm{C}} = \frac{v}{k}\left[1 - \sqrt{1 - \frac{4}{z}}\right]^{-1}.$$

(See Fowler and Guggenheim, *Statistical Thermodynamics,* p. 574; J. Kirkwood: J. Chem. Phys. **6** (1938).)

20. Calculate first the partition function Z according to the given Hamiltonian:

$$Z(H, H', T) =$$
$$= \sum_{\sigma_0=\pm1}\sum_{\sigma_1=\pm1}\cdots\sum_{\sigma_z=\pm1}\exp\left[\frac{1}{kT}\{\mu_0 H\sigma_0 + \mu_0(H+H')\textstyle\sum\sigma_j + J\sum\sigma_0\sigma_j\}\right]$$
$$\equiv \textstyle\sum\exp\left[\alpha\sigma_0 + (\alpha+\alpha')\sum\sigma_j + \beta\sum\sigma_0\sigma_j\right] \tag{1}$$

where the following abbreviations are made:

$$\alpha = \frac{\mu_0 H}{kT}, \qquad \alpha' = \frac{\mu_0 H'}{kT}, \qquad \beta = \frac{J}{kT}. \tag{2}$$

Equation (1) can be divided into two parts according to $\sigma_0 = \pm 1$: namely,

$$Z = Z_+ + Z_-,$$
$$Z_\pm = \sum_{\sigma_1=\pm1}\cdots\sum_{\sigma_z=\pm1}\exp\{\pm\alpha + (\alpha+\alpha'\pm\beta)\textstyle\sum_j\sigma_j\}$$
$$= e^{\pm\alpha}\left[2\cosh(\alpha+\alpha'\pm\beta)\right]^z. \tag{3}$$

Then one can calculate the averages of the spins in the following way:

$$\bar{\sigma}_0 = (Z_+ - Z_-)/Z, \tag{4}$$

and

$$\begin{aligned}\bar{\sigma}_j &= \frac{1}{z}\sum_{j=1}^{z} \bar{\sigma}_j = \frac{1}{z}\frac{\partial Z}{\partial \alpha'}\Big/ Z \\ &= [Z_+ \tanh(\alpha + \alpha' + \beta) + Z_- \tanh(\alpha + \alpha' - \beta)]/Z.\end{aligned} \tag{5}$$

Equating (4) with (5) gives

$$Z_+ \{1 - \tanh(\alpha + \alpha' + \beta)\} = Z_- \{1 + \tanh(\alpha + \alpha' - \beta)\}; \tag{6}$$

substitution of (3) into this expression finally gives

$$\left[\frac{\cosh(\alpha + \alpha' + \beta)}{\cosh(\alpha + \alpha' - \beta)}\right]^{z-1} = e^{2\alpha'}. \tag{7}$$

This is the condition of self-consistency which determines α'. In order to find the Curie temperature, one uses equation (7) with $\alpha = 0$ ($H = 0$), or,

$$\frac{\cosh(\alpha' + \beta)}{\cosh(\alpha' - \beta)} = e^{2\alpha'/(z-1)} \qquad \frac{\alpha'}{z-1} = \tfrac{1}{2}\log\frac{\cosh(\beta + \alpha')}{\cosh(\beta - \alpha')}. \tag{8}$$

$\alpha' = 0$ is always a solution of this equation. In order to find out whether another solution $\alpha' \neq 0$ exists at small values, one expands the right hand side of equation (8):

$$\begin{aligned}\frac{\alpha'}{z-1} &= \tfrac{1}{2}\log\frac{\cosh\beta + \alpha'\sinh\beta + \tfrac{1}{2}\alpha'^2\cosh\beta + \cdots}{\cosh\beta - \alpha'\sinh\beta + \tfrac{1}{2}\alpha'^2\cosh\beta + \cdots} \\ &= \tanh\beta\,\{\alpha' - \tfrac{1}{3}\alpha'^3/\cosh^2\beta + \cdots\}.\end{aligned}$$

Therefore, at temperatures lower than T_C, defined by

$$\tanh\beta_C \equiv \tanh\frac{J}{kT_C} = \frac{1}{z-1}, \tag{9}$$

there exists a non-zero real solution α', which is defined by

$$\alpha'^2 = 3\frac{\cosh^3\beta}{\sinh\beta}\left\{\tanh\beta - \frac{1}{z-1} + \cdots\right\}. \tag{10}$$

Thus T_C, as defined by (9), is an approximate value of the Curie temperature.

NOTE: If $z \gg 1$, the solution of (9) reduces to $J/kT_C \simeq (z-1)^{-1} \simeq z^{-1}$

and agrees with (5.10), which is the answer in the simplest molecular field approximation. Though $z = 1$ gives no solution for T_C, and $z = 2$ gives $T_C = 0$, T_C, as given by (9), is generally lower than that of (5.10) for the case $z > 2$, and is a better value. A discussion of this approximation, primarily in the case of alloys, is given by Fowler; *Statistical Mechanics*, p. 800; Fowler-Guggenheim, *Statistical Thermodynamics*, p. 576; Bethe, Proc. Roy. Soc. A **150** (1935) 552.

21. If in calculating the partition function, given as (1) in the solution of the preceding problem, one performs the summation over ± 1 of the variables $\sigma_2, \ldots, \sigma_z$ with the exception of σ_0 and σ_1, one obtains

$$Z = \sum_{\sigma_0 = \pm 1} \sum_{\sigma_1 = \pm 1} \exp\{\alpha\sigma_0 + (\alpha + \alpha')\sigma_1 + \beta\sigma_0\sigma_1\} \cdot [2\cosh(\alpha + \alpha' + \beta\sigma_0)]^{z-1}. \tag{1}$$

Dividing this into three parts according the pair of values of σ_0 and σ_1:

$$Z = Z_{++} + Z_{+-} + Z_{--}, \tag{2}$$

one has

$$\bar{N}_{++} : \bar{N}_{+-} : \bar{N}_{--} = Z_{++} : Z_{+-} : Z_{--}. \tag{3}$$

Performing this program explicitly and taking account of (7) in the previous solution, one obtains

$$\bar{N}_{++} \propto e^{2\alpha + \alpha' + \beta}[2\cosh(\alpha + \alpha' + \beta)]^{z-1},$$

$$\bar{N}_{+-} \propto e^{-\alpha' - \beta}[2\cosh(\alpha + \alpha' + \beta)]^{z-1} + e^{\alpha' - \beta}[2\cosh(\alpha + \alpha' - \beta)]^{z-1} =$$
$$= 2e^{-\alpha' - \beta}[2\cosh(\alpha + \alpha' + \beta)]^{z-1},$$

$$\bar{N}_{--} \propto e^{-2\alpha - 3\alpha' + \beta}[2\cosh(\alpha + \alpha' + \beta)]^{z-1}$$

From these it follows immediately that

$$\frac{\bar{N}_{+-}^2}{\bar{N}_{++}\bar{N}_{--}} = 4e^{-4\beta} = 4\exp(-4J/kT).$$

NOTE: If one exchanges the two atoms between the two pairs $(+\ +)$ and $(-\ -)$ of atoms, one will obtain two $(+\ -)$ pairs as a result. One may regard this process as a sort of chemical reaction. Then the relation obtained here will represent an approximation to the equilibrium condition of the reaction (law of mass action). On the other hand, there are evidently the following relations between the quantities $\bar{N}_{++}$, $\bar{N}_{+-}$ and $\bar{N}_{--}$:

$$2\bar{N}_{++} + \bar{N}_{+-} = z\bar{N}_{+}, \qquad 2\bar{N}_{--} + \bar{N}_{+-} = z\bar{N}_{-}.$$

With the use of these one can represent $\bar{N}_{++}$ and $\bar{N}_{--}$ in terms of $\bar{N}_{+}$, $\bar{N}_{-}$ and $\bar{N}_{+-}$ and solve the relation above for $\bar{N}_{+-}$ as a function of $\bar{N}_{+}$ and $\bar{N}_{--}$. This is another interpretation of Bethe's approximation and was called the *quasi-chemical method* by Fowler and Guggenheim (c.f. *Statistical Thermodynamics,* p. 576).

22. The equation derived in example 4 now becomes

$$\frac{\partial}{\partial \boldsymbol{r}_i} g_s(\boldsymbol{r}_1, \cdots, \boldsymbol{r}_s) + \beta \frac{\partial U\{s\}}{\partial \boldsymbol{r}_i} g_s(\boldsymbol{r}_1, \cdots, \boldsymbol{r}_s) + \\ + \frac{\beta}{v} \int \frac{\partial \phi(\boldsymbol{r}_i, \boldsymbol{r}_{s+1})}{\partial \boldsymbol{r}_i} g_{s+1}(\boldsymbol{r}_1, \cdots, \boldsymbol{r}_{s+1}) \mathrm{d}\boldsymbol{r}_{s+1} = 0. \quad (1)$$

Substituting the expansion for g_s (in powers of v^{-1}) in the foregoing and putting the coefficient of each power equal to zero, one obtains

$$\frac{\partial}{\partial \boldsymbol{r}_i} g_s^{(0)} + \beta \frac{\partial U\{s\}}{\partial \boldsymbol{r}_i} g_s^{(0)} = 0, \quad (2)$$

$$\frac{\partial}{\partial \boldsymbol{r}_i} g_s^{(1)} + \beta \frac{\partial U\{s\}}{\partial \boldsymbol{r}_i} g_s^{(1)} + \beta \int \frac{\partial \phi(\boldsymbol{r}_i, \boldsymbol{r}_{s+1})}{\partial \boldsymbol{r}_i} g_{s+1}^{(0)}(\boldsymbol{r}_1, \cdots, \boldsymbol{r}_{s+1}) \mathrm{d}\boldsymbol{r}_{s+1} = 0 \quad (3)$$

and so on. Putting

$$g_s^{(0)} = C_s^{(0)}(\boldsymbol{r}_1, \cdots, \boldsymbol{r}_s) \exp(-\beta U\{s\}) \quad (4)$$

and substituting it in (2) reduces the latter to $\partial C_s^0/\partial \boldsymbol{r}_1 = 0$. Since $C_s^{(0)}$ has to be symmetric with respect to $\boldsymbol{r}_1, \ldots, \boldsymbol{r}_s$, $C_s^{(0)}$ = constant. In particular $C_1^{(0)} = 1$. The condition $|\boldsymbol{r}_{ij}| \to \infty$, which is given in the text of the problem is equivalent to

$$\lim g_s(\boldsymbol{r}_1, \cdots, \boldsymbol{r}_s) \to 1 \qquad (|\boldsymbol{r}_{ij}| \to \infty). \quad (5)$$

This in turn requires, since $U\{s\} \to 0$ as all $|\boldsymbol{r}_{ij}| \to \infty$, that

$$C_s^{(0)}(\boldsymbol{r}_1, \ldots, \boldsymbol{r}_s) \to 1, \text{ as all } |\boldsymbol{r}_{ij}| \to \infty .$$

From this and the constancy of $C_s^{(0)}$, derived above, it follow that $C_s^{(0)} = 1$ and hence (4) reduces to

$$g_s^{(0)} = \exp(-\beta U\{s\}). \quad (6)$$

Next, on substituting (6) in (3) one finds

$$\frac{\partial g_s^{(1)}}{\partial \boldsymbol{r}_i} + \beta \frac{\partial U\{s\}}{\partial \boldsymbol{r}_i} g_s^{(1)} + \beta \int \frac{\partial \phi(\boldsymbol{r}_{i,s+1})}{\partial \boldsymbol{r}_i} \exp(-\beta U\{s+1\}) \mathrm{d}\boldsymbol{r}_{s+1} = 0. \quad (7)$$

One can transform the integrand in (7) in the following way:

$$\beta\left(\frac{\partial}{\partial \boldsymbol{r}_i}\phi(\boldsymbol{r}_{i,s+1})\right)\exp(-\beta U\{s+1\}) =$$

$$= -\exp(-\beta U\{s\})\frac{\partial}{\partial \boldsymbol{r}_i}\exp\left\{-\beta\sum_{j=1}^{s}\phi(\boldsymbol{r}_{j,s+1})\right\}$$

$$= -\exp(-\beta U\{s\})\frac{\partial}{\partial \boldsymbol{r}_i}\prod_{j=1}^{s}\{1+f(\boldsymbol{r}_{j,s+1})\},$$

where

$$f(r) = \mathrm{e}^{-\beta\phi(r)} - 1 .$$

Therefore, (7) can be written as

$$\frac{\partial g_s^{(1)}}{\partial \boldsymbol{r}_i} + \beta\frac{\partial U\{s\}}{\partial \boldsymbol{r}_i}g_s^{(1)} = \mathrm{e}^{-\beta U\{s\}}\frac{\partial}{\partial \boldsymbol{r}_i}\int\prod_{j=1}^{s}\{1+f(\boldsymbol{r}_{j,s+1})\}\,\mathrm{d}\boldsymbol{r}_{s+1}. \tag{8}$$

Examining the form of this equation, one again puts

$$g_s^{(1)}(\boldsymbol{r}_1,\cdots,\boldsymbol{r}_s) \equiv C_s^{(1)}(\boldsymbol{r}_1,\cdots,\boldsymbol{r}_s)\mathrm{e}^{-\beta U\{s\}} \tag{9}$$

and reduces (8) to an equation for $C_s^{(1)}$, which turns out to be

$$\frac{\partial C_s^{(1)}}{\partial \boldsymbol{r}_i} = \frac{\partial}{\partial \boldsymbol{r}_i}\int\prod_{1\leqq j\leqq s}\{1+f(\boldsymbol{r}_{j,s+1})\}\,\mathrm{d}\boldsymbol{r}_{s+1}. \tag{10}$$

The solution of the foregoing is readily seen to be

$$C_s^{(1)} = \int\prod_{1\leqq j\leqq s}\{1+f(\boldsymbol{r}_{j,s+1})\}\,\mathrm{d}\boldsymbol{r}_{s+1} + k_s, \tag{11}$$

where k_s is a constant. To determine the value of k_s we proceed in the following way: Let $r_{ij} \to \infty$ for all pairs i, j $(1 < i < j < s)$. Then, since $f(\boldsymbol{r}_{j,s+1})$ gets small rapidly as $|\boldsymbol{r}_{j,s+1}|$ increases, terms up to the first order in f remain in the expansion of the product in (11) and hence

$$g_s^{(1)} \to C_s^{(1)} \to \int\{1+\sum_{j=1}^{s}f(\boldsymbol{r}_{i,s+1})\}\,\mathrm{d}\boldsymbol{r}_{s+1} + k_s. \tag{12}$$

However from (5) and (6) it is necessary that

$$g_s^{(1)} \to 0, \qquad \text{as all} \qquad |\boldsymbol{r}_{ij}| \to \infty .$$

This, then determines the constant k_s. Thus we finally find that

$$g_s^{(1)} = \mathrm{e}^{-\beta U\{s\}}\int\{\prod_{j=1}^{s}(1+f(\boldsymbol{r}_{j,s+1})) - 1 - \sum_{j=1}^{s}f(\boldsymbol{r}_{j,s+1})\}\,\mathrm{d}\boldsymbol{r}_{s+1}. \tag{13}$$

In particular, from (6) and (13)

$$g_2 = e^{-\beta\phi(r_{12})}\left\{1 + \frac{1}{v}\int f(r_{13})f(r_{23})\,d\boldsymbol{r}_3 + \cdots\right\}. \tag{14}$$

Substituting the foregoing in

$$\frac{p}{kT} = \frac{1}{v}\left\{1 - \frac{\beta}{6v}\int r_{12}\frac{d\phi}{dr_{12}}g(r_{12})\,d\boldsymbol{r}_{12}\right\} \tag{15}$$

which was derived in problem 12, gives the virial expansion

$$\frac{p}{kT} = \frac{1}{v}\left\{1 - \frac{\beta_1}{2v} - \frac{2}{3}\frac{\beta_2}{v^2} - \cdots\right\}, \tag{16}$$

where

$$\begin{aligned}
\beta_1 &= \tfrac{1}{3}\beta\int r_{12}\frac{d\phi}{dr_{12}}e^{-\beta\phi(r_{12})}\,d\boldsymbol{r}_{12} = -\tfrac{1}{3}\int r_{12}\frac{d}{dr_{12}}e^{-\beta\phi}\,d\boldsymbol{r}_{12}\\
&= -\tfrac{1}{3}\int r_{12}\frac{d}{dr_{12}}f(r_{12})\,d\boldsymbol{r}_{12} = \int f(r_{12})\,d\boldsymbol{r}_{12},\\
\beta_2 &= \tfrac{1}{4}\beta\int\int r_{12}\frac{d\phi(r_{12})}{dr_{12}}e^{-\beta\phi(r_{12})}f(r_{13})f(r_{23})\,d\boldsymbol{r}_{12}\,d\boldsymbol{r}_3\\
&= \tfrac{1}{2}\int\int f(r_{12})f(r_{23})f(r_{31})\,d\boldsymbol{r}_2\,d\boldsymbol{r}_3.
\end{aligned}$$

The transformation to the last line is achieved through an appropriate integration by parts, in much the same way as in the case of β_1.

CHAPTER 6

FLUCTUATIONS AND KINETIC THEORIES

So far we have been mainly considering the averages of physical quantities. It is, however, also one of the principal objects of statistical mechanics to find the laws of fluctuations of observed values from the averages. If we go a step further beyond the statistical mechanics of equilibrium states, we are faced with non-equilibrium problems, that is to say, irreversible processes which involve deviations from equilibrium. These may be stationary or non-stationary, but in general depend on the time either explicitly or implicitly. Traditional methods for treating such problems are known as kinetic methods, and their prototype is found in the kinetic theory of gases. It would be beyond the scope of this book to go too far into these theories, but a few introductory problems are given in this chapter since a fundamental understanding of these subjects is desirable.

Fundamental Topics

§ 6.1. FLUCTUATIONS

Statistical mechanics can tell us the probability distribution law of microscopic states of a given system under certain prescribed conditions. For instance, it will be a micro-canonical distribution law if the energy is prescribed, a canonical distribution if the temperature is prescribed, and so on. Once such a distribution law is known, the distribution law for any physical quantity of the system may be calculated. Therefore, there is no new principle required in the calculation of fluctuations.

Thermodynamical theory of fluctuations: It should be noticed, however, that the probability of fluctuation of macroscopic quantities can be expressed in terms of thermodynamic variables. Let $\alpha = (\alpha_1, \alpha_2, \ldots)$ be a set of such quantities whose equilibrium values are $\alpha^* = (\alpha_1^*, \alpha_2^*, \ldots)$. The probability $P(\alpha')$ that a deviation from equilibrium, $\alpha' = (\alpha_1', \alpha_2', \ldots)$, is realized has been already given by (1.68) and (1.69) in Chapter 1. In particular, when these variables are those for a subsystem of a large system, this probability can be expressed as

$$P(\alpha') = C \exp\{- W_{\min}(\alpha^*, \alpha')/kT^*\}, \tag{6.1}$$

where $W_{\min}(\alpha^*, \alpha')$ is the minimum work required to bring the subsystem from an equilibrium state α^* to another state $\alpha^* + \alpha'$ under the interaction with the rest of the whole system (environment). T^* here is the equilibrium temperature. For the derivation see example 1. This kind of thermodynamic consideration is often very useful.

§ 6.2. COLLISION FREQUENCY

In kinetic theories of gases or electrons in solids, it is necessary to find the average frequency of collisions which a particular particle makes with other particles in the system or with obstacles (the wall of the container, or impurities in a crystal). A few definitions are first introduced:

Duration time of collision means the time which the particle under consideration spends in the force range of the partner in the collision. The average of this time over all possible collisions is the average (duration) time of collision.

Free time of flight is the time between two successive collisions. Mean free flight time (average free flight time) is the average of the free flight time over all possible collisions. In ordinary kinetic treatment, it is always assumed that a collision takes place instantaneously and so the particles are free most of the time. That is, it should be possible to choose a time interval Δt satisfying the condition:

$$\text{mean free flight time} \gg \Delta t \gg \text{average collision time.} \tag{6.2}$$

Assumption for calculating the number of collisions (Stosszahlansatz): To calculate the number of collisions per unit time, the following assumption is usually introduced. Consider a scatterer (a partner in the collision) with which the particles are colliding. In order that a particle makes a collision of a certain type with this scatterer during a time interval Δt, it must have been

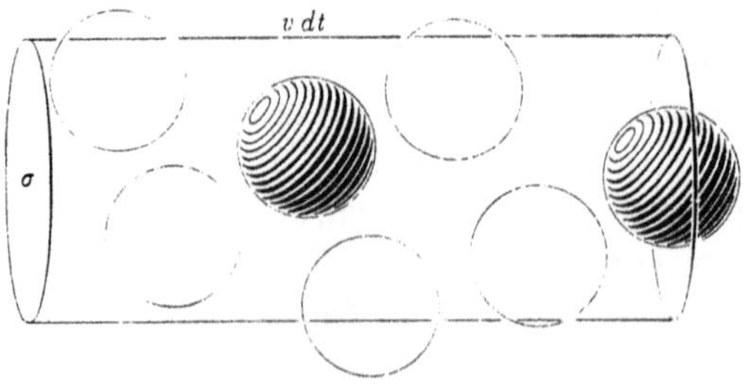

Fig. 6.1.

in a certain region of the space in the immediate neighborhood of the scatterer (see Fig. 6.1). Then it is assumed that

$$\begin{pmatrix}\text{number of}\\ \text{collisions}\\ \text{in } \Delta t\end{pmatrix} = \Delta V \cdot \begin{pmatrix}\text{probability that the}\\ \text{particles are found}\\ \text{in unit volume of space}\end{pmatrix} \cdot \begin{pmatrix}\text{the number}\\ \text{of scatterers}\end{pmatrix}, \tag{6.3}$$

where the second factor on the right hand side means the distribution function given by (1.9).

For example, when there are s scatterers per unit volume, each having a cross section σ, the number of particles which collide per unit time with these scatterers with velocities in the interval between $\boldsymbol{v}$ and $\boldsymbol{v} + \mathrm{d}\boldsymbol{v}$ is given by

$$\frac{\mathrm{d}\mathfrak{N}}{\mathrm{d}t} = \sigma v \cdot n f(\boldsymbol{v})\, \mathrm{d}\boldsymbol{v} \cdot s. \tag{6.4}$$

Here $\boldsymbol{n}$ is the particle number density, $f(\boldsymbol{v})$ is the velocity distribution function (normalized as $\int f(\boldsymbol{v})\, \mathrm{d}\boldsymbol{v} = 1$). This can be seen from Figure 6.1 which shows a cylinder of volume $v\mathrm{d}t \times \sigma$.

The assumption of molecular chaos means that the collision frequency can be calculated by means of (6.2) as long as time intervals Δt which satisfy the condition (6.2) are considered. Upon this assumption it is possible to set up the Boltzmann equation (see the following section). In other words, the dynamic relations between successive collisions are easily lost in chaos because the scatterers are so numerous and randomly distributed in space.

§ 6.3. BOLTZMANN TRANSPORT EQUATION

Viscous forces in a streaming gas are related to the transport of momentum of molecules, heat conduction to that of energy, and diffusion processes to that of mass. These phenomena and similar phenomena such as electric conduction are called *transport phenomena*. The viscosity coefficient, heat conductivity, diffusion constant and electric conductivity are generally called the *transport coefficients* of these phenomena. One of the main objects of kinetic theories is the calculation of such transport coefficients from the microscopic motion of particles.

In kinetic theories one usually considers the distribution function of molecules in μ-space (see § 1.2). It is convenient to define the distribution function f in such a way that $f(\boldsymbol{x}, \boldsymbol{p}, t)\, \mathrm{d}\boldsymbol{x}\, \mathrm{d}\boldsymbol{p}$ represents the number of molecules which are in the volume element $\mathrm{d}\boldsymbol{x}$ around the coordinate $\boldsymbol{x}$ and which have momenta between $\boldsymbol{p}$ and $\boldsymbol{p} + \mathrm{d}\boldsymbol{p}$. The equation governing the temporal change of this distribution function is called the Boltzmann equation, which has the form

$$\frac{\partial f}{\partial t} + \boldsymbol{v}\cdot\frac{\partial f}{\partial \boldsymbol{x}} + \boldsymbol{F}\cdot\frac{\partial f}{\partial \boldsymbol{p}} = \left(\frac{\partial f}{\partial t}\right)_{\text{coll.}}, \tag{6.5}$$

where $\boldsymbol{x}$ is the coordinate, $\boldsymbol{p}$ the momentum, and $\boldsymbol{F}$ the external force acting on a particle.

Drift term: Equation (6.5) is written as

$$\frac{\partial f}{\partial t} = \left(\frac{\partial f}{\partial t}\right)_{\text{drift}} + \left(\frac{\partial f}{\partial t}\right)_{\text{coll.}}, \qquad \left(\frac{\partial f}{\partial t}\right)_{\text{drift}} \equiv -\boldsymbol{v}\cdot\frac{\partial f}{\partial \boldsymbol{x}} - \boldsymbol{F}\cdot\frac{\partial f}{\partial \boldsymbol{p}} \tag{6.6}$$

$(\partial f/\partial t)_{\text{drift}}$ is called the drift term, which represents the rate of change of the distribution function f due to the motion of particles without collisions. Such motion causes the changes

$$\boldsymbol{x} \to \boldsymbol{x}' = \boldsymbol{x} + \boldsymbol{v}\,\mathrm{d}t, \qquad \boldsymbol{p} \to \boldsymbol{p}' = \boldsymbol{p} + \dot{\boldsymbol{p}}\,\mathrm{d}t = \boldsymbol{p} + \boldsymbol{F}\,\mathrm{d}t$$

in a time interval $\mathrm{d}t$. This is a flow in μ-space. The change of f must satisfy the continuity condition

$$f(\boldsymbol{x}', \boldsymbol{p}', t + \mathrm{d}t) = f(\boldsymbol{x}, \boldsymbol{p}, t)$$

or

$$f(\boldsymbol{x} + \boldsymbol{v}\,\mathrm{d}t, \boldsymbol{p} + \boldsymbol{F}\,\mathrm{d}t, t + \mathrm{d}t) = f(\boldsymbol{x}, \boldsymbol{p}, t). \tag{6.7}$$

The second equation of (6.6) is easily obtained from (6.7).

Collision term: The right hand side of (6.5) represents the effect of collisions between the particles and the scatterers or between the particles themselves. In simple theories this term is obtained in a more or less intuitive manner by making the assumptions mentioned in the previous section. To give a few examples:

(1) Classical particles (for example, non-degenerate electrons) scattered by foreign particles (scatterers). Let $\sigma_\alpha(\boldsymbol{p}, \boldsymbol{p}')$ be the differential cross section of the scattering of a particle by one of the scatterers of the α-th kind. This means that the frequency of the process in which an incident particle with momentum $\boldsymbol{p}$ makes per unit time and per unit volume a collision with a scatterer of type α and is scattered with a momentum between $\boldsymbol{p}'$ and $\boldsymbol{p}' + \mathrm{d}\boldsymbol{p}'$ is given by

$$s_\alpha \sigma_\alpha(\boldsymbol{p}, \boldsymbol{p}')\,\mathrm{d}\boldsymbol{p}' \cdot v(\boldsymbol{p}) f(\boldsymbol{p}), \tag{6.8}$$

s_α being the number density of the scatterers of type α. The collision term in this case can be written as

$$\left(\frac{\partial f}{\partial t}\right)_{\text{coll.}} = -\sum_\alpha s_\alpha \int \sigma_\alpha(\boldsymbol{p}, \boldsymbol{p}')\,\mathrm{d}\boldsymbol{p}' \cdot v(\boldsymbol{p}) f(\boldsymbol{p}) + \sum s_\alpha \int \sigma_\alpha(\boldsymbol{p}', \boldsymbol{p}) v(\boldsymbol{p}') f(\boldsymbol{p}')\,\mathrm{d}\boldsymbol{p}' . \tag{6.9}$$

The first term on the right hand side means the rate of loss from the state $\boldsymbol{p}$ and the second term is the gain from other states with all possible $\boldsymbol{p}'$s.

(2) Degenerate particles scattered by foreign scatterers. In this case the scattering probability (6.8) must be amended as

$$s_\alpha \sigma_\alpha(\boldsymbol{p}, \boldsymbol{p}')\,\mathrm{d}\boldsymbol{p}' v(\boldsymbol{p}) f(\boldsymbol{p}) \{1 - f(\boldsymbol{p}')\} . \tag{6.10}$$

By the Pauli principle, a scattering process ($\boldsymbol{p} \to \boldsymbol{p}'$) can only take place when the state $\boldsymbol{p}'$ is vacant. This restriction makes collisions less frequent. This effect is taken into consideration by the last factor of (6.10). (If the electrons are non-degenerate, this factor is not important since $f \ll 1$.) Accordingly (6.9) is now modified as

$$\left(\frac{\partial f}{\partial t}\right)_{\text{coll.}} = -\sum_\alpha s_\alpha \int \sigma_\alpha(\boldsymbol{p}, \boldsymbol{p}') v(\boldsymbol{p}) f(\boldsymbol{p}) \{1 - f(\boldsymbol{p}')\}\,\mathrm{d}\boldsymbol{p}' + \sum_\alpha s_\alpha \int \sigma_\alpha(\boldsymbol{p}', \boldsymbol{p}) v(\boldsymbol{p}') f(\boldsymbol{p}') \{1 - f(\boldsymbol{p})\}\,\mathrm{d}\boldsymbol{p}' . \tag{6.11}$$

(3) Collisions between the particles themselves (classical). In this case, let $\Phi(\boldsymbol{p}_1, \boldsymbol{p}_2; \boldsymbol{p}_1', \boldsymbol{p}_2')$ be the probability density describing the process in which two particles with momenta $\boldsymbol{p}_1$ and $\boldsymbol{p}_2$ collide and make transitions to the momenta $\boldsymbol{p}_1'$ and $\boldsymbol{p}_2'$. Then the collision term may be written as

$$\left(\frac{\partial f}{\partial t}\right)_{\text{coll.}} = -\iiint \Phi(\boldsymbol{p}, \boldsymbol{p}_2; \boldsymbol{p}_1', \boldsymbol{p}_2') f(\boldsymbol{p}) f(\boldsymbol{p}_2)\,\mathrm{d}\boldsymbol{p}_2\,\mathrm{d}\boldsymbol{p}_1'\,\mathrm{d}\boldsymbol{p}_2' + \iiint \Phi(\boldsymbol{p}_1', \boldsymbol{p}_2'; \boldsymbol{p}, \boldsymbol{p}_2) f(\boldsymbol{p}_1') f(\boldsymbol{p}_2')\,\mathrm{d}\boldsymbol{p}_1'\,\mathrm{d}\boldsymbol{p}_2'\,\mathrm{d}\boldsymbol{p}_2 . \tag{6.12}$$

The collision term of the Boltzmann equation is now written simply as

$$\left(\frac{\partial f}{\partial t}\right)_{\text{coll.}} \equiv \Gamma[f], \tag{6.13}$$

where $\Gamma[f]$ indicates the action of an integral operator, which may be linear or non-linear on the function f. As long as no creation or annihilation occurs by collision processes, the particle number does not change, so that we must have

$$\int \Gamma[f]\,\mathrm{d}\boldsymbol{p} = 0 . \tag{6.14}$$

Also, if f_e is the equilibrium distribution we must have

$$\Gamma[f_e] \equiv 0, \tag{6.15}$$

because the collision process is, in fact, a microscopic mechanism to bring the particles into equilibrium.

The general problem of a kinetic theory which uses a Boltzmann equation is to solve (6.5) for a given collision mechanism, and for a given external force $\boldsymbol{F}$ under certain initial and boundary conditions. This equation is generally an integro-differential equation and is usually non-linear except for some fortunate cases like (6.9). Therefore, a rigorous solution is very difficult to obtain. However, linearization is possible if the deviation from the equilibrium state is small, because then one may put

$$f = f_e + g \tag{6.16}$$

and assume that the deviation g is sufficiently small. By this procedure a Boltzmann equation becomes a linear integro-differential equation.

Furthermore, the collision term is often replaced by a simple approximation

$$\left(\frac{\partial f}{\partial t}\right)_{\text{coll.}} = -\frac{f - f_e}{\tau} \equiv -g/\tau, \tag{6.17}$$

which means that the deviation of the distribution function from equilibrium decays simply in time due to collision processes. Here, τ is sometimes assumed simply as a constant or more generally a function of $\boldsymbol{p}$. It measures the rate of recovery from a non-equilibrium state and so is often called the relaxation time. It is approximately of the same order as the mean free flight time, but it is not exactly equal to it. This simplification is not rigorous but is useful to obtain a qualitative understanding of many problems.

Examples

1. Show that, if in (1.68) the (semi-) macroscopic variable α characterizes the state of a small subsystem I of the system under consideration, the probability distribution governing its fluctuation can be written in the form

$$P(\alpha') = C \exp\left\{-\frac{W_{\min}(\alpha^*, \alpha')}{kT^*}\right\},$$

where $W_{\min}(\alpha^*, \alpha')$ stands for the minimum work required to transfer the subsystem I from the equilibrium state α^* to the state $\alpha^* + \alpha'$ under action of the remaining part II of the system, and T^* is the equilibrium temperature, i.e. the temperature of the subsystem II.

SOLUTION

Equation (1.68) is valid for the composite system I + II. Let the total system I + II be transferred from the state A^* into a state $A^* + A'$, when the subsystem I is transferred from the state α^* to the state $\alpha^* + \alpha'$, external work being done. Then, since whenever work is done in a quasistatic (or reversible) way it becomes a minimum, we obtain

$$W_{\text{min}}(\alpha^*, \alpha') = \int_{A^*}^{A^*+A'} (\mathrm{d}U - T\,\mathrm{d}S), \tag{1}$$

where U and S denote the internal energy and the entropy of the composite system I + II respectively, and T is the temperature at each intermediate stage of the process. Since α is the change in a sufficiently small subsystem I, we may regard A' as a very small deviation for the total system I + II, and therefore approximate T by the value T^* at the state A^*. When we discuss the fluctuation, the composite system is considered to be isolated, so that the internal energy at the state $A^* + A'$ must be equal to the value E in the state A^*. In this case, equation (1) may be written as

$$W_{\text{min}}(\alpha^*, \alpha') = -T^*\{S(E, N, V, \alpha^* + \alpha') - S(E, N, V, \alpha^*)\}. \tag{2}$$

Substituting this into (1.68), we obtain the required expression.

2. For a rarefied gas obeying the Maxwellian velocity distribution law, find the mean free flight time, i.e. the mean time between two successive collisions suffered by one particular molecule, regarding the molecule as an elastic sphere of radius a. Find the mean free path, i.e. the mean distance traversed by a molecule during the mean free flight time.

Using the resulting formulae, calculate the respective values at standard conditions (0° C, 1 atm) for N_2 gas (using $a = 1.90$ Å).

SOLUTION

In order to clarify our calculations, let us assume a mixture of a gas composed of molecules with mass m_1 and another gas composed of molecules with mass m_2, and consider a collision between molecules m_1 and m_2. Let us denote their radii by a_1 and a_2, and their velocities by $\boldsymbol{v}_1$ and $\boldsymbol{v}_2$, respectively. Then the velocity of their center of mass $\boldsymbol{V}$ and the relative velocity $\boldsymbol{u}$ are given by

$$\boldsymbol{V} = \frac{m_1 \boldsymbol{v}_1 + m_2 \boldsymbol{v}_2}{m_1 + m_2}, \qquad \boldsymbol{u} = \boldsymbol{v}_2 - \boldsymbol{v}_1 . \tag{1}$$

Collisions of the type shown in Figure 6.2 occur

$$f_1 \, \mathrm{d}\boldsymbol{v}_1 f_2 \, \mathrm{d}\boldsymbol{v}_2 \cdot \pi(a_1 + a_2)^2 \{\sin^2(\theta + \mathrm{d}\theta) - \sin^2\theta\} u \tag{2}$$

times per unit time and per unit volume of the gas. Here, f_1 and f_2 are the Maxwellian distribution function, i.e. expression (1.100) multiplied by the number density of molecules. From (1) we get

$$m_1 \boldsymbol{v}_1^2 + m_2 \boldsymbol{v}_2^2 = (m_1 + m_2) \boldsymbol{V}^2 + \mu \boldsymbol{u}^2, (\mu^{-1} = m_1^{-1} + m_2^{-1}), \tag{3}$$

$$\mathrm{d}\boldsymbol{v}_1 \, \mathrm{d}\boldsymbol{v}_2 = \left| \frac{\partial(\boldsymbol{v}_1, \boldsymbol{v}_2)}{\partial(\boldsymbol{V}, \boldsymbol{u})} \right| \mathrm{d}\boldsymbol{V} \mathrm{d}\boldsymbol{u} = \mathrm{d}\boldsymbol{V} \mathrm{d}\boldsymbol{u} \tag{4}$$

and by making use of these relations, we can integrate expression (2) over all possible orientations of the line $(0 < \theta < \frac{1}{2}\pi)$ to obtain

$$\begin{aligned} N_{12} &= \iint \mathrm{d}\boldsymbol{V} \mathrm{d}\boldsymbol{u} \, n_1 n_2 \frac{(m_1 m_2)^{\frac{3}{2}}}{(2\pi kT)^3} \\ &\quad \times \exp\left\{ - \frac{(m_1 + m_2)\boldsymbol{V}^2 + \mu \boldsymbol{u}^2}{2kT} \right\} \\ &\quad \times \pi(a_1 + a_2)^2 u \int_0^{\frac{1}{2}\pi} 2 \sin\theta \cos\theta \, \mathrm{d}\theta \\ &= n_1 n_2 \left(\frac{\mu}{2\pi kT} \right)^{\frac{3}{2}} \int_0^\infty 4\pi u^2 \, \mathrm{d}u \, \mathrm{e}^{-\mu u^2/2kT} \pi(a_1 + a_2)^2 u \\ &= 2 n_1 n_2 (a_1 + a_2)^2 \sqrt{\frac{2\pi kT}{\mu}} . \end{aligned} \tag{5}$$

This gives the mean number of m_1 - m_2 collisions occurring per unit time and per unit volume of the gas. The mean number of collisions involving one particular molecule is given by (5) divided by n_1. Since that number of collisions occurs per unit time, the mean time interval between collisions becomes

$$\tau = \left\{ 4n\pi d^2 \sqrt{\frac{kT}{\pi m}} \right\}^{-1} = \frac{1}{4n\sigma} \sqrt{\frac{\pi m}{kT}}, \tag{6}$$

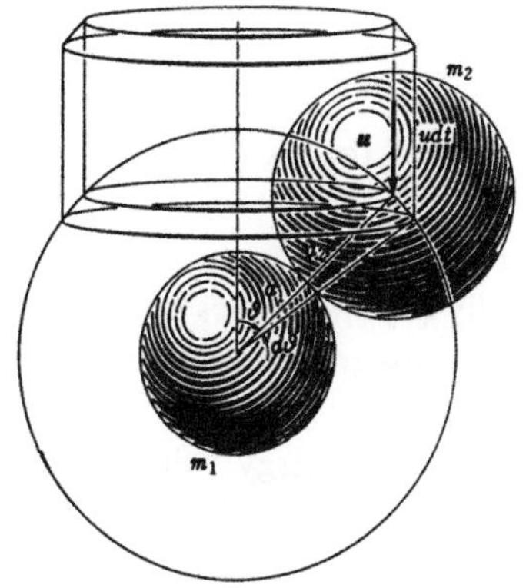

Fig. 6.2.

Fig. 6.3. N_2 molecule.

where we have put $m_1 = m_2 = m$, $n_1 = n_2 = n$, and $a_1 = a_2 = \frac{1}{2}d$, while $\sigma = \pi d^2$ is just the cross-section. The distance traversed by the molecule during this time interval may be estimated as

$$l = \tau \bar{v} = \tau \sqrt{\frac{8kT}{\pi m}} = \frac{1}{n\sigma\sqrt{2}}. \tag{7}$$

The number density can be determined from the equation of state for the ideal gas $p = nkT$. At $p = 1$ atm $= 1.013 \times 16^6$ dynes cm^{-2} and $T = 273°$ K, we obtain $n = 2.69 \times 10^{19}$ cm^{-3}. The cross-section $\sigma = \pi d^2 = 4\pi a^2$ becomes 45.4×10^{-16} cm^2, if we take $a = 1.90 \times 10^{-8}$ cm. The mass of the N_2-molecule is given by $m = 28.0 \times 1.660 \times 10^{-24} = 46.5 \times 10^{-24}$ g, where 28.0 is the molecular weight. Thus we get $\bar{v} = \sqrt{8kT/\pi m} = 453$ m sec^{-1}.

From (7) we obtain, first, $l = 5.79 \times 10^{-6} = 5.8 \times 10^{-6}$ cm. The mean free flight time τ is given by dividing this value by $\bar{v}$, and is $\tau = 1.29 \times 10^{-10}$ sec.

NOTE 1: In equation (7), which is called Maxwell's equation, it is not obvious from the above context that we should use for $\bar{v}$ the mean velocity.

NOTE 2: The mean distance between two molecules in the gas $1/n^{\frac{1}{3}}$ is $\frac{1}{3} \times 10^{-6}$ cm at 0° C and 1 atm. Notice that the value of l obtained above is more than 17 times this mean distance.

NOTE 3: As the value of the radius of the N_2 molecule we have used $a = 1.90$ Å, which is the average of the values estimated from experimental values of the coefficients of viscosity, of thermal conduction, and of diffusion. The actual shape of the molecule, e.g., as pictured by using the van der Waals radius of atoms (1.5 Å for N) and the nuclear distance (1.09 Å for N_2), is different from the sphere considered above, which is called the kinetic theory shell.

3. Estimate the value of thermal conductivity of the electron gas in a metal based on the following assumptions:

a. all electrons move with the same speed v;

b. each electron is scattered isotropically (by impurities contained in the metal, say) at every time interval τ, which is assumed to be the same for all electrons;

c. after scattering, each electron has just the energy value equal to the internal energy (per electron) of the electron gas u at the point where the scattering occurred.

SOLUTION

Let us suppose a weak temperature gradient in the direction of the z-axis: the variation of temperature $T(z)$ over a distance of the order of mean free path $l = \tau v$ is assumed to be small compared with the value of T itself. The internal energy u depends on the coordinate z through the temperature T: in this sense we write it as a function $u(z)$.

The heat flux q is defined as the energy transferred per unit time through unit area on the plane $z = 0$, and the heat conductivity λ by the relation

$$q = -\lambda \frac{\partial T}{\partial z}. \tag{1}$$

Since an electron passing through the plane $z = 0$ should move freely during a time of the order of the mean free flight time τ, let us assume, for the sake of obtaining an estimate that each electron moves freely during exactly the

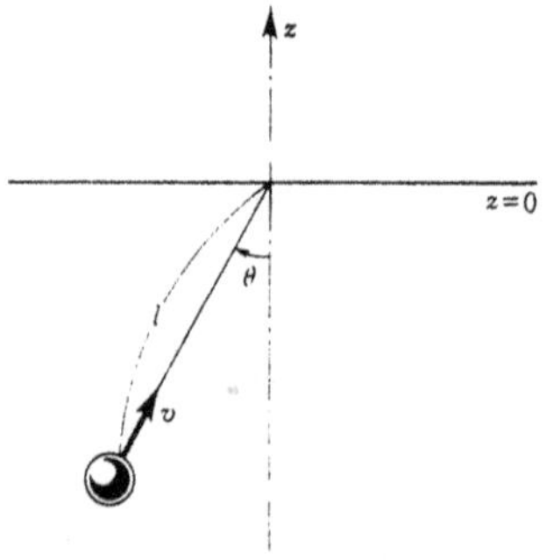

Fig. 6.4.

time τ before reaching the plane $z = 0$. An electron whose path makes an angle θ with the $-z$ axis should transfer internal energy of the amount $u(-l\cos\theta)$ by virture of the assumption c. Since the temperature gradient is assumed to be weak, we may put

$$u(-l\cos\theta) \risingdotseq u(0) - l\cos\theta\left(\frac{\partial u}{\partial z}\right)_{z=0}. \tag{2}$$

According to assumption b, the probability of finding an electron path making an angle within the interval $(\theta, \theta + \mathrm{d}\theta)$ with the $-z$ axis is given by $2\pi \sin\theta \, \mathrm{d}\theta/4\pi$, and hence the number of electrons passing through a unit area on the plane $z = 0$ per unit time at such an angle is given by

$$nv\cos\theta \cdot \tfrac{1}{2}\sin\theta \, \mathrm{d}\theta, \tag{3}$$

n being the number density of electrons. Thus, integrating over θ from 0 to π, we obtain the net flow of energy through the plane $z = 0$, i.e. the heat flux:

$$q = -\tfrac{1}{2} n v l \frac{\partial u}{\partial z}\int_0^\pi \cos^2\theta \sin\theta \, \mathrm{d}\theta = -\tfrac{1}{3} nvl \frac{\partial u}{\partial z}. \tag{4}$$

Since $\partial u/\partial z = (\partial u/\partial T)\,\partial T/\partial z$, and since $\partial u/\partial T$ is the heat capacity per electron and becomes the heat capacity per unit volume if multiplied by n, we obtain, on comparing (4) with (1),

$$\lambda = \tfrac{1}{3} Cvl. \tag{5}$$

NOTE: The condition of zero electric current is necessary, and is, in fact, mathematically satisfied by our model, as is seen by integrating (3) from 0 to π. However, for this condition to be satisfied, the pressure p must be uniform, and since the temperature T is a function of z, the number density n must be also a function of z. The internal energy u is, in general, a function of n and T, so that the kind of heat capacity entering in the derivative $\partial u/\partial z$ is not obvious in the above context.

The model used in the problem is, however, a fairly good approximation for a strongly degenerate electron gas.

4. Find the electric conductivity of a homogeneous metal at temperature T by assuming the Boltzmann equation for strongly degenerate conduction electrons in the form

$$\frac{\partial f}{\partial t} + \boldsymbol{v}\cdot\frac{\partial f}{\partial x} - e\boldsymbol{E}\cdot\frac{\partial f}{\partial \boldsymbol{p}} = -\frac{f - f_0}{\tau}$$

where $-e$ denotes the electric charge of an electron, and $\boldsymbol{p}$ the momentum. The equilibrium distribution function f_0 is given by the Fermi function $f_0(\varepsilon)$, $\varepsilon = \varepsilon(\boldsymbol{p})$ being the energy of an electron. The electric field $\boldsymbol{E}$ is assumed to be uniform, and the relaxation time τ to be a function of T and ε only (i.e. depending on $\boldsymbol{p}$ through ε).

In particular, calculate the conductivity when $\varepsilon(\boldsymbol{p})$ is a quadratic form $p^2/(2m^*)$, m^* being the effective mass.

SOLUTION

It is sufficient to find the current density in the form of a term proportional to the electric field. Under the assumptions of steadiness and uniformity, the Boltzmann equation reduces to

$$-e\boldsymbol{E}\cdot\frac{\partial f}{\partial \boldsymbol{p}} = -\frac{f-f_0}{\tau}. \qquad (1)$$

In order to determine an expression correct to first order in $\boldsymbol{E}$, the distribution function f on the right-hand side may be replaced by the zeroth approximation f_0, i.e. the solution in the case of $\boldsymbol{E} = 0$. Noticing that f_0 is a function of ε, one obtains

$$f = f_0 + \frac{\partial f_0}{\partial \varepsilon}\tau e\boldsymbol{v}\cdot\boldsymbol{E} \qquad (\partial\varepsilon/\partial\boldsymbol{p} = \boldsymbol{v} = \text{velocity}). \qquad (2)$$

The current density is obtained by multiplying (2) by $-e\boldsymbol{v}$ and integrating over all values of the momentum:

$$\boldsymbol{j} = -e\int \boldsymbol{v}f\frac{2\,\mathrm{d}\boldsymbol{p}}{h^3} = e^2\int\left(-\frac{\partial f_0}{\partial\varepsilon}\right)\tau\boldsymbol{v}\boldsymbol{v}\cdot\boldsymbol{E}\frac{2\,\mathrm{d}\boldsymbol{p}}{h^3}. \qquad (3)$$

Here $\mathrm{d}\boldsymbol{p}$ stands for $\mathrm{d}p_x\,\mathrm{d}p_y\,\mathrm{d}p_z$, and the factor 2 accounts for the weight due to the electron spin. Thus the components of the electric conductivity tensor become

$$\sigma_{\alpha\beta} = e^2\int\tau v_\alpha v_\beta\left(-\frac{\partial f_0}{\partial\varepsilon}\right)\frac{2\,\mathrm{d}\boldsymbol{p}}{h^3} = e^2\int\tau\overline{v_\alpha v_\beta}\left(-\frac{\partial f_0}{\partial\varepsilon}\right)D(\varepsilon)\,\mathrm{d}\varepsilon, \quad (\alpha,\beta = x,y,z) \qquad (4)$$

where $D(\varepsilon)$ denotes the state density of the conduction band, and $\overline{v_\alpha v_\beta}$ the average of $v_\alpha v_\beta$ taken over the energy surface ε.

If the electron system is strongly degenerate, $-\partial f_0/\partial\varepsilon$ has a sharp maximum at $\varepsilon = \mu$ (μ being the chemical potential), so that (4) may be approximated as

$$\sigma_{\alpha\beta} = e^2\tau(\mu)\overline{v_\alpha v_\beta}^{\mathrm{F}}D(\mu), \qquad (5)$$

where $\overline{v_\alpha v_\beta}^{\mathrm{F}}$ stands for the value of $\overline{v_\alpha v_\beta}$ on the Fermi surface $\varepsilon = \mu$.

In the case of a spherical energy surface $\varepsilon(\boldsymbol{p}) = p^2/2m^*$, one has $\boldsymbol{v} = \boldsymbol{p}/m^*$, $\overline{v_\alpha v_\beta} = \delta_{\alpha\beta}\, 2\varepsilon/3m^*$ and $D(\varepsilon) = \sqrt{2\varepsilon}\, m^{*\frac{3}{2}}/\pi^2 \hbar^3$. Thus

$$\sigma = \tfrac{1}{3} e^2 \tau(\mu) \frac{(2m^*\mu)^{\frac{3}{2}}}{m^* \pi^2 \hbar^3} = \frac{ne^2}{m^*} \tau(\mu). \tag{6}$$

Here $n = (2m^*\mu)^{\frac{3}{2}}/3\pi^2\hbar^3$ is the number density of the electrons.

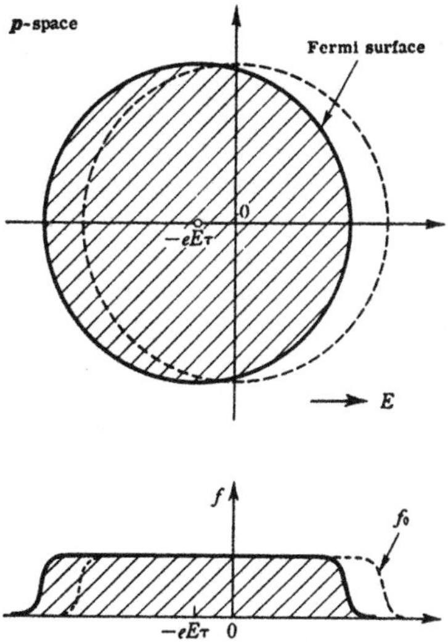

Fig. 6.5.

NOTE: According to (2), one sees that the electric current is produced by a shift of the center of the Fermi distribution. This is most clearly seen in the case of $\varepsilon(\boldsymbol{p}) = p^2/2m^*$:

$$f \doteq f_0(\varepsilon + \tau e \boldsymbol{v} \cdot \boldsymbol{E}) \doteq f_0\left(\frac{1}{2m^*} |\boldsymbol{p} + \tau e \boldsymbol{E}|^2\right). \tag{7}$$

Here $-e\, \boldsymbol{E}\tau$ gives the shift of the center.

DIVERTISSEMENT 13

Onsager's reciprocity relations and the thermodynamics of irreversible processes. Many physical phenomena show a certain symmetry which is often called the reciprocity. When a signal is sent out from a point A and received at another point B, it is just as equally receivable as a signal sent in the reverse way. The transmission of signals is reciprocal. In irreversible processes, which we here suppose to be stationary, for the sake of simplicity, we usually are dealing with various kinds of flow such as heat current, electric current, particle flow and so forth. These currents are driven by forces which are produced by the general tendency of nature to recover equilibrium once a system is shifted from it. So far as the state of a system is not too far from equilibrium we may assume linear relations between flows and forces; namely

$$J_i = \sum_k L_{ik} X_k$$

where J_i ($i = 1, 2, \ldots$) denote flows of various sorts and the X_i's are forces, X_i being the force conjugate to J_i. The coefficients L_{ik} are called *kinetic coefficients*, examples of which are heat conductivity, diffusion constant, electric conductivity and so on. Generally speaking, the off-diagonal elements of the matrix (L_{ik}) will not vanish; they are responsible for cross effects. It had been long recognized empirically that reciprocal relations such as $L_{ik} = L_{ki}$ exist. For example, the tensor of heat conductivity in an anisotropic crystal is symmetric. Why is this so? This reciprocity looks somewhat different from the above mentioned example of signal transmission, which is due to the dynamical laws of electromagnetic waves or sound waves. (To be more exact, the reciprocity of signal transmissions is also a case of Onsager's reciprocity.) The reciprocity of kinetic coefficients is not a direct consequence of such dynamical laws. L. Onsager raised this question and gave the answer to it. His proof, appearing in 1931 (Phys. Rev. **37**, 405; **38**, 2265), was based upon the analysis of fluctuation processes and the reversibility of dynamical laws governing the microscopic processes behind observable macroscopic phenomena.

The fundamental importance of Onsager's reciprocity law has only fully recognized more recently. During and after World War II, the so-called quasi-thermodynamics or thermodynamics of irreversible processes was developed mainly in Europe. This is a phenomenological approach to non-equilibrium phenomena which aims at establishing internal relations between irreversible processes under different but related conditions. The usefulness of this approach would be very much restricted if we did not have Onsager's reciprocity relations which form in fact, the most essential basis of the whole theory. For the subject of quasi-thermodynamics we refer to the following books: S. R. de Groot, *Thermodynamics of Irreversible Processes* (North-Holland Publ. Co., Amsterdam, 1952); I. Prigogine, *Introduction to Thermodynamics of Irreversible Processes* (Thomas, Springfield, Ill., 1955); R. Becker, *Theorie der Wärme* (Springer, 1955) Chapter VII; S. R. de Groot and P. Mazur, *Non-Equilibrium Thermodynamics* (North-Holland Publ. Co., Amsterdam, 1962).

It must also be mentioned that Onsager's reciprocity law has a deep connection with the so-called dissipation-fluctuation theorem and the recent development of statistical-mechanics of irreversible processes (see Divertissement 14).

Problems

A

1. One may use the result of example 1 to discuss fluctuations of thermodynamical variables (α) characterizing a subsystem containing a certain amount of mass (certain number of molecules) of a homogeneous large system.

(i) Show that the probability of finding a departure α' of α from its equilibrium value α^* may generally be written in the form

$$P(\alpha') = C \exp\left[\frac{1}{2kT^*}(\Delta p \Delta V - \Delta T \Delta S)\right]$$

T^* being the equilibrium temperature, provided that α' is not very large and the probability may be regarded as Gaussian.

Hint: Use the equation for the minimum work,

$$W_{\min} = \Delta U - T^* \Delta S + p^* \Delta V$$

for a process in which the changes ΔU, ΔS and ΔV are given, T^* and p^* being the equilibrium temperature and pressure. Then, expand ΔU up to the second order terms in ΔS and ΔV. See volume I, (2.24) and (3.32b).

(ii) Take especially V and T as α, and find $\overline{\Delta V^2}$, $\overline{\Delta V \Delta T}$ and $\overline{\Delta T^2}$.

(iii) Find Δp .

2. The probability distribution of energy E and volume V of the subsystem considered in the previous problem may be regarded as the $T-p$ distribution given in § 1.13 if the remaining part of the total system is treated as an environment of constant temperature T^* and constant pressure p^* for the subsystem. Based on this, derive the probability distribution discussed in (i) of the previous problem, and find $\overline{\Delta E^2}$, $\overline{\Delta V^2}$ and $\overline{\Delta E \Delta V}$.

3. Discuss the fluctuation in the number of solute molecules N_1 contained in a small part of a two-component solution which contains a constant number of solvent molecules N_0. Applying the result to a dilute solution, derive the relation between the fluctuation in N_1 and the osmotic pressure. The temperature may be treated as constant, and its fluctuation may be neglected. (*Hint:* Extend the solution to problem 1 to include a change in N_1, or, which amounts to the same thing, use the $T-\mu$ distribution.)

4. Show that in a stationary state, the pressures p_A and p_B of an very dilute ideal gas kept in two compartments A and B separated by a porous wall satisfy the relation

$$\frac{p_A}{p_B} = \sqrt{\frac{T_A}{T_B}}$$

when the temperatures of those two compartments are held at T_A and T_B respectively. Discuss what happens if the two compartments are connected by

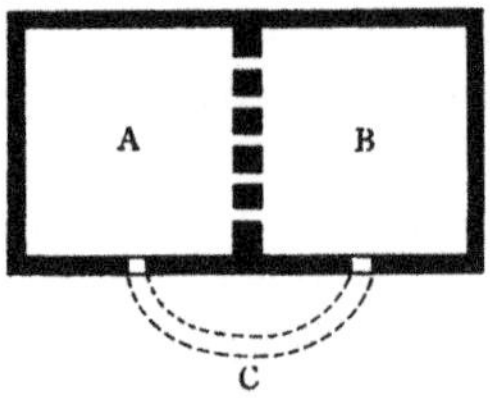

a tube C, as shown in the figure. (*Hint:* consider the balance in the gas flowing through small holes in the porous wall.)

5. Estimate the coefficient of viscosity of a rarefied gas based on the following assumptions:

a. all the molecules are in thermal motion with the same constant speed;

b. each molecule is isotropically scattered as a result of encounters with other molecules, and this occurs regularly at time intervals τ;

c. the average speed of a molecule immediately after an encounter is equal to the flow velocity V of the gas at the point of encounter. (*Hint:* the velocity of a molecule is the vector sum of the flow velocity and the velocity of thermal agitation. The origin of viscous force is the momentum transfer between layers with different flow velocities. Calculate the momentum transfer using the method of example 3.)

6. Show that a mirror, much smaller in size than the mean free path of a molecule, suspended by means of a very thin wire in a rarefied gas of temperature T and pressure p, and rotating with an angular velocity ω, experiences a couple equal to $-\zeta\omega$, where the friction constant ζ is given by:

$$\zeta = \frac{2m\bar{v}pI}{\sigma kT}.$$

Here, $\bar{v} = \sqrt{8kT/\pi m}$, I is the moment of inertia, σ is the mass per unit area of the mirror, and m is the mass of a gas molecule.

7. Find the electric conductivity of a metal by calculating the average velocity of an electron on the assumption that each electron is scattered isotropically (e.g. by impurities in the metal) at constant time intervals τ.

8. For non-degenerate electrons in a semiconductor, f_0 appearing in the Boltzmann equation given in example 4 may be put equal to the Maxwell-Boltzmann distribution. Calculate the electric conductivity by assuming $\varepsilon(\boldsymbol{p}) = p^2/2m^*$ and $\tau = A\,|\,\boldsymbol{v}\,|^s$ ($A > 0$ and $s > -7$ are constants).

9. Calculate the electric conductivity tensor of a homogeneous metal under action of a uniform static magnetic field $\boldsymbol{H}$, assuming a Boltzmann equation of the type treated in example 4 and the quadratic law $\varepsilon(\boldsymbol{p}) = p^2/2m^*$, m^* being the effective mass.

B

10. Find the average fluctuation of the deviation of each point of a string stretched between two fixed pins with constant stress, by assuming that the deviation is small compared with the length of string. (*Hint:* expand the deviation $\xi(x)$

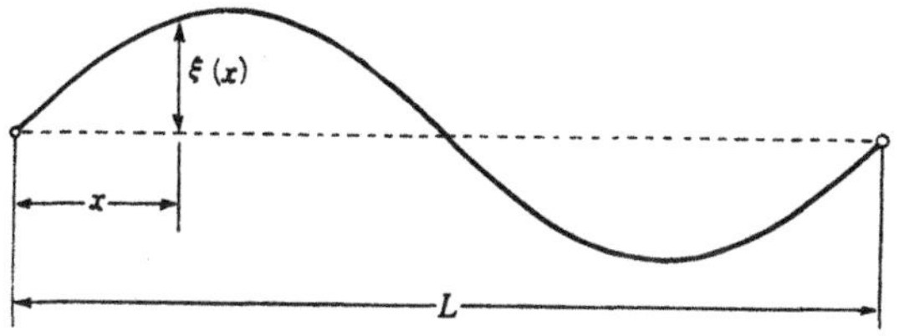

into a Fourier series, and regard the expansion coefficients $A_1, A_2, \ldots$ as the coordinates describing the deviation.)

11. Show that, for the number of molecules N_A contained in a macroscopic region V_A within a liquid, the relation

$$\langle (N_A - \langle N_A \rangle)^2 \rangle = \langle N_A \rangle \{1 + n \int \{g(R) - 1\}\, \mathrm{d}\boldsymbol{R}\}$$

holds, where n stands for the (average) number density of molecules, and $g(R)$ is the radial distribution function defined in equation (5.36). Furthermore, show that

$$n \int \{g(R) - 1\}\, \mathrm{d}\boldsymbol{R} = nkT\kappa_T - 1 \qquad (\kappa_T = \text{isothermal compressibility}).$$

12. The probability that a particle moving along the x-axis, in such a way that it is displaced by either $+a$ or $-a$ with equal probability during each time interval τ, reaches a point ya at a time $t = n\tau$ after it starts from a point xa, satisfies the Smoluchowski equation

$$P_n(x \mid y) = \sum_{z=-\infty}^{\infty} P_{n-1}(x \mid z) P_1(z \mid y), \qquad n \geqq 1,$$

where
$$P_1(x \mid y) = \tfrac{1}{2}\delta_{y,x-1} + \tfrac{1}{2}\delta_{y,x+1}, \qquad P_0(x \mid y) = \delta_{x,y}.$$

Find $P_n(x \mid y)$ by solving the Smoluchowski equation. In particular, discuss the asymptotic form of the solution when n is very large. (*Hint:* put

$$\sum_{y=-\infty}^{\infty} P_n(x \mid y)\xi^y = Q_n(\xi)$$

and solve the equation for Q_n, and then determine the $P_n(x \mid y)$ as the expansion coefficients of Q_n in powers of ξ.)

13. A velocity component v of a microscopic particle (e.g. colloidal particle) of mass m suspended in a liquid of temperature T satisfies the equation of motion

$$m\frac{dv}{dt} = -\zeta v + F(t),$$

where ζ denotes the friction constant, and $F(t)$ the fluctuating part of the force exerted by liquid molecules on the particle. Assuming that the fluctuating force is a stochastic variable and that its average and correlation function are given by
$$\overline{F(t)} = 0, \qquad \overline{F(t)F(t')} = 2\zeta kT\delta(t-t')$$

respectively, find the average and the correlation function of the velocity component v of the particle. Furthermore, calculate the mean-square displacement of the particle for a sufficiently long time interval ($t \gg m/\zeta$). (*Hint:* solve the given differential equation and express v in terms of F.)

NOTE: The above-mentioned equation of motion for a microscopic particle is called the *Langevin equation*, and the thermal motion of such a microscopic particle is called *Brownian motion*.

14. Calculate the heat conductivity of a metal by assuming the Boltzmann equation treated in example 4, in which the electric field vanishes and instead a uniform stationary temperature gradient is introduced. Consider the case of the quadratic law $\varepsilon(p) = p^2/2m^*$ (m^* being an effective mass) and $\tau = Av^s$ ($A > 0$ and $s > -7$ are constants).

C

15. Find the resistive force suffered by a sphere, which moves at constant velocity in a rarefied gas in equilibrium at temperature T and which has a radius large compared with the radius of the gas molecules but small compared with the mean free path of the gas molecules. Assume that the collisions between the sphere and the gas molecules are completely elastic (the surface of the sphere is smooth).

16. The temporal variation of the orientational distribution function $f(\theta,\varphi,t)$ of electric dipoles of molecules with effective radius a and permanent dipole moment μ suspended in a liquid of viscosity coefficient η under action of an electric field $E(t)$ is governed by the equation

$$\frac{\partial f}{\partial t} = \frac{B}{\sin\theta}\frac{\partial}{\partial\theta}\left\{\sin\theta\left(\frac{\partial f}{\partial\theta} + \frac{\mu E(t)\sin\theta}{kT}f\right)\right\} + \frac{B}{\sin^2\theta}\frac{\partial^2 f}{\partial\varphi^2},$$

where the polar angle $(\theta,\ \varphi)$ describes an orientation referred to the electric field as the polar axis, and the orientation diffusion coefficient B is given by $kT/8\pi\eta a^3$, T being the temperature of the liquid. Find the complex dielectric constant based on the assumption $|\mu E(t)| \ll kT$. (*Hint:* the *complex dielectric constant* $\varepsilon(x)$ is defined as the Fourier transform of the relaxation function $\varphi(t)$:

$$\varepsilon(\omega) = \varepsilon_\infty + \int_0^\infty \varphi(t)\,\mathrm{e}^{-\mathrm{i}\omega t}\,\mathrm{d}t$$

provided that the electric displacement $D(t)$ is given as a linear equation of the electric field $E(t)$ (assumed to vanish at the infinite past) in the form

$$D(t) = \varepsilon_\infty E(t) + \int_{-\infty}^{t} \varphi(t-t')E(t')\,\mathrm{d}t'.$$

Solve the given equation for f, and calculate the electric polarization).

17. Derive expressions for the viscosity coefficients (bulk and shear) and the thermal conductivity of a rarefied gas, for which the Boltzmann equation has the form

$$\frac{\partial f}{\partial t} + v_x\frac{\partial f}{\partial x} + v_y\frac{\partial f}{\partial y} + v_z\frac{\partial f}{\partial z} = -\frac{f-f_0}{\tau},$$

where f_0 stands for the local equilibrium distribution, i.e. the Maxwellian distribution corresponding to the local number density $n(\boldsymbol{x})$ and the local temperature $T(\boldsymbol{x})$ and extending around the local flow velocity $V(\boldsymbol{x})$. Here $\boldsymbol{x} = (x, y, z)$ denotes the radius vector, and $\boldsymbol{v} = (v_x, v_y, v_z)$ the velocity vector. The relaxation time τ is assumed to be a function of $|\boldsymbol{v} - \boldsymbol{V}|$. Calculate these expressions in the case of $\tau = A|\boldsymbol{v} - \boldsymbol{V}|^s$ ($A > 0$ and $s > -9$ are constants).

18. The motion of an electron belonging to a molecule in a rarefied gas may, in some cases, be replaced by that of a harmonic oscillator: it is determined by

$$\frac{\mathrm{d}\boldsymbol{x}}{\mathrm{d}t} = \frac{\boldsymbol{p}}{m}, \qquad \frac{\mathrm{d}\boldsymbol{p}}{\mathrm{d}t} = -m\omega_0^2\boldsymbol{x} - eE(t),$$

where $\boldsymbol{x}$ and $\boldsymbol{p}$ denote the radius vector and the momentum of the electron within the molecule respectively, m the mass, $-e$ the electric charge, ω_0 the characteristic angular frequency, and $E(t)$ an external electric field. Show that the average $\bar{f}(\boldsymbol{x}, \boldsymbol{p}, t)$ of the electron distribution function $f(\boldsymbol{x}, \boldsymbol{p}, t)$, taken over all possible values of the time and of the position at which collisions occur, obeys the equation

$$\frac{\partial \bar{f}}{\partial t} + \frac{\boldsymbol{p}}{m}\cdot\frac{\partial \bar{f}}{\partial \boldsymbol{x}} + \{-m\omega_0^2\boldsymbol{x} - e\boldsymbol{E}(t)\}\cdot\frac{\partial \bar{f}}{\partial \boldsymbol{p}} = -\frac{\bar{f} - f_0}{\tau}$$

by assuming that collisions between molecules occur with a mean free flight time τ, and that the electron distribution function immediately after a collision reduces to a given distribution function $f_0(\boldsymbol{x}, \boldsymbol{p}, t)$.

19. Assume that the collision term in the Boltzmann equation of an electron system is of the form (6.9), and that the operator D is defined by

$$\left(\frac{\partial f}{\partial t}\right)_{\text{coll.}} \equiv -Df = -f(\boldsymbol{v})\int W(\boldsymbol{v}, \boldsymbol{v}')\mathrm{d}\boldsymbol{v}' + \int W(\boldsymbol{v}', \boldsymbol{v})f(\boldsymbol{v}')\mathrm{d}\boldsymbol{v}'$$

where $W(\boldsymbol{v}, \boldsymbol{v}')\,\mathrm{d}\boldsymbol{v}'$ denotes the transition rate of electrons with velocity $\boldsymbol{v}$ into a state with velocity in the interval between $\boldsymbol{v}'$ and $\boldsymbol{v}' + \mathrm{d}\boldsymbol{v}'$ (for simplicity, the collisions are assumed to be elastic). Solve the time-dependent Boltzmann equation up through a term linear in $\boldsymbol{E}$, with the initial condition that $\boldsymbol{E}$ vanishes at $t = -\infty$ and that the electron distribution is the equilibrium distribution f_0, and show that the electric current density at a time t is generally given in the form

$$j_i(t) = \sum_l \int_{-\infty}^{t} E_l(t')\mathrm{d}t'\Phi_{il}(t - t') \qquad (i, l = x, y, z)$$

with

$$\Phi_{il}(t) = \frac{\langle j_i(t) j_l(0)\rangle}{kT}.$$

Here $\langle j_i(t) j_i(0) \rangle$ is the correlation function of the electric current, which occurs as a fluctuation in an electron system at equilibrium. Based on this result, express the static and dynamical electric conductivities in terms of the correlation function.

Solutions

1. (i) Let ΔU, ΔS and ΔV be changes in U, S and V, respectively, which follow α', a change in α. The minimum work required is given as

$$W_{\min} = \Delta U - T^* \Delta S + p^* \Delta V. \tag{1}$$

Here T^* and p^* are the equilibrium temperature and pressure, respectively. We expand ΔU in powers of ΔS and ΔV, to obtain

$$\Delta U = T^* \Delta S - p^* \Delta V + \\ + \frac{1}{2}\left\{\left(\frac{\partial^2 U}{\partial S^2}\right)^* (\Delta S)^2 + 2\left(\frac{\partial^2 U}{\partial S \partial V}\right)^* \Delta S \Delta V + \left(\frac{\partial^2 U}{\partial V^2}\right)^* (\Delta V)^2\right\}, \tag{2}$$

where quantities marked with an asterisk(*) represent their equilibrium values. Inserting the equations

$$\Delta T = \Delta\left(\frac{\partial U}{\partial S}\right) = \left(\frac{\partial^2 U}{\partial S^2}\right)^* \Delta S + \left(\frac{\partial^2 U}{\partial V \partial S}\right)^* \Delta V,$$

$$-\Delta p = \Delta\left(\frac{\partial U}{\partial V}\right) = \left(\frac{\partial^2 U}{\partial S \partial V}\right)^* \Delta S + \left(\frac{\partial^2 U}{\partial V^2}\right)^* \Delta V$$

into (2), we obtain

$$W_{\min} = \tfrac{1}{2}(\Delta T \Delta S - \Delta p \Delta V). \tag{3}$$

Substitution of this equation into the equation in example 1 yields

$$P(\alpha') = C \exp\left[\frac{1}{2kT^*}\{\Delta p \Delta V - \Delta T \Delta S\}\right], \quad (C = \text{normalization constant}). \tag{4}$$

(ii) To obtain the answer, we need to express the right hand side of (3) as a quadratic form in ΔT and ΔV. If we substitute the relations

$$\Delta p = \left(\frac{\partial p}{\partial V}\right)_T^* \Delta V + \left(\frac{\partial p}{\partial T}\right)_V^* \Delta T, \qquad \Delta S = \left(\frac{\partial p}{\partial T}\right)_V^* \Delta V + \frac{C_V^*}{T^*} \Delta T \tag{5}$$

(where we have used both Maxwell's relation $(\partial S/\partial V)_T = (\partial p/\partial T)_V$ and $(\partial S/\partial T)_V = C_V/T$) into (3), we find that the term proportional to $\Delta V \Delta T$ vanishes and we obtain

$$P(\Delta V, \Delta T) = C \exp\left[\frac{1}{2kT^*}\left(\frac{\partial p}{\partial V}\right)_T^* (\Delta V)^2 - \frac{C_V^*}{2kT^{*2}} (\Delta T)^2\right]. \tag{6}$$

From (6), it is apparent that ΔV and ΔT are independent of each other (hence $\overline{\Delta V \Delta T} = 0$), and we have

$$\overline{(\Delta V)^2} = -kT^*\left(\frac{\partial V}{\partial p}\right)_T^* = -kT^* V\kappa_T, \tag{7a}$$

(κ_T = isothermal compressibility)

$$\overline{(\Delta T)^2} = kT^{*2}/C_V^*. \tag{7b}$$

(iii) Calculation becomes tedious or easy according to which is chosen as an independent variable besides p. If we take p and S as independent variables, expressing ΔV and ΔT in terms of Δp and ΔS, as in (5), and inserting them into (3), we find that the term proportional to $\Delta p \Delta S$ fortunately cancels [with the aid of Maxwell's relation $(\partial V/\partial S)_p = (\partial T/\partial p)_S$, which can be obtained from the derivatives of enthalpy H, $(\partial H/\partial S)_p = T$ and $(\partial H/\partial p)_S = V$]. Finally, (4) takes the form

$$P(\Delta p, \Delta S) = C' \exp\left\{\frac{1}{2kT^*}\left(\frac{\partial V}{\partial p}\right)^* (\Delta p)^2 - \frac{1}{2kC_p^*}(\Delta S)^2\right\}. \tag{8}$$

Therefore, we have

$$\overline{(\Delta p)^2} = -kT^*\left(\frac{\partial p}{\partial V}\right)^* = \frac{kT^*}{V^*\kappa_S}, \qquad (\kappa_S = \text{adiabatic compressibility}). \tag{9}$$

NOTE: Instead of p and S, we may take p and T (or p and V) as independent variables. In this case, however, the term proportional to $\Delta p \Delta T$ (or $\Delta p \Delta V$) does not cancel out.

2. Let the density of states be denoted by $\Omega(E, V)$. Then the entropy is expressed as $S = k \log \Omega(E, V)$. According to (1.76), the probability distribution of E and V is given as

$$P(E, V) = C \exp\left[-\frac{1}{kT^*}(E + p^*V)\right]\Omega(E, V) =$$

$$= C \exp\left[-\frac{1}{kT^*}(E + p^*V - T^*S)\right]. \tag{1}$$

The equilibrium values of E and V, E^* and V^*, respectively, are determined from the requirement that the argument of the exponential function should be maximum, i.e.,

$$\frac{\partial S}{\partial E} = \frac{1}{T^*}, \qquad \frac{\partial S}{\partial E} = \frac{p^*}{T^*}. \tag{2}$$

Putting $E = E^* + \Delta E$ and $V = V^* + \Delta V$, and retaining terms quadratic in ΔE and ΔV, we have for (1)

$$P(\Delta E, \Delta V) = C' \exp\left[\frac{1}{2k}\left\{\left(\frac{\partial^2 S}{\partial E^2}\right)^* \Delta E^2 + 2\left(\frac{\partial^2 S}{\partial E \partial V}\right)^* \Delta E \Delta V + \left(\frac{\partial^2 S}{\partial V^2}\right)^* \Delta V^2\right\}\right]. \tag{3}$$

As in the last problem, by virture of the equations $T^{-1} = \partial S/\partial E$ and $p/T = \partial S/\partial V$, the argument of the exponential function can be transformed as

$$\begin{aligned}\Delta E \Delta\left(\frac{1}{T}\right) + \Delta V \Delta\left(\frac{p}{T}\right) &= \frac{1}{T^*}\left\{\Delta p \Delta V - \frac{p^*}{T^*}\Delta T \Delta V - \frac{1}{T^*}\Delta T \Delta E\right\} \\ &= \frac{1}{T^*}\{\Delta p \Delta V - \Delta T \Delta S\}.\end{aligned} \tag{4}$$

Next, the covariance matrix for the normal distribution (3) of ΔE and ΔV is given as

$$\begin{pmatrix} \overline{\Delta E^2} & \overline{\Delta E \Delta V} \\ \overline{\Delta V \Delta E} & \overline{\Delta V^2} \end{pmatrix} =$$

$$= k\begin{pmatrix} -\dfrac{\partial^2 S}{\partial E^2} & -\dfrac{\partial^2 S}{\partial E \partial V} \\ -\dfrac{\partial^2 S}{\partial V \partial E} & -\dfrac{\partial^2 S}{\partial V^2} \end{pmatrix}^{-1} = k\begin{pmatrix} -\dfrac{\partial \theta}{\partial E} & -\dfrac{\partial \theta}{\partial V} \\ -\dfrac{\partial \pi}{\partial E} & -\dfrac{\partial \pi}{\partial V} \end{pmatrix}^{-1}, \tag{5}$$

the proof of which will be found in the note for this solution. In the above, we have put $\theta = 1/T$ and $\pi = p/T$. Hereafter, we shall omit asterisks (*) and simply write the equilibrium values as p, T, and so on. With the help of a formula for the Jacobian, (5) can be transformed as follows (note that $\partial\theta/\partial V = \partial\pi/\partial E$):

$$\text{the right hand side of (5)} = \frac{k}{\dfrac{\partial(\theta, \pi)}{\partial(E, V)}}\begin{pmatrix} -\dfrac{\partial \pi}{\partial V} & \dfrac{\partial \theta}{\partial V} \\ \dfrac{\partial \theta}{\partial V} & -\dfrac{\partial \theta}{\partial E} \end{pmatrix} =$$

$$= k\frac{\partial(E, V)}{\partial(\theta, \pi)}\begin{pmatrix} -\dfrac{\partial \pi}{\partial V} & \dfrac{\partial \theta}{\partial V} \\ \dfrac{\partial \theta}{\partial V} & -\dfrac{\partial \theta}{\partial E} \end{pmatrix} = k\begin{pmatrix} -\left(\dfrac{\partial E}{\partial \theta}\right)_\pi & -\left(\dfrac{\partial E}{\partial \pi}\right)_\theta \\ -\left(\dfrac{\partial V}{\partial \theta}\right)_\pi & -\left(\dfrac{\partial V}{\partial \pi}\right)_\theta \end{pmatrix} \tag{6}$$

(note that $(\partial E/\partial \pi)_\theta = (\partial V/\partial \theta)_\pi$). In the last step, we have used relations such as

$$\frac{\partial(E,V)}{\partial(\theta,\pi)}\left(\frac{\partial \pi}{\partial V}\right)_E = \frac{\partial(E,V)}{\partial(\theta,\pi)}\frac{\partial(\pi,E)}{\partial(V,E)} = \frac{\partial(E,\pi)}{\partial(\theta,\pi)} = \left(\frac{\partial E}{\partial \theta}\right)_\pi .$$

From (6), we immediately obtain

$$\overline{\Delta V^2} = -k\left(\frac{\partial V}{\partial(p/T)}\right)_T = -kT\left(\frac{\partial V}{\partial p}\right)_T, \tag{7}$$

$$\overline{\Delta E \Delta V} = -k\left(\frac{\partial E}{\partial \pi}\right)_\theta = -kT\left(\frac{\partial E}{\partial p}\right)_T = kT\left(\frac{\partial E}{\partial V}\right)_T \Big/ \left(\frac{\partial p}{\partial V}\right)_T$$

$$= kT\left\{p - T\left(\frac{\partial p}{\partial T}\right)_V\right\} \Big/ \left(\frac{\partial p}{\partial V}\right)_T = kT\left\{T\left(\frac{\partial V}{\partial T}\right)_p + p\left(\frac{\partial V}{\partial p}\right)_T\right\}. \tag{8}$$

Also, $\overline{\Delta E^2}$ is obtained and rewritten as follows:

$$\overline{\Delta E^2} = -k\left(\frac{\partial E}{\partial \theta}\right)_\pi = k\left(\frac{\partial \pi}{\partial \theta}\right)_E \Big/ \left(\frac{\partial \pi}{\partial E}\right)_\theta$$

$$= kT\left\{p - T\left(\frac{\partial p}{\partial T}\right)_E\right\} \Big/ \left(\frac{\partial p}{\partial E}\right)_T = kT\left\{p\left(\frac{\partial E}{\partial p}\right)_T + T\left(\frac{\partial E}{\partial T}\right)_p\right\}.$$

If, in the above, we use the relation

$$\left(\frac{\partial E}{\partial T}\right)_p = \frac{\partial(E,p)}{\partial(T,V)} \Big/ \frac{\partial(T,p)}{\partial(T,V)} = \left(\frac{\partial E}{\partial T}\right)_V - \left(\frac{\partial E}{\partial V}\right)_T \left(\frac{\partial p}{\partial T}\right)_V \Big/ \left(\frac{\partial p}{\partial V}\right)_T$$

and the identity for $(\partial E/\partial p)_T$ which has been used in the calculation of (8), we have

$$\overline{\Delta E^2} = kT\left\{TC_V - \left(\frac{\partial E}{\partial V}\right)_T^2 \left(\frac{\partial V}{\partial p}\right)_T\right\}, \tag{9}$$

where $(\partial E/\partial V)_T$ is given, of course, by (3.21b).

NOTE: *Covariance matrix for n-dimensional normal distribution:* For a two-dimensional normal distribution, by virtue of the formula

$$\iint \exp\left[-\tfrac{1}{2}(ax^2 + 2bxy + cy^2)\right] \mathrm{d}x\,\mathrm{d}y = \frac{2\pi}{\sqrt{ac - b^2}}$$

(which is obtained by rewriting the argument of the exponential function as $ax^2 + 2bxy + cy^2 = a[x + (b/a)y]^2 + [c - (b/a)]y^2$ and by integrating with respect to x first, and to y next), we can easily evaluate $\langle x^2 \rangle$, $\langle xy \rangle$ and

$\langle y^2 \rangle$ (by taking the logarithmic derivatives of the above formula with respect to a, b and c, respectively), i.e.,

$$\begin{pmatrix} \langle x^2 \rangle & \langle xy \rangle \\ \langle xy \rangle & \langle y^2 \rangle \end{pmatrix} = \frac{1}{ac - b^2} \begin{pmatrix} c & -b \\ -b & a \end{pmatrix}. \tag{10}$$

More generally, for the n-dimensional normal distribution

$$P(x_1, \cdots, x_n) = C \exp\left[-\tfrac{1}{2} \sum_i \sum_j a_{jk} x_j x_k\right] \tag{11}$$

the matrix consisting of the mean values $\langle x_i \, x_k \rangle$ as its elements is called *the covariance matrix*. To obtain this matrix, it is convenient to introduce the characteristic function:

$$\begin{aligned} G(\xi_1, \cdots, \xi_n) &\equiv \langle \exp(\mathrm{i} \sum_{j=1}^{n} \xi_j x_j) \rangle \\ &= C \int \cdots \int \exp\{-\tfrac{1}{2} \sum_j^n \sum_k^n a_{jk} x_j x_k + \mathrm{i} \sum_j^n \xi_j x_j\} \, \mathrm{d}x_1 \cdots \mathrm{d}x_n . \end{aligned} \tag{12}$$

The integral in (12) is the same as given in (1) of the solution for problem 30, Chapter 2, i.e.,

$$G = C \cdot \frac{(2\Delta)^{\frac{1}{2}n}}{[\det A]^{\frac{1}{2}}} \exp\left[-\tfrac{1}{2} \sum A_{jk}^{-1} \xi_j \xi_k\right].$$

If we set $\xi_1 = \ldots = \xi_n = 0$, obviously we have $G = 1$ from (12). Hence the normalization constant C cancels the factor in front of the exponential function. Therefore, we get

$$G(\xi_1, \cdots, \xi_n) = \exp\left[-\tfrac{1}{2} \sum A_{jk}^{-1} \xi_j \xi_k\right], \tag{13}$$

where $A_{jk}{}^{-1}$ is the jk element of A^{-1}, the inverse matrix to the matrix $(a_{jk}) = A$. Differentiation of (12) with respect to ξ yields

$$\langle x_j x_k \rangle = \frac{1}{\mathrm{i}^2} \left(\frac{\partial^2 G}{\partial \xi_j \partial \xi_k} \right)_{\xi_1 = \cdots = \xi_n = 0},$$

whence $\langle x_j \, x_k \rangle = A_{jk}^{-1}$, i.e.

$$(\langle x_j x_k \rangle) = A^{-1}. \tag{14}$$

3. In problem 1, the change in the number of molecules was not considered. If this is taken into account, the minimum work required is given as

$$W_{\min} = \Delta U - T \Delta S + p \Delta V - \mu_1 \Delta N_1 , \tag{1}$$

where μ_1 and N_1 are the chemical potential and the number of solute molecules. A transformation just like that in problem 1 yields

$$W_{\min} = \tfrac{1}{2}(\Delta T\Delta S - \Delta p\Delta V + \Delta\mu_1\Delta N_1). \tag{2}$$

Assuming $\Delta T = 0$, and rewriting (3) in terms of Δp and ΔN_1, we find that the term proportional to $\Delta p\Delta N$ vanishes on account of Maxwell's relation $(\partial V/\partial N_1)_{T,p} \equiv (\partial\mu_1/\partial p)_{T,N_1}$ (which can be deduced from $V=(\partial G/\partial p)_{T,N}$ and $\mu_1 = (\partial G/\partial N_1)_{T,p}$) and that (2) becomes

$$W_{\min} = \frac{1}{2}\left\{-\left(\frac{\partial V}{\partial p}\right)_{T,N}(\Delta p)^2 + \left(\frac{\partial\mu_1}{\partial N_1}\right)_{T,p}(\Delta N_1)^2\right\}. \tag{3}$$

Hence a normal distribution for ΔN_1

$$P(\Delta N_1) = C\exp\left[-\frac{1}{2kT}\left(\frac{\partial\mu_1}{\partial N_1}\right)_{T,p}(\Delta N_1)^2\right] \tag{4}$$

is obtained. [We can proceed in the same way as in problem 2 by considering the system in contact with the environment with constant p and u, and by taking the canonical distribution $\exp[-(1/kT)\{F(T,V) + pV - \mu_1 N_1\}]$. In this way, (2) above and also (4) can easily be derived.] From (4), we get

$$\overline{(\Delta N_1)^2} = kT\bigg/\left(\frac{\partial\mu_1}{\partial N_1}\right)_{T,p}. \tag{5}$$

On the other hand, the Gibbs-Duhem relation, $\sum \mu_j\,\mathrm{d}\mu_j = 0$, tells us that

$$\begin{aligned}\frac{\partial\mu_1}{\partial N_1} &= -\frac{N_0}{N_1}\frac{\partial\mu_0}{\partial N_1} = -\frac{N_0}{N_1}\frac{\partial}{\partial N_1}(kT\log a_0)\\ &= \frac{N_0}{N_1}\frac{\bar{V}_0}{N_A}\frac{\partial\pi}{\partial N_1} \simeq \frac{\bar{V}_0}{N_A N_0 x}\frac{\partial\pi}{\partial x},\end{aligned} \tag{6}$$

where N_A = the Avogadro number, $\bar{V}_0$ = the molar volume of solvent, x = mol fraction $\simeq N_1/N_0$, and π = osmotic pressure. In the above, a_0 is the activity of the solvent defined by the equation

$$\mu_0 = \mu_0^0 + kT\log a_0$$

where μ_0^0 is the chemical potential of a solvent molecule in its pure state (see (4.42), Vol. 1). Note that the osmotic pressure π is related to this activity by the equation

$$\pi = -\frac{RT}{\bar{V}_0}\log a\,,$$

(see example 6, Chapter 4, Vol. 1). Substituting (6) into (5), we finally obtain

$$\overline{(\Delta N_1)^2} = RTN_0 x \Big/ \left(\bar{V}_0 \frac{\partial \pi}{\partial x} \right). \tag{7}$$

NOTE: *Experiment on the Scattering of Light:* When there exists a fluctuation in the concentration of a solution, the refractive index also fluctuates and gives rise to a scattering of light. This scattering is observed as turbidity. Let the refractive index be ν, which is given as $\nu = \nu_0 + \Delta\nu$ and

$$\Delta\nu = \frac{\partial \nu}{\partial x} \Delta x = \frac{1}{N_0} \left(\frac{\partial \nu}{\partial x} \right)_0 \Delta N_1 .$$

Then $\overline{(\Delta\nu)^2}$ can be evaluated with the aid of (7) as

$$\begin{aligned} \overline{(\Delta\nu)^2} &= \frac{1}{N_0^2} \left(\frac{\partial \nu}{\partial x} \right)_0^2 \overline{(\Delta N_1)^2} = \frac{RT}{N_0} \left(\frac{\partial \nu}{\partial x} \right)_0^2 x \Big/ \left(\bar{V}_0 \frac{\partial \pi}{\partial x} \right) \\ &\simeq \frac{1}{V} \left(\frac{\partial \nu}{\partial x} \right)_0^2 M_0 RT x \Big/ \left(\rho N_A \bar{V}_0 \frac{\partial \pi}{\partial x} \right). \end{aligned} \tag{8}$$

Here, V is the volume of the part under consideration, ρ the density, and M_0 the molecular weight of the solvent. The turbidity, τ, is defined as a quantity representing the strength of scattering and is proportional to $V\overline{(\Delta\nu)^2}$. Therefore, measurements of τ and $\partial\nu/\partial x$ as a function of the concentration give us $\partial\pi/\partial x$. When π is expressed in terms of the weight concentration $c\,(\simeq M_1 x/\bar{V}_0$, M_1 = the molecular weight of solute), the molecular weight M_1 is obtained from the relation

$$\frac{\partial \pi}{\partial c} = \frac{RT}{M_1} + \alpha c + \beta c^2 + \cdots \qquad (\alpha \text{ and } \beta \text{ being constants}).$$

These measurements afford an important method for obtaining molecular weight, especially for polymer solutions [c.f. P. J. Flory, *Principles of Polymer Chemistry* (Cornell Univ. Press, Ithaca, 1953)].

4. The number of molecules incident per unit time upon a hole with cross section S is given, according to the Maxwellian velocity distribution law, as follows: If the direction of the outward normal of S is chosen as that of the

x-axis, the velocity components of molecules are denoted by (v_x, v_y, v_z), and the mass is m:

$$J = n\left(\frac{m}{2\pi kT}\right)^{\frac{3}{2}} \int_{-\infty}^{\infty}\int_{-\infty}^{\infty} \mathrm{d}v_y \mathrm{d}v_z \int_0^{\infty} \mathrm{d}v_x \exp\left\{-\frac{m}{2kT}(v_x^2 + v_y^2 + v_z^2)\right\} v_x S$$

$$= Sn\left(\frac{m}{2\pi kT}\right)^{\frac{1}{2}} \int_0^{\infty} v_x \mathrm{e}^{-mv_x^2/(2kT)} \mathrm{d}v_x = Sn\sqrt{\frac{kT}{2\pi m}}, \tag{1}$$

where n is the number density of gas molecules and T is the absolute temperature. Since the pressure of an ideal gas is given by $p = nkT$, (1) can be transformed into the following form:

$$J = S\frac{p}{\sqrt{2\pi mkT}} \propto \frac{p}{\sqrt{T}}. \tag{2}$$

Now the first part of the problem is answered, if we consider that in a stationary state the flow per unit time through a tiny hole on the wall from A to B is equal to that from B to A, which is expressed by the condition $p_A/\sqrt{T_A} = p_B/\sqrt{T_B}$. The answer to the second part is as follows: When $p_A = p_B$, the flow across the wall from A to B becomes smaller or larger than that from B to A, according as $T_A > T_B$ or $T_A < T_B$. Thus, a stationary flow through the tube C is set up from A to B, or from B to A. (This is analogous to thermoelectricity.)

NOTE: In reality, the above theory applies only when the collision between molecules is not essential, i.e., when the gas is so rarefied that the dimensions of the holes are comparable with the mean free path of molecules. When a gas is in such a state of dilution, it is said to be in a Knudsen region.

5. We can proceed just as in example 3. Assume that the flow is parallel to the

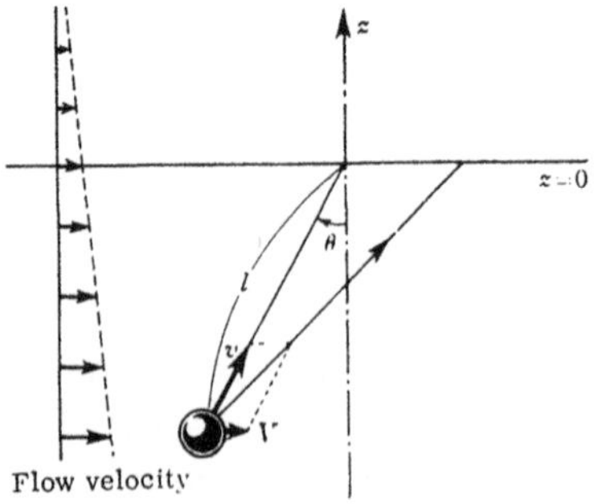

x-axis and that the flow velocity V is a function of the z coordinate only. The velocity gradient is assumed to be so weak that it changes only slightly as z changes by a distance of the order of mean free path $l = v\tau$. To facilitate calculation, a molecule which passes through the plane $z = 0$ is assumed to arrive at the plane $z = 0$ after a time of flight τ. If the angle θ is measured as in the figure, the mean velocity of the molecules is given, according to assumption c as:

$$V(-l\cos\theta) \fallingdotseq V(0) - l\cos\theta\left(\frac{\partial V}{\partial z}\right)_{z=0}. \tag{1}$$

The number of molecules which pass through unit area on the plane $z = 0$ per unit time is given by (3) in the solution to example 3. The momentum transfered per unit time from the side $z < 0$ to the side $z > 0$ through unit area on the plane $z = 0$ becomes:

$$-\tfrac{1}{2}mn\,vl\frac{\partial V}{\partial z}\int_0^\pi \cos^2\theta\sin\theta\,\mathrm{d}\theta = -\tfrac{1}{3}mnvl\frac{\partial V}{\partial z} \tag{2}$$

where m is the molecular mass. This quantity is equal to the force acting along the x-direction in the plane normal to the z-axis, p_{xz} (the xz-component of the pressure tensor). The viscosity coefficient is defined as the ratio of this quantity to the velocity gradient, i.e.,

$$P_{xz} = -\eta\frac{\partial V}{\partial z}. \tag{3}$$

Hence we have

$$\eta = \tfrac{1}{3}\rho v l, \tag{4}$$

$\rho = mn$ being the density of the gas.

NOTE: The calculation in example 3 is valid also for a rarefied gas. If we notice that the heat capacity per unit volume, C, divided by the density ρ, is equal to the specific heat c, the ratio of the heat conductivity to ηc can be seen to be

$$\lambda/\eta c = 1. \tag{5}$$

Measured ratios give values of about 2, which shows that the above argument is essentially valid, though a factor of the order of one is uncertain on account of the crude approximations used.

6. Let us take an infinitesimal portion $\mathrm{d}S$ of the mirror. If the distance from the axis of rotation of the mirror is denoted by ρ, and the rotation velocity of the

mirror by ω, $\mathrm{d}S$ moves with a velocity $\rho\omega$ in the normal direction, which is taken as the x-axis. We need to calculate the average torque exerted upon the mirror when ω is given. Since, presumably, the gas is in thermal equilibrium and is sufficiently rarefied, the Maxwellian velocity distribution law can be assumed to apply. The number of molecules striking $\mathrm{d}S$ on the $-x$ side per unit time with velocity $\boldsymbol{v}$ is

$$n\left(\frac{m}{2\pi kT}\right)^{\frac{3}{2}} \mathrm{e}^{-mv^2/(2kT)}\,\mathrm{d}v_x\,\mathrm{d}v_y\,\mathrm{d}v_z\cdot(v_x - \mu\omega)\,\mathrm{d}S, \tag{1}$$

n being the number density of the molecules. If we assume that the collisions are perfectly elastic and that the mirror is sufficiently heavy, the molecule gives

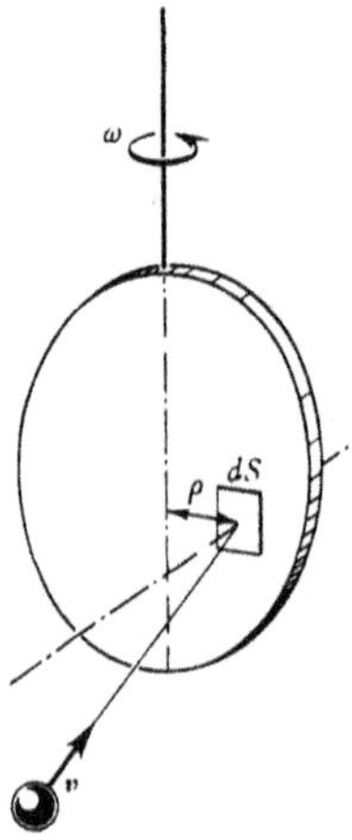

a momentum $2m(v_x - \rho\omega)$ to the mirror on impact. Hence, multiplying (1) by $2m(v_x - \rho\omega)\rho$, we have the torque exerted by the group of molecules with velocity $\boldsymbol{v}$. Summing up contributions from molecules over the velocity region in which molecules collide with $\mathrm{d}S$ on the $-x$ side, i.e., $v_x > \rho\omega$, we obtain the torque exerted by the molecules hitting $\mathrm{d}S$ on the side of $-x$:

$$\mathrm{d}M_- = \mathrm{d}S\cdot n\cdot\left(\frac{m}{2\pi kT}\right)^{\frac{3}{2}} 2m\rho \int_{\rho\omega}^{\infty}\mathrm{d}v_x \int_{-\infty}^{\infty}\int_{-\infty}^{\infty}\mathrm{d}v_y\,\mathrm{d}v_z(v_x - \rho\omega)^2\,\mathrm{e}^{-mv^2/(2kT)}. \tag{2}$$

Now, we consider the case when ω is small and carry out the calculation retaining terms up to first order in ω. We use the approximation $(v_x - \rho\omega)^2 \simeq v_x^2 - 2v_x\rho\omega$ and replace the lower limit of integration by 0 for simplicity, noting that the error involved is only O (ω^2). Thus, we easily arrive at the expression

$$dM_- = p\,dS\cdot\rho - nm\bar{v}\rho\omega\,dS\cdot\rho\,. \tag{3}$$

Here we have used the equation $p = nkT$. Similarly, we obtain the torque exerted by the molecules hitting dS on the $+x$ side as

$$dM_+ = -\,p\,dS\cdot\rho - nm\bar{v}\rho\omega\,dS\cdot\rho\,. \tag{4}$$

Adding (3) and (4), and collecting contributions from all dS, we have the average torque acting on the mirror when ω is given:

$$\overline{M(t)}^{\omega} = \int(dM_- + dM_+) = -\,2nm\bar{v}\int\rho^2\,dS\cdot\omega = -\,2nm\bar{v}(I/\sigma)\omega\,,$$

or

$$\overline{M(t)}^{\omega} = -\,\zeta\omega\,. \tag{5}$$

7. If a uniform electric field $\boldsymbol{E}$ is applied along the x-axis, a force $-e\boldsymbol{E}$ is exerted upon the electrons, which undergo an acceleration $-e\boldsymbol{E}/m$. Although the electrons are in thermal motion, their mean velocity is zero in the absence of the electric field. The effect of the electric field is to raise the mean velocity of electrons by virtue of the acceleration $-e\boldsymbol{E}/m$. Since the mean velocity of the electrons just after each collision is zero, according to the assumption, the mean velocity increases in proportion to time and reaches $-e\boldsymbol{E}\tau/m$ after the electrons have been accelerated for a time τ. Then the electrons are scattered again and the mean velocity becomes zero. Therefore, if averaged over time (since each electron is scattered at a different time), the mean velocity of the electron system is of the order $\bar{v}_x = -e\boldsymbol{E}\tau/2m$. Since each electron carries an electric charge $-e$, the current density is given as

$$j_x = -\,en\bar{v}_x = \frac{ne^2}{2m}\,\tau\,\boldsymbol{E} \tag{1}$$

where n is the number density of electrons. The electric conductivity is defined by the relation $j_x = \sigma\boldsymbol{E}$, and its approximate value is given by

$$\sigma = \frac{ne^2}{2m}\,\tau\,. \tag{2}$$

NOTE: The above is a theory due to Drude. Compare it with the result of example 4. If the assumption is made that the scattering occurs randomly at a mean free flight time interval τ instead of the assumption of scattering at constant time intervals, it is to be noted that according to (2) in the solution to problem 18, the mean velocity becomes

$$-\int_0^\infty \frac{eE}{m} t e^{-t/\tau} \mathrm{d}t/\tau = -\frac{eE}{m}\tau .$$

8. Considerations leading to (4) in the solution to example 4 equally apply when f_0 is the Maxwell-Boltzmann distribution. For the Maxwell-Boltzmann distribution, the identity

$$-\frac{\partial f_0}{\partial \varepsilon} = \frac{1}{kT} f_0 \qquad (f_0 = e^{(\mu-\varepsilon)/kT}) \tag{1}$$

holds, and we see that the electric conductivity is

$$\sigma = \frac{e^2}{3kT}\int \tau \overline{v^2} f_0(\varepsilon) D(\varepsilon)\, \mathrm{d}\varepsilon . \tag{2}$$

In the case where $\varepsilon(\boldsymbol{p}) = \boldsymbol{p}^2/2m^*$, and $\tau = A\,|\,\boldsymbol{v}\,|^s$, we obtain $D(\varepsilon) = \sqrt{2\varepsilon}\, m^{*\frac{3}{2}}/\pi^2\hbar^3$ and $|\boldsymbol{v}| = \sqrt{2\varepsilon/m^*}$, so that the conductivity becomes

$$\sigma = \frac{e^2}{3kT}\int_0^\infty A\left(\frac{2\varepsilon}{m^*}\right)^{\frac{1}{2}s} e^{(\mu-\varepsilon)/kT} \frac{(2\varepsilon m^*)^{\frac{3}{2}}}{m^*\pi^2\hbar^3}\, \mathrm{d}\varepsilon . \tag{3}$$

If we put $\varepsilon/kT = x$, the integral is seen to be expressed in terms of the gamma function:

$$\sigma = \frac{e^2}{m^*} A\left(\frac{2kT}{m^*}\right)^{\frac{1}{2}s} \frac{(2kTm^*)^{\frac{3}{2}}}{3\pi^2\hbar^3} e^{\mu/kT}\, \Gamma\left(\frac{s+5}{2}\right). \tag{4}$$

Since the number density of electrons is given by

$$n = \int_0^\infty f_0(\varepsilon) D(\varepsilon)\, \mathrm{d}\varepsilon = \frac{(kTm^*)^{\frac{3}{2}}}{\pi^{\frac{3}{2}}\hbar^3} e^{\mu/kT}, \tag{5}$$

(4) may be expressed in the following form:

$$\sigma = \frac{ne^2}{m^*} A\left(\frac{2kT}{m^*}\right)^{\frac{1}{2}s} \frac{2\sqrt{2}}{3\sqrt{\pi}}\, \Gamma\left(\frac{s+5}{2}\right). \tag{6}$$

When τ = const., $s = 0$, it becomes

$$\sigma = \frac{ne^2}{m^*}\frac{\tau}{\sqrt{2}} \tag{7}$$

and when the mean free path $l = \tau v = \text{const.}$, i.e., $s = -1$, we have

$$\sigma = \frac{ne^2}{m^*} \frac{l}{\bar{v}} \frac{4\sqrt{2}}{3\pi} \tag{8}$$

where $\bar{v} = \sqrt{8kT/\pi m^*}$ is the mean speed of the electrons.

9. When both a uniform electric field $\boldsymbol{E}$ and a uniform magnetic field $\boldsymbol{H}$ are applied, the Boltzmann equation for the stationary state takes the form:

$$-e\left(\boldsymbol{E} + \frac{\boldsymbol{v}}{c} \times \boldsymbol{H}\right)\cdot\frac{\partial f}{\partial \boldsymbol{p}} = -\frac{f - f_0}{\tau}. \tag{1}$$

Here, c is the velocity of light. Since only the first order term in the electric field is to be considered, f in the coefficient of $\boldsymbol{E}$ may be approximated by f_0. Noting that the coefficient of $\boldsymbol{H}$ vanishes if f_0 is substituted for f (hence f_0 is the solution when $\boldsymbol{E} = 0$), we are led to the equation

$$-e\boldsymbol{v}\cdot\boldsymbol{E}\frac{\partial f_0}{\partial \varepsilon} - \frac{e}{c}(\boldsymbol{v} \times \boldsymbol{H})\cdot\frac{\partial (f - f_0)}{\partial \boldsymbol{p}} = -\frac{f - f_0}{\tau}. \tag{2}$$

In the case where $\varepsilon(\boldsymbol{p}) = \boldsymbol{p}^2/2m^*$, we have $\boldsymbol{p} = m^*\boldsymbol{v}$ and the above equation is solved if the solution is assumed to be of the form

$$f = f_0 - \boldsymbol{v}\cdot\boldsymbol{c}(\varepsilon)\frac{\partial f_0}{\partial \varepsilon}. \tag{3}$$

From (2), it is seen that the unknown vector $\boldsymbol{c}(\varepsilon)$ satisfies the equation

$$-e\boldsymbol{v}\cdot\boldsymbol{E} + \frac{e}{m^* c}(\boldsymbol{v} \times \boldsymbol{H})\cdot\boldsymbol{c} = \frac{\boldsymbol{v}\cdot\boldsymbol{c}}{\tau}. \tag{4}$$

Noting that $(\boldsymbol{v} \times \boldsymbol{H})\cdot \boldsymbol{c} = (\boldsymbol{H} \times \boldsymbol{c})\cdot \boldsymbol{v}$, and comparing the coefficients of $\boldsymbol{v}$, we obtain

$$-e\boldsymbol{E} + \boldsymbol{\omega} \times \boldsymbol{c} = \boldsymbol{c}/\tau, \tag{5}$$

where $\boldsymbol{\omega} = e\boldsymbol{H}/m^*c$ is a vector having the direction of the magnetic field and the magnitude $\omega = eH/m^*c$. To solve (5), we first take the scalar product of (5) and $\boldsymbol{\omega}$. Making use of the relation $\boldsymbol{\omega}\cdot(\boldsymbol{\omega} \times \boldsymbol{c}) = 0$, we deduce that

$$\boldsymbol{\omega}\cdot\boldsymbol{c} = -e\tau\boldsymbol{\omega}\cdot\boldsymbol{E}. \tag{6}$$

Next we construct the vector product of (5) and $\boldsymbol{\omega}$. If we note that $\boldsymbol{\omega} \times (\boldsymbol{\omega} \times \boldsymbol{c}) = (\boldsymbol{\omega} \cdot \boldsymbol{c})\boldsymbol{\omega} - \omega^2 \boldsymbol{c}$,* another relation is obtained:

$$\boldsymbol{\omega} \times \boldsymbol{c} = - e\tau\boldsymbol{\omega} \times \boldsymbol{E} + \tau(\boldsymbol{\omega}\cdot\boldsymbol{c})\boldsymbol{\omega} - \tau\omega^2\boldsymbol{c}\,. \tag{7}$$

Eliminating $\boldsymbol{\omega} \cdot \boldsymbol{c}$ and $\boldsymbol{\omega} \times \boldsymbol{c}$ from (5), (6) and (7), we get an equation for $\boldsymbol{c}$:

$$(1 + \omega^2\tau^2)\boldsymbol{c} = - e\tau\,\{\boldsymbol{E} + \tau^2(\boldsymbol{\omega}\cdot\boldsymbol{E})\boldsymbol{\omega} + \tau\boldsymbol{\omega} \times \boldsymbol{E}\}\,, \tag{8}$$

thus

$$f = f_0 + \frac{e\tau}{1 + \omega^2\tau^2}\,\{\boldsymbol{v} + \tau^2(\boldsymbol{v}\cdot\boldsymbol{\omega})\boldsymbol{\omega} + \tau(\boldsymbol{v} \times \boldsymbol{\omega})\}\cdot\boldsymbol{E}\frac{\partial f_0}{\partial\varepsilon}\,. \tag{9}$$

The current density is given as

$$\begin{aligned} \boldsymbol{j} &= - e\int \boldsymbol{v} f\frac{2\,\mathrm{d}\boldsymbol{p}}{h^3} \\ &= e^2\int\left(-\frac{\partial f_0}{\partial\varepsilon}\right)\frac{\tau}{1 + \omega^2\tau^2}\,\{\boldsymbol{v}\boldsymbol{v} + \tau^2(\boldsymbol{v}\cdot\boldsymbol{\omega})\boldsymbol{v}\boldsymbol{\omega} + \tau\boldsymbol{v}(\boldsymbol{v} \times \boldsymbol{\omega})\}\cdot\boldsymbol{E}\frac{2\,\mathrm{d}\boldsymbol{p}}{h^3}\,. \end{aligned} \tag{10}$$

Noting the spherical symmetry of $\varepsilon(\boldsymbol{p})$, we obtain for the components of the electric conductivity tensor

$$\begin{aligned} \sigma_{\alpha\beta} &= e^2\int\left(-\frac{\partial f_0}{\partial\varepsilon}\right)\frac{\tau}{1 + \omega^2\tau^2}\frac{v^2}{3}\,\{\delta_{\alpha\beta} + \tau^2\omega_\alpha\omega_\beta \pm (1 - \delta_{\alpha\beta})\tau\omega_\gamma\}\,\frac{2\,\mathrm{d}\boldsymbol{p}}{h^3} \\ &= \frac{ne^2}{m^*}\frac{\tau(\mu)}{1 + \omega^2\tau^2(\mu)}\,\{\delta_{\alpha\beta} + \tau^2(\mu)\omega_\alpha\omega_\beta \pm (1 - \delta_{\alpha\beta})\tau(\mu)\omega_\gamma\}\,. \end{aligned} \tag{11}$$

In the above, the double sign in front of the last term in the brackets should be understood to represent + or − according as $\alpha\beta\gamma$ is an odd or even permutation of xyz, respectively. Though the above expression is of the most general form, the direction of the magnetic field can conveniently be chosen as the z-axis without loss of generality, because $\varepsilon(\boldsymbol{p})$ is spherically symmetric. Referred to this coordinate system, the electric conductivity is expressed as follows:

$$\left.\begin{aligned} &\sigma_{xx} = \sigma_{yy} = \frac{ne^2}{m^*}\frac{\tau(\mu)}{1 + \omega^2\tau^2(\mu)}\,, \qquad \sigma_{zz} = \frac{ne^2}{m^*}\tau(\mu)\,, \\ &-\sigma_{xy} = \sigma_{yx} = \frac{ne^2}{m^*}\frac{\omega\tau^2(\mu)}{1 + \omega^2\tau^2(\mu)}\,, \qquad \sigma_{xz} = \sigma_{zx} = \sigma_{yz} = \sigma_{zy} = 0\,. \end{aligned}\right\} \tag{12}$$

---

* $\boldsymbol{A} \times (\boldsymbol{B} \times \boldsymbol{C}) = (\boldsymbol{A} \cdot \boldsymbol{C})\boldsymbol{B} - (\boldsymbol{A} \cdot \boldsymbol{B})\boldsymbol{C}$.

10. If $\xi(x)$ denotes the deviation at a point distant by x from one end of the string, $\xi = 0$ at both ends of the string, i.e., at $x = 0$ and L, and $\xi(x)$ can be expressed by a Fourier series of the following form:

$$\xi(x) = \sum_{n=1}^{\infty} A_n \sqrt{\frac{2}{L}} \sin\left(n\pi \frac{x}{L}\right). \tag{1}$$

When the string has the shape represented by $\xi(x)$ and the stress on the string is F, the elastic energy is, in case the deviation is small, equal to

$$\begin{aligned} \Phi &= \tfrac{1}{2} F \int_0^L \left(\frac{\partial \xi}{\partial x}\right)^2 \mathrm{d}x \\ &= \tfrac{1}{2} F \sum_{n,n'=1}^{\infty} A_n A_{n'} \frac{n\pi}{L} \frac{n'\pi}{L} \frac{2}{L} \int_0^L \cos\left(n\pi \frac{x}{L}\right) \cos\left(n'\pi \frac{x}{L}\right) \mathrm{d}x \\ &= \tfrac{1}{2} F \sum_{n=1}^{\infty} A_n^2 \left(\frac{n\pi}{L}\right)^2. \end{aligned} \tag{2}$$

The shape of the string is specified by the coefficients $(A_1, A_2, \ldots)$, and the potential energy is given by (2) when these coefficients take on particular values $(A_1, A_2, \ldots)$. Therefore, the fluctuations of the coordinates at absolute temperature T are

$$\begin{aligned} \overline{A_n A_{n'}} &= \int_{-\infty}^{\infty} \cdots \int_{-\infty}^{\infty} \mathrm{d}A_1\, \mathrm{d}A_2 \cdots \mathrm{e}^{-\Phi/kT} A_n A_{n'} \Big/ \int_{-\infty}^{\infty} \cdots \int_{-\infty}^{\infty} \mathrm{d}A_1\, \mathrm{d}A_2 \cdots \mathrm{e}^{-\Phi/kT} \\ &= \delta_{n,n'} \frac{kT}{F} \Big/ \left(\frac{n\pi}{L}\right)^2. \end{aligned} \tag{3}$$

According to (1), we have

$$\begin{aligned} \overline{\xi^2(x)} &= \sum_{n,n'=1}^{\infty} \overline{A_n A_{n'}} \frac{2}{L} \sin\left(n\pi \frac{x}{L}\right) \sin\left(n'\pi \frac{x}{L}\right) \\ &= \frac{kT}{FL} 2 \sum_{n=1}^{\infty} \sin^2\left(\frac{n\pi}{L} x\right) \Big/ \left(\frac{n\pi}{L}\right)^2 = \frac{kT}{FL} x(L - x). \end{aligned} \tag{4}$$

ALTERNATIVE SOLUTION

If the shape of the string, $\xi(x)$, is taken as in the figure to be a triangle for which $\xi = \xi_0$ at $x = x_0$, the deviation is given by

$$\xi(x) = \begin{cases} \xi_0 x/x_0 , & x < x_0 \\ \xi_0(L-x)/(L-x_0), & x > x_0 , \end{cases}$$

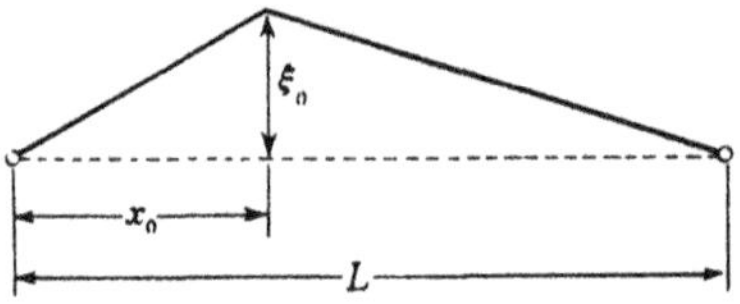

and according to (2) the elastic energy can immediately be written as

$$\Phi = \frac{F}{2}\left\{x_0\left(\frac{\xi_0}{x_0}\right)^2 + (L-x_0)\left(\frac{\xi_0}{L-x_0}\right)^2\right\}$$

$$= \frac{F}{2}\xi_0^2\left\{\frac{1}{x_0} + \frac{1}{L-x_0}\right\}. \tag{5}$$

Regarding this as the potential energy for the coordinate ξ_0, we can determine the fluctuation of ξ_0 as

$$\overline{\xi_0^2} = \int_{-\infty}^{\infty} \xi_0^2 e^{-\Phi/kT} d\xi_0 / \int_{-\infty}^{\infty} e^{-\Phi/kT} d\xi_0 = \frac{kT}{FL} x_0(L-x_0), \tag{6}$$

which is the same as (2) if we substitute there $x = x_0$. Generally, in order to obtain the fluctuation of a deviation ξ_0 at $x = x_0$ according to example 1, we determine the equilibrium shape of the string $\xi(x)$ under the condition $\xi = \xi_0$ at $x = x_0$, and calculate the minimum work $W_{\min} = \Phi$ required to form this equilibrium shape. In the present problem, the minimum work is

$$W_{\min} = F(\sqrt{x_0^2+\xi_0^2} - x_0) + F\{\sqrt{(L-x_0)^2+\xi_0^2} - (L-x_0)\}, \tag{7}$$

which agrees with (5) when ξ_0 is small compared with x_0 and $L - x_0$.

NOTE: In the same way as above, we can calculate the correlation of the deviations at two points from (1) and (3) with the help of the formula for Fourier series. Calculation gives the result

$$\overline{\xi(x)\xi(x')} = \frac{kT}{FL} G(x,x'), \qquad G(x,x') = \begin{cases} x(L-x'), & x \leqq x' \\ x'(L-x), & x \geqq x' . \end{cases} \tag{8}$$

Here, the function $G(x, x')$ is equal to what is called Green's function for an eigenvalue equation $u''(x) = -\lambda u$, with $u(0) = u(L) = 0$. Since Φ is an integral form quadratic in $\xi(x)$, the calculation of $\overline{\xi(x)\xi(x')}$ is equivalent to calculating the inverse to this integral kernel. (See note for problem 2. There, the treatment is limited to a matrix of finite dimensions: however, extension to the continuous case can easily be made by analogy.) This is nothing but Green's function (8). With regard to Green's function applied to quantum statistics, see, for example, D. J. Thouless, *The Quantum Mechanics of Many-Body Systems* (Academic Press, New York, 1961) p. 59.

11. Let us introduce a function which has the following property:

$$m(\boldsymbol{r}) = \begin{cases} 1, & \text{if } \boldsymbol{r} \text{ lies within the region } V_{\mathrm{A}}. \\ 0, & \text{otherwise}. \end{cases}$$

This function can be used to express the number of particles contained in V_{A} as $N_{\mathrm{A}} = \sum_{i=1}^{N} m(\boldsymbol{r}_i)$. Hence

$$\begin{aligned} \langle N_{\mathrm{A}} \rangle &= \sum_{i=1}^{N} \int \cdots \int m(\boldsymbol{r}_i) F\{N\} \, \mathrm{d}\{N\} \\ &= N \int \cdots \int m(\boldsymbol{r}_1) \, \mathrm{d}\boldsymbol{r}_1 \, F\{N\} \, \mathrm{d}\{N-1\} \\ &= \int m(\boldsymbol{r}) F_1(\boldsymbol{r}) \, \mathrm{d}\boldsymbol{r} . \end{aligned}$$

If we take the limit $N \to \infty$ under the condition $N/V = n = $ constant, we have $F_1(\boldsymbol{r}) = n$ and accordingly

$$\langle N_{\mathrm{A}} \rangle = n V_{\mathrm{A}} . \tag{1}$$

In a similar fashion, we can calculate

$$\langle N_{\mathrm{A}}^2 \rangle = \langle \sum_i \sum_j m(\boldsymbol{r}_i) m(\boldsymbol{r}_j) \rangle = \langle \{ \sum_{i \neq j} \sum m(\boldsymbol{r}_i) m(\boldsymbol{r}_j) + \sum_{i=1}^{N} m(\boldsymbol{r}_i) \} \rangle .$$

To get the last form of the above equation, we have used the property $m(\boldsymbol{r}) m(\boldsymbol{r}) = m(\boldsymbol{r})$. Calculation gives

$$\begin{aligned} \langle N_{\mathrm{A}}^2 \rangle &= n V_{\mathrm{A}} + \sum_{i \neq j} \sum \int_V \cdots \int m(\boldsymbol{r}_i) m(\boldsymbol{r}_j) F\{N\} \, \mathrm{d}\{N\} \\ &= n V_{\mathrm{A}} + N(N-1) \int_V \cdots \int m(\boldsymbol{r}_1) m(\boldsymbol{r}_2) \, \mathrm{d}\boldsymbol{r}_1 \, \mathrm{d}\boldsymbol{r}_2 F\{N\} \, \mathrm{d}\{N-2\} \\ &= n V_{\mathrm{A}} + \iint_V m(\boldsymbol{r}_1) m(\boldsymbol{r}_2) F_2(\boldsymbol{r}_1, \boldsymbol{r}_2) \, \mathrm{d}\boldsymbol{r}_1 \, \mathrm{d}\boldsymbol{r}_2 \\ &= n V_{\mathrm{A}} + n^2 \iint_{V_{\mathrm{A}}} g(\boldsymbol{r}_1, \boldsymbol{r}_2) \, \mathrm{d}\boldsymbol{r}_1 \, \mathrm{d}\boldsymbol{r}_2 . \end{aligned} \tag{2}$$

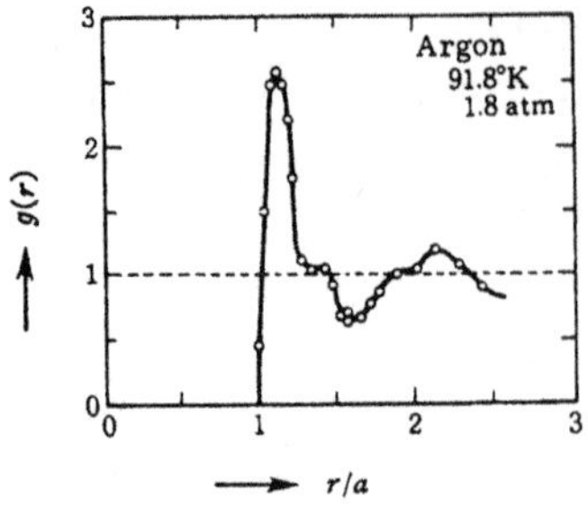

Therefore,

$$\frac{\langle (N_A - \langle N_A \rangle)^2 \rangle}{\langle N_A \rangle} = \frac{\langle N_A^2 \rangle - \langle N_A \rangle^2}{\langle N_A \rangle} =$$

$$= 1 + n^2 \frac{1}{\langle N_A \rangle} \iint_{V_A} g(\boldsymbol{r}_1, \boldsymbol{r}_2)\, d\boldsymbol{r}_1\, d\boldsymbol{r}_2 - \langle N_A \rangle$$

$$= 1 + n \frac{1}{V_A} \iint_{V_A} \{g(\boldsymbol{r}_1, \boldsymbol{r}_2) - 1\}\, d\boldsymbol{r}_1\, d\boldsymbol{r}_2 . \quad (3)$$

The function $g(\boldsymbol{r}_1, \boldsymbol{r}_2) = g(|\boldsymbol{r}_1 - \boldsymbol{r}_2|)$ is of the form shown in the figure, and is nearly equal to unity when $|\boldsymbol{r}_2 - \boldsymbol{r}_1| = r$ is larger than a certain range l. If V_A is taken much larger than l^3 so that the contribution to the integral from a surface layer of width l may be neglected, integration with respect to $\boldsymbol{r}_1$ in (3) can be performed, which results in cancelling the factor $1/V_A$ and yields

$$\frac{\langle (N_A - \langle N_A \rangle)^2 \rangle}{\langle N_A \rangle} = n \int \{g(\boldsymbol{R}) - 1\}\, d\boldsymbol{R} + 1 . \quad (4)$$

In order to calculate the fluctuation in the particle number, we can use various methods. If we regard the surroundings as an environment having a temperature T and a chemical potential μ, the T-μ distribution

$$P(N) = C \exp\left[-\frac{1}{kT} \{F_N(V, T) - N\mu\} \right]$$

determines the distribution of N. The grand partition function,

$$\Xi = \sum_N \exp\left[-\frac{1}{kT} \{F_N(V, T) - N\mu\} \right]$$

can be used to calculate the following averages

$$\langle N \rangle = kT \frac{\partial \log \Xi}{\partial \mu}, \qquad \langle N^2 \rangle - \langle N \rangle^2 = (kT)^2 \frac{\partial^2 \log \Xi}{\partial \mu^2} . \quad (5)$$

From $kT \log \Xi = pV$, we have

$$\langle N^2 \rangle - \langle N \rangle^2 = VkT \left(\frac{\partial^2 p}{\partial \mu^2} \right)_T . \quad (6)$$

Using the relation $d\mu = v dp - s dT$ ($v = V/N$, $s = S/N$), we obtain

$$\left(\frac{\partial^2 p}{\partial \mu^2}\right)_T = \left(\frac{\partial}{\partial \mu}\frac{1}{v}\right)_T = -\frac{1}{v^2}\left(\frac{\partial v}{\partial \mu}\right)_T = -\frac{1}{v^2}\left(\frac{\partial p}{\partial \mu}\right)_T\left(\frac{\partial v}{\partial p}\right)_T = -\frac{1}{v^3}\left(\frac{\partial v}{\partial p}\right)_T,$$

or

$$\frac{\langle N^2\rangle - \langle N\rangle^2}{\langle N\rangle} = nkT\kappa_T \qquad (\kappa_T = \text{isothermal compressibility}). \tag{7}$$

Substitution of this identity into (4) will complete the proof.

12. In this problem, $P_n(x_0 \mid x)$ satisfies

$$P_n(x \mid y) = \tfrac{1}{2} P_{n-1}(x \mid y+1) + \tfrac{1}{2} P_{n-1}(x \mid y-1). \tag{1}$$

Multiplying this by ξ^y and summing with respect to y, we easily obtain

$$Q_n(\xi) = \tfrac{1}{2}\left(\xi + \frac{1}{\xi}\right) Q_{n-1}(\xi). \tag{2}$$

Since

$$Q_0(\xi) = \sum_y P_0(x \mid y)\xi^y = \sum_y \delta_{x,y}\xi^y = \xi^x,$$

the solution to equation (2) is given as

$$Q_n(\xi) = [\tfrac{1}{2}(\xi + 1/\xi)]^n \xi^x. \tag{3}$$

Comparing the expanded form

$$Q_n(\xi) = \frac{1}{2^n}\sum_{m=-n}^{n} \frac{n!}{\frac{1}{2}(n+m)!\frac{1}{2}(n-m)!}\xi^{x+m}$$

with the definition of $Q_n(\xi)$, we get

$$P_n(x \mid y) = \begin{cases} \dfrac{n!}{\frac{1}{2}(n+y-x)!\frac{1}{2}(n-y+x)!} \times 2^{-n}, & |y-x| \leqq n, \\ = 0, & |y-x| > n. \end{cases} \tag{4}$$

This is nothing but a binomial distribution. For $n \gg 1$, use of the extended Stirling's formula, which was given in (8a) of note 2 for example 2, Chapter 1, leads to

$$\log P_n(x \mid y) \sim (n + \tfrac{1}{2})\log n - \tfrac{1}{2}(n + y - x + 1)\log \tfrac{1}{2}n\left(1 + \frac{y-x}{n}\right)$$
$$- \tfrac{1}{2}(n - y + x + 1)\log \tfrac{1}{2}n\left(1 - \frac{y-x}{n}\right) - \log\sqrt{2\pi}. \tag{5}$$

Moreover, if we assume $| x - y | \ll n$, we arrive at

$$P_n(x \mid y) \sim \frac{1}{\sqrt{2\pi n}} \exp\left[- \frac{(x-y)^2}{2n} \right], \tag{6}$$

which alternatively can be written as

$$\frac{1}{a} P_n(x \mid y) \sim \frac{1}{\sqrt{4\pi Dt}} \exp\left[- \frac{(x-y)^2}{4Dt} \right], \qquad D = a^2/2\tau. \tag{7}$$

Here D is the diffusion coefficient.

NOTE: A problem of this type is called a *random walk problem*. Example 4 and problem 13, Chapter 1 are equivalent to the present problem, and example 4 and problem 17, Chapter 2 are elementary examples of three-dimensional random walk problems. The groggy steps of a drunkard, the flight of a particle which suffers occasional scattering, and the Brownian motion of a colloidal particle are all problems in the same category [cf. *Fundamental Problems in Statistical Mechanics,* compiled by E. G. D. Cohen (North-Holland Publ. Co., Amsterdam, 1962) p. 33, and S. Chandrasekhar, Rev. Mod. Phys. **15** (1943) 1].

13. Putting $\tau = m/\zeta$, and regarding $F(t)$ as a given function of t, we integrate the equation of motion as

$$v(t) = v(t_0)\,\mathrm{e}^{-(t-t_0)/\tau} + \frac{1}{m}\int_{t_0}^{t} \mathrm{e}^{-(t-s)/\tau} F(s)\,\mathrm{d}s, \qquad (t \geqq t_0). \tag{1}$$

If we keep $v(t_0)$ constant, and take the average of $F(s)$, the above equality shows that the average velocity decays with a relaxation time τ. For simplicity, we take $v(t_0)$ finite and $t_0 \to -\infty$ in the following; then equation (1) gives

$$v(t) = \frac{1}{m}\int_{-\infty}^{t} \mathrm{e}^{-(t-s)/\tau} F(s)\,\mathrm{d}s, \quad \text{thus} \quad \overline{v(t)} = 0. \tag{2}$$

The latter identity is obtained from the former by taking the average, because $\overline{F(s)} = 0$.

In order to calculate the correlation function, we multiply two equations which correspond to (2) for t and t', respectively, and take the average using $\overline{F(t)F(t')} = 2\zeta kT\,\delta(t-t')$:

$$\overline{v(t)v(t')} = \frac{1}{m^2}\int_{-\infty}^{t} \mathrm{d}s\,\mathrm{e}^{-(t-s)/\tau}\int_{-\infty}^{t'} \mathrm{d}s'\,\mathrm{e}^{-(t'-s')/\tau}\,2\zeta kT\delta(s-s')$$

$$= \frac{1}{m^2}\int_0^{\infty} \mathrm{d}u\,\mathrm{e}^{-u/\tau}\int_0^{\infty} \mathrm{d}u'\,\mathrm{e}^{-u'/\tau}\,2\zeta kT\delta(t-t'-u+u').$$

Applying the formula $\int f(x)\delta(x-a)\,\mathrm{d}x = f(a)\int\delta(x-a)\,\mathrm{d}x$ to the integral with respect to u', we obtain

$$\overline{v(t)v(t')} = \frac{2kT}{m\tau}\mathrm{e}^{(t-t')/\tau}\int_0^{\infty}\mathrm{d}u\,\mathrm{e}^{-2u/\tau}\int_0^{\infty}\mathrm{d}u'\delta(t-t'-u+u').$$

Noticing that if $t-t'>0$ the u'-integral is zero for $u<t-t'$, we are led to

$$\overline{v(t)v(t')} = \frac{kT}{m}\frac{2}{\tau}\mathrm{e}^{(t-t')/\tau}\int_{\max(0,\,t-t')}^{\infty}\mathrm{e}^{-2u/\tau}\mathrm{d}u,$$

or

$$\overline{v(t)v(t')} = \frac{kT}{m}\mathrm{e}^{-|t-t'|/\tau}. \qquad (3)$$

The mean square of the displacement can be calculated as

$$\overline{\{x(t)-x(0)\}^2} = \int_0^t \mathrm{d}t_1\int_0^t \mathrm{d}t_2\overline{v(t_1)v(t_2)} = \frac{kT}{m}\int_0^t \mathrm{d}t_1\int_0^t \mathrm{d}t_2\exp\{-|t_1-t_2|/\tau\}. \quad (4)$$

The integral region is the square shown in the figure. The integral is equal to twice the contribution from the lower triangular region. Transforming the integration variables to $t_1-t_2=t'$ and t_2 $(0<t_2<t-t')$, we calculate the integral as follows:

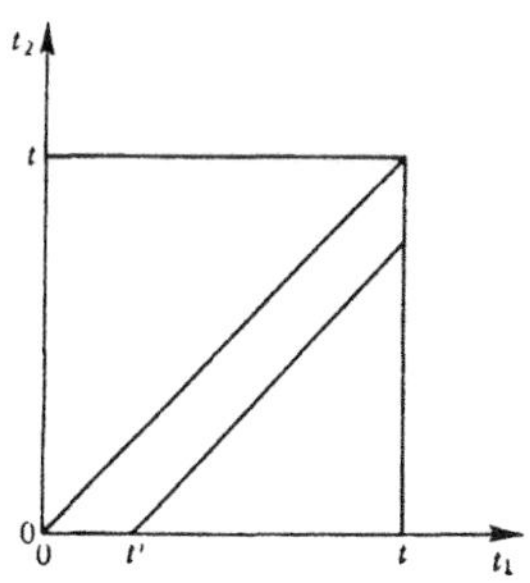

$$2\int_0^t (t-t')\exp(-t'/\tau)\,\mathrm{d}t' =$$

$$= 2\left\{\frac{t}{\alpha}(1-\mathrm{e}^{-\alpha t}) + \frac{\mathrm{d}}{\mathrm{d}\alpha}\frac{1}{\alpha}(1-\mathrm{e}^{-\alpha t})\right\} \qquad (\alpha = 1/\tau)$$

$$= 2\{\tau t - \tau^2(1-\mathrm{e}^{-t/\tau})\},$$

whence for $t \gg \tau$,

$$\overline{\{x(t)-x(0)\}^2} \to 2\frac{kT}{m}\tau t = 2\frac{kT}{\zeta}t. \qquad (5)$$

NOTE: When (5) is rewritten as

$$\overline{(\Delta x)^2} = 2Dt \qquad (D = kT/\zeta) \tag{6}$$

D is called *the diffusion coefficient.* If we observe the distribution of particles which undergo Brownian motion, it obeys the diffusion equation

$$\frac{\partial f}{\partial t} = D\left\{\frac{\partial^2 f}{\partial x^2} + \frac{\partial^2 f}{\partial y^2} + \frac{\partial^2 f}{\partial z^2}\right\}$$

in a time scale longer than τ. On the other hand, when a constant force F_0 is applied, the particle moves with a constant velocity $v = F_0/\zeta$. The proportionality constant $1/\zeta = \mu$ (or e/ζ in case of ions) is called *the mobility.* Therefore, the following identity (*Einstein's relation*)

$$D = \mu kT \qquad (\text{or } D = \mu kT/e) \tag{7}$$

holds. [For a general discussion of the Einstein relation, see R. Kubo, J. Phys. Soc. Japan **12** (1957) 570; R. Kubo, *Some Aspects of the Statistical-Mechanical Theory of Irreversible Processes, Lectures in Theoretical Physics,* Vol. I., ed. by Brittin and Dunham (Interscience Publ., New York, 1959).]

14. It suffices to consider terms up to first order in the temperature gradient, $\partial T/\partial \boldsymbol{x}$. Since, in the absence of a temperature gradient, the stationary solution is $f = f_0$, $f - f_0$ begins with the first order term in the temperature gradient. Therefore, in the equation

$$\boldsymbol{v}\cdot\frac{\partial f}{\partial \boldsymbol{x}} = -\frac{f - f_0}{\tau} \tag{1}$$

f on the left hand side may be replaced by f_0. Although τ generally depends on the temperature T, it may be evaluated at the temperature existing in the absence of a temperature gradient, i.e., the average temperature. Noting that

$$f_0 = \{e^{(\varepsilon-\mu)/kT} + 1\}^{-1}, \tag{2}$$

we have

$$\boldsymbol{v}\cdot\frac{\partial f_0}{\partial \boldsymbol{x}} = \frac{\partial f_0}{\partial \varepsilon} T\boldsymbol{v}\cdot\frac{\partial}{\partial \boldsymbol{x}}\left(\frac{\varepsilon - \mu}{T}\right) = \frac{\partial f_0}{\partial \varepsilon}\boldsymbol{v}\cdot\left\{-\varepsilon\frac{\partial \log T}{\partial \boldsymbol{x}} - T\frac{\partial}{\partial \boldsymbol{x}}\left(\frac{\mu}{T}\right)\right\}, \tag{3}$$

$$f = f_0 - \frac{\partial f_0}{\partial \varepsilon}\tau\boldsymbol{v}\cdot\left\{-\varepsilon\frac{\partial \log T}{\partial \boldsymbol{x}} - T\frac{\partial}{\partial \boldsymbol{x}}\left(\frac{\mu}{T}\right)\right\}. \tag{4}$$

In order to obtain the thermal conductivity, the thermal current should be calculated under the condition that no electric current flows. For this purpose, it is necessary to choose properly the gradient of the chemical potential, μ. The flow of electrons is given by

$$\boldsymbol{J} = \int \boldsymbol{v} f \frac{2\,\mathrm{d}\boldsymbol{p}}{h^3} = \int \left(-\frac{\partial f_0}{\partial \varepsilon}\right) \tau \overline{\boldsymbol{v}\boldsymbol{v}} \cdot \left\{ -\varepsilon \frac{\partial \log T}{\partial \boldsymbol{x}} - T \frac{\partial}{\partial \boldsymbol{x}}\left(\frac{\mu}{T}\right)\right\} D(\varepsilon)\,\mathrm{d}\varepsilon . \tag{5}$$

Since we take $\varepsilon(\boldsymbol{p}) = p^2/2m^*$, we have, similarly to example 4, $\overline{v_\alpha v_\beta} = \delta_{\alpha\beta} 2\varepsilon/3m^*$, $D(\varepsilon) = \sqrt{2\varepsilon} m^{*\frac{3}{2}}/\pi^2\hbar^3$, and also

$$\boldsymbol{J} = L_1\left(-\frac{\partial \log T}{\partial \boldsymbol{x}}\right) + L_0\left\{-T\frac{\partial}{\partial \boldsymbol{x}}\left(\frac{\mu}{T}\right)\right\}; \tag{6}$$

$$L_\nu = \int \left(-\frac{\partial f_0}{\partial \varepsilon}\right) \tau \frac{\overline{v^2}}{3} \varepsilon^s D(\varepsilon)\,\mathrm{d}\varepsilon = \frac{A}{m^*}\left(\frac{2}{m^*}\right)^{\frac{1}{2}s} \frac{(2m^*)^{\frac{3}{2}}}{3\pi^2\hbar^3} \int_0^\infty \left(-\frac{\partial f_0}{\partial \varepsilon}\right) \varepsilon^{\nu + \frac{1}{2}(s+3)}\,\mathrm{d}\varepsilon$$

$$= \frac{n}{m^*} \mu_0^\nu A \left(\frac{2\mu_0}{m^*}\right)^{\frac{1}{2}s} \left\{1 + \tfrac{1}{6}\left[\nu + \tfrac{1}{2}(s+3)\right](\nu + \tfrac{1}{2}s)\left(\frac{\pi k T}{\mu_0}\right)^2 + \cdots\right\}. \tag{7}$$

If we put $\boldsymbol{J} = 0$ in (6), the gradient of the chemical potential can be expressed in terms of the temperature gradient as

$$-T\frac{\partial}{\partial \boldsymbol{x}}\left(\frac{\mu}{T}\right) = -\frac{L_1}{L_0}\left(-\frac{\partial \log T}{\partial \boldsymbol{x}}\right). \tag{8}$$

Hence the energy flow is given as

$$\boldsymbol{q} = \int \varepsilon \boldsymbol{v} f \frac{2\,\mathrm{d}\boldsymbol{p}}{h^3} = \int \left(-\frac{\partial f_0}{\partial \varepsilon}\right) \tau \varepsilon \overline{\boldsymbol{v}\boldsymbol{v}} \cdot \left\{ -\varepsilon \frac{\partial \log T}{\partial \boldsymbol{x}} - T \frac{\partial}{\partial \boldsymbol{x}}\left(\frac{\mu}{T}\right)\right\} D(\varepsilon)\,\mathrm{d}\varepsilon , \tag{9}$$

or

$$\boldsymbol{q} = L_2\left(-\frac{\partial \log T}{\partial \boldsymbol{x}}\right) + L_1\left\{-T\frac{\partial}{\partial \boldsymbol{x}}\left(\frac{\mu}{T}\right)\right\}. \tag{10}$$

The energy flow in the case $\boldsymbol{J} = 0$ is equal to the thermal current. Substitution of (10) into (8) leads to

$$\boldsymbol{q} = -\frac{L_0 L_2 - L_1 L_1}{T L_0}\frac{\partial T}{\partial \boldsymbol{x}}, \quad \text{thus} \quad \lambda = \frac{L_0 L_2 - L_1^2}{T L_0}, \tag{11}$$

where λ is the thermal conductivity. If we take only the first term of the expansion (7) as in example 4, (11) vanishes. In order to obtain the answer to even the lowest approximation, we must retain the second term of (7). In this way, we obtain

$$\lambda = \frac{n}{m^*} A \left(\frac{2\mu_0}{m^*} \right)^{\frac{1}{2}s} \frac{\mu_0^2}{3T} \left(\frac{\pi k T}{\mu_0} \right)^2 = \frac{n}{m^*} \tau(\mu_0) \tfrac{1}{3} \pi^2 k^2 T . \tag{12}$$

NOTE 1: The thermal current is produced by a higher order effect which results from the deformation in shape of the electronic distribution from the Fermi distribution without shifting its center.

NOTE 2: If we compare the thermal conductivity with the electric conductivity obtained in example 4 (where we may replace μ by μ_0), we have the relation

$$\frac{\lambda}{\sigma T} = \frac{\pi^2}{3} \left(\frac{k}{e} \right)^2 . \tag{13}$$

The *Wiedemann-Franz law* is an empirical statement that at a given temperature the ratio λ/σ is a constant irrespective of the material. Also, the rule that λ/σ is proportional to T is called *Lorentz' law*. The quantity on the left hand side of (13) is named the *Lorentz number*. According to Drude's theory, which was described in example 3 and problem 7, we have $vl = v^2\tau = \tau 2kT/m$, $C = \frac{3}{2}kn$ so that

$$\frac{\lambda}{\sigma T} = 3 \left(\frac{k}{e} \right)^2 . \tag{14}$$

For standard metals near room temperature, (13) gives roughly the right value. Measured values of the Lorentz number, however, exhibit a dependence on temperature, and also vary somewhat from one metal to another.

NOTE 3: When an electric field $\boldsymbol{E}$ is present, a term $-e\boldsymbol{E}$ is added inside the brackets of the second term in (6) and (10). The fact that the same coefficient L_1 appears in (6) and (10) is a consequence of *Onsager's reciprocal theorem*. This theorem states the reciprocity of interference effects between irreversible processes. In the present case, it states that the two effects are closely correlated, the one being that a flow of electrons is produced by a temperature gradient and the other that thermal flow is generated by an electric field.

15. Let the velocity of the spherical body be $\boldsymbol{V}$ and the velocity of a molecule $\boldsymbol{v}$. Then the relative velocity is given by $\boldsymbol{u} = \boldsymbol{v} - \boldsymbol{V}$. Referred to a coordinate system which translates with the spherical body, the molecule makes a perfectly elastic collision at the surface of the spherical body. Take a polar coordinate system which is fixed in space and has the direction of $\boldsymbol{V}$ as its polar axis: let the direction of $-\boldsymbol{u}$ be specified by (θ,φ) in this coordinate.

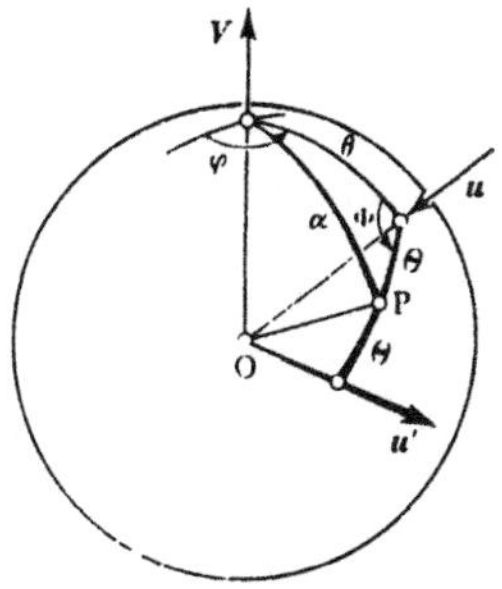

Let the direction of the point P where the molecule collides with the spherical body be represented by (Θ, Φ) in a polar coordinate system which is fixed with $-\boldsymbol{u}$ as its polar axis and with the plane determined by $-\boldsymbol{u}$ and $\boldsymbol{V}$ as the reference plane for azimuthal angle. Then the angle α between $\overrightarrow{\mathrm{OP}}$ and $\boldsymbol{V}$ is given by

$$\cos\alpha = \cos\theta\cos\Theta + \sin\theta\sin\Theta\cos\Phi . \tag{1}$$

The relative velocity after the collision, $\boldsymbol{u}'$, has a magnitude equal to that of $\boldsymbol{u}$, makes an angle Θ with $\overrightarrow{\mathrm{OP}}$, and lies in the plane determined by $-\boldsymbol{u}$ and $\overrightarrow{\mathrm{OP}}$. Hence the momentum received by the spherical body upon an impact with a molecule is parallel to $\overrightarrow{\mathrm{OP}}$ and has a magnitude $2mu\cdot\cos\Theta$. The component of this momentum in the direction of $\boldsymbol{V}$ is

$$-2mu\cos\Theta\cos\alpha . \tag{2}$$

From the symmetry of the problem, it is obvious that the net force exerted upon the spherical body is parallel to $\boldsymbol{V}$.

Since the gas is rarefied, the Maxwell velocity distribution applies, the number of molecules moving in the direction within a solid angle $\sin\theta\,\mathrm{d}\theta\,\mathrm{d}\varphi$ and with speed between $(u, u+\mathrm{d}u)$ per unit volume being given by

$$n\left(\frac{m}{2\pi kT}\right)^{\frac{3}{2}}\exp\left\{-\frac{m}{2kT}(u^2+V^2-2uV\cos\theta)\right\}u^2\,\mathrm{d}u\sin\theta\,\mathrm{d}\theta\,\mathrm{d}\varphi, \tag{3}$$

where n is the number density of molecules. If we take the radius of the spherical body as R and that of the molecule as a, the number of collisions occurring per unit time such that the impact point P lies within a solid angle

$\sin \Theta \, d\Theta \, d\Phi$ in the direction (Θ, Φ), is equal to the number of molecules contained in the volume

$$(R + a)^2 \sin \Theta \, d\Theta \, d\Phi \cdot u \cos \Theta . \tag{4}$$

Therefore, the net force experienced by the spherical body due to the gas molecules is obtained on multiplication of (2), (3) and (4) and summation over the configuration $0 \leqq \Theta < \frac{1}{2}\pi$ for which the collisions take place. Substitution of (1) into (2) yields

$$K = -2mn\left(\frac{m}{2\pi kT}\right)^{\frac{3}{2}} (R + a)^2 \int_0^\infty u^4 \, du \int_0^\pi 2\pi \sin\theta \, d\theta \times \exp\left\{-\frac{m}{2kT}(u^2 + V^2 - 2uV\cos\theta)\right\}$$
$$\times \int_0^{\frac{1}{2}\pi} \cos^2\Theta \sin\Theta \, d\Theta \int_0^{2\pi} d\Phi (\cos\theta\cos\Theta + \sin\theta\sin\Theta\cos\Phi)$$
$$= -2m\pi^2 n\left(\frac{m}{2\pi kT}\right)^{\frac{3}{2}} (R + a)^2 \int_0^\infty u^4 \, du \int_0^\pi \cos\theta \sin\theta \, d\theta \times \exp\left\{-\frac{m}{2kT}(u^2 + V^2 - 2uV\cos\theta)\right\}$$
$$= -2\pi^2 (R + a)^2 n\left(\frac{m}{2\pi kT}\right)^{\frac{3}{2}} \frac{(kT)^2}{mV^2} \int_0^\infty u^2 \left\{\left(\frac{mV}{kT}u - 1\right) e^{-m(u-V)^2/(2kT)} + \left(\frac{mV}{kT}u + 1\right) e^{-m(u+V)^2/(2kT)}\right\} du . \tag{5}$$

The last integral can be transformed into an error integral by the substitution $x = (u \mp V)\sqrt{m/2kT}$. Putting $x_0 = V\sqrt{m/2kT}$, we obtain

$$K = -nkT\pi(R + a)^2 \frac{1}{x_0^2}\left\{\frac{2}{\sqrt{\pi}} \int_{-x_0}^\infty (x + x_0)^2 (xx_0 + x_0^2 - \tfrac{1}{2}) e^{-x^2} dx + \frac{2}{\sqrt{\pi}} \int_{+x_0}^\infty (x - x_0)^2 (xx_0 - x_0^2 + \tfrac{1}{2}) e^{-x^2} dx\right\} . \tag{6}$$

With the help of the formulae

$$\frac{2}{\sqrt{\pi}}\int_{\pm x_0}^{\infty} e^{-x^2}dx = 1 \mp \Phi(x_0), \qquad \frac{2}{\sqrt{\pi}}\int_{\pm x_0}^{\infty} x\,e^{-x^2}dx = \frac{1}{\sqrt{\pi}}e^{-x_0{}^2},$$

$$\frac{2}{\sqrt{\pi}}\int_{\pm x_0}^{\infty} x^2 e^{-x^2}dx = \tfrac{1}{2} \pm \left\{\frac{x_0}{\sqrt{\pi}}e^{-x_0{}^2} - \tfrac{1}{2}\Phi(x_0)\right\},$$

$$\frac{2}{\sqrt{\pi}}\int_{\pm x_0}^{\infty} x^3 e^{-x^2}dx = \frac{x_0^2+1}{\sqrt{\pi}}e^{-x_0{}^2}, \qquad \Phi(x_0) = \frac{2}{\sqrt{\pi}}\int_0^{x_0} e^{-x^2}dx$$

and on introducing the pressure of the gas $p = nkT$, we are led to

$$K = -p\pi(R+a)^2\left\{\frac{e^{-x_0{}^2}}{\sqrt{\pi}x_0}(1+2x_0^2) + \left(2x_0^2 + 2 - \frac{1}{2x_0^2}\right)\Phi(x_0)\right\}. \tag{7}$$

The quantity x_0 is of the order of the ratio between V, the velocity of the spherical body, and the mean velocity of the molecules. For the case $x_0 \ll 1$, the expansion

$$\Phi(x_0) = \frac{2}{\sqrt{\pi}}\left(x_0 - \frac{x_0^3}{1!3} + \frac{x_0^5}{2!5} - \cdots\right) \tag{8}$$

can be applied to expand (7) in powers of x_0, the expansion beginning with a term proportional to x_0:

$$K = -p\pi(R+a)^2\cdot\frac{16}{3\sqrt{\pi}}x_0(1 + \tfrac{1}{5}x_0^2 + \cdots). \tag{9}$$

Since $R \gg a$, we finally obtain the following result:

$$K = -p\pi R^2\cdot\frac{16}{3}\sqrt{\frac{m}{2\pi kT}}\,V\left(1 + \frac{1}{5}\frac{mV^2}{2kT} + \cdots\right). \tag{10}$$

This result shows that if $mV^2 \ll kT$, the spherical body is acted upon by a resistive force which is proportional to V, the speed of the spherical body.

16. The electric displacement is written as $D(t) = E(t) + 4\pi P(t)$, where the electric polarization is given by the following equation:

$$P(t) = \int_0^{\pi} \mu\cos\theta f(\theta,\varphi,t)2\pi\sin\theta\,d\theta. \tag{1}$$

If the electric field does not depend on time, the orientational distribution is given by the Boltzmann factor at equilibrium, i.e.,

$$f(\theta, \varphi, t) = \frac{n}{4\pi}\left(1 + \frac{\mu E}{kT}\cos\theta\right), \qquad (|\mu E| \ll kT) \tag{2}$$

where n is the number density of molecules. Then we assume that the required orientational distribution takes the following form:

$$f(\theta, \varphi, t) = \frac{n}{4\pi}\left\{1 + \frac{\mu F(t)}{kT}\cos\theta\right\}, \qquad (|\mu F| \ll kT). \tag{3}$$

In the above, an unknown function $F(t)$ is assumed to be linear in the electric field. Since we may only keep linear terms in $\mu F(t)/kT$ and $\mu E(t)/kT$, the equation determining $F(t)$ is

$$\frac{\mathrm{d}F(t)}{\mathrm{d}t} + 2BF(t) = 2BE(t). \tag{4}$$

A solution which gives $P(t) \to 0$ when $t = -\infty$ also gives $F(t) \to 0$. Hence we have

$$F(t) = 2B\int_{-\infty}^{t} \mathrm{e}^{-2B(t-t')}E(t')\,\mathrm{d}t'. \tag{5}$$

Substitution of (3) into (1) leads to

$$P(t) = \frac{n\mu_2}{3kT}F(t), \quad \text{thus} \quad D(t) = E(t) + \frac{4\pi n\mu^2}{3kT}F(t). \tag{6}$$

Inserting (5) into the above equation, and comparing it with the equation in the Hint, we can determine ε_∞ and the relaxation function as

$$\varepsilon_\infty = 1, \qquad \varphi(t) = \frac{4\pi n\mu^2}{3kT}2B\mathrm{e}^{-2Bt}, \tag{7}$$

thus

$$\varepsilon(\omega) = 1 + \frac{4\pi n\mu^2}{3kT}\frac{1}{1 + \mathrm{i}\omega\tau}. \tag{8}$$

Here, the relaxation time τ is given by the following equation:

$$\tau = \frac{1}{2B} = \frac{4\pi\eta a^3}{kT}. \tag{9}$$

NOTE 1: By putting $\omega = 0$, we obtain the static dielectric constant as $\varepsilon_s = 1 + 4\pi n\mu^2/3kT$. Hence (8) can be rewritten in the following form:

$$\varepsilon(\omega) = \varepsilon'(\omega) + \mathrm{i}\varepsilon''(\omega); \quad \varepsilon'(\omega) - \varepsilon_\infty = \frac{\varepsilon_s - \varepsilon_\infty}{1 + \omega^2\tau^2}, \quad \varepsilon''(\omega) = \frac{(\varepsilon_s - \varepsilon_\infty)\omega\tau}{1 + \omega^2\tau^2}. \tag{10}$$

In general, an equation of this type is called *Debye's equation.* It has the characteristic that the curve which is drawn through points with ε' and ε'' as their rectangular coordinates (ω being a parameter) becomes a semi-circle. This curve is called the Cole-Cole plot.

NOTE 2: Equation (3) is not a general solution even in the approximation under consideration. Nevertheless, it can be shown that the solution is sufficiently precise for the present problem. It is advisable for the reader to show this by making use of the expansion in terms of spherical harmonics.

NOTE 3: On substitution of (6) into (4), the equation satisfied by the electric polarization can be found to be

$$\frac{\mathrm{d}P(t)}{\mathrm{d}t} = -\frac{1}{\tau}\{P(t) - P_0\}, \quad P_0 = \frac{n\mu_2}{3kT}E(t). \tag{11}$$

Here, P_0 is the equilibrium value expected for the instantaneous value of the electric field.

17. Let the mass of the molecules be m, and $\mathrm{d}v_x\,\mathrm{d}v_y\,\mathrm{d}v_z = \mathrm{d}\boldsymbol{v}$. Then the density of the gas ρ, the flow velocity V, and the internal energy density u are given by the following equations, respectively:

$$\begin{aligned} \rho = mn = \int mf\,\mathrm{d}\boldsymbol{v} = \int mf_0\,\mathrm{d}\boldsymbol{v}, \quad \rho V = \int m\boldsymbol{v}f\,\mathrm{d}\boldsymbol{v} = \int m\boldsymbol{v}f_0\,\mathrm{d}\boldsymbol{v}, \\ \rho u = \tfrac{1}{2}\int m(\boldsymbol{v} - V)^2 f\,\mathrm{d}\boldsymbol{v} = \int \tfrac{1}{2}m\boldsymbol{v}^2 f\,\mathrm{d}\boldsymbol{v} - \tfrac{1}{2}\rho V^2. \end{aligned} \tag{1}$$

Therefore, multiplying the Boltzmann equation by m, $m\boldsymbol{v}$, and $\frac{1}{2}m\boldsymbol{v}^2$ respectively, and integrating over all $\boldsymbol{v}$, we obtain*

$$\frac{\partial \rho}{\partial t} + \frac{\partial}{\partial \boldsymbol{x}}\cdot(\rho V) = 0, \qquad \text{or} \qquad \frac{\mathrm{d}\rho}{\mathrm{d}t} = -\rho \operatorname{div} V; \tag{2}$$

$$\frac{\partial(\rho V)}{\partial t} + \frac{\partial}{\partial \boldsymbol{x}}\cdot(\rho V V + \mathbf{P}) = 0, \qquad \text{or} \qquad \rho\frac{\mathrm{d}V}{\mathrm{d}t} = -\operatorname{Div}\mathbf{P}; \tag{3}$$

$$\frac{\partial}{\partial t}(\rho u + \tfrac{1}{2}\rho V^2) + \frac{\partial}{\partial \boldsymbol{x}}\cdot\{V(\rho u + \tfrac{1}{2}\rho V^2) + \mathbf{P}\cdot V\} = 0, \qquad \text{or}$$

$$\rho\frac{\mathrm{d}u}{\mathrm{d}t} = -\mathbf{P}:\frac{\partial}{\partial \boldsymbol{x}}V - \operatorname{div}\boldsymbol{q}. \tag{4}$$

---

* $\operatorname{Div} V = \dfrac{\partial}{\partial \boldsymbol{x}}\cdot V = \dfrac{\partial V_x}{\partial x} + \dfrac{\partial V_y}{\partial y} + \dfrac{\partial V_z}{\partial z}$,

$(\operatorname{Div}\mathbf{P})_\mu = \left(\dfrac{\partial}{\partial \boldsymbol{x}}\cdot\mathbf{P}\right)_\mu = \dfrac{\partial P_{x\mu}}{\partial x} + \dfrac{\partial P_{y\mu}}{\partial y} + \dfrac{\partial P_{z\mu}}{\partial z}$, $\quad \mathbf{P}:\dfrac{\partial}{\partial \boldsymbol{x}}V = \displaystyle\sum_{\mu,\nu=x,y,z} P_{\mu\nu}\frac{\partial V_\mu}{\partial \nu}$.

In the above, $\mathrm{d}/\mathrm{d}t = \partial/\partial t + \boldsymbol{V}\cdot\partial/\partial\boldsymbol{x}$ denotes the temporal variation one sees when one moves with the flow, and

$$P_{\mu\nu} = \int m(v_\mu - V_\mu)(v_\nu - V_\nu) f\,\mathrm{d}\boldsymbol{v}, \qquad q_\mu = \int \tfrac{1}{2} m(\boldsymbol{v} - \boldsymbol{V})^2 (v_\mu - V_\mu) f\,\mathrm{d}\boldsymbol{v}. \quad (5)$$

Since (2) is the equation of continuity, (3) that of motion and (4) the first law of thermodynamics, $\boldsymbol{P}$ represents the pressure tensor and $\boldsymbol{q}$ the thermal flow vector. For a rarefied gas in which velocity gradient and temperature gradient are moderate, such relations as

$$P_{\mu\nu} = \{p + (\tfrac{2}{3}\eta - \kappa)\operatorname{div}\boldsymbol{V}\}\,\delta_{\mu\nu} - \eta\left(\frac{\partial V_\mu}{\partial \nu} + \frac{\partial V_\nu}{\partial \mu}\right), \qquad q_\mu = -\lambda\frac{\partial T}{\partial \boldsymbol{x}}, \quad (6)$$

$$(\mu, \nu = x, y, z)$$

are expected to hold, where the pressure is denoted by $p = nkT$. In (6), η and κ are the coefficients of viscosity (shear viscosity and bulk viscosity, respectively), λ is the coefficient of thermal conductivity, and $\delta_{\eta\nu}$ is Kronecker's delta.

In order to calculate the coefficients of viscosity and of thermal conductivity, the deviation of the distribution function f from its local equilibrium value f_0 should be obtained up to terms linear in the velocity gradient and the temperature gradient. For this purpose, f may be approximated by f_0 in the left hand side of the Boltzmann equation:

$$\frac{\partial f_0}{\partial t} + \boldsymbol{v}\cdot\frac{\partial f_0}{\partial \boldsymbol{x}} = \frac{\mathrm{d}f_0}{\partial t} \fallingdotseq -\frac{f - f_0}{\tau}. \quad (7)$$

Furthermore, for the calculation of $\mathrm{d}n/\mathrm{d}t$, $\mathrm{d}\boldsymbol{V}/\mathrm{d}t$, and $\mathrm{d}T/\mathrm{d}t$ which occur in the expression for $\mathrm{d}f_0/\mathrm{d}t$, the quantities u, $\mathbf{P}$, and $\boldsymbol{q}$ in (2), (3), and (4) may be replaced by their values calculated using f_0:

$$u \fallingdotseq \tfrac{3}{2}kT, \qquad P_{\mu\nu} \fallingdotseq p\delta_{\mu\nu}, \qquad \mathbf{q}_\mu \fallingdotseq 0. \quad (8)$$

As a result, putting $\boldsymbol{u} = \boldsymbol{v} - \boldsymbol{V}$, we obtain the equation

$$f_0\left\{\left(\frac{m\boldsymbol{u}\boldsymbol{u}}{kT} - \frac{m\boldsymbol{u}^2}{3kT}\mathbf{1}\right) : \frac{\partial}{\partial \boldsymbol{x}}\boldsymbol{V} + \left(\frac{m\boldsymbol{u}^2}{2kT} - \frac{5}{2}\right)\boldsymbol{u}\cdot\frac{\partial \log T}{\partial \boldsymbol{x}}\right\} \fallingdotseq -\frac{f - f_0}{\tau}, \quad (9)$$

where $\mathbf{1}$ denotes the unit tensor. Since f_0 is a function of $\boldsymbol{u}^2$ only, we see, on substitution of (9) into (5), that only the velocity gradient term in (9) contributes to $\mathbf{P}$, while only the temperature gradient term contributes to $\boldsymbol{q}$:

$$P_{\mu\nu} = p\delta_{\mu\nu} - \sum_{\gamma,\beta = x,y,z} \int f_0 \tau m u_\mu u_\nu \left(\frac{m u_\beta u_\gamma}{kT} - \frac{m\boldsymbol{u}^2}{3kT}\delta_{\gamma\beta}\right)\mathrm{d}\boldsymbol{u}\cdot\frac{\partial V_\beta}{\partial \gamma}, \quad (10)$$

$$q_\mu = -\sum_{\gamma=x,y,z} \int f_0 \tau \tfrac{1}{2} m u^2 u_\mu \left(\frac{mu^2}{2kT} - \frac{5}{2}\right) u_\gamma \, \mathrm{d}\boldsymbol{u} \cdot \frac{\partial \log T}{\partial \gamma}. \tag{11}$$

If we take a component with $\mu \neq \nu$ in (10), the only terms which contributes occurs when $u_\beta u_\gamma = u_\mu u_\nu$. Thus we have

$$\eta = \int f_0 \tau \frac{(m u_\mu u_\nu)^2}{kT} \mathrm{d}\boldsymbol{u} = \tfrac{1}{15} \int f_0 \tau \frac{(mu^2)^2}{kT} \mathrm{d}\boldsymbol{u} \tag{12}$$

$$(\because \int_0^{2\pi}\int_0^{\pi} (\sin^2\theta \cos\varphi \sin\varphi)^2 \sin\theta \, \mathrm{d}\theta \, \mathrm{d}\varphi = \tfrac{4}{15}\pi). \tag{13}$$

In the case where $\mu = \nu$, only a term with $\beta = \gamma$ is left, and with the help of (13), the result

$$\kappa = 0 \tag{14}$$

can be derived. In the summation in (11), a term with $\gamma = \mu$ remains, and the following result is obtained:

$$\lambda = \frac{k}{3} \int f_0 \tau \frac{mu^2}{2kT} u^2 \left(\frac{mu^2}{2kT} - \frac{5}{2}\right) \mathrm{d}\boldsymbol{u}. \tag{15}$$

For the case $\tau = Au^s$, we can calculate from (12)

$$\eta = \frac{4\pi}{15} \frac{Am^2}{kT} n \left(\frac{m}{2\pi kT}\right)^{\frac{3}{2}} \int_0^\infty u^{6+s} \mathrm{e}^{-mu^2/(2kT)} \mathrm{d}u$$

$$= AnkT \left(\frac{2kT}{m}\right)^{\frac{1}{2}s} \frac{\Gamma(\frac{1}{2}[7+s])}{\Gamma(\frac{7}{2})}. \tag{16}$$

In particular, when $s = 0$, i.e., $\tau = $ constant, (16) becomes $\eta = nkT\tau = p\tau$. When $s = -1$, i.e., the mean free path $l = \tau u = $ constant, it follows that

$$\eta = \tfrac{16}{15} An \left(\frac{mkT}{2\pi}\right)^{\frac{1}{2}}.$$

Similarly, we obtain from (15)

$$\lambda = \frac{4\pi k}{3} An \left(\frac{m}{2\pi kT}\right)^{\frac{3}{2}} \int_0^\infty \frac{mu^{6+s}}{2kT} \left(\frac{mu^2}{2kT} - \frac{5}{2}\right) \mathrm{e}^{-mu^2/(2kT)} \mathrm{d}u$$

$$= Ank \left(\frac{2kT}{m}\right)^{1+\frac{1}{2}s} \frac{s+2}{3\sqrt{\pi}} \Gamma(\tfrac{1}{2}[7+s]), \tag{17}$$

which becomes $\lambda = 5\tau nk^2T/2m = 5p\tau k/2m$ when τ = constant, and $\lambda = \frac{4}{3}Ank\sqrt{2kT/\pi m}$ when l = constant. From (16) and (17), remembering that the specific heat at constant volume is given $c_V = 3k/2m$, we have

$$\frac{\lambda}{\eta c_V} = \frac{5}{6}(s+2). \tag{18}$$

NOTE: If we use the Boltzmann equation in which collisions between mole-

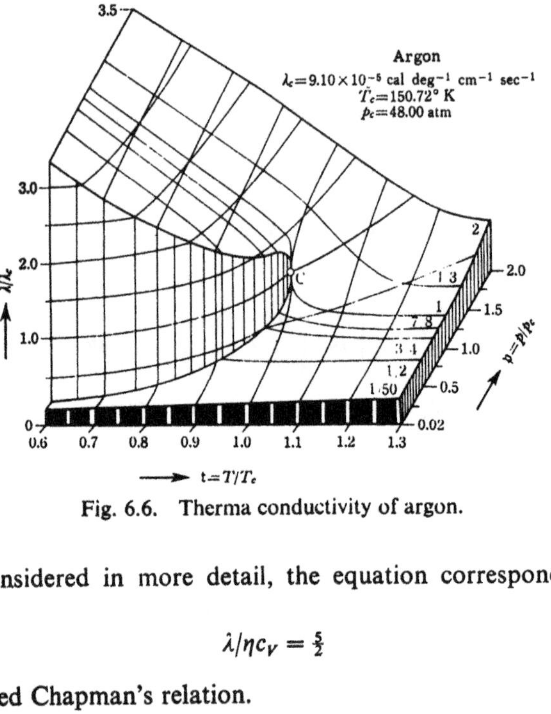

Fig. 6.6. Therma conductivity of argon.

cules are considered in more detail, the equation corresponding to (18) becomes

$$\lambda/\eta c_V = \frac{5}{2} \tag{19}$$

which is called Chapman's relation.

18. When collisions between molecules are not considered, the variation of the distribution function is determined by Liouville's equation

$$\frac{\partial f}{\partial t} + \frac{\boldsymbol{p}}{m}\cdot\frac{\partial f}{\partial \boldsymbol{x}} + \{-m\omega_0^2\boldsymbol{x} - eE(t)\}\cdot\frac{\partial f}{\partial \boldsymbol{p}} = 0. \tag{1}$$

Let t_c be the time during which one molecule collides with another one, and assume that no collision occurs again until a time t. Then the solution $f_{t_c}(\boldsymbol{x}, \boldsymbol{p}, t)$, which is obtained upon solution of (1) with the initial condition that $f = f_0$ at $t = t_c$, can be used between t_c and t. The distribution of free flight time $\theta = t - t_c$

can be calculated in the following way. The probability that a collision takes place within a time interval $(\theta, \theta + d\theta)$ is $d\theta/\tau$ according to the definition of the mean free flight time τ. Hence, if the probability that the molecule does not collide with any other one within $(\theta, \theta + d\theta)$ is denoted by $P(\theta)\, d\theta$, we have $P(\theta + d\theta) = P(\theta)(1 - d\theta/\tau)$, and

$$\frac{\partial P(\theta)}{\partial \theta} = -\frac{P(\theta)}{\tau}, \qquad \text{or} \quad P(\theta) = \frac{1}{\tau} e^{-\theta/\tau}. \tag{2}$$

When the time t is fixed, this expression gives the probability distribution for the time t_c when the last collision took place. Therefore,

$$\bar{f}(\boldsymbol{x}, \boldsymbol{p}, t) = \int_{-\infty}^{t} f_{t_c}(\boldsymbol{x}, \boldsymbol{p}, t) \frac{e^{-(t-t_c)/\tau}}{\tau}\, dt_c \tag{3}$$

is the required average of the distribution function. As a factor $1/\tau$ is contained in (3), $P(\theta)$ is properly normalized. Performing partial differentiation of (3) with respect to t, inserting (1) into this result, and noticing that $f_t(\boldsymbol{x}, \boldsymbol{p}, t) = f_0(\boldsymbol{x}, \boldsymbol{p}, t)$, we finally obtain

$$\begin{aligned}\frac{\partial \bar{f}}{\partial t} &= f_t(\boldsymbol{x}, \boldsymbol{p}, t)\frac{1}{\tau} + \int_{-\infty}^{t} \frac{\partial f_{t_c}}{\partial t} \frac{e^{-(t-t_c)/\tau}}{\tau}\, dt_c - \frac{1}{\tau}\int_{-\infty}^{t} f_{t_c} \frac{e^{-(t-t_c)/\tau}}{\tau}\, dt_c \\ &= \frac{f_0}{\tau} - \left[\frac{\boldsymbol{p}}{m}\cdot\frac{\partial \bar{f}}{\partial \boldsymbol{x}} + \{-m\omega_0^2 \boldsymbol{x} - e\boldsymbol{E}(t)\}\cdot\frac{\partial \bar{f}}{\partial \boldsymbol{p}}\right] - \frac{\bar{f}}{\tau},\end{aligned} \tag{4}$$

which is the required equation.

NOTE: The above derivation of the Boltzmann equation makes clear its intuitive meaning. The assumption used concerning the nature of the collisions is called the assumption of *strong collisions*. The function f_0 may be given in various ways.

19. The Boltzmann equation (10.5) (with $\partial f/\partial \boldsymbol{x} = 0$) is

$$\frac{\partial f}{\partial t} - e\boldsymbol{E}(t)\cdot\frac{\partial f}{\partial \boldsymbol{p}} = -Df. \tag{1}$$

In the approximation, retaining terms up to first order in $\boldsymbol{E}$, the value of f in the second term on the left hand side may be replaced by the equilibrium distribution $f_0(\varepsilon)$:

$$\frac{\partial f_0}{\partial \boldsymbol{p}} = \frac{\partial \varepsilon}{\partial \boldsymbol{p}}\frac{\partial f_0}{\partial \varepsilon} = \boldsymbol{v}\frac{\partial f_0}{\partial \varepsilon} = -\frac{\boldsymbol{v}}{kT} f_0. \tag{2}$$

Putting $f = f_0 + g$, and regarding g as a quantity which is linear in $\boldsymbol{E}$, we have an approximate equation valid to this order:

$$\frac{\partial g}{\partial t} + Dg = -\frac{e\boldsymbol{E}(t)\cdot\boldsymbol{v}f_0}{kT}. \tag{3}$$

A formal solution with $\boldsymbol{E}(-\infty) = 0$ and $g(-\infty) = 0$ is easily obtained (by the method of variation of constants) as

$$g(\boldsymbol{v}, t) = -\int_{-\infty}^{t} \mathrm{e}^{-D(t-t')}\mathrm{d}t' \frac{e\boldsymbol{E}(t')\cdot\boldsymbol{v}f_0}{kT}. \tag{4}$$

Therefore, the components of the electric current can be expressed in the form

$$j_i(t) = -e\int v_i g(\boldsymbol{v}, t) 2\frac{\mathrm{d}\boldsymbol{p}}{h^3} = \frac{e^2}{kT}\sum_l \int_{-\infty}^{t} E_l(t')\mathrm{d}t' \int \{\mathrm{e}^{-D(t-t')}v_l f_0\}\, v_i \cdot 2\,\mathrm{d}\boldsymbol{p}/h^3. \tag{5}$$

Since we have assumed that the energy does not change on scattering, we can write

$$\mathrm{e}^{-D(t-t')}v_l f_0 = \{\mathrm{e}^{-D(t-t')}v_l\} f_0 .$$

Here, D is the operator defined in the problem. With the notation

$$\mathrm{e}^{-D\tau}\delta(\boldsymbol{v} - \boldsymbol{v}_0) = \Omega(\boldsymbol{v}, \boldsymbol{v}_0 \,|\, \tau), \tag{6}$$

the integral representation of $\mathrm{e}^{-D\tau}$ is given by

$$\mathrm{e}^{-D\tau}\boldsymbol{v} = \int \Omega(\boldsymbol{v}, \boldsymbol{v}' \,|\, r)\boldsymbol{v}'\,\mathrm{d}\boldsymbol{v}' .$$

If we use this expression, (5) can be rewritten in the form

$$j_i(t) = \sum_l \int_{-\infty}^{t} E_l(t')\mathrm{d}t' \Phi_{li}(t - t'), \tag{7}$$

with

$$\begin{aligned}\Phi_{li}(t - t') &= \frac{e^2}{kT}\int \{\mathrm{e}^{-D(t-t')}v_l\} f_0 v_i \cdot 2\,\mathrm{d}\boldsymbol{p}/h^3 \\ &= \frac{e^2}{kT}\int\int v_i \Omega(\boldsymbol{v}, \boldsymbol{v}' \,|\, t - t')v_l' f_0 2\,\mathrm{d}\boldsymbol{p}/h^3 \\ &= \frac{e^2 n}{kT}\langle v_i(t) v_l(t')\rangle \qquad (n = \text{density of electrons}).\end{aligned} \tag{8}$$

In the last form of (8), $\langle v_i(t) v_l(t')\rangle$ represents the correlations between $\boldsymbol{v}(t')$, the velocity of an electron at time t', and $\boldsymbol{v}(t)$, that at time t. This is because Ω in (6) represents the velocity distribution, at time τ, of an electron originally moving with velocity $\boldsymbol{v}_0$ (which is guaranteed by the delta function) at $t = 0$, i.e., the transition probability from $\boldsymbol{v}_0$ to $\boldsymbol{v}$. Alternatively, (8) can be written as

$$\Phi_{li}(t - t') = \frac{\langle j_i(t) j_l(t') \rangle}{kT} = \frac{\langle j_i(t - t') j_l(0) \rangle}{kT}. \tag{9}$$

Here, $j_i(t)$ is the electric current which occurs as a fluctuation in conductors in equilibrium, and (9) is proportional to its correlation. For a system in equilibrium, this correlation is dependent only on the time difference $t - t'$. Equation (9) can easily be derived from (8) on the assumption that the n electrons per unit volume are independent of each other.

In particular, let us consider the case $\boldsymbol{E} = (E, 0, 0)$ (where we take $E_x(t) = E\mathrm{e}^{\varepsilon t}$ and let $\varepsilon \to 0$). Then the static conductivity

$$\sigma_{xx} = \frac{1}{kT} \int_0^\infty \langle j_x(t) j_x(0) \rangle \, \mathrm{d}t \tag{10}$$

is obtained from (7), (8) and (9). If we set $\boldsymbol{E} = (E \cos \omega t, 0, 0)$, the conductivity for an oscillating electric field,

$$\sigma_{xx}(\omega) = \frac{1}{kT} \Re \int_0^\infty \langle j_x(t) j_x(0) \rangle \, \mathrm{e}^{-\mathrm{i}\omega t} \mathrm{d}t \tag{11}$$

can be derived (on substitution of $E_x(t) = \Re E\, \mathrm{e}^{\mathrm{i}\omega t}$ into (7)). Since

$$\langle j_x(t) j_x(0) \rangle = \langle j_x(0) j_x(-t) \rangle = (\text{even function of } t), \tag{12}$$

(11) may be expressed as

$$\sigma_{xx}(\omega) = \frac{1}{2kT} \int_{-\infty}^{\infty} \langle j_x(t) j_x(0) \rangle \, \mathrm{e}^{-\mathrm{i}\omega t} \mathrm{d}t. \tag{13}$$

NOTE: From (13), we can obtain the inverse relation

$$\langle j_x(t) j_x(t') \rangle = \frac{kT}{\pi} \int_{-\infty}^{\infty} \sigma_{xx}(\omega) \, \mathrm{e}^{\mathrm{i}\omega(t - t')} \mathrm{d}\omega. \tag{14}$$

In particular, if $\sigma_{xx}(\omega)$ is assumed to be independent of ω, we obtain, with the aid of

$$\int_{-\infty}^{\infty} \mathrm{e}^{\mathrm{i}\omega t} \mathrm{d}\omega = 2\pi \delta(t),$$

the result

$$\langle j_x(t) j_x(t') \rangle = 2kT\sigma\delta(t-t'). \tag{15}$$

When a bar conductor of cross-section S and length L is at temperature T, an electric potential difference $V(t)$ appears as a fluctuation between the two ends of the bar. If we use the relation $V(t) = Lj_x(t)/\sigma$, we may rewrite (15) in terms of $V(t)$ as

$$\langle V(t)V(t') \rangle = 2RkT \cdot LS\delta(t-t'),$$

where $R = L/S$ is the resistance of the bar. The expression per unit volume is

$$\langle v(t)v(t') \rangle = 2RkT\delta(t-t').$$

If we take the Fourier component of $v(t)$ as

$$v(\omega) = (\sqrt{2}/2\pi) \int_{-\infty}^{\infty} v(t) e^{i\omega t} dt,$$

we can write

$$\langle v(\omega)v(\omega') \rangle = \frac{2}{\pi} RkT\delta(\omega-\omega').$$

This relation is called *Nyquist's equation.*

Nyquist's equation can be shown to be valid in its more general form (14) or (13) under conditions much more general than those considered here. The general theory is one of the most important foundations in the statistical mechanics of irreversible processes (cf. H. Nyquist, Phys. Rev. **32** (1928) 110; H. B. Callen and T. A. Welton, Phys. Rev. **83** (1951) 34; R. Kubo, J. Phys. Soc. Japan **12** (1957) 570; R. Kubo: *Some Aspects of the Statistical-Mechanical Theory of Irreversible Processes, Lectures in Theoretical Physics,* Vol. I ed. by Brittin and Dunham (Interscience Publ., New York, 1959).

DIVERTISSEMENT 14

Fluctuation-dissipation theorem. When you are listening to a radio, you can faintly hear a noise which the electrons at work in the apparatus produce through the irregularity of their motion. Nyquist was the first who recognized the important relationship between such thermal noise and the impedance of a resistor across which an irregular voltage difference is incessantly induced as the result of the thermal motion of the electrons. The average power of thermal noise in a given frequency band is essentially proportional to the temperature (more precisely, to the average energy of a harmonic oscillator with that frequency) and to the impedance function of the resistor.

This phenomenon is, by its nature, very closely related to Brownian motion, and so Nyquist's theorem can be greatly generalized. Such a generalization has been stated by many authors, for example, by H. Takahashi (J. Phys. Soc. Japan **7** (1952) 439); H. B. Callen and T. A. Welton (Phys. Rev. **83** (1951) 34); and R. Kubo (J. Phys. Soc. Japan **12** (1957) 570). The generalized theorem is now called the *fluctuation-dissipation theorem,* because it relates in a most general way the fluctuations of a physical quantity of a system in equilibrium to a dissipation process which can be realized when the system is subject to an external force driving it away from equilibrium.

It has been recognized that this theorem makes it possible to find the non-equilibrium properties of a system, for example an impedance function defined with respect to a given external disturbance, from the analysis of thermal fluctuations of the system in equilibrium states. Therefore, the statistical mechanics of irreversible processes may, in a sense, be reduced to that of equilibrium states, which however has to deal with time-dependent fluctuation processes. Thus a new method of approach is now opened to the theory of irreversible processes. The reader is referred to the following article: R. Kubo, *Some Aspects of the Statistical-Mechanical Theory of Irreversible Processes, Lectures in Theoretical Physics,* Vol. I, ed. by Brittin and Dunham (Interscience Publ., New York, 1959).

INDEX